Principles of Mathematics

principles of

Carl B. Allendoerfer
Professor of Mathematics | University of Washington

Cletus O. Oakley
Professor and Department Head
Department of Mathematics | Haverford College

International Student Edition

McGraw-Hill Book Company, Inc.
New York | San Francisco | Toronto | London
Kōgakusha Company, Ltd.
Tokyo

mathematics

second edition

Principles of Mathematics

INTERNATIONAL STUDENT EDITION

I

***Library of Congress Catalog Card Number* 63-12123**

TOSHO PRINTING CO., LTD., TOKYO, JAPAN

Preface

In our preface to the first edition of this book we wrote: "This book has been written with the conviction that large parts of the standard undergraduate curriculum in mathematics are obsolete and that it is high time that our courses take due advantage of the remarkable advances that have been made in mathematics during the past century. All other branches of science manage to incorporate modern knowledge into their elementary courses, but mathematicians hesitate to teach their elementary students anything more modern than the works of Descartes and Euler. It should be granted that mathematics is a cumulative subject and that one cannot run until he has learned to walk. Thus it is not realistic to start our students off with Functions of a Complex Variable, or other higher branches of our subject. The authors believe, however, that some of the content and much of the spirit of modern mathematics can be incorporated in courses given to our beginning students. This book is designed to do just that."

Since the publication of the first edition this point of view has been adopted by many leading writers of mathematics textbooks and by groups such as the Commission on Mathematics of the College Entrance Examination Board, the School Mathematics Study Group, and the Committee on the Undergraduate Program of the Mathematical Association of America. There is now a considerable reservoir of experience with materials of this kind, and the methods for teaching them can no longer be considered to be experimental.

This second edition incorporates much that we have learned from our own experience and from the experience of others. The book is intended for students who have completed Intermediate Algebra and a first course in Trigonometry and carries them to the point at which they can begin a serious university course in Calculus. It is therefore appropriate for students in the twelfth grade in high school or for college freshmen who have entered with only three years of preparatory mathematics. It is also useful for preservice or in-service courses for secondary school teachers of mathematics.

The chief changes from the first edition are the following:

(1) The number of problems has been greatly expanded. There are now over 2,700 in the text. Answers to odd-numbered problems appear at the end of this volume; answers to even-numbered problems are included in the Teachers' Manual. The more difficult problems have been marked with asterisks (*). A few problems are marked "BT", which means "Booby Trap", "Use your head", "Be careful", or "Don't make a fool of yourself".

(2) The language of sets has been used systematically wherever it is appropriate. The basic ideas of sets and logic are presented jointly in an expanded first chapter, which forms the foundation of the rest of the book.

(3) The treatment of Mathematical Induction has been clarified and enlarged, and new material on Number Theory has been included.

(4) Methods for the solution of equations and inequalities have been brought together in a new chapter which treats these in a systematic and parallel fashion.

(5) The chapter on Trigonometry is focused on the trigonometry of real numbers and the analytic properties of the trigonometric functions. This is intended to supplement the brief treatment of these topics in many high school courses which are centered around numerical trigonometry and the solution of triangles.

(6) There is a completely new chapter on Probability, which replaces the former discussion of Statistics and Probability. This chapter depends heavily on set theory and incorporates many of the most recent approaches to the subject. It can be studied at any time after Chapter 1 has been completed.

(7) The material on Boolean Algebra has been completely rewritten and appears in the final chapter. This chapter is intended as an "extra" for the student who wishes greater competence with the algebra of sets, or experience with an abstract system significantly different from ordinary algebra. It, too, can be studied at any time after Chapter 1, but perhaps should be deferred until after Chapter 4.

Outlines for possible courses using the book and various teaching suggestions are included in the Teachers' Manual, which is published separately.

"It is hoped that the book is relatively free of errors, but each author blames the other for those that may be discovered."

Carl B. Allendoerfer
Cletus O. Oakley

Contents

The Integers | 3

Groups | 4

Equations and Inequalities | 5

Analytic Geometry | 9

Limits | 10

The Calculus | 11

Probability | 12

Boolean Algebra | 13

Appendix

Logic and Sets | 1

1.1 The Nature of Mathematics

Mathematics is so widely spread throughout our culture that it presents different faces to different people. A description of the nature of mathematics is therefore likely to depend upon the relationship of the author to the subject, and we may well find ourselves in a situation similar to that of the three blind men who tried to describe an elephant. In all probability your own experience with mathematics has introduced you into only certain of its aspects, which may differ from those we are going to examine in this book. So that we may all be ready to start together, let us begin by presenting a somewhat oversimplified version of the nature of mathematics, into which you can fit your own experience and out of which this book will grow.

Virtually all the mathematics with which you are familiar had its roots somewhere in nature. Arithmetic and algebra grew out of men's needs for counting, financial management, and other simple operations of daily life; geometry and trigonometry developed from problems of land measurement, surveying, and astronomy; and calculus was invented to assist in the solution of certain basic problems in physics. In recent years new forms of mathematics have been invented to help us cope with problems in social science, business, biology, and warfare; and new mathematical subjects are sure to arise from other portions of human endeavor. Let us lump all these sources of mathematical ideas together and call them *Nature*.

At first our approach to nature is descriptive, but as we learn more about it and perceive relationships between its parts, we begin to construct a *Mathematical Model* of nature. This is a highly creative step which requires deep insight and inspiration. In doing this we settle on the meanings of the words that we shall be using and develop axioms which will form the foundation of our mathematical theory. Perhaps you are familiar with this sort of process through your study of geometry, in which the axioms form an abstract description of what man saw when he began to measure the earth. In later sections we shall discuss axioms and related matters in some detail, but we hope that you have at least an elementary idea about them at this time.

The next step in the process is to deduce the consequences of our collection of axioms. By applying logical methods of deduction we then arrive at theorems. These theorems are nothing more than logical conclusions from our axioms and must not be assumed to be firm statements about relationships which are necessarily true in nature. In order to prove these theorems we must know the rules of logic and the methods by which mathematical proofs are constructed. Probably you already know something about logic and have proved theorems in geometry, but now

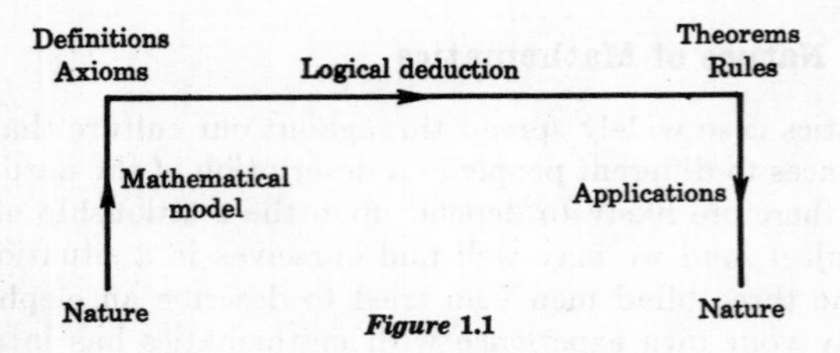

Figure 1.1

it is time to be rather systematic and to be certain that you really know what you are doing. Since much of this book is devoted to proving theorems, our first chapter will be spent on a brief treatment of logic and mathematical proof.

After the theorems are proved, the mathematician then looks for applications of them to nature. He solves problems and obtains numerical answers to questions raised in his investigations. You are probably most familiar with this portion of mathematics, since it has been traditionally the main content of school mathematics. Knowing how to solve problems is very important, but you must also know a good deal about other portions of mathematics, such as model building and the logical proof of theorems.

We recognize this structure of mathematics when we look back over a mature mathematical subject, but most people do not learn mathematics in this precise way. They are more likely to begin with the descriptive aspects of nature and proceed to theorems and rules by *intuition* rather than by proof. Having obtained these rules, they then apply them to nature and convince themselves that "mathematics is good for something". This process is indicated in Fig. 1.2.

After these intuitive theorems and rules have been invented, the investigator will often be faced with a confusing and disorganized set of statements which are hard to remember and apply. In his attempt to organize and simplify these, he will go back and establish the necessary axioms and derive his theorems from them. In this way he will build the structure of Fig. 1.1.

In writing this book, we shall assume that you have already been through the intuitive phases of algebra and geometry and that you are ready for a more organized approach to these subjects. If your intuitive background seems to be weak, be sure to review your school mathematics before trying to understand our more abstract development.

The above account of the nature of mathematics is oversimplified in one important way. There are numerous branches of modern mathematics that do not appear to be rooted in nature. Systems of axioms

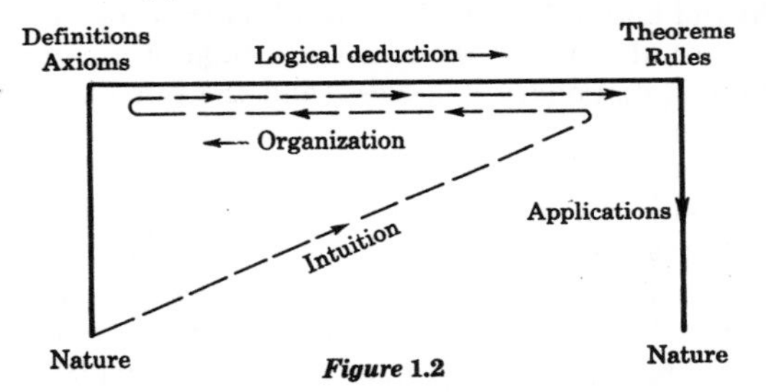

Figure 1.2

have been developed which seem valid and meaningful to mathematicians but which are in no way descriptions of anything in man's experience. The theorems developed from these axioms are often of great interest to mathematicians and frequently have unexpected applications. As you study more advanced mathematics, you will encounter such theories, but for our purposes in this book we shall stay reasonably close to those portions of our subject which grew out of man's direct needs.

1.2 Definitions

In this section we shall begin the study of the logical development of a mathematical system. The first requirement for an understanding of any branch of knowledge is that you know the meanings of the words you use. In school you were taught to look in a dictionary for definitions of unfamiliar words. What you find, of course, is a statement of the meaning of the new word in terms of other words. But what do these new words mean? If we look them up, they are defined in terms of still more words. If we carry on this process indefinitely, we shall eventually start going in circles, as in the following illustrations.

Illustration 1

(a) *dead:* lifeless
lifeless: deprived of *life*
life: the quality distinguishing an animal or plant from organic or *dead* bodies

(**b**) *point:* the common part of two intersecting *lines*
line: the figure traced by a *point* which moves along the shortest path between two *points*

Exercise A. Choose an ordinary, nontechnical word and build a circular chain of definitions from this word back to itself. Use a standard dictionary for your definitions. Do not put simple connectives such as *the, and, in, is,* etc., in your chain.

The evils of this circular process can be seen by supposing that we try to learn a foreign language, say French, by using only an ordinary French dictionary—not a French-English dictionary. We look up a particular French word and find it described in more French words, and we find ourselves no further ahead. Without a knowledge of a certain amount of French, a French dictionary is useless; and the same is true of our own English dictionary.

In this book we are starting to learn something about mathematics, and we shall soon find that mathematics has many similarities to a foreign language. We must learn what its words mean and must build up a basic mathematical vocabulary. The thing that will appear strange to you is that in mathematics we shall not allow our chain of definitions to become circular. We shall take a small number of words which are *undefined.* Then other words are defined in terms of these undefined words, or in terms of words which themselves are defined in terms of the undefined words. This process is not in general use in most fields of knowledge, but it is essential for the construction of any logical system, and it is universally applied in mathematics.

It may come as a shock to you to find that we do not define certain words, and indeed that we *cannot* define them; therefore let us take some examples from plane geometry. In this subject it is customary to take *point* and *line* as undefined words, and this may not seem natural at first. Why not take the dictionary definition of *point:*

point: that which has neither parts nor extent, but position only (Webster)

Think this over and see whether it really helps very much. One may say that a point is an infinitely small dot on a piece of paper, but if this matter is pressed, it becomes hard to ascribe meaning to an *infinitely small dot.* Perhaps we can do better with *line:*

line: that which has length but not breadth or thickness (Webster)

If we knew anything about length, breadth, and thickness, this might help. But an investigation of the meanings of these only leads us further into difficulties.

Thus we begin our logical structure with a set of *undefined mathematical words.* All other mathematical words will be defined in terms of these, with the understanding that our definitions may also contain common English words (*is, and, the*, etc.) which have no special mathematical meanings. It is not easy to decide which words should be left undefined and which should be defined in terms of the undefined words. Many choices can be made, and the final decision is largely based upon considerations of simplicity and elegance.

Let us suppose that *point, line*, and *between* are undefined. Then we may define:

line segment: that portion of a line contained between two given points on a line

The words in this definition other than *point, line*, and *between* are words without special technical meanings, and thus may be used freely.

CAUTION: Although *line* is undefined, you doubtless have an intuitive picture of a *line* in mind. The picture which will correspond to our use of the word *line* is that of a straight, infinite line extending indefinitely in both directions. We use *line segment* to denote a finite portion of such a line. Please keep this in mind; there is often confusion between the concepts of *line* and *line segment* in elementary textbooks.

In forming definitions, we must be careful about a few matters if we are to do anything useful. In the first place, we must make sure that our definition is *meaningful* and *consistent*. Examples which violate this requirement are:

triangle: a line line line point point point
square: a four-sided polygon having five equal sides
(It is assumed here that *polygon, side* of a polygon, and *equal* have been previously defined.)

Further, when we formulate a definition, we always have something in mind which we are trying to describe, and we must make certain that our definition *includes* all cases which we have in mind and *excludes* all others. The definition:

square: a four-sided polygon all of whose sides are equal

is satisfied by a rhombus and hence includes too much. On the other hand, suppose we define:

rectangle: a four-sided polygon all of whose vertex angles are right angles and whose sides are equal

This is unsatisfactory because it excludes nonsquare rectangles and hence does not include all the desired cases.

A better way of putting this idea is to define:

square: a four-sided polygon is a square if and only if its sides are all equal and its vertex angles are all right angles

The *if* in this statement says that we are to include all the cases which follow in the next clause, and the *only if* tells us to exclude all other cases.

1.3 Propositions

When we have built up an adequate vocabulary of technical words—both undefined and defined—we are ready to form sentences using these words and also nontechnical English words. If we examine our intuitive reasoning processes, we see at once that there is an important class of sentences which will play a key role in our subject. This is the class of sentences to which one and only one of the terms *true* or *false* "can be meaningly applied".

Illustration 1. The following sentences are members of this class of sentences:

The sum of the measures of the interior angles of any plane triangle is equal to 180. (True)
$2 + 6 = 7$. (False)

Illustration 2. The following sentences are not members of this class:

Mathematics is green. (Nonsense)
A triangle consists of three lines and three points. (Vague)
It is raining. (Requires qualifications as to place)
$x + 5 = 0$. (Truth cannot be determined until x is given a specific value)

Although our intuition is usually a safe guide for deciding whether a sentence belongs to this class, we can give no clear-cut rules which will enable us to decide this matter in general. One of the difficulties is that the words *true* or *false* have connotations in common usage which are based upon our particular philosophy or experience, and hence cannot be made sufficiently precise for our purposes. Another difficulty is that the phrase *can be meaningly applied* requires a subjective decision whose outcome may well differ from person to person or culture to culture. As a result, the best that we can do is to give an intuitive description of the type of sentences under consideration. A sentence of this type will be called a *proposition*.

Intuitive description. A proposition is a sentence to which one and only one of the terms true or false can be meaningly applied.

Although intuition is essential, we cannot base a rigorous theory on it and must proceed more formally. In the formal theory, *proposition*, *true*, and *false* are undefined words. In order to work with these we assume the axioms:

Axiom 1. There is given a class C of propositions. Hereafter, all propositions which we mention must be members of this class.

Axiom 2. To each proposition in C we can assign one and only one of the words *true* or *false*.

Notation. We shall represent propositions by small letters such as p, q, r, s, etc., and shall let T and F stand for *true* and *false*, respectively. T and F are called *truth values*, so that each proposition p has a truth value which is either T or F but not both.

In our discussion of logic we shall work wholly in this formal theory, but we shall nevertheless be greatly interested in its application to mathematical reasoning. In order to do this we shall carry two trains of thought simultaneously, and it is important that you understand their relationship:

(**1**) The formal theory based on Axioms 1 and 2. This is an abstract model of mathematical reasoning which is constructed to agree with our logical intuition.
(**2**) Mathematical reasoning, as such. This is an intuitive theory from which we shall draw illustrations. We must use our intuition to decide whether the conclusions of the formal theory are appropriate guides for mathematical reasoning. As you can imagine, however, we should not bother to develop the formal theory in this book unless it was an excellent model of our intuition. Virtually all mathematicians agree that the conclusions of the formal theory do give valid rules for mathematical reasoning, and we shall proceed on this basis without further mention of the logical distinction between the two theories.

One of our chief tasks in mathematics is to determine which truth value, *true* or *false*, is to be applied to a given proposition. The usual procedure is to take a collection of propositions whose truth values we know already and then by means of logic to derive the truth value of the given proposition from these. We cannot begin this process unless we assume that certain initial propositions are *true*, and we call such propositions *axioms*.

Definition: An *axiom* is an initial proposition which is assumed to be true.

Exercise A. What is the logical status in our political philosophy of the statement:

> We hold these truths to be self-evident, that all men are created equal, that . . . . (Declaration of Independence)

There are two reasonable questions which you may ask at this time:

(1) How do we choose our axioms? This choice is a creative rather than a logical matter. We may choose axioms that reflect observed properties of nature which we are describing in our mathematical model of it. Thus the axiom:

There is one and only one line that passes through two distinct points

is derived from our physical experience in drawing lines on paper. On the other hand, we may choose axioms which are not derived from nature, so that we can investigate some interesting new mathematical theory. For example, we can write the axiom:

The sum of the interior angles of every triangle is less than 180°

even when all our observations suggest that this sum is equal to 180°.

Some choices of axioms turn out to be useful or interesting, and others do not. Thus the selection of axioms is a highly personal matter and requires the creative talents of the best mathematicians.

(2) How do we know that our axioms are true? The answer is that we have no such knowledge. We assume them to be true and proceed from this assumption in assigning truth values to other propositions.

1.4 Sets

The concept of a class of propositions introduced above is an illustration of a wider notion which is of great importance in mathematics. This is the concept of a *set*, which we introduce in this section.

We think of a set as a collection of objects: pencils, trees, numbers, men, points, propositions, etc. The individual components of the set are called its *elements*. As an example consider the set consisting of four boys named: John, Joe, Jerry, Jim. This set has four elements. Sets may be of any size. We may think of the set of all particles of sand on a beach; this has a finite number of elements, but this number is certainly very large. A set, however, may have infinitely many elements. An example of an infinite set is the set of all positive integers: 1, 2, 3, 4, 5, ... Indeed, a set may contain no element, in which case we call it the *empty*, or *null*, *set*.

We can describe sets in this way, but *set* is a primitive notion which cannot be defined. Hence we take *set* and *element* to be undefined. The statement: "a is an element of the set A", which we write "$a \in A$", is similarly an undefined relationship.

Examples and notation. We give below some typical examples of sets occurring in mathematics and indicate the notations appropriate for them. Note that, with some exceptions, we regularly use braces { } to represent a set.

(1) $\emptyset$, the empty, or *null, set,* containing no elements.
(2) $\{3\}$, the finite set, of which 3 is the only element. Note that this *set* is quite different from the *number* 3.
(3) $\{2, 7, 15, 36\}$, a finite set of four elements.
(4) $\{2, 4, 6, 8, 10, \ldots\}$, the infinite set of all even positive integers.
(5) X, the set of all real numbers. At this stage we assume that you have a rough idea of what a real number is. We shall go into details in Chap. 2.
(6) $X \times Y$, the set of all ordered pairs of real numbers (x,y). The adjective *ordered* means that the pairs (a,b) and (b,a) are understood to be distinct.
(7) $\{x \text{ (men)} \mid x \text{ was a President of the United States}\}$
This expression should be read: "The set of all men x such that x was a President of the United States", the vertical line standing for *such that.*
(8) $\{x \text{ (triangles)} \mid x \text{ is isosceles}\}$
This is the set of all isosceles traingles.
(9) $\{x \text{ (real numbers)} \mid x + 3 = 5\}$
This set can also be represented by the symbol $\{2\}$.
(10) The set C of all propositions.

Universal set. In examples (7), (8), and (9) above we had to specify that x stood, respectively, for men, triangles, and real numbers. In most portions of mathematics we omit references of this kind, for we usually know from the context what kind of an object x is. In other words, there is given in advance a large set, called the *universal set* U, to which x is supposed to belong. This set may vary from situation to situation, as in the examples above, but is fixed for the duration of a particular discussion. Statements like (7), (8), and (9) then define *subsets* of the set U which is involved.

Definition: A set A is a *subset* of a set B if and only if every element of A is an element of B.

Notation. We write this relationship $A \subseteq B$, read "A is a subset of B", or "A is included in B".

Definition: A set A is a *proper subset* of B if and only if A is a subset of B and at least one element of B is not an element of A.

Notation. We write this relationship $A \subset B$, read "A is a proper subset of B", or "A is properly included in B".

By convention, the null set $\emptyset$ is a subset of every set and is a proper subset of every set except itself.

Illustration 1

(a) Given the set $\{1, 2, 3\}$, its subsets are $\{1, 2, 3\}$, $\{1, 2\}$, $\{1, 3\}$, $\{2, 3\}$, $\{1\}$, $\{2\}$, $\{3\}$, $\emptyset$. Its proper subsets are the above excepting for $\{1, 2, 3\}$.
(b) The set $\{1, 2, 3, 4, 5, \ldots\}$ is a proper subset of the set of all real numbers.
(c) The set $\{2, 4, 6, 8, 10, \ldots\}$ is a proper subset of the set $\{1, 2, 3, 4, 5, \ldots\}$.

Subsets of a set A are often defined as containing those elements of A which have some property in common. If S represents the universal set consisting of all the students at your university, we may be interested in the following subsets:

$\{s \mid s \text{ is a girl}\}$
$\{s \mid s \text{ is a football player}\}$
$\{s \mid s \text{ is a member of the } \Sigma O \Sigma \text{ fraternity}\}$
$\{s \mid s \text{ is a graduate student}\}$

Illustration 2. Let U be the set X of real numbers. Then:

(a) $\{x \mid x - 3 = 0\}$ is the set $\{3\}$.
(b) $\{x \mid x^2 - 5x + 6 = 0\}$ is the set $\{2, 3\}$.
(c) $\{x \mid x^2 + 1 = 0\}$ is the null set $\emptyset$.
(d) $\{x \mid x^2 = 0\}$ is the set $\{0\}$. Even though 0 is a double root of the expression $x^2 = 0$, we do not write symbols like $\{0, 0\}$.

There are two other elementary relationships between sets which we shall need in the future. One of these is the notion of *identity*.

Definition: Two sets are said to be *identical* if and only if every element of each is an element of the other. When A and B are identical, we write $A = B$.

Illustration 3. The sets $\{1, 2, 3\}$, $\{3, 2, 1\}$, and $\{2, 3, 1\}$ are identical since they consist of the same elements. The order in which we write these elements is unimportant.

Definition: Two sets $A = \{a_1, a_2, \ldots\}$ and $B = \{b_1, b_2, \ldots\}$ are said to be in 1 *to* 1 *correspondence* when there exists a pairing of the a's and the b's such that each a corresponds to one and only one b and each b corresponds to one and only one a.

Illustration 4. Establish a 1 to 1 correspondence between the set of numbers $\{1, 2, \ldots, 26\}$ and the set of letters of the alphabet $\{a, b, \ldots, z\}$.

We make the pairing:

$$\begin{matrix} 1 & 2 & \cdots & 26 \\ a & b & \cdots & z \end{matrix}$$

However, there are many other possible pairings such as:

2	3	$\cdots$	26	1
a	b	$\cdots$	y	z

Illustration 5. Establish a 1 to 1 correspondence between the set $\{1, 2, 3, 4, 5, \ldots\}$ and the set $\{2, 4, 6, 8, 10, \ldots\}$.

Let n represent an element of $\{1, 2, 3, 4, 5, \ldots\}$. Then the pairing $n \leftrightarrow 2n$ gives the required correspondence, examples of which are:

1	2	$\cdots$	50	$\cdots$	100	$\cdots$
2	4	$\cdots$	100	$\cdots$	200	$\cdots$

Exercise A. Establish a 1 to 1 correspondence between the sets {John, Joe, Jerry, Jim} and {Mildred, Marcia, Ruth, Sandra}.

Exercise B. Establish a 1 to 1 correspondence between the sets $\{1, 2, 3, 4, 5, \ldots\}$ and $\{3, 6, 9, 12, 15, \ldots\}$.

Problems 1.4

In Probs. 1 to 10 state which of the given definitions are satisfactory.

1. A *triangle* is a geometric figure with three edges.
2. A *square* is a four-sided rectangle.
3. Two lines in the plane are *parallel* if and only if they do not intersect.
4. Two line segments are *congruent* if and only if their lengths are equal.
5. A *regular pentagon* is a polygon with five sides.
6. An *octagon* is a polygon with eight sides.
7. A *cube* is a rectangular parallelopiped.
8. A *tetrahedron* is a polyhedron with triangular faces.
9. A *polynomial* is an algebraic expression of the form $2x^3 - x^2 + x - 1$.
10. A *real number* is one which is not complex.

In Probs. 11 to 20 state which of the given sentences are propositions.

11. Tomorrow will be Tuesday.
12. John Doe's hair is red.
13. Some students are poor.
14. Chalk is white.
15. $x^2 + 3x + 2 = 0$.
16. $x^2 - 16 = (x + 4)(x - 4)$.
17. All ripe apples are red.
18. A triangle is isosceles.
19. This triangle is isosceles.
20. $5 = 3 + 2$, and $7 = 2 + 4$.

In Probs. 21 to 30 write a simple expression for the given set. The universal set is X, the set of real numbers.

21. $\{x \mid 2x = 4\}$.
22. $\{x \mid 4x^2 = 16\}$.
23. $\{x \mid x^2 + 2x + 1 = 0\}$.
24. $\{x \mid x^2 + x + 1 = 0\}$.
25. $\{x \mid x^2 < 0\}$.
26. $\{x \mid x + 1 = x + 1\}$.
27. $\{x \mid x = -x\}$.
28. $\{x \mid x^3 = 1\}$.
29. $\{x \mid x \text{ is a prime and } x \text{ is even}\}$.
30. $\{x \mid x \text{ is negative and } x^2 = 4\}$.

In Probs. 31 to 44 establish a 1 to 1 correspondence between the given two sets whenever this is possible.

31. The set of negative integers; the set of positive integers.
32. $\{2, 4, 6, 8, 10, \ldots\}$; $\{3, 6, 9, 12, 15, \ldots\}$.
33. The set of married men; the set of married women.
34. The set of Chevrolets in operation; the set of Fords in operation.
35. The set of all students in your university; the set of mathematics professors in your university.
36. The set of all French words; the set of all English words.
37. The set of positions on a baseball team; the set of players of one team as listed in their batting order.
38. The set of elective offices in your state government; the set of elected officials.
39. The set of all integers; the set of positive integers.
40. The set of real numbers in the interval $[0,1]$; the set of real numbers in the interval $[0,2]$.
41*. The set of real numbers in the interval $0 < x < 1$; the set of all real numbers.
42*. The set of positive integers; the set of rational numbers. (See Courant and Robbins, "What Is Mathematics?" pp. 79–80.)
43*. The set of positive integers; the set of real numbers. (See Courant and Robbins, "What Is Mathematics?" pp. 81–83.)
44*. The set of rational numbers, the set of ordered pairs of rational numbers.

In Probs. 45 to 50 list all subsets of the given set. Which of these are *proper* subsets?

45. $\{2, 6\}$.
46. $\{a,b\}$.
47. $\{3, 5, 7\}$.
48. {John, James, Jerry}.
49. $\{a, b, c, d\}$.
50. $\{1, 2, 3, 4\}$.

51*. Count the number of subsets in each of Probs. 45 to 50 that you have worked. Now guess a general formula for the number of subsets of a given, finite set. Prove that your guess is correct.
52. Show that, if $A \subseteq B$ and $B \subseteq A$, then A and B are identical.

1.5 Open Sentences and Quantifiers

We have previously seen that a statement such as

$$x^2 + 3x + 2 = 0$$

is not a proposition, since a truth value cannot be assigned to it. If, however, we assume that the universal set U is the set X of real numbers, we can substitute any element of X for x into the above statement and obtain a proposition. If we put $x = 0$, we obtain the proposition: $0^2 + 3 \times 0 + 2 = 0$, which is false. On the other hand, if we put $x = -1$, we obtain $(-1)^2 + 3 \times (-1) + 2 = 0$, which is true. Let us define some terms.

Definition: A *variable* is a symbol for which any element of the universal set may be substituted.

Notation. It is customary to represent variables by letters such as x, y, and z.

Definition: An *open sentence* is a statement which contains a variable and which becomes a proposition when an element of the universal set is substituted for this variable. If the variable occurs more than once in the open sentence, the *same* element of the universal set must be substituted for it at each occurrence.

Notation. We shall represent open sentences by symbols such as p_x, q_x, etc., where x is the variable involved. The symbol p_x should be read "p sub x".

Remark. Although we shall simplify our treatment by considering only open sentences which contain a single variable, open sentences containing several variables frequently occur. For example,

$$x^2y + 3y^2 + 2x = 0$$

is an open sentence with two variables. The variables x and y can refer to different universal sets, and we are perfectly free to substitute one element for x and a different element for y. We must, however, substitute the same element for x each time that it appears, and similarly the same element for y each tie that it appears.

Exercise A. Define an open sentence containing two variables.

Illustration 1

(a) Let the universal set be the set X of real numbers. Then:

$$x^2 - 5x + 4 = 0 \qquad \text{and} \qquad x + 1 = 0$$

are open sentences.

(b) Let the universal set be the set of triangles in the plane. Then:

x is equilateral and *The area of x is 10 in.^{2}*

are open sentences.

We shall be especially interested in an element $x \in U$ if its substitution in p_x results in a true proposition, and we give the set of such elements a special name, the *truth set* of p_x.

Definition: An element $x \in U$ is an element of the *truth set* of an open sentence p_x if and only if its substitution for x in p_x results in a true proposition.

Notation. We shall write P, Q, etc., for the truth sets of p_x, q_x, etc. Whenever this notation is not sufficiently clear, we shall use longer expressions such as:

$$\{x \mid x^2 = 1\}$$

Strictly speaking, we should write this in the form

$$\{x \mid x^2 = 1 \text{ is true}\}$$

but it is customary to omit the phrase *is true* on the assumption that the meaning is clear. Examples (7), (8), and (9) and Illustration 2 of Sec. 1.4 can now be interpreted as representing the truth sets of certain open sentences.

In many mathematical situations we wish to consider the possibility that the truth set of an open sentence is the whole of the universal set U. To do this we invent the symbol $\forall_x$ (read "for all x") and write the proposition: $\forall_x p_x$.

Definition: $\forall_x p_x$ is true if and only if P is the set U. $\forall_x p_x$ is false if and only if P is not the set U; that is, there is at least one element of U which does not belong to P.

In a similar fashion, we may be interested in the possibility that the truth set of an open sentence contains at least one element of U, that is, that this truth set is not the null set. We invent the symbol $\exists_x$ (read "for some x" or "there exists an x such that") and write the proposition: $\exists_x p_x$.

Definition: $\exists_x p_x$ is true if and only if P is not the set $\emptyset$. $\exists_x p_x$ is false if and only if P is the set $\emptyset$.

Definition: The symbols $\forall_x$ and $\exists_x$ defined above are called *quantifiers*, a term which specifies "how many".

One of the difficulties in reading mathematics is that quantifiers are frequently omitted and must be understood from the context. You will be expected to supply them as indicated in the illustrations below.

Illustration 2

(a) Let the universal set be the set X of real numbers. Then the open sentence

$$x^2 + 2x + 1 = (x + 1)^2$$

is understood to have the quantifier $\forall_x$. Such expressions are often called *identities*.

(b) Let the universal set be the set X of real numbers. Then the open sentence

$$x^2 + 2x + 1 = 0$$

is understood to have the quantifier $\exists_x$. Such expressions are often called *equations of condition*, or *conditional equations*.

(c) Let the universal set be the set of all triangles in a plane. Then the open sentence

If x is isosceles, then its base angles are equal

is understood to have the quantifier $\forall_x$. This is typical of most theorems in geometry.

1.6 Conjunction and Intersection

Whenever we begin a chain of reasoning, we must have available to us a collection of propositions with known truth values. These may be axioms whose truth values we have assumed, or they may be theorems whose truth has been established by the methods to be explained in this chapter. In either case we wish to combine these given propositions into new propositions to which we must assign truth values. The purpose of the next few sections of this chapter is to define operations for combining propositions and to give the rules for assigning truth values to the resulting compound propositions.

The simplest combination of two propositions is their *conjunction*, in which we put them together with an *and* in between. For example, the conjunction of

Jones is a farmer   Smith is a plumber

is

Jones is a farmer, *and* Smith is a plumber.

Intuitively, the truth of a conjunction means that its two component parts are both true. Also, it is reasonable to assume that a conjunction is false if either or both of its components is false. In order to express these ideas more formally, we must introduce some notation.

We recall that the letters p, q, r, etc., are used to represent propositions. The capital letters T and F will represent the truth values, true and false, which are assigned to these propositions. The symbol $p \wedge q$ (read "p and q") represents the conjunction of the two unspecified propositions p, q. The truth value of $p \wedge q$ depends upon the truth values of the specific propositions which are substituted for p, q. This dependence is given in the *truth table* below.

Conjunction

p	q	$p \wedge q$
T	T	T
T	F	F
F	T	F
F	F	F

Corresponding values are on the same horizontal line; for example, the second line in the body of the table says that: if p is true and q is false, then $p \wedge q$ is false.

Exercise A. Show that $p \wedge q$ and $q \wedge p$ have the same truth values.

By analogy with the conjunction of two propositions we can discuss the conjunction of two open sentences, which we shall write: $p_x \wedge q_x$. In order for this to qualify as an open sentence, we must specify the proposition which we will obtain if we substitute a particular element of U for x in $p_x \wedge q_x$. By definition this proposition is the conjunction of the two propositions obtained by substituting this element of U in p_x and q_x separately.

Illustration 1. Let p_x be $x^2 + 3x + 1 = 0$; q_x be $x + 3 = 0$.
Then $p_x \wedge q_x$ is $(x^2 + 3x + 1 = 0) \wedge (x + 3 = 0)$.
When $x = 2$, this becomes the proposition:

$$(2^2 + 3 \times 2 + 1 = 0) \wedge (2 + 3 = 0)$$

which is false since each component is false.

Illustration 2. Let p_x be $x^2 - 3x + 3 = 0$; q_x be $x - 2 = 0$.
Then $p_x \wedge q_x$ is $(x^2 - 3x + 2 = 0) \wedge (x - 2 = 0)$.
When $x = 2$, this becomes the proposition:

$$(2^2 - 3 \times 2 + 2 = 0) \wedge (2 - 2 = 0)$$

which is true since both components are true.

From this discussion it is clear that the truth set of $p_x \wedge q_x$, which we write $\{x \mid p_x \wedge q_x\}$, consists of those elements $x \in U$ which are elements of both P and Q. (Recall that we write P for the truth set of p_x.) A set which is obtained from two other sets in this fashion is called the *intersection* of the two sets.

Definition: The *intersection* of two sets A and B, written $A \cap B$, is the set of elements which belong to both A and B. (See Fig. 1.3, in which $A \cap B$ is shaded.)

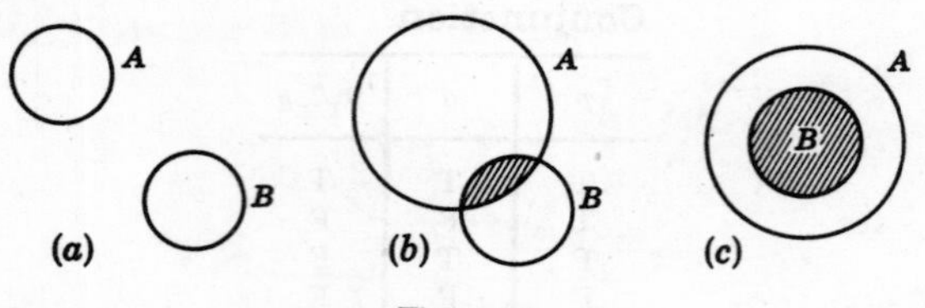

Figure 1.3

In view of this definition, we can summarize the above discussion in the following theorem.

Theorem 1. $\{x \mid p_x \wedge q_x\} = P \cap Q$.

Illustration 3

(**a**) Let A be the set of red haired men, and
B be the set of blue eyed men.
Then $A \cap B$ is the set of men who have both red hair and blue eyes.

(**b**) Let A be the set of rectangles in the plane, and
B be the set of rhombuses in the plane.
Then $A \cap B$ is the set of squares in the plane.

(**c**) Let A be the set of points on line a of Fig. 1.4, and
B be the set of points on line b.
Then $A \cap B$ is the point N, which is common to both lines.

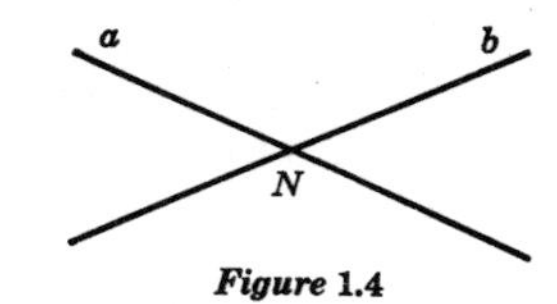

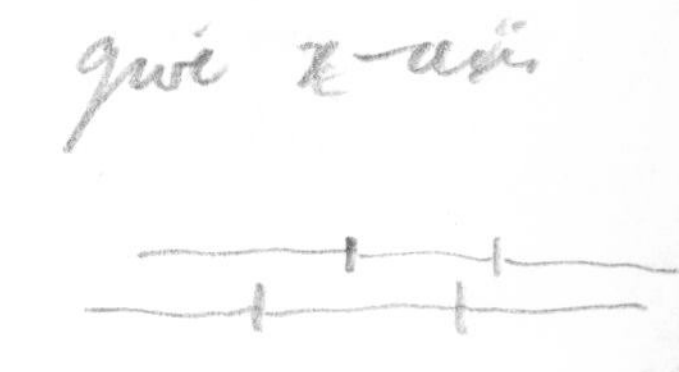

Figure 1.4

(**d**) Let A be the set of odd integers, and
B be the set of even integers.
Then $A \cap B$ is the empty set $\emptyset$.

(**e**) Let A be the set of real numbers, and
B be the set of positive real numbers.
Then $A \cap B$ is the set of positive real numbers.

1.7 Disjunction and Union

Nearly as simple as the notion of a conjunction is that of a *disjunction*. In a disjunction we put the two propositions together with an *or* between them and write symbolically $p \vee q$. In trying to arrive at the proper truth values of a disjunction, we run into an ambiguity in English regarding the meaning of *or*. When we say that "p or q is true", we can mean one of two things:

(**1**) Either p is true, or q is true, but not both are true.
(**2**) Either p is true, or q is true, or both are true.

The first of these is called the *exclusive or*, and the second is the *inclusive or*. The inclusive *or* is often expressed in legal language by the wording *and/or*. When we use *or* as a disjunction, we shall hereafter mean it in the inclusive sense. We can now write down the truth table for a disjunction.

Disjunction

p	q	$p \vee q$
T	T	T
T	F	T
F	T	T
F	F	F

Exercise A. Show that $p \vee q$ and $q \vee p$ have the same truth values.

On several occasions, we shall need a symbol for the exclusive *or*. To express this idea we shall write symbolically $p \veebar q$.

Exercise B. Write down the truth table for $p \veebar q$.

Again we wish to extend this definition to that of the disjunction of two open sentences. We follow the procedure which we used in the case of conjunctions and arrive at the conclusion that the elements of the truth set of $p_x \vee q_x$ are those which are elements of either or both of the truth sets P and Q. A set which is obtained from two other sets in this fashion is called the *union* of the two sets.

Definition: The *union* of two sets A and B, written $A \cup B$, is the set of elements which belong to A or to B or to both A and B. (See Fig. 1.5, in

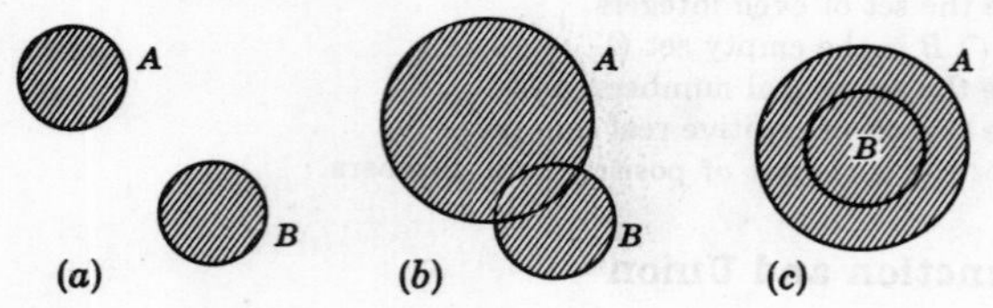

Figure 1.5

which $A \cup B$ is shaded.)

This definition permits us to state the theorem:

Theorem 2. $\{x \mid p_x \vee q_x\} = P \cup Q.$

Illustration 1

(a) Let A be the set of even integers, and
B be the set of odd integers.
Then $A \cup B$ is the set of all integers.

(b) Let A be the set of real numbers greater than 2, and
B be the set of real numbers less than 5.
Then $A \cup B$ is the set of all real numbers.

(c) Let A be the set of all students in your university and
B be the set of those students in your university who play football.
Then $A \cup B$ is the set of all students in your university.

(d) Let A be the set of all rectangles in the plane, and
B be the set of all squares in the plane.
Then $A \cup B$ is the set of all rectangles in the plane.

Problems 1.7

In Probs. 1 to 10 state what quantifiers are implied, if any, which make the given statements true.

1. $(x - 3)(x - 4) = x^2 - 7x + 12$.
2. $x^2 - 7x + 12 = 0$.
3. The integer 11 is a prime.
4. In a right triangle, the square of the length of the hypotenuse is equal to the sum of the squares of the lengths of the two legs.
5. Vertical angles are equal.
6. The area of this square is 12 in.^{2}
7. If x is an even integer, then x^2 is an even integer.
8. 5 is a square root of 25.
9. $x + 2 = 5$.
10. It is impossible to trisect an angle with the use of straight-edge and compass alone. (What if the angle is a right angle?)

In Probs. 11 to 14 write the conjunction and disjunction of the given propositions.

11. John is a student. Mary is beautiful.
12. All summers are hot. Some winters are cold.
13. All lines are straight. All circles are round.
14. These lines are parallel. These lines intersect.

In Probs. 15 to 24 write the conjunction and disjunction of the given open sentences. Find the truth sets of the resulting compound open sentences. The universal set is the set X of real numbers.

15. $x + 1 = 0 \quad x^2 - 1 = 0$.
16. $x^2 = 0 \quad x = 1$.
17. $(x + 2)^2 = 0 \quad x + 2 = 0$.
18. $x^2 - 8x + 15 = 0 \quad x^2 - 7x + 12 = 0$.
19. $x > 3 \quad x > 5$.
20. $x^2 + 4 = 0 \quad x^2 - 4 = 0$.
21. $x - 7 = 0 \quad x > 5$.
22. $x^3 - 1 = 0 \quad x^2 + x + 1 = 0$.
23. $x^2 - 16 = (x + 4)(g - 4) \quad x^2 - 9 = (x + 3)(x - 3)$.
24. $3x + 4 = 2 + 3x + 2 \quad x^2 < 0$.

In Probs. 25 to 32 find the set defined by the given operation. The universal set is the set of real numbers.

25. $\{1, 3, 7\} \cap \{3, 8, 10\}$.
26. $\{a, h, q\} \cup \{b, h, r\}$.
27. $\{1, 5, 9, 11, 23\} \cup \{2, 4, 8, 11, 60\}$.
28. $\{2, 7, 6, 1, 12\} \cap \{15, 2, 3\}$.
29. $\{x \mid x \text{ is an integer}\} \cap \{x \mid x > 0\}$.
30. $\{x \mid x \text{ is an even integer}\} \cup \{x \mid x \text{ is an odd integer}\}$.
31. $\{x \mid x \text{ is positive}\} \cup \{x \mid x \text{ is negative}\}$.
32. $\{x \mid x^2 - 25 = 0\} \cap \{x \mid x^2 - 9x + 20 = 0\}$.
33. Show that $\{x \mid x^2 - 7x + 12 = 0\} = \{x \mid x - 3 = 0\} \cup \{x \mid x - 4 = 0\}$.
34. Show that $\{x \mid x^2 + 2x - 24 = 0\} = \{x \mid x + 6 = 0\} \cup \{x \mid x - 4 = 0\}$.
35. Show that $A \cap B = B \cap A$ for any sets A and B. (Use the definition of $\cap$.)
36. Show that $A \cup B = B \cup A$ for any sets A and B. (Use the definition of $\cup$.)
37. Show that if $A \subseteq B$, then $A \cap B = A$.
38. Show that if $A \subseteq B$, then $A \cup B = B$.

1.8 Negation and Complement

Our next operation on propositions differs from those just given in that it applies to a single proposition alone. The negation of a proposition is a new proposition which has the opposite truth value; i.e., if p is true, the negation of p is false, and if p is false, the negation of p is true. It will be convenient to write $\sim p$ (read "not p") for the negation of p. The appropriate truth table is:

Negation

p	$\sim p$
T	F
F	T

As we shall see later, the process of forming $\sim p$ from p requires considerable care. For the present we consider only the case in which p is a simple proposition which does not contain any quantifiers. In this situation the negation can be formed by inserting a *not* in a suitable place.

Illustration 1

(a) p: The number 3 is a perfect square.
$\sim p$: The number 3 is not a perfect square.

(b) p: The sum of the measures of the interior angles of a given triangle is 180°.
$\sim p$: The sum of the measures of the interior angles of a given triangle is not 180°.

When p_x is an open sentence, we form the new open sentence $\sim p_x$ and must examine its truth set. If the substitution of an element of U in p_x yields a true proposition, the substitution of this element in $\sim p_x$ yields a false proposition. Similarly, if the substitution of an element of U in p_x yields a false proposition, the substitution of this element in $\sim p_x$ yields a true proposition. Thus we see that the truth set of $\sim p_x$ consists of those elements of U which are not elements of the truth set of p_x. A set which is formed in this way from another set is called the *complement* of the given set.

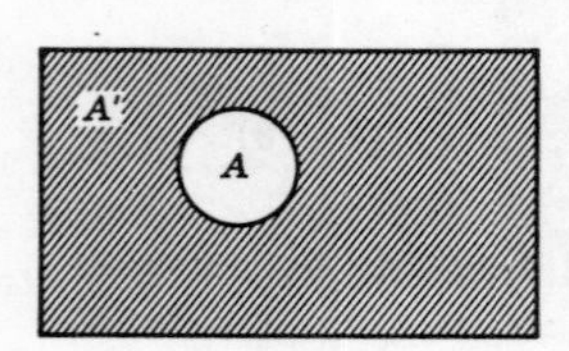

Figure 1.6

Definition: Let a set A be a subset of a universal set U. Then the *complement* of A, written A', is the set of those elements of U which are not elements of A. (See Fig. 1.6, where A' is shaded.)

Exercise A. Show that A is the complement of A'.

Remark. The complement of a set is defined relative to a particular universal set. If the universal set is changed, the complement must be changed accordingly.

In view of this definition we have the theorem:

Theorem 3. $\{x \mid \sim p_x\} = P'$.

Illustration 2

(a) The complement of the set $\{1, 3, 6\}$ relative to the universal set $\{1, 2, 3, 4, 5, 6, 7\}$ is the set $\{2, 4, 5, 7\}$.
(b) The complement of the set $\{1, 3, 6\}$ relative to the universal set $\{1, 3, 6, 9, 12\}$ is the set $\{9, 12\}$.

Illustration 3. Let U be the set X of real numbers.

(a) Let A be the set consisting of 0 and all the positive real numbers. Then A' is the set of all negative real numbers.
(b) Let A be the set $\{0\}$. Then A' is the set of all positive and negative real numbers.
(c) Let A be the empty set $\emptyset$. Then A' is the set U.
(d) Let A be the set U. Then A' is the set $\emptyset$.

Illustration 4. Let U be the set of all real numbers, and p_x be the open sentence: $x^2 - 9 = 0$. Then the truth set P is $\{-3, 3\}$, and its complement P' is the set $\{x \mid x \neq -3 \text{ or } 3\}$. Hence $\{x \mid \sim p_x\} = \{x \mid x \neq -3 \text{ or } 3\}$.

1.9 Implication

An implication is a proposition formed from two propositions as an *if—then* statement. For example, two implications can be formed from the propositions:

Seattle is beautiful. Philadelphia is old.

They are:

If Seattle is beautiful, then Philadelphia is old

and

If Philadelphia is old, then Seattle is beautiful.

You will notice that in our example there is no relationship between the two clauses, and none is required in an implication. For the most part, however, useful implications have clauses which have some connection with each other, for example:

If I jump off a cliff, I shall be injured.

The clause which follows the *if* is called the *antecedent*, or *hypothesis*, and the clause which follows the *then* is called the *consequent*, or *conclusion*.

We should like to know the truth value of the implication when we know the truth values of the two clauses. Although the final procedure is a matter of definition, it is based upon the intuitive grounds of correct reasoning. Let us examine all the possible cases.

(1) If a true hypothesis leads to a true conclusion, we believe that we have reasoned correctly and wish to call the implication true.

(2) If a true hypothesis leads to a false conclusion, we have surely made a mistake is reasoning and should call the implication false.

(3) If we start from a false hypothesis, correct reasoning can bring us to a true conclusion and we should call the implication true. As an example of such a procedure let us take as our hypothesis:

$$\begin{array}{lcl} & 1 = 2 & \text{(false hypothesis)} \\ \text{Then} & \underline{2 = 1} & \\ \text{Adding, we have} & 3 = 3 & \text{(true conclusion)} \end{array}$$

(4) If we start from a false hypothesis, correct reasoning can equally well bring us to a false conclusion and we should call the implication true. For example:

$$\begin{array}{cl} 1 = 2 & \text{(false hypothesis)} \\ \underline{3 = 3} & \\ 4 = 5 & \text{(false conclusion)} \end{array}$$

We write the implication *if p, then q* in the form $p \to q$ and construct the truth table:

Implication

p	q	$p \to q$
T	T	T
T	F	F
F	T	T
F	F	T

Exercise A. Do $p \to q$ and $q \to p$ have the same truth values?

Next we write the implication of two open sentences: $p_x \to q_x$ and wish to determine its truth set. In the previous cases of conjunction and disjunction we invented new symbols, $\cap$ and $\cup$, respectively, which we used to represent the truth sets $P \cap Q$ and $P \cup Q$. It would now be reasonable to invent a new symbol to help us write the truth set of an implication, but we shall find this unnecessary, for the symbols already defined are sufficient for the job. When we examine the truth table for an implication, we find that we can make $p \to q$ true by making p false or by making q true. This suggests the correct conclusion that:

Theorem 4. $\{x \mid p_x \to q_x\} = P' \cup Q$.

After we have developed some more machinery (Sec. 1.11, Prob. 26), we can give a formal proof of this result.

Illustration 1. We give some examples of true and false implications.

	Implication	**Truth value**
(a)	If $3 + 5 = 8$, then $2 + 4 = 6$	T
(b)	If $3 + 5 = 8$, then $2 + 4 = 5$	F
(c)	If $2 + 3 = 6$, then $4 + 8 = 12$	T
(d)	If $2 + 3 = 6$, then $4 + 8 = 5$	T

Illustration 2

(a) The truth set of "If $x^2 = 1$, then $x = 1$" is $\{x \mid x \neq -1\}$; for $P = \{-1, 1\}$, $P' = \{x \mid x \neq -1 \text{ or } 1\}$, $Q = \{1\}$. Hence $P' \cup Q = \{x \mid x \neq -1\}$.

(b) The truth set of: "If $x = 1$, then $x^2 = 1$" is the universal set; for $P = \{1\}$, $P' = \{x \mid x \neq 1\}$; $Q = \{-1, 1\}$. Hence $P' \cup Q$ is the universal set.

1.10 Equivalence

In the further development of the laws of logic we shall not care about the exact form of a proposition, but shall be concerned only with its truth value. For this reason it is desirable to develop the notion of the *equivalence* of two propositions. We introduce the compound proposition $p \leftrightarrow q$, read "p is equivalent to q", which is to be true when p and q have the same truth values and false when they have opposite truth values. This leads us to the truth table:

Equivalence

p	q	$p \leftrightarrow q$
T	T	T
T	F	F
F	T	F
F	F	T

Since the equivalences available to us at this point are all very trivial, the following illustrations may seem too easy. They are worth mastering, however, so that you will understand more complicated ones in later sections.

Illustration 1

(a) The equivalence $(2^2 = 4) \leftrightarrow (3^2 = 9)$ is true, since both propositions are true.

(b) The equivalence $(2^2 = 4) \leftrightarrow (3 + 5 = 7)$ is false, since one proposition is true and one is false.

(c) The equivalence $(2^2 = 5) \leftrightarrow (3 + 5 = 7)$ is true, since both propositions are false.

From the definition of equivalence it is easy to verify the following three statements:

(1) p is equivalent to p.
(2) If p is equivalent to q, then q is equivalent to p.
(3) If p is equivalent to q and if q is equivalent to r, then p is equivalent to r.

These follow immediately, for in each of them all of the propositions which occur have the same truth values.

As defined here, *equivalence* refers solely to the equivalence of two propositions, but later on we shall have to define the equivalence of other kinds of mathematical concepts. In such general contexts we shall make use of the notion of a *relation* between two mathematical entities. These entities may be numbers, points, lines, triangles, or other mathematical objects, and we shall let a and b, p and q, AB and CD, etc., be pairs of these of the same type. We indicate how these are related by placing a mathematical symbol between them as in the examples:

$$p \leftrightarrow q \qquad AB \parallel CD$$
$$a = b \qquad AB \perp CD$$
$$a < b \qquad \triangle ABC \cong \triangle XYZ$$
$$a \neq b \qquad \{1, 2\} = \{x \mid x^2 - 3x + 2 = 0\}$$

When we wish to refer to a relation without specifying the meanings of its component parts we write $a * b$, read "a is related to b".†

Our concern here is with equivalence relations as defined below.

Definition: A relation $a * b$ between two mathematical entities is an *equivalence relation* if and only if it has the properties:

(1) $a * a$.
(2) If $a * b$, then $b * a$.
(3) If $a * b$ and if $b * c$, then $a * c$.

Exercise A. Show that the equivalence $p \leftrightarrow q$ is an equivalence relation.

Illustration 2. Throughout mathematics the equal sign "=" is used to mean that the symbols on the two sides are the same or different names for the same thing. For example, $5 = 2 + 3$ means that 5 and $2 + 3$ are different symbols for the same number "five". Similarly, the equality $\{x \mid x^2 = 9\} = \{-3, 3\}$ means that the two sets mentioned are identical. In view of this definition, equality has the properties:

(1) $a = a$.
(2) If $a = b$, then $b = a$.
(3) If $a = b$, and if $b = c$, then $a = c$.

Consequently, *equality* is an equivalence relation in the sense of the above definition.

† More generally, a relation is the truth set of an open sentence in two variables. See Sec. 6.4.

Remark. This definition of equality permits us to carry out certain familiar operations without further ado. For example, from $a = b$ we can conclude that $a + c = b + c$. The reason is that a and b are two names for the same number, and hence that $a + c$ and $b + c$ are merely two different names for another number. Hence $a + c$ and $b + c$ are equal. Similarly, from $a = b$ we can conclude that $a - c = b - c$, $ac = bc$, and $a/c = b/c$ (if $c \neq 0$).

Illustration 3. The following relations are not equivalence relations for the reasons given.

(**a**) $a < b$, violates (1) and (2).
(**b**) $a \leq b$, violates (2).
(**c**) $a \neq b$, violates (1) and (3).
(**d**) $AB \perp CD$, violates (1) and (3).

As before, we can extend this notion of equivalence to that of open sentences and wish to discuss the truth set of the equivalence $p_x \leftrightarrow q_x$. If we substitute an element of U which belongs to $P \cap Q$, we obtain a true proposition since both sides of the equivalence are now true propositions. If we substitute an element of U which belongs to $P' \cap Q'$ we again obtain a true proposition since both sides of the equivalence are false propositions. The substitution of any other element of U gives us a false proposition since one side of the equivalence will be true and the other side will be false. This discussion amounts to a proof of the theorem:

Theorem 5. $\{x \mid p_x \leftrightarrow q_x\} = (P \cap Q) \cup (P' \cap Q')$.

In Sec. 1.11, Illustration 2, we shall give another proof of this theorem. In specific cases, however, we can often obtain the truth set of an equivalence more rapidly by proceeding directly instead of by using this theorem.

Illustration 4

(**a**) The truth set of $(x^2 = 1) \leftrightarrow (x = 1)$ is the set $\{x \mid x \neq -1\}$, since -1 is the only value of x for which both sides do not have the same truth values. We can obtain this result from Theorem 5 as follows:

Let p_x be $x^2 = 1$, and q_x be $x = 1$. Then $P = \{-1, 1\}$ and $Q = \{1\}$. Hence $P' = \{x \mid x \neq -1 \text{ or } 1\}$ and $Q' = \{x \mid x \neq 1\}$. Combining these results we see that $P \cap Q = \{1\}$ and $P' \cap Q' = \{x \mid x \neq -1 \text{ or } 1\}$ and that $(P \cap Q) \cup (P' \cap Q') = \{x \mid x \neq -1\}$.

(**b**) The truth set of the equivalence:

Triangle x is equilateral $\leftrightarrow$ Triangle x is equiangular

is the universal set consisting of all triangles, since for every triangle the two sides of the equivalence have equal truth values.

Problems 1.10

In Probs. 1 to 6 find the negation of the given propositions.

1. Jones is a carpenter.
2. Smith is a merchant.
3. 7 is a prime number.
4. 16 is a perfect square.
5. These two lines are parallel.
6. The area of this circle is 25 in.^{2}

In Probs. 7 to 16 find the complement of the given set relative to the stated universal set U.

7. $\{4, 6, 8\}$ $U = \{1, 2, 3, 4, 5, 6, 7, 8\}$.
8. $\{4, 6, 8\}$ $U = \{2, 4, 6, 8, 10\}$.
9. $\{1, 3, 5, 7, \ldots\}$ $U =$ the set of all positive integers.
10. $\{2, 4, 6, 8, \ldots\}$ $U =$ the set of all positive integers.
11. $\{x \mid x^2 = 1\}$ $U = X$, the set of all real numbers.
12. $\{x \mid (x^2 = 1) \wedge (x > 0)\}$ $U =$ the set of all positive real numbers.
13. $\{x \mid x > 3\}$ $U = X$.
14. $\{x \mid x \leq 4\}$ $U = X$.
15. $\{x \mid x = 1\} \cap \{x \mid x = 2\}$ $U = X$.
16. $\{x \mid x = 5\} \cup \{x \mid x = 7\}$ $U = X$.

In Probs. 17 to 26 find $P' = \{x \mid \sim p_x\}$ for the given open sentences p_x. In all cases $U = X$.

17. $x = 1$.
18. $x = -4$.
19. $x + 7 = 15$.
20. $2x + 5 = 13$.
21. $x^2 = 16$.
22. $x^2 - 5x + 6 = 0$.
23. $x^2 + 1 = 0$.
24. $x^3 + 1 = 0$.
25. $x^2 \geq 4$.
26. $x^2 \leq 9$.

In Probs. 27 to 36 form two implications from the two given propositions and give their truth values.

27. $3 + 7 = 10$ $4 + 9 = 13$.
28. $6 - 2 = 3$ $4^2 = 15$.
29. $2 + 5 = 7$ $3 + 3 = 8$.
30. $22 + 1 = 221$ $60 = 2 \cdot 30$.
31. This circle of radius 5 in. has an area of 25π in.^{2} This circle of radius 5 in. has a circumference of 10π in.
32. This square of side 4 in. has an area of 16 in.^{2} This square of side 4 in. has a perimeter of 8 in.
33. 24 is a perfect square. 13 is a composite number.
34. "Two in One" is shoe polish. "Three in One" is oil.
35. Philadelphia is in New Jersey. Atlanta is in Georgia.
36. Brooklyn is in Los Angeles. New York is in San Francisco.

In Probs. 37 to 46 derive the truth set of the given implications from the formula of Theorem 4. U is the set X.

37. If $x^2 = 1$, then $x = 1$ or -1.
38. If $x^2 + 7x + 12 = 0$, then $x = -3$ or -4.
39. If $x = 4$, then $x^2 = 16$.
40. If $x^2 - 2x + 1 = 0$, then $x = 1$.
41. If $x^2 = 9$, then $x = -3$.
42. If $x^2 - 6x + 8 = 0$, then $x = 2$.
43. If $x = 3$, then $x \neq 5$.
44. If $x = 4$, then $x \neq 4$.
45. If $x^2 \geq 0$, then $x^4 < 0$.
46. If $x \geq 0$, then $x^2 < 0$.

In Probs. 47 to 54 state whether or not the given propositions are equivalent.

47. $2 + 3 = 5$; $1 + 7 = 8$.
48. $3 - 2 = 1$; $4^2 = 16$.
49. $6 + 4 = 10$; $15^2 = 125$.
50. $5 + 1 = 51$; $26 = 62$.

51. December 7, 1941, was Sunday; Mt. Rainier is in Oregon.
52. Independence Hall is in Boston; the White House is in Washington, D.C.
53. 4 is divisible by 2; 4 is an even number.
54. 16 is a composite number; $16 = 2 \cdot 8$.

In Probs. 55 to 64 find the truth set os the given equivalence. If possible, obtain the answer by inspection, but in any case derive it from Theorem 5. U is the set X.

55. $(x = 1) \leftrightarrow (x = -1)$.
56. $(x^2 = 1) \leftrightarrow [(x = 1) \vee (x = -1)]$.
57. $(x^2 = 0) \leftrightarrow (x = 0)$.
58. $(3x = 6) \leftrightarrow (x = 3)$.
59. $(2x = 8) \leftrightarrow (x = 4)$.
60. $(3x = 9) \leftrightarrow (3x + 2 = 11)$.
61. $(x \neq 1) \leftrightarrow (x^2 \neq 1)$.
62. $(x \geq 0) \leftrightarrow (x \leq 0)$.
63. $(x = 5) \leftrightarrow (x \neq 5)$.
64. $(x^2 + 4 = 0) \leftrightarrow [(x + 2)^2 = 0]$.

1.11 Tautology

In the previous sections we have studied expressions such as $p \wedge q$, $p \vee q$, $\sim p$, $p \rightarrow q$, and $p \leftrightarrow q$. If we wish, we can combine these operations and write longer expressions such as

$$[p \wedge (p \rightarrow q)] \rightarrow q \qquad \text{and} \qquad (p \wedge q) \rightarrow (\sim p)$$

When p and q are specified propositions, the expressions above are also propositions, and we have developed techniques for finding their truth values. It is often useful, however, to talk about these expressions without specifying what propositions p and q stand for and to see what conclusions can be drawn under these circumstances. In order to be definite, let us define a new term.

Definition: An open propositional formula (or *formula*, for short) is an expression containing a finite number of variables, p, q, r, etc., and a finite number of the logical operations: $\wedge$, $\vee$, $\sim$, $\rightarrow$, and $\leftrightarrow$, which becomes a proposition when specific propositions are substituted for the variables.

In general, a formula will not have a truth value, for this will depend upon the truth values of the propositions which are substituted for its variables. There are formulas, however, which result in true propositions when any propositions whatsoever are substituted for their variables. We call such formulas *tautologies*.

Definition: A formula is called a *tautology* if and only if it becomes a true proposition when any propositions whatsoever are substituted for its variables.

In our development of the laws of logic we shall have frequent occasion to prove that certain formulas are tautologies, and so we shall now explain how to do this by means of an extended truth table.

Theorem 6. $p \vee (\sim p)$ is a tautology.

Proof: We write the headings

$$p \qquad \sim p \qquad p \vee (\sim p)$$

Then we form the table

p	$\sim p$	$p \vee (\sim p)$
T	F	T
F	T	T

In this table the first column lists the possible truth values of p. The second column is obtained from the first by use of the truth table for a negation. The third column is derived from the first two and the truth table for a disjunction. Since all entries in the third column are T, we have shown that the given formula is a tautology. This tautology is called the *Law of the Excluded Middle.*

Theorem 7. $\sim[p \wedge (\sim p)]$ is a tautology.

Proof:

p	$\sim p$	$p \wedge (\sim p)$	$\sim[p \wedge (\sim p)]$
T	F	F	T
F	T	F	T

This tautology is called the *Law of Contradiction.*

An important special case is that of an equivalence which is a tautology. We suppose that r and s are two formulas each of which has variables p, q, etc., and that $r \leftrightarrow s$ is a tautology. This implies that, when any propositions are substituted for the variables p, q, etc., in both the formulas r and s, the truth value of r will be equal to the truth value of s. This relationship between two formulas is extremely important and deserves a special name.

Definition: Two formulas r and s are called *equivalent* if and only if the truth value of r is equal to the truth value of s for each choice of propositions which are substituted for their respective variables.

Because of this property a formula r may be replaced in any logical argument by a formula s which is equivalent to it. From the discussion above we also have the result:

Theorem 8. Two formulas r and s are equivalent if and only if $r \leftrightarrow s$ is a tautology.

Exercise A. Prove that the equivalence of formulas has properties 1, 2, and 3 of Sec. 1.10 and hence is a proper equivalence relation.

As an illustration of the equivalence of two formulas, let us prove the next theorem.

Theorem 9. $p \leftrightarrow q$ is equivalent to $[p \wedge q] \vee [(\sim p) \wedge (\sim q)]$.

Proof:

p	q	$\overset{*}{p \leftrightarrow q}$	$p \wedge q$	$\sim p$	$\sim q$	$(\sim p) \wedge (\sim q)$	$\overset{*}{(p \wedge q] \vee [(\sim p) \wedge (\sim q)]}$
T	T	T	T	F	F	F	T
T	F	F	F	F	T	F	F
F	T	F	F	T	F	F	F
F	F	T	F	T	T	T	T

The equivalence follows from the equality of entries in the two starred columns.

Tautologies have important consequences for open sentences which are derived from them. Suppose that we have a tautology t with variables p, q, etc. Let us substitute open sentences p_x, q_x, etc., for p, q, etc., and thus obtain an open sentence t_x. From the fact that t is a tautology we see immediately that t_x is true when any element of the universal set is substituted for x. Let us state this formally.

Theorem 10. Let t be a formula which is a tautology. Then the derived open sentence t_x is true for all $x \in U$.

We may express this conclusion in the alternative forms:

$$\forall_x t_x \text{ is true} \qquad \{x \mid t_x\} = U$$

Illustration 1. Let us apply Theorem 10 to the tautologies of Theorems 6 and 7.

(**a**) From Theorem 6 we have the alternative conclusions:

$$\forall_x[p_x \vee (\sim p_x)] \text{ is true}$$

and

$$\{x \mid p_x \vee (\sim p_x)\} = U$$

Using Theorem 2 of Sec. 1.7, we can rewrite the second of these in the form $\{x \mid p_x\} \cup \{x \mid \sim p_x\} = U$, and using Theorem 3 of Sec. 1.8, we finally write: $P \cup P' = U$, where $P = \{x \mid p_x\}$.

(**b**) From Theorem 7 we similarly derive the conclusion that $(P \cap P')' = U$ or that $P \cap P' = \emptyset$.

Finally, consider the special case of Theorem 10, in which the tautology t is an equivalence $r \leftrightarrow s$, that is, r and s are equivalent formulas. The derived open sentence $r_x \leftrightarrow s_x$ is then true for all $x \in U$. This means that for each x, r_x and s_x have the same truth values; that is, $\{x \mid r_x\} = \{x \mid s_x\}$.

Definition: Two open sentences are said to be *equivalent* if and only if their truth sets are identical.

In view of this definition and the above discussion we can state the most important theorems:

Theorem 11. Two open sentences r_x and s_x are equivalent if they are derived from formulas r and s which are equivalent.

Theorem 12. If r_x and s_x are equivalent, we may replace r_x with s_x in any logical argument.

Exercise B. Show that the notions of the equivalence of two formulas and of two compound open sentences satisfy the three properties of a general equivalence relationship which are given in Sec. 1.10.

Remark: Theorem 11 gives us an effective procedure for finding alternative expressions for the truth sets of compound propositions. This method is best explained by an illustration.

Illustration 2. Find another expression for the truth set of an equivalence of two open sentences, that is, $\{x \mid p_x \leftrightarrow q_x\}$.

Solution: First we look for a propositional formula with variables p and q which involves only the operations $\wedge$, $\vee$, and $\sim$ and which is equivalent to $p \leftrightarrow q$. Theorem 9 gives us just such a compound proposition, namely:

$$[p \wedge q] \vee [(\sim p) \wedge (\sim q)]$$

If we now replace p and q, respectively, with p_x and q_x and use Theorem 11 and the definition preceding it, we conclude that:

$$\{x \mid p_x \leftrightarrow q_x\} = \{x \mid [p_x \wedge q_x] \vee [(\sim p_x) \wedge (\sim q_x)]\}$$

Using Theorems 1, 2, and 3 in turn and writing P and Q for the truth sets of p_x and q_x, respectively, we finally reduce the right-hand side to: $(P \cap Q) \cup (P' \cap Q')$. In this way we have given a formal proof of Theorem 5 of Sec. 1.10.

Problems 1.11

In Probs. 1 to 25 use truth tables to show that the given propositions are tautologies.

1. $[p \rightarrow q] \leftrightarrow [(\sim p) \vee q]$ [Law of Equivalence for Implication and Disjunction].
2. $[\sim(p \wedge q)] \leftrightarrow [(\sim p) \vee (\sim q)]$ [DeMorgan's First Law].
3. $[\sim(p \vee q)] \leftrightarrow [(\sim p) \wedge (\sim q)]$ [DeMorgan's Second Law].
4. $[\sim(\sim p)] \leftrightarrow p$ [Law of Double Negation].
5. $[\sim(p \rightarrow q)] \leftrightarrow [p \wedge (\sim q)]$ [Law of Negation for Implication].
6. $[\sim(p \leftrightarrow q)] \leftrightarrow [(\sim p) \leftrightarrow q]$ [Law of Negation for Equivalence].
7. $[\sim(p \leftrightarrow q)] \leftrightarrow [p \leftrightarrow (\sim q)]$ [Law of Negation for Equivalence].

8. $[p \leftrightarrow q] \leftrightarrow [(p \rightarrow q) \wedge (q \rightarrow p)]$.
9. $[p \wedge (p \rightarrow q)] \rightarrow q$.
10. $[(\sim q) \wedge (p \rightarrow q)] \rightarrow (\sim p)$.
11. $[(\sim p) \wedge (p \vee q)] \rightarrow q$.
12. $(p \wedge q) \rightarrow p$.
13. $[(p \rightarrow q) \wedge (q \rightarrow r)] \rightarrow (p \rightarrow r)$.
14. $[p \rightarrow (q \rightarrow r)] \rightarrow [(p \wedge q) \rightarrow r]$.
15. $[(p \wedge q) \rightarrow r] \rightarrow [p \rightarrow (q \rightarrow r)]$.
16. $[(p \rightarrow q) \wedge (\sim q)] \rightarrow (\sim p)$.
17. $p \rightarrow (p \vee q)$.
18. $p \vee (q \vee r) \leftrightarrow (p \vee q) \vee r$.
19. $p \wedge (q \wedge r) \leftrightarrow (p \wedge q) \wedge r$.
20. $p \wedge (q \vee r) \leftrightarrow (p \wedge q) \vee (p \wedge r)$.
21. $p \vee (q \wedge r) \leftrightarrow (p \vee q) \wedge (p \vee r)$.

22. Let t be a true proposition; then $[p \wedge t] \leftrightarrow p$.
23. Let f be a false proposition; then $[p \vee f] \leftrightarrow p$.
24. $[p \vee p] \leftrightarrow p$.
25. $[p \wedge p] \leftrightarrow p$.

In Probs. 26 to 30 find alternative expressions for the truth sets of the given open sentences. You may use the results of Probs. 1 to 25 in your solutions.

26. $\{x \mid p_x \rightarrow q_x\}$ (see Sec. 1.9, Theorem 4).
27. $\{x \mid p_x \vee (q_x \vee r_x)\}$.
28. $\{x \mid p_x \wedge (q_x \wedge r_x)\}$.
29. $\{x \mid p_x \wedge (q_x \vee r_x)\}$.
30. $\{x \mid p_x \vee (q_x \wedge r_x)\}$.

31. Use the result of Prob. 8 to find an expression for $\{x \mid p_x \leftrightarrow q_x\}$ different from that of Theorem 5.

1.12 Applications to Negations

We now have enough machinery to discuss the negations of propositions which are compound or which involve quantifiers. In Probs. 1.11 we derived a number of tautologies involving the negations of compound propositions, and here we state these results in the following theorem.

Theorem 13. The negations of the basic propositional formulas can be written in the forms below:

	Formula	Negation
Conjunction	$p \wedge q$	$(\sim p) \vee (\sim q)$
Disjunction	$p \vee q$	$(\sim p) \wedge (\sim q)$
Equivalence	$p \leftrightarrow q$	$(\sim p) \leftrightarrow q$ or $p \leftrightarrow (\sim q)$
Implication	$p \rightarrow q$	$p \wedge (\sim q)$

Illustration 1

Proposition	**Negation**
(a) The sky is blue, and the grass is green.	The sky is not blue, or the grass is not green.
(b) The woods are wet, or there are forest fires.	The woods are not wet, and there are not forest fires.
(c) $4^2 = 16$ is equivalent to $3 \times 5 = 15$.	$4^2 \neq 16$ is equivalent to $3 \times 5 = 15$.
(d) If I am working, then I am happy.	I am working, and I am not happy.

Let us now consider the negations of propositions involving quantifiers. Careful analysis suggests the following negations:

Proposition	**Negation**
All men are hungry.	Some men are not hungry.
Some men are soldiers.	All men are not soldiers; i.e., No men are soldiers.

You should carefully examine the reasons for rejecting the following statements as negations for the above propositions.

Incorrect negations

All men are not hungry.
Some men are not soldiers.
Not all men are soldiers.

Our logical problems here are severely complicated by the imprecise usage of "All . . . not . . ." and "Not all . . ." in the English language. H. W. Fowler has a delightful discussion of this point in his estimable book "A Dictionary of Modern English Usage", but we fear that Mr. Fowler is a better linguist than he is a logician. He regrets the use of "All men do not speak German", and prefers "Not all men speak German", when what he really wants to say is "Some men do not speak German", or "It is false that all men speak German". Notice also that the common proverb "All that glitters is not gold" is a logical misstatement of "Some things that glitter are not gold".

For this reason a careful writer will avoid "All . . . not . . ." and "Not all . . ." entirely if he wishes his meaning to be precise. In particular, it is wise to replace the logically correct "All x are not . . ." by "No x are . . ." when we take the negation of "Some x are . . ." On the other hand, the negation of "All x are . . ." is "Some x are not . . .", and there is no ambiguity.

These examples lead us to state the theorem:

Theorem 14. The negations of propositions involving quantifiers are given by the following equivalences:

$$\sim(\forall_x p_x) \leftrightarrow \exists_x(\sim p_x)$$
$$\sim(\exists_x p_x) \leftrightarrow \forall_x(\sim p_x)$$

Proof: We prove only the first of these. Let us first find the truth values of the left-hand side. By definition (Sec. 1.5),

(1) $\begin{cases} \forall_x p_x \text{ is true if and only if } P \text{ is the set } U. \\ \forall_x p_x \text{ is false if and only if } P \text{ is not the set } U. \end{cases}$

Taking the negation of (1) and reversing truth values, we obtain:

(2) $\begin{cases} \sim(\forall_x p_x) \text{ is true if and only if } P \text{ is not the set } U. \\ \sim(\forall_x p_x) \text{ is false if and only if P is the set } U. \end{cases}$

Turning now to the right-hand side, we have, by definition:

(3) $\begin{cases} \exists_x p_x \text{ is true if and only if } P \text{ is not the set } \emptyset. \\ \exists_x p_x \text{ is false if and only if } P \text{ is the set } \emptyset. \end{cases}$

In (3) we substitute $\sim p_x$ for p_x and obtain:

(4) $\begin{cases} \exists_x(\sim p_x) \text{ is true if and only if } P' \text{ is not the set } \emptyset\text{; that is, } P \text{ is not the set } U. \\ \exists_x(\sim p_x) \text{ is false if and only if } P' \text{ is the set } \emptyset\text{; that is, } P \text{ is the set } U. \end{cases}$

The first of the above two equivalences is now verified since (2) and (4) are identical.

Exercise A. Prove the second equivalence in a similar manner.

In forming the negations of compound propositions which involve quantifiers, we must apply Theorems 13 and 14 in turn as in the following illustrations.

Illustration 2

(a) Write the negation of:

Some sailing is dangerous, and all fishing is tedious.

Solution: First we recognize that this is a conjunction and write the negation in the preliminary form:

[$\sim$(Some sailing is dangerous)] or [$\sim$(all fishing is tedious)].

Now we apply Theorem 14 to the separate clauses and obtain:

All sailing is not dangerous, or some fishing is not tedious

or No sailing is dangerous, or some fishing is not tedious.

(b) Write the negation of:

If the nation is prosperous, then some people are rich.

Solution: First we recognize that this is an implication and write the negation in the preliminary form:

[The nation is prosperous] and [$\sim$(some people are rich)].

Now we apply Theorem 14 to the second clause and obtain:

The nation is prosperous, and all people are not rich

or The nation is prosperous, and no person is rich.

(c) Write the negation of:

$$\forall_x[(x \neq 0) \rightarrow (x^2 > 0)]$$

Solution: First we must apply Theorem 14 to handle the quantifier. This gives:

$$\exists_x(\sim[(x \neq 0) \rightarrow (x^2 > 0)])$$

Now apply Theorem 13 inside the quantifier and obtain:

$$\exists_x[(x \neq 0) \wedge (x^2 \leq 0)]$$

(d) Write the negation of:

If two triangles are congruent, their medians are equal.

Solution: This implication is actually shorthand for the following implication involving a quantifier:

$\forall_x$ [If the triangles of x are congruent, the medians of the triangles of x are equal], where U is the set of pairs of triangles in the plane.

This is of the form $\forall_x p_x$, whose negation is $\exists_x[\sim p_x]$. Since p_x is an implication, the correct negation is

$\exists_x$ [The triangles of x are congruent, and the medians of the triangles of x are not equal.]

A smoother statement of this is:

There exists a pair of triangles which are congruent and whose medians are not equal.

Problems 1.12

In Probs. 1 to 10 write the negation of each of the given formulas. In each case reduce the result to one in which the $\sim$ sign is applied only to the symbols p, q, r.

1. $p \rightarrow (\sim q)$.
2. $(\sim p) \vee q$.
3. $(p \wedge q) \vee r$.
4. $r \wedge (p \rightarrow q)$.
5. $[p \wedge (\sim r)] \rightarrow q$.
6. $p \rightarrow (q \rightarrow r)$.
7. $(p \vee r) \vee (\sim q)$.
8. $(q \vee r) \rightarrow (p \wedge r)$.
9. $(p \vee q) \rightarrow (\sim q \vee r)$.
10. $p \vee (q \vee r)$.

In Probs. 11 to 26 find the negations of the given propositions.

11. Some numbers are even.
12. Some professors are bald.
13. All rooms are taken.
14. All students are young.
15. $\exists_x(x^2 < 0)$.
16. $\exists_x(x^2 + 5x + 4 = 0)$.
17. $\forall_x(x^2 \geq 0)$.
18. $\forall_x[x^2 + 25 = (x - 5)(x + 5)]$.
19. For all triangles, the sum of the measures of the interior angles is 180°.
20. For some circles the circumference is not π times the diameter.
21. If some people are rich, some people are poor.
22. If all newspapers are bad, then some people are misinformed.
23. All roads lead to Rome, and some roads are impassable.
24. Some houses are expensive, and some houses have views.
25. All windows are open, or I am too hot.
26. Some men are soldiers, or all men are slaves.

In Probs. 27 to 32 insert the implied quantifiers and find the negations.

27. If two angles are congruent, their measures are equal.
28. If p is a prime, there is a prime larger than p.
29. Let x be an integer. If x^2 is even, then x is even.
30. If point A is between points B and C, then C is not between A and B.
31. If two triangles are congruent, their pairs of corresponding altitudes are equal.
32. If the bases of two rectangles are equal, their areas are proportional to their altitudes.

1.13 Applications to Set Theory

The tautologies proved in Sec. 1.11 permit us to derive a number of important formulas in set theory. We assume a given universal set and that all sets mentioned in this section are subsets of U. We shall demonstrate our general method by giving the details in a typical example.

Let us begin with the tautology

$$[p \wedge (q \wedge r)] \leftrightarrow [(p \wedge q) \wedge r]$$

which is Prob. 19 in Sec. 1.11. If we substitute open sentences for the variables of this tautology, we obtain the following open sentence which is true for all $x \varepsilon U$:

$$[p_x \wedge (q_x \wedge r_x)] \rightarrow [(p_x \wedge q_x) \wedge r_x]$$

Hence we know that

$$\{x \mid p_x \wedge (q_x \wedge r_x)\} = \{x \mid (p_x \wedge q_x) \wedge r_x\}$$

By repeated application of Theorem 1, we find that

$$\{x \mid p_x \wedge (q_x \wedge r_x)\} = P \cap (Q \cap R)$$

and that

$$\{x \mid (p_x \wedge q_x) \wedge r_x\} = (P \cap Q) \cap R$$

Hence

$$P \cap (Q \cap R) = (P \cap Q) \cap R$$

This formula has therefore been proved for all P, Q, and R, which are the truth sets of open sentences. Is it true for all subsets of U?

In order to answer this question let us prove Theorem 15.

Theorem 15. Every subset of U is the truth set of an open sentence.

Proof: Let A be a subset of U. Then A is the truth set of the open sentence: $x \varepsilon A$.

Theorem 15 removes the restriction on our formula which required P, Q, and R to be truth sets. Hence we have proved the result:

Theorem 16. For any P, Q, and R which are subsets of U,

$$P \cap (Q \cap R) = (P \cap Q) \cap R$$

This method can be applied quite generally, as stated in the following theorem.

Theorem 17. Suppose that we are given two formulas whose variables are p, q, r, etc., which are equivalent and which involve only the operators $\wedge$, $\vee$, and $\sim$. Then we shall obtain a true relationship among arbitrary subsets P, Q, R, etc., of U by replacing p by P, q by Q, r by R, etc., $\wedge$ by $\cap$, $\vee$ by $\cup$, $\sim$ by $'$, and $\leftrightarrow$ by $=$.

Illustration 1

(a) From the tautology $\sim(p \wedge q) \leftrightarrow (\sim p) \vee (\sim q)$ we obtain the equality $(P \cap Q)' = P' \cup Q'$.

(b) From the tautology $[p \wedge (q \vee r)] \leftrightarrow [(p \wedge q) \vee (p \wedge r)]$ we obtain the equality $P \cap (Q \cup R) = (P \cap Q) \cup (P \cap R)$.

Finally, we shall develop an equivalence between set inclusion and implication. Let p_x and q_x be open sentences with truth sets P and Q. Then $P \subseteq Q$ and $\forall_x(p_x \to q_x)$ are propositions which may be true or false. We shall prove the result:

Theorem 18. $(P \subseteq Q) \leftrightarrow [\forall_x(p_x \to q_x)]$.

It suffices to show that when $P \subseteq Q$ is false, then $\forall_x(p_x \to q_x)$ is false, and that when $\forall_x(p_x \to q_x)$ is false, then $P \subseteq Q$ is false.

Assume first that $P \subseteq Q$ is false. Then there is an x for which p_x is true and q_x is false. For such an x, $p_x \to q_x$ is false, and hence $\forall_x(p_x \to q_x)$ is false.

Conversely, assume $\forall_x(p_x \to q_x)$ is false. Then there is some x for which $p_x \to q_x$ is false. This can happen if and only if for this x: p_x is true and q_x is false. The existence of such an x, however, shows that $P \subseteq Q$ is false.

Exercise A. Show that $[\forall_x(p_x \to q_x)] \leftrightarrow [(P \cap Q') = \emptyset]$ and hence that $(P \subseteq Q) \leftrightarrow [(P \cap Q') = \emptyset]$.

Theorem 18 permits us to give a pictorial representation for theorems in mathematics which are expressed as implications that are asserted to be true for all objects under consideration.

Illustration 2. The theorem:

If x is a prime larger than 2, then it is odd

can be illustrated as follows. Let the plane represent the universal set (all positive integers), and let circular disks represent the truth sets of the two clauses of this implication. Namely, let one circular disk represent the set of odd numbers, and another circular disk represent the set of primes larger than 2. The implication is thus equivalent to the inclusion of the second of these in the first as illustrated in Fig. 1.7.

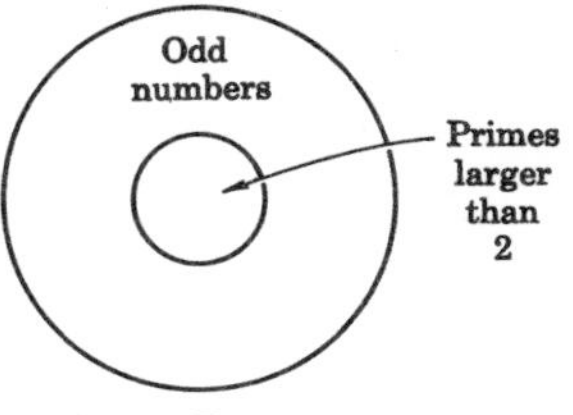

Figure 1.7

1.14 Implications in Mathematics

Theorems in mathematics fall into two groups, which we shall call *specific* and *general*. A specific theorem states a fact about a particular mathematical object, such as:

7 is a prime number.
3.1416 is an approximate value for π.

General theorems express relationships which are true for all members of a stated set of objects, such as:

The base angles of an isosceles triangle are equal.

For any two real numbers, a and b, $a + b = b + a$.

In this section we shall be interested in those general theorems which have the form $\forall_x(p_x \rightarrow q_x)$ or which can be rewritten in this form. For example, we can rephrase the first of the above general theorems as follows:

> $\forall_x$ (triangles): If x is isosceles, then the base angles of x are equal.

If we start from the implication $\forall_x(p_x \rightarrow q_x)$, we can state two other implications which are related to it.

Converse: $\forall_x(q_x \rightarrow p_x)$
Contrapositive: $\forall_x[(\sim q_x) \rightarrow (\sim p_x)]$

Suppose that we know that $\forall_x(p_x \rightarrow q_x)$ is true, what can we say about the truth of its converse and its contrapositive? A simple approach is to consider the diagrams formed according to Theorem 18.

From Fig. 1.8*a* and *b* it is evident that an implication and its converse

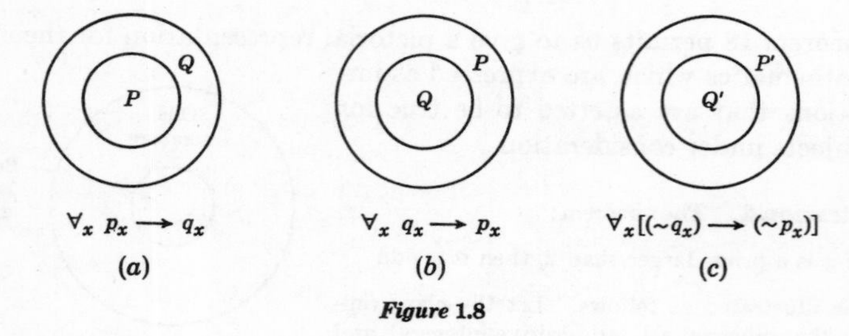

***Figure* 1.8**

are quite separate statements and that the truth of one cannot be concluded from the truth of the other.

Illustration 1

Implication: If two triangles are congruent, then they are similar. (True)
Converse: If two triangles are similar, then they are congruent. (False)

On the other hand, the converse of a true implication may well be true. This clearly happens if and only if the sets P and Q in Fig. 1.8 are identical.

Illustration 2

Implication: If $x = 3$, then $x + 5 = 8$. (True)
Converse: If $x + 5 = 8$, then $x = 3$. (True)

Note that the quantifier $\forall_x$ is implied here so that a full statement of the implication is: $\forall_x$ (real numbers), [If $x = 3$, then $x + 5 = 8$]. As was stated earlier, it is customary to omit such quantifiers since they can easily be supplied by the reader.

Exercise A. Rewrite the implications of Illustration 1 so that the proper quantifier is stated explicitly.

From Fig. 1.8 we can conclude immediately that an implication and its contrapositive are equivalent. In Fig. 1.8*a* the region outside the circle enclosing P is the set P', and the region outside the circle enclosing Q is the set Q'. Hence $Q' \subseteq P'$, which is the configuration represented by Fig. 1.8*c*. By reversing the argument from Fig. 1.8*c* to Fig. 1.8*a*, we complete the proof of the theorem:

Theorem 19. An implication of the form $\forall_x(p_x \rightarrow q_x)$ and its contrapositive are equivalent propositions.

Illustration 3

Implication: If $x = 3$, then $x^2 = 9$.
Contrapositive: If $x^2 \neq 9$, then $x \neq 3$.

Because of this theorem we can prove the truth of a theorem by proving the truth of its contrapositive. We shall proceed in this way whenever the proof of the contrapositive is easier than the proof of the given theorem.

Mathematicians frequently express implications in language different from that used above, and consequently you must learn to recognize implications even when they are disguised in a fashion which may seem confusing at first.

Definition: The following six statements all carry the same meaning.

$\forall_x(p_x \rightarrow q_x)$.
$P \subseteq Q$.
p_x is sufficient for q_x.
q_x is necessary for p_x.
If p_x, then q_x.
Only if q_x, then p_x.

The last four statements can be understood by referring to Fig. 1.8*a*. From this we see that:

For x to be in Q, it is sufficient that it be in P.
For x to be in P, it is necessary that it be in Q.
If x is in P, then x is in Q.
Only if x is in Q is it in P.

Illustration 4. The implication, "If a polygon is a square, then it is a rectangle", can be rewritten in the following equivalent ways:

The fact that a polygon is a square is a sufficient condition that it is a rectangle.
The fact that a polygon is a rectangle is a necessary condition that it is a square.
A polygon is a square only if it is a rectangle.
The first two of these may be rephrased:
A sufficient condition that a polygon is a rectangle is that it is a square.
A necessary condition that a polygon is a square is that it is a rectangle.

Remark. From the definition above it follows that the converse of an implication can be obtained in any of the following ways, depending on how the given implication is phrased:

Interchange p_x and q_x.
Interchange P and Q.
Replace "necessary" by "sufficient".
Replace "sufficient" by "necessary".
Replace "if" by "only if".
Replace "only if" by "if".

In a similar fashion, the equivalence $\forall_x(p_x \leftrightarrow q_x)$ can be expressed in a number of alternative ways.

Definition: The following six statements all carry the same meaning:

$\forall_x(p_x \leftrightarrow q_x)$.
$P = Q$.
p_x is necessary and sufficient for q_x.
q_x is necessary and sufficient for p_x.
p_x if and only if q_x.
q_x if and only if p_x.

Illustration 5. The equivalence, "A triangle is equilateral if and only if it is equiangular", can be rewritten in the following equivalent ways:

A triangle is equiangular if and only if it is equilateral.
A necessary and sufficient condition that a triangle is equilateral is that it is equiangular.
A necessary and sufficient condition that a triangle is equiangular is that it is equilateral.

Problems 1.14

In Probs. 1 to 10 write the relationship between sets which is derived from the given equivalence.

1. $[\sim(p \wedge q)] \leftrightarrow [(\sim p) \vee (\sim q)]$.
2. $[\sim(p \vee q)] \leftrightarrow [(\sim p) \wedge (\sim q)]$.
3. $[\sim(\sim p)] \leftrightarrow p$.
4. $p \vee (q \vee r) \leftrightarrow (p \vee q) \vee r$.
5. $p \wedge (q \wedge r) \leftrightarrow (p \wedge q) \wedge r$.
6. $p \vee (q \wedge r) \leftrightarrow (p \vee q) \wedge (p \vee r)$.
7. $(p \vee p) \leftrightarrow p$.
8. $(p \wedge p) \leftrightarrow p$.
9. $(p \wedge t) \leftrightarrow p$ (t is always true).
10. $(p \vee f) \leftrightarrow p$ (f is always false).

In Probs. 11 to 14 draw a diagram like Fig. 1.7 which illustrates the given implication.

11. If two triangles are congruent, then they are similar.
12. If a triangle is equilateral, then it is isosceles.
13. If a polygon is a square, then it is a rectangle.
14. If an integer is a multiple of 4, then it is even.

In Probs. 15 to 20 state the converse and the contrapositive of the given implication.

15. If a is divisible by 4, then $2a$ is divisible by 8.
16. If the angles of a triangle are all equal, then the triangle is equilateral.
17. If a quadrilateral is a parallelogram, then its diagonals bisect each other.
18. If $x \neq 0$, then $x^2 > 0$.
19. If $a > b$, then $a - c > b - c$.
20. If r is a solution of $a + x = b$, then $-r$ is a solution of $b + x = a$.

21. Using truth tables, show that $p \to q$ is not equivalent to $q \to p$.
22. Using truth tables show that $(p \to q) \leftrightarrow [(\sim q) \to (\sim p)]$.
23. Write the contrapositive of the converse of $\forall_x(p_x \to q_x)$. This is called the "inverse".
24. Write the converse of the contrapositive of $\forall_x(p_x \to q_x)$. Show that this is the "inverse" (see Prob. 23).

In Probs. 25 to 30, write the given implication using the "sufficient condition" language.

25. If the base angles of a triangle are equal, the triangle is isosceles.
26. If two triangles are congruent, their corresponding altitudes are equal.
27. If two lines are perpendicular to the same line, they are parallel.
28. If two spherical triangles have their corresponding angles equal, they are congruent.
29. If $3x + 2 = x + 4$, then $x = 1$.
30. If $x^2 = 0$, then $x = 0$.

In Probs. 31 to 36, write the given implication, using the "necessary condition" language.

31. If a triangle is inscribed in a semicircle, then it is a right triangle.
32. If $x = 3$, then $x^2 = 9$.
33. If a body is in static equilibrium, the vector sum of all forces acting on it is zero.
34. If a body is in static equilibrium, the vector sum of the moments of all forces acting on it is zero.
35. If two forces are in equilibrium, they are equal, opposite, and collinear.
36. If three nonparallel forces are in equilibrium, their lines of action are concurrent.

In Probs. 37 to 42, write the given implication using the phrase "only if".

37. The implication of Prob. 31.
38. The implication of Prob. 32.
39. The implication of Prob. 33.
40. The implication of Prob. 34.
41. The implication of Prob. 35.
42. The implication of Prob. 36.

In Probs. 43 to 48 write the converse of the given implication using "necessary" and then "sufficient" language. Give two answers to each problem.

43. The implication of Prob. 25.
44. The implication of Prob. 26.
45. The implication of Prob. 27.
46. The implication of Prob. 28.
47. The implication of Prob. 29.
48. The implication of Prob. 30.

In Probs. 49 to 54, write the converse of the given implication using the phrase "only if"

49. The implication of Prob. 31.
50. The implication of Prob. 32.
51. The implication of Prob. 33.
52. The implication of Prob. 34.
53. The implication of Prob. 35.
54. The implication of Prob. 36.

In Probs. 55 to 58, write the given equivalence in "necessary and sufficient" language.

55. Two lines are parallel if and only if they are equidistant.
56. An integer is even if and only if it is divisible by 2.
57. Three concurrent forces are in equilibrium if and only if their vector sum is zero.
58. A lever is balanced if and only if the algebraic sum of all moments about its fulcrum is zero.
59. A man promised his girl: "I will marry you only if I get a job." He got the job and refused to marry her. She sued for breach of promise. Can she logically win her suit? Why?

1.15 Methods of Proof (Direct Proof)

We shall first extract from our previous discussion two statements which are commonly used in constructing mathematical proofs.

Law of Detachment. Let $\forall_x(p_x \rightarrow q_x)$ be true, and let a be in the truth set P of p_x. Then a is in the truth set Q of q_x.

This is nothing but a rephrasing of Theorem 18.

Law of Substitution. At any stage of a proof we may replace a proposition or an open sentence by an equivalent one.

There are two important applications of this law:

(1) At any stage of a proof we may replace an implication by its contrapositive

(2) Instead of proving that $\forall_x(p_x \rightarrow q_x)$ is true, we may prove the truth of the equivalent proposition: $P \subseteq Q$.

The usual proofs of theorems, indeed, make essential use of this second application. Suppose that you are proving the following proposition: $\forall_x$: if x is even, then x^2 is even. Immediately you convert this to the proposition: $\{x \mid x \text{ is even}\} \subseteq \{x \mid x^2 \text{ is even}\}$. Usually you are not even aware of this conversion and begin the argument with the statement: "x is even by hypothesis." Then you proceed to prove that x^2 is even. This proof, however, is nothing more than the proof of the set inclusion written above.

Direct proof proceeds in a great variety of fashions, but the following outline is quite typical. Suppose that the following implications are known to be true by virtue of our axioms, definitions, or previously proved theorems:

Given: (1) $\forall_x(p_x \rightarrow q_x)$
(2) $\forall_x[(\sim r_x) \rightarrow (\sim q_x)]$
(3) $\forall_x(r_x \leftrightarrow s_x)$

Then we are asked to prove:

(4) $\forall_x(p_x \rightarrow s_x)$, or $P \subseteq S$

Proof:
(1) Assume $x \in P$, or "p_x is true by hypothesis".
(2) $\forall_x(p_x \rightarrow q_x)$ [Given]
(3) q_x is true for $x \in P$ [Detachment]
(4) $\forall_x[(\sim r_x) \rightarrow (\sim q_x)]$ [Given]
(5) $\forall_x(q_x \rightarrow r_x)$ [Substitution]
(6) r_x is true for $x \in P$ [Detachment]
(7) $\forall_x(r_x \leftrightarrow s_x)$ [Given]
(8) s_x is true for $x \in P$ [Detachment]
(9) $P \subseteq S$ [Definition]

The problem of constructing a proof is to arrange a sequence of steps of this kind which leads to the desired conclusion. There is no automatic way of doing this; you must develop skill through experience and the use of your originality.

Illustration 1. Prove that the square of an odd number is odd.
Let us rephrase this as follows:

$\forall_x$ (integers): If x is odd, then x^2 is odd.

By the Law of Substitution we write this:

$\{x \mid x \text{ is odd}\} \subseteq \{x \mid x^2 \text{ is odd}\}$

Now we construct the proof.

Proof:
(1) x is odd. [Hypothesis]
(2) There is an integer a such that $x = 2a + 1$. [Definition]
(3) $x^2 = (2a + 1)^2$ [Theorems of algebra]
$= 4a^2 + 4a + 1$
$= 2(2a^2 + 2a) + 1.$
(4) x^2 is odd. [Definition]

1.16 Other Methods of Proof

(*a*) *Indirect Proof.* If you have difficulty in constructing a direct proof, you can sometimes make progress by using other tactics. The method of "indirect proof" relies on the fact that if $\sim p$ is false, then p is true. Hence, to prove that p is true, we attempt to show that $\sim p$ is false. The best way to accomplish this is to show that $\sim p$ is not consistent with the given statements. In other words, we assume that $\sim p$ is true and add it to the list of known statements and attempt to show that this augmented list of statements leads to a contradiction. When the contradiction is reached, we know that the truth of $\sim p$ is not consistent with our given true statements and that hence it is false. Therefore p is true.

Let us apply this method to prove that the implication $\forall_x(p_x \rightarrow q_x)$ is true. We assume that $\sim[\forall_x(p_x \rightarrow q_x)]$ is true.

First we must rewrite this using the rules for forming negations. The result is:

$$\exists_x[p_x \wedge (\sim q_x)] \text{ is true.}$$

This means that there is an x for which both p_x and $\sim q_x$ are true. Consequently, we suppose that there is an example for which the hypothesis of the implication is true and its conclusion is false and proceed to find a contradiction. Such a contradiction shows that

$$\exists_x[p_x \wedge (\sim q_x)] \text{ is false} \qquad \text{and} \qquad \forall_x(p_x \rightarrow q_x) \text{ is true.}$$

Illustration 1. Prove: If two lines are cut by a transversal so that a pair of alternate interior angles are equal, the lines are parallel.

We assume the existence of an example consisting of two lines AB and CD such that $\angle 1 = \angle 2$ and AB is not parallel to CD. Let AB and CD intersect at O. Then, in ΔMNO, $\angle 1$ is an exterior angle and $\angle 2$ is an interior angle. But then $\angle 1$ is greater than $\angle 2$ by a known theorem. Hence there is a contradiction with the given fact that $\angle 1 = \angle 2$. This contradiction establishes the proof of the theorem.

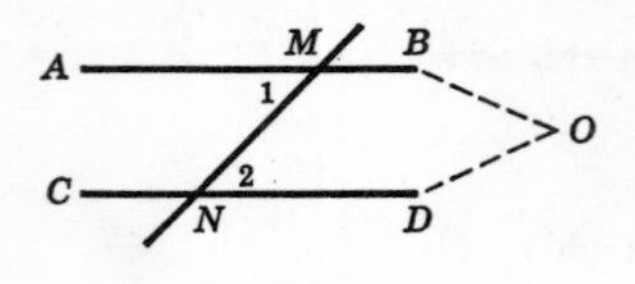

Figure 1.9

(*b*) *Use of the Contrapositive.* Instead of proving $\forall_x(p_x \rightarrow q_x)$, we can just as well prove $\forall_x[(\sim q_x) \rightarrow (\sim p_x)]$, the contrapositive. Often there are great similarities between this method and indirect proof. Let us consider the theorem of Illustration 1.

Illustration 2. The contrapositive of the given implication is: If two lines are not parallel, then the alternate interior angles obtained by cutting these lines by a transversal are not equal.

We establish this by the precise argument used in Illustration 1. Hence the given implication is true.

(*c*) *Proof of Existence.* Before you spend a lot of time and money (on a high-speed computer, say) trying to solve a problem, it is a good idea to determine in advance that the problem actually does have a solution. You have probably never seen problems that do not have solutions, for most textbooks and teachers consider it to be bad form to ask students to do something which is impossible. In actual practice, however, such problems may arise, and it is a good idea to know how to recognize them. A very simple example of such a problem is the following:

Find all integers x which satisfy the equation

$$7x + 5 = 2x + 9$$

There is no solution.

In order to reassure you that you are working on problems that do have solutions, mathematicians have developed a number of *existence theorems.* These are statements of the form: $\exists_x p_x$; that is,

There exists a number x which has a given property.

An important example of such a theorem is this one:

$\exists_x$ (x a real number): [If a and b are real numbers with $a \neq 0$, then $ax + b = 0$].

The best way of proving such a theorem is to exhibit a number x with the required property. The proof of the above theorem amounts to checking that $x = -(b/a)$ satisfies the given equation.

Although there are other forms of existence proofs, a constructive proof of this kind is considered to be of greater merit, and this method is widely used in establishing the existence of solutions of various types of equations.

1.17 Methods of Disproof

If you have tried unsuccessfully to prove a conjectured theorem, you may well spend some time trying to disprove it. There are two standard methods for disproving such statements.

(*a*) *Disproof by Contradiction.* In this case we assume that the given statement is true and then proceed to derive consequences from it. If we succeed in arriving at a consequence which contradicts a known theorem, we have shown that the given statement is false.

Illustration 1. Disprove the statement: "The square of every odd number is even."

Of course, this immediately contradicts our previous result (Sec. 1.15, Illustration 1) that the square of every odd number is odd. But let us disprove it from first principles. Since every odd number can be written in the form $2a + 1$, where a is an integer, and since every even number can be written in the form $2b$, where b is an integer, the given statement implies that:

$$(2a + 1)^2 = 2b \qquad \text{for some } a \text{ and } b$$

or

$$4a^2 + 4a + 1 = 2b$$

Both sides are supposed to represent the same integer, but the left-hand side *is not* divisible by 2, while the right-hand side *is* divisible by 2. This is surely a contradiction, and so the given statement is false.

(b) *Disproof by Counterexample.* This method is effective in disproving statements of the form: $\forall_x p_x$. An example is the following:

$$\forall_x[x^2 + 16 = (x + 4)(x - 4)]$$

In order to disprove such an assertion, we proceed to find a "counterexample". In other words, we look for *one* value of x for which the statement is false; and since the statement was supposed to be true for *all* values of x, this single counterexample is the end of the matter. In the above example, $x = 0$ does the job.

Illustration 2. Disprove the statement: "The square of every odd number is even."

All that we have to do is to find a single odd number whose square is odd. Since $3^2 = 9$, we have established the disproof.

We close with this warning: Although *disproof* by counterexample is a valid procedure, theorems are not to be *proved* by verifying them in a number of special cases. Be sure that you do not confuse these two ideas.

Problems 1.17

In Probs. 1 to 18 you are given a series of mathematical statements, some of which are true and some of which are false. Prove those which are true, and disprove those which are false.

1. The sum of three odd integers is odd.
2. The product of two odd integers is odd.
3. $\forall_x: 3x^2 - 5x - 2 = (3x + 1)(x - 2)$.
4. $\exists_x: 2x + 3 = 4x - 2 - 2x$.
5. $\exists_x: 2^x = 16$.
6. $\forall_x: 2^{x+3} = 8(2^x)$.
7. The sum of the roots of: $x^2 + 5x + 4 = 0$ is equal to 5.
8. The product of the roots of: $x^2 - 6x - 16 = 0$ is equal to 16.
9. $\forall_x: (x + 2)^2 = x + 4$.
10. $\forall_x: (x + 4)^3 = x^3 + 64$.
11. Two triangles are congruent if two sides and the angle opposite one of these of one triangle are equal, respectively, to the corresponding parts of the other triangle.
12. If two triangles are similar, then they have the same area.
13. A sufficient condition that $ax^2 + bx + c = 0$ $(a \neq 0)$ have a real root is that $b^2 - 4ac > 5$.

14. A necessary condition that $ax^2 + bx + c = 0$ $(a \neq 0)$ have a real root is that $b^2 - 4ac = 0$.
15. Provided that $a \neq 0$, every equation of the form: $ax + b = c$ has a solution.
16. Provided that $a \neq b$, every equation of the form: $(x - a)(x - b) = 0$ has two distinct solutions.
17. The sum of the exterior angles of any triangle is equal to 180°.

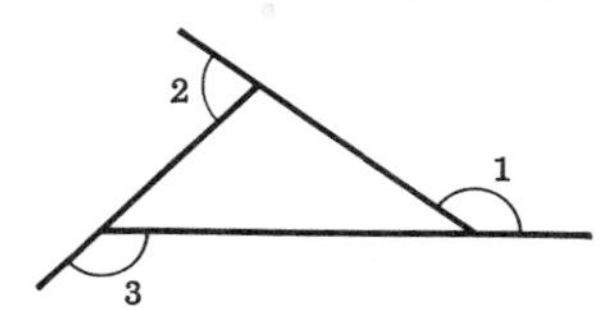

Figure 1.10

18. Any two medians of a triangle bisect each other.
19. You are given the following axiom: "One and only one line can be drawn through any two points." Prove: "Any two distinct lines meet in at most one point." HINT: Use indirect proof.
20. You are given the theorems: "(1) If $ax = 0$ and $a \neq 0$, then $x = 0$." "(2) Provided that $a \neq 0$, every equation of the form $ax + b = 0$ has a solution." Prove that every equation of this form has at most one solution. HINT: Use indirect proof.
21. You are given the theorem: "Every equation of the form $a + x = b$ has a solution." Prove that every equation of this form has at most one solution. HINT: Use indirect proof.
22. You are given the theorem: "At most one circle can be drawn through three, distinct points." Prove that two distinct circles can intersect in at most two points. HINT: Use indirect proof.
23. Give an indirect proof of the theorem: "There exist an infinite number of primes." If you are unable to do so, consult Courant and Robbins, "What Is Mathematics?" p. 22. This theorem is due to Euclid.

References

Carroll, Lewis: "Logical Nonsense", Putnam, New York (1934).
Courant, Richard, and Herbert Robbins: "What is Mathematics?" Oxford, New York (1941).
Stabler, E. R.: "An Introduction to Mathematical Thought", Addison-Wesley, Reading, Mass. (1953).
Stoll, Robert R: "Sets, Logic, and Axiomatic Theories", Freeman, San Francisco (1961).
Suppes, Patrick: "Introduction to Logic", Van Nostrand, Princeton, N.J. (1957).
Tarski, Alfred: "Introduction to Logic", Oxford, New York (1946).

Also consult the following articles in the *American Mathematical Monthly:*

Hempel, C. G.: Geometry and Empirical Science, vol. 52, p. 7 (1945).
Hempel, C. G.: On the Nature of Mathematical Proof, vol, 52, p. 543 (1945).
MacLane, Saunders: Symbolic Logic, vol. 46, p. 289 (1939).
Wilder, R. L.: The Nature of Mathematical Proof, vol. 51, p. 309 (1944).

2 | Number Fields

2.1 Introduction

Since numbers are basic ideas in mathematics, we shall devote this chapter to a discussion of the most important properties of our number system. We do not give a complete account of this subject, and you are likely to study it in more detail when you take more advanced courses in mathematics. Numerous suggestions for further reading are given at the end of the chapter.

Let us retrace briefly the development of numbers as it is usually presented in schools. As a young child you first learned to count, and thus became acquainted with the *natural numbers* 1, 2, 3, ... In your early study of arithmetic you learned how to add, subtract, multiply, and divide pairs of natural numbers. Although some divisions such as $6 \div 3 = 2$ were possible, it soon developed that new numbers had to be invented so as to give meaning to expressions like $7 \div 2$ and $3 \div 5$. To handle such situations, fractions were introduced, and the arithmetic of fractions was developed.

It should be noted that the invention of fractions was a major step in the development of mathematics. In the early days many strange practices were followed. The Babylonians considered only fractions whose denominators were 60, the Romans only those whose denominators were 12. The Egyptians insisted that the numerators must be 1, and wrote $\frac{1}{3} + \frac{1}{15}$ instead of $\frac{2}{5}$. Our modern notation dates from Leonardo of Pisa (also called Fibonacci), whose great work *Liber Abaci* was published in A.D. 1202.

Later on you became acquainted with zero and negative numbers such as -7, -3, $-\frac{5}{3}$, $-4\frac{1}{5}$, etc., and you learned how to calculate with these. The entire collection consisting of the positive and negative integers, zero, and the positive and negative fractions is called the system of *rational numbers.* The advantage of using this system in contrast to the system of purely positive numbers is that it is possible to subtract any rational number from any rational number. With only positive numbers available, 3 − 5, for instance, is meaningless. It is interesting to note that it took many years before negative numbers were permanently established in mathematics. Although they were used to some extent by the early Chinese, Indians, and Arabs, it was not until the beginning of the seventeenth century that mathematicians accepted negative numbers on an even footing with positive numbers.

When you were introduced to *irrational* numbers such as $\sqrt{2}$ and π, you were told that these could not be expressed as ordinary fractions. Instead, they are written as infinite decimal expansions such as 1.4142 . . . and 3.1415 . . . . The decimal expansions of the rational numbers are also infinite; for example,

$$\begin{aligned} \tfrac{1}{4} &= 0.25000\ \ldots \\ \tfrac{1}{3} &= 0.33333\ \ldots \\ 2 &= 2.00000\ \ldots \\ \tfrac{1}{7} &= 0.142857142857\ \ldots \end{aligned}$$

These, however, repeat after a certain point, whereas the irrationals do not have this property. The collection of all these, the rationals plus the irrationals, is called the system of *real* numbers. It is quite difficult to give a completely satisfactory definition of a real number, but for our present purposes the following will suffice:

Definition: A *real number* is a number which can be represented by an infinite decimal expansion.

If you wish a more subtle definition of a real number, read Courant and Robbins, "What Is Mathematics?", chap. 2.

Although a real number is a definite mathematical object, we can express such a number in a great variety of notations. For example, we can write 7 in the following ways:

$$\text{VII},\ 111_{\text{two}}\ (\text{base two}),\ \tfrac{21}{3},\ 7.000\ \ldots,\ 9 - 2$$

The rational number usually written $\frac{1}{2}$ can also be written:

$$\frac{2}{4}, \frac{8\pi}{16\pi}, 0.5000\ \cdots, \frac{\frac{1}{8}}{\frac{1}{4}}, \frac{1}{4} + \frac{1}{4}, \left(\frac{1}{\sqrt{2}}\right)^2, \sqrt{\frac{1}{4}}$$

For each real number it is customary to adopt a "simplest" expression which is commonly used to represent it (7 and $\frac{1}{2}$ in the examples above), but we shall not hesitate to use other representations when they are more convenient.

2.2 Addition of Real Numbers

Addition is defined for *pairs* of real numbers such as $2 + 3 = 5$, $-3 + 2\frac{1}{2} = -\frac{1}{2}$, etc. Indeed, the sum of every pair of real numbers is defined as a third real number. We give this property the name "closure" and write the following law.

Closure law of addition. The sum $a + b$ of any two real numbers is a unique real number c.

This property of closure may seem so simple that you wonder why we mention it at all. Let us consider some situations where closure is not true.

Illustration 1

(**a**) The sum of two *odd* numbers is *not* an odd number.
(**b**) The sum of two *irrational* numbers is *not necessarily* irrational, for $(2 + \sqrt{3}) + (4 - \sqrt{3}) = 6$; i.e., the irrational numbers are not closed under addition.
(**c**) The sum of two *prime* numbers is *not necessarily* a prime, for $7 + 11 = 18$; i.e., the prime numbers are not closed under addition.

You are very familiar with the fact that the order of addition is not important. For instance, $2 + 4 = 4 + 2$, $-3 + \pi = \pi + (-3)$, etc. To describe this property, we say that addition is "commutative" and write the following law.

Commutative law of addition. $a + b = b + a$.

It is slightly more difficult to add three numbers such as $2 + 4 + 7$, for addition is defined for *pairs* of real numbers and not for *triples*. Normally, we first add $2 + 4 = 6$, and then add $6 + 7 = 13$. But we could just as well have added $4 + 7 = 11$ and then $2 + 11 = 13$. That is, $(2 + 4) + 7 = 2 + (4 + 7)$. To describe this property, we say that addition is "associative" and write the following law.

Associative law of addition. $(a + b) + c = a + (b + c)$.

Actually, the sum $a + b + c$ of three real numbers needs to be defined; for originally we knew only how to add two numbers, $a + b$. Therefore we make the following definition:

Definition: $a + b + c$ is defined to be the sum $(a + b) + c$.

We now prove a theorem which illustrates the fact that the sum of three real numbers is the same regardless of the order in which the addition is performed.

Theorem 1. $a + b + c = c + b + a$.

Proof:

$$\begin{aligned} a + b + c &= (a + b) + c && \text{[Definition]} \\ &= (b + a) + c && \text{[Commutative Law]} \\ &= c + (b + a) && \text{[Commutative Law]} \\ &= (c + b) + a && \text{[Associative Law]} \\ &= c + b + a && \text{[Definition]} \end{aligned}$$

In a similar fashion we can define the sum of four real numbers.

Definition: $a + b + c + d$ is defined to be the sum

$$(a + b + c) + d$$

The number zero plays a special role in addition; the sum of zero and any real number a is a itself:

$$a + 0 = 0 + a = a$$

Since this leaves a identically as it was before the addition, we make the following definition:

Definition: The real number *zero* is called the *identity element* in the addition of real numbers. This statement is equivalent to the statement: "For any real number a, $a + 0 = 0 + a = a$".

The set of real numbers is the union of three subsets, namely:

The set of positive real numbers
The set of negative real numbers
The set with the single element, *zero*, which is neither positive nor negative

If p is a positive real number and $-p$ is its negative, then $p + (-p) = 0$ and $(-p) + p = 0$. This enables us to conclude that, for every real number a, there is another real number ^-a, called its *additive inverse*, such that $a + {}^-a = 0$ and $^-a + a = 0$. For when a is positive so that $a = p$ (where p is positive), then $^-a = -p$, and when a is negative so that $a = -p$, then $^-a = p$. Finally, when $a = 0$, then $^-a = 0$. Putting all these together, we have the definition:

Definition: The *additive inverse* of a real number a is the real number ^-a having the property that

$$a + {}^-a = {}^-a + a = 0$$

The use of the term *inverse* may be motivated as follows: We start at 0 and add a, thus obtaining a. We now wish to retrace our steps and return to 0; hence we must add ^{-}a to a. The operation of adding ^{-}a undoes the operation of adding a and thus is said to be the *inverse* operation.

Exercise A. Show that $^{-}(^{-}a) = 0$.

We must further define the *difference* of two real numbers. Of course, this is familiar when a and b are both positive and $a < b$. Other cases, however, must be treated, and we include these in the definition below. We introduce the symbol $a - b$ to denote subtraction and define it as follows:

Definition: Let a and b be real numbers. Then, by definition,

$$a - b = a + {}^{-}b$$

In other words, in order to subtract b from a, add ^{-}b to a.

You will notice that the symbol "$-$" is used in two distinct ways: (1) to represent a negative number such as -2 and (2) to represent the operation of subtraction as in $a - b$. It is further customary to give it a third meaning: (3) to represent the additive inverse ^{-}a by $-a$. In order for these three uses of the symbol "$-$" to be consistent, we must assume the conventions:

For all real numbers a and b:

$$-(-a) = a \qquad \text{and} \qquad a - b = a + (-b)$$

With this understanding we shall drop the notation ^{-}a and use minus signs freely without specifying the sense in which they are employed.

We shall have frequent occasion to refer to the absolute value of a real number a. This is written $|a|$ and is defined as follows:

Definition: The *absolute value* of a real number a, $|a|$, is the real number such that:

(1) If a is positive or zero, then $|a| = a$.
(2) If a is negative, then $|a| = -a$.

Illustration 2. $|5| = 5 \quad |-6| = 6 \quad |0| = 0$.

2.3 Multiplication of Real Numbers

Now that the essential laws of addition are before us, the laws of multiplication are easy to learn; they are almost the same, with "product" written in the place of "sum".

Closure law of multiplication. The product $a \times b$ of any two real numbers is a unique real number c.

Commutative law of multiplication. $a \times b = b \times a$.

Associative law of multiplication. $(a \times b) \times c = a \times (b \times c)$.

We now ask: "What is the identity element for multiplication?" It should be the number b such that, for any a, $a \times b = a$. In other words, multiplication by b leaves a unchanged, just as in addition the addition of 0 to a leaves a unchanged. Clearly, the correct choice for the identity element is 1.

Definition: The real number 1 is called the *identity element* in the multiplication of real numbers. This statement is equivalent to the statement: "For any real number a, $a \times 1 = 1 \times a = a$."

If a is a real number different from zero, we know that

$$a \times (1/a) = (1/a) \times a = 1$$

This enables us to conclude that for every real number a ($\neq 0$), there is another real number a', called its *multiplicative inverse*, such that

$$a \times a' = a' \times a = 1$$

Definition: The *multiplicative inverse* or a real number a ($\neq 0$) is the real number a' having the property that

$$a \times a' = a' \times a = 1$$

Exercise A. Show that $(a')' = a$ if $a \neq 0$.

Next let us define *division*. Just as the difference of a and b is defined to be the sum of a and the additive inverse of b, the quotient of a by b is defined to be the product of a and the multiplicative inverse of b:

Definition: Let a and b be real numbers, and let $b \neq 0$. Then the *quotient* of a by b $\left(\text{written } a/b \text{ or } \frac{a}{b}\right)$ is defined to be:

$$a/b = a \times b'$$

Note that division by zero is not defined.

Up to this point we have distinguished the quotient $1/a = 1 \times a'$ from the multiplicative inverse a', but there is no longer any reason for doing so. Hereafter we shall write $1/a$ for the multiplicative inverse of

a with the understanding that the slant (/) is used in the two senses: (1) to represent the multiplicative inverse $1/a$, and (2) to represent the quotient a/b. Since these uses are compatible, this will cause no confusion.

Exercise B. Show that $b \times (a/b) = a$, where $b \neq 0$.

Exercise C. Show that if $a/b = c$, then $a = b \times c$, where $b \neq 0$.

Remark. The multiplicative and divisibility properties of zero are among the more troublesome parts of the study of the real numbers. In Sec. 2.4, Theorem 2, we shall show that for any real number a, $a \times 0 = 0$; but let us assume this result here. There are three situations to consider.

(1) Let $a \neq 0$, then $0/a = 0$. For $0/a = 0 \times (1/a) = 0$ by the above definition and Theorem 2.
(2) Let $a \neq 0$, then $a/0$ is *meaningless*. Suppose that we were to write $a/0 = c$, and that the result of Exercise C above still holds true. Then $a = c \times 0 = 0$, which cannot be the case since $a \neq 0$.
(3) The symbol $0/0$ is *indeterminate*. Suppose that $0/0 = c$, or that $0 = c \times 0$. Since any real number c satisfies this equation, the value of c is indeterminate. Do not confuse $0/0$ with a/a ($a \neq 0$), which is equal to 1.

In summary, we note that *zero may never appear in the denominator of a fraction:*

NEVER DIVIDE BY ZERO

2.4 The Distributive Law

There is one final law; this connects multiplication and addition. You are used to writing $4(2 + 3) = (4 \times 2) + (4 \times 3)$; $2(x + y) = 2x + 2y$; etc. Or probably you did the reverse in factoring when you wrote $3x + 6y = 3(x + 2y)$. These are illustrations of the following law.

Distributive law. $a \times (b + c) = (a \times b) + (a \times c)$.

This law is the basis for many familiar operations. For example, the usual way of multiplying 15×23 is

$$\begin{array}{r} 15 \\ \underline{23} \\ 45 \\ \underline{30} \\ 345 \end{array}$$

But this really amounts to the statement that

$$\begin{aligned}15 \times 23 &= 15 \times (20 + 3)\\ &= (15 \times 20) + (15 \times 3)\\ &= 300 + 45\\ &= 345\end{aligned}$$

As a more complicated example, consider the following illustration.

Illustration 1. Show that $(a + b)(c + d) = ac + bc + ad + bd$.

$$\begin{aligned}(a + b)(c + d) &= (a + b)c + (a + b)d && \text{[Distributive Law]}\\ &= c(a + b) + d(a + b) && \text{[Commutative Law]}\\ &= (ca + cb) + (da + db) && \text{[Distributive Law]}\\ &= ca + cb + da + db && \text{[Property of Addition]}\\ &= ac + bc + ad + bd && \text{[Commutative Law]}\end{aligned}$$

The distributive law has a number of important consequences. The first of these states the multiplicative property of zero.

Theorem 2. Let a be any real number; then $a \times 0 = 0$.

Proof:

(1) $0 = 0 + 0$ [Definition, Sec. 2.2]
(2) $a \times 0 = a \times (0 + 0)$
(3) $a \times 0 = (a \times 0) + (a \times 0)$ [Distributive Law]
(4) $a \times 0 = a \times 0$ [Identity]

Subtracting (4) from (3), we obtain:

(5) $0 = a \times 0$.

From this theorem we conclude the following useful result.

Theorem 3. If a and b are two real numbers such that $ab = 0$, then $a = 0$, or $b = 0$.

Proof: If $a = 0$, the theorem is immediately verified (Theorem 2).
If $a \neq 0$, then $1/a$ is defined. Hence we may write:

$$(1/a)(ab) = (1/a)(0)$$

Using the associative law for multiplication and Theorem 2, we find that $b = 0$, which proves the theorem.

This theorem has very many applications, especially in the solution of equations.

Illustration 1. Solve: $x^2 - 5x + 6 = 0$.
By factoring we find that: $(x - 2)(x - 3) = 0$.
From Theorem 2 we see that:

Either $x - 2 = 0$ and $x = 2$
or $x - 3 = 0$ and $x = 3$

Hence 2 and 3 are solutions of the given equation.

THE PRODUCT OF TWO REAL NUMBERS IS ZERO IF AND ONLY IF AT LEAST ONE OF THE TWO FACTORS IS ZERO.

A second consequence of the distributive law is the set of rules for multiplying signed numbers. These are easily derived from the theorem:

Theorem 4. For any real number a, $(-1) \times a = -a$.

Proof:

(1) $1 + (-1) = 0$ [Definition of additive inverse]
(2) $1 \times a + (-1) \times a = 0 \times a$ [Distributive law]
(3) $0 \times a = 0$ [Theorem 2]
(4) $1 \times a = a$ [Definition of 1]
(5) $a + (-1) \times a = 0$ [(2), (3), and (4)]
(6) $a + (-a) = 0$ [Definition of additive inverse]
(7) $a + (-1) \times a = a + (-a)$ [(5) and (6)]
(8) $(-1) \times a = -a$ [Subtraction of a from both sides]

Corollary. $(-1) \times (-1) = 1$.
Put $a = -1$ in Theorem 4 and apply the convention that $-(-a) = a$.
Now we can prove the usual rules as follows:

Theorem 5. Let p and q be any positive real numbers. Then:

(1) $p \times (-q) = -(pq)$
(2) $(-p) \times (-q) = pq$

Proof: Write $(-p) = (-1) \times p$; $(-q) = (-1) \times q$; and

$$-(pq) = (-1) \times p \times q$$

Then each of the identities of the theorem follows from the associative and commutative properties of multiplication and the corollary to Theorem 4.

2.5 Formal Properties of Real Numbers

In summary of Secs. 2.2 to 2.4, we state the following properties of the arithmetic of real numbers. The letters a, b, c stand for arbitrary real numbers.

Addition

R1. $a + b$ is a unique real number [Closure Law]
R2. $(a + b) + c = a + (b + c)$ [Associative Law]
R3. $a + 0 = 0 + a = a$ [Identity Law]
R4. $a + (-a) = (-a) + a = 0$ [Inverse Law]
R5. $a + b = b + a$ [Commutative Law]

Multiplication

R6. $a \times b$ is a unique real number [Closure Law]
R7. $(a \times b) \times c = a \times (b \times c)$ [Associative Law]
R8. $a \times 1 = 1 \times a = a$ [Identity Law]
R9. $a \times \frac{1}{a} = \frac{1}{a} \times a = 1$ for $a \neq 0$ [Inverse Law]
R10. $a \times b = b \times a$ [Commutative Law]

Distributive Law

R11. $a \times (b + c) = (a \times b) + (a \times c)$

These 11 laws form the foundations of the entire subjects of arithmetic and ordinary algebra. They should be carefully memorized. Later in this chapter they are taken to be the axioms of an abstract system called a *field*. Hence we may say that the real numbers form a field.

Problems 2.5

Addition

In Probs. 1 to 4 use the commutative and associative laws to establish the truth of the given statement. Model your proofs on that given for Theorem 1.

1. $4 + 2 + 7 = 2 + 7 + 4$.
2. $1 + 5 + 9 = 9 + 5 + 1$.
3. $a + b + c = c + a + b$.
4. $a + b + c = b + c + a$.

5. Define $a + b + c + d$.
6. Assuming that $a + b + c + d$ has been defined (Prob. 5), define $a + b + c + d + e$.
7. Find the additive inverse of each of the following: 4, -2, π, 0, $-\sqrt{5}$.
8. Find the additive inverse of each of the following: -6, 2, $\frac{3}{4}$, -0, 3.

9. Find the absolute value of each of the following: $7, -3, 0, \frac{2}{3}, -\frac{1}{2}$.
10. Find the absolute value of each of the following: $-8, -0, \pi^2, \frac{3}{5}, -\sqrt{5}$.

Multiplication

11. Define $a \times b \times c$.
12. Assuming that $a \times b \times c$ is defined (Prob. 11), define $a \times b \times c \times d$.

In Probs. 13 to 16 use the commutative and associative laws to establish the truth of the given statement.

13. $4 \times 5 \times 7 = 7 \times 5 \times 4$.
14. $6 \times 2 \times 3 = 6 \times 3 \times 2$.
15. $a \times b \times c = b \times a \times c$.
16. $a \times b \times c = c \times a \times b$.

17. Find the multiplicative inverse of each of the following: $\frac{1}{3}, -4, -\frac{3}{4}, 1, 0$.
18. Find the multiplicative inverse of each of the following: $2, -\pi, \frac{1}{2}, \sqrt{5}, \frac{7}{3}$.

Subtraction and Division

19. Does the commutative law hold for the subtraction of real numbers?
20. Does the commutative law hold for the division of real numbers?
21. Does the associative law hold for the subtraction of real numbers?
22. Does the associative law hold for the division of real numbers?
23. Is there an identity element for subtraction? If so, what is it?
24. Is there an identity element for division? If so, what is it?

Zero

25. What meaning is to be attached to each of the following?

$$\frac{4}{0}, \frac{0}{4}, \frac{4}{4}, \frac{0}{\frac{1}{4}}, \frac{0}{0}$$

26. What meaning is to be attached to each of the following?

$$\frac{2}{0}, \frac{0}{2}, \frac{2}{2}, \frac{0}{\frac{1}{2}}, \frac{0}{0}$$

27. For what real values of x are the following expressions meaningless?

$$\frac{2x-1}{x+1}, \frac{5}{x}, \frac{x+4}{x-2}, \frac{0}{x^2+5}, \frac{4}{x^2-5x+4}$$

28. For what real values of x are the following expressions meaningless?

$$\frac{x}{3}, \frac{3x+2}{x-1}, \frac{x-3}{x^2+8}, \frac{3}{x}, \frac{2}{x^2+4x+4}$$

29. For what real values of x are the following expressions indeterminate?

$$\frac{x}{2x^2}, \frac{x+2}{2x+4}, \frac{1+x^2}{3+x^2}, \frac{0}{x}, \frac{x-3}{x^2-9}$$

30. For what real values of x are the following expressions indeterminate?

$$\frac{2x+6}{x+3}, \quad \frac{0}{x^2}, \quad \frac{x^4}{2x^3}, \quad \frac{3+x^2}{1+x^4}, \quad \frac{2x+5}{4x^2-25}$$

In Probs. 31 to 34 factor and solve for x.

31. $x^2 + 5x + 6 = 0$.
32. $x^2 - 8x + 12 = 0$.
33. $x^2 - 5 = 1$.
34. $x^2 - 7 = 18$.

Proofs

In Probs. 35 to 40 prove or disprove the given statement. You may use R1 to R11 as given axioms.

35. $(a + b) \times c = (a \times c) + (b \times c)$.
36. $a + (b \times c) = (a + b) \times (a + c)$.
37. $a \div (b + c) = (a \div b) + (a \div c)$.
38. $(a - b) + b = a$.

39. If $a \neq 0$, $ax + b = 0$ has a unique solution.
40. To any real number a there corresponds a real number x such that $0x = a$.
41. Let "addiplication" be defined (with symbol $\odot$) as follows:

$$a \odot b = (a + b) + (a \times b)$$

Under addiplication are the real numbers closed? Is addiplication commutative; associative? Is there an identity; an addiplicative inverse?

2.6 Rational Numbers

Before treating the properties of fields, we wish to develop your intuition by showing you some other examples of fields. The first of these is a subset of the real numbers called the *rational numbers*. A rational number is really nothing but a fraction whose numerator and denominator are both integers. Let us give a formal definition:

Definition: A *rational number* is a real number which can be expressed in the form a/b where a and b are integers and $b \neq 0$.

As we know from the example: $\frac{1}{2} = \frac{2}{4}$, there are many expressions of the form a/b which represent the same rational number. So that we can identify such cases easily, we need the following definition:

Definition: The expressions a/b and c/d where a, b, c, and d are integers and $b \neq 0$, $d \neq 0$ represent the *same rational number* if and only if $ad = bc$.

We can now state the following theorem, which expresses the most important and useful property of rational numbers.

Theorem 6. Given any pair of integers a and b ($\neq 0$), there exists a rational number x such that $bx = a$. Moreover, any two rational numbers x_1 and x_2 with this property are equal.

Proof: The existence of a solution is immediate, for $x = a/b$ has the required property. In order to establish the second part of the theorem, we suppose that x_1 and x_2 both satisfy $bx = a$. Then:

$$bx_1 = a$$
$$bx_2 = a$$

Subtracting, we have:

$$b(x_1 - x_2) = 0$$

or $$x_1 - x_2 = 0 \qquad \text{[Theorem 3, since } b \neq 0\text{]}$$

Finally we must remind you of the rules for adding and multiplying rational numbers. These are given by the theorems below.

Theorem 7. $$\frac{a}{b} + \frac{c}{d} = \frac{ad + bc}{bd}$$

Proof: Let $x = a/b$; then $bx = a$
Let $y = c/d$; then $dy = c$
From these two equations we obtain:

$$bdx = ad$$
$$bdy = bc$$

Adding and using the distributive law, we get

$$bd(x + y) = ad + bc$$

So $$x + y = \frac{ad + bc}{bd}$$

Theorem 8. $(a/b) \times (c/d) = ac/bd.$

Proof: Using the notation in the proof of Theorem 7, we have again

$$bx = a$$
$$dy = c$$

Multiplying the left-hand sides and the right-hand sides separately, we have:

$$(bd)(xy) = ac$$

Therefore
$$xy = \frac{ac}{bd}$$

With these concepts of addition and multiplication we can now check to see how many of R1 to R11 are satisfied by the rational numbers. As a matter of fact we find that all of these are satisfied. This means that the arithmetic of rational numbers is just like that of the real numbers. This might lead us to believe that there is no difference between the real numbers and their special case, the rational numbers. However, we shall see that real numbers such as $\sqrt{2}$ are not rational and hence that a distinction must be made.

Problems 2.6

1. Verify that R1 to R11 are satisfied when $a = 3$, $b = \frac{1}{2}$, $c = -2$.
2. Verify that R1 to R11 are satisfied when $a = \frac{1}{3}$, $b = -4$, $c = 5$.
3. Prove: For any pair of rational numbers a and b, there exists a rational number x such that $a + x = b$. Moreover, any two rationals x_1 and x_2 with this property are equal.
4. Prove: For any three rational numbers a, b, and c, where $a \neq 0$, there exists a rational number x such that $ax + b = c$. Moreover, any two rationals x_1 and x_2 with this property are equal.
5. Prove: For any two rational numbers a/b and c/d, the quotient $\dfrac{a/b}{c/d} = \dfrac{ad}{bc}$.
   HINT: Consider $(c/d)x = a/b$.
6. Prove: The two rational numbers: $(-a)/b$ and $-(a/b)$ are equal. HINT: Show that $a/b + (-a)/b = 0$ and $a/b + [-(a/b)] = 0$.

2.7 Complex Numbers

Our third example of a field is one which includes the real numbers as a subset; it is known as the field of complex numbers. This field was invented to permit us to solve certain problems which had no solution if we were limited to real numbers. The simplest of these is the equation $x^2 = -1$, for which there is no real solution for x.

In order to handle such situations the new symbol i is introduced, which, by definition, is to have the property that $i^2 = -1$. We then write expressions like $a + bi$ where a and b are real numbers and call these expressions *complex numbers*. The number a is the *real part of* $a + bi$, and bi is its *imaginary part*. So that we can treat these like numbers, we must define the usual arithmetic operations on them.

Definition: The arithmetic operations on complex numbers are defined as follows:

Equality: $a + bi = c + di$ if and only if $a = c$ and $b = d$.

Addition: $(a + bi) + (c + di) = (a + c) + (b + d)i$.

Multiplication: $(a + bi) \times (c + di) = (ac - bd) + (bc + ad)i$.

Note that the definition of multiplication is consistent with the property that $i^2 = -1$. For we can multiply $(a + bi)(c + di)$ by ordinary algebra and obtain $ac + i(bc + ad) + i^2(ad)$. When we replace i^2 with -1 and rearrange, we obtain the formula in the definition.

Illustration 1

(**a**) $(3 + 6i) + (2 - 3i) = 5 + 3i$.
(**b**) $(7 + 5i) - (1 + 2i) = 6 + 3i$.
(**c**) $(5 + 7i)(3 + 4i) = 15 + 41i + 28i^2 = (15 - 28) + 41i = -13 + 41i$.
(**d**) $(2 - 3i)(-1 + 4i) = -2 + 11i - 12i^2 = (-2 + 12) + 11i = 10 + 11i$.

We also must consider division. We wish to express $1/(a + bi)$ as a complex number. This is best approached through the use of the conjugate complex number $a - bi$.

Definition: The complex numbers $a + bi$ and $a - bi$ are called *conjugates*.

We write:

$$\frac{1}{a + bi} = \left(\frac{1}{a + bi}\right)\left(\frac{a - bi}{a - bi}\right) = \frac{a - bi}{a^2 + b^2} = \frac{a}{a^2 + b^2} + \frac{-b}{a^2 + b^2}\, i$$

which is the required complex number equal to $1/(a + bi)$. By an extension of this method we can evaluate general quotients $(a + bi)/(c + di)$:

$$\frac{a + bi}{c + di} = \left(\frac{a + bi}{c + di}\right)\left(\frac{c - di}{c - di}\right) = \frac{(ac + bd) + (bc - ad)i}{c^2 + d^2}$$

Hence we have the rule for division:

Division. In order to form the quotient $(a + bi)/(c + di)$, multiply numerator and denominator by the conjugate complex number $c - di$ and simplify the result.

Illustration 2

$$\frac{4 + i}{2 - 3i} = \frac{4 + i}{2 - 3i} \times \frac{2 + 3i}{2 + 3i} = \frac{(4 + i)(2 + 3i)}{(2 - 3i)(2 + 3i)} = \frac{5 + 14i}{13}$$

We could write this answer as $\frac{5}{13} + \frac{14}{13}i$, but this is an unnecessary refinement.

Finally, let us solve some equations involving complex numbers. The general method is suggested by the illustration below:

Illustration 3. Solve: $(x + yi)(2 - 3i) = 4 + i$. We could do this by writing $x + yi = (4 + i)/(2 - 3i)$ and evaluating the quotient on the right. But let us use another method. If we multiply out the left-hand side, we get:

$$(2x + 3y) + (-3x + 2y)i = 4 + i$$

From our definition of the equality of two complex numbers, the *real parts* of both sides must be equal, and similarly the *imaginary* parts must be equal. Hence:

$$\begin{aligned} 2x + 3y &= 4 \\ -3x + 2y &= 1 \end{aligned}$$

We can solve these simultaneous equations and obtain: $x = \frac{5}{13}$ and $y = \frac{14}{13}$.

This method of equating real and imaginary parts is of great importance in the application of complex numbers to engineering, and you should be certain that you understand it.

There are a number of other important properties of complex numbers that need to be discussed. Since these depend upon a knowledge of trigonometry, we defer their treatment to Chap. 8.

The above definition of a complex number is somewhat lacking in intuitive appeal. We have said that there is no real number x such that $x^2 = -1$, but immediately we introduce i with this property. What, then, is i? Mathematicians were sufficiently disturbed about this to call i an *imaginary* number and $a + bi$ *complex* numbers; by contrast, other numbers in our system are *real*. Our purpose now is to give an alternative development of the complex numbers in a logical and nonimaginary fashion.

Definitions:

Complex Number. A *complex number* is an ordered pair of real numbers (a,b).

Real Part of a Complex Number. The complex number $(a,0)$ is called the *real part* of the complex number (a,b). We shall see that the pairs $(a,0)$ can be identified with the real numbers a in a natural fashion.

Imaginary Part of a Complex Number. The complex number $(0,b)$ is called the *imaginary part* of the complex number (a,b). We shall also call complex numbers of this form *pure imaginary numbers*.

The arithmetic of complex numbers is given by the following basic definitions:

Definitions:

Equality. Two complex numbers (a,b) and (c,d) are said to be *equal* if and only if $a = c$ and $b = d$.

Addition. $(a,b) + (c,d) = (a + c, b + d)$.
Multiplication. $(a,b) \times (c,d) = (ac - bd, bc + ad)$.

It is evident that there is a 1 to 1 correspondence between the complex numbers $(a,0)$ and the real numbers a which is defined by $(a,0) \leftrightarrow a$. This is a particularly useful correspondence, for under it sums correspond to sums and products to products. That is:

$$\begin{array}{ccccc} (a,0) & + & (c,0) & = & (a + c,\ 0) \\ \updownarrow & & \updownarrow & & \updownarrow \\ a & + & c & = & a + c \end{array} \qquad \begin{array}{ccccc} (a,0) & \times & (c,0) & = & (ac,0) \\ \updownarrow & & \updownarrow & & \updownarrow \\ a & \times & c & = & ac \end{array}$$

Such a correspondence is called an *isomorphism*, and we say that the set of complex numbers $(a,0)$ is isomorphic to the set of real numbers a relative to addition and multiplication. We are therefore justified in identifying these two symbols and in calling $(a,0)$ a real number when there is no source of confusion.

Although the complex numbers $(a,0)$ are really nothing new, the pure imaginaries $(0,b)$ *are* something new. Their arithmetic, as derived from the definitions, is given by the rules:
Addition. $(0,b) + (0,d) = (0, b + d)$.
Multiplication. $(0,b) \times (0,d) = (-bd,0)$.

It is important to note that the product of two pure imaginaries is a real number. In particular,

$$(0,1) \times (0,1) = (-1,0)$$

We now recall that our motivation for introducing the complex numbers was our inability to solve the equation $x^2 = -1$ in terms of real numbers. Let us see how the introduction of complex numbers enables us to provide such a solution. By means of the isomorphism above, we see that the equation $x^2 = -1$ corresponds to the equation

$$(x,y)^2 = (x,y) \times (x,y) = (-1,0)$$

As we have noted, $(x,y) = (0,1)$ is a solution of this equation, and we also see that $(x,y) = (0,-1)$ is another solution. Therefore our introduction of complex numbers permits us to solve equations of this type, which had no solution in terms of real numbers.

In order to complete our discussion we need to show the correspondence between our two definitions of complex numbers. In preparation for this we note the following identities:

$$(0,b) = (b,0) \times (0,1)$$
$$(a,b) = (a,0) + [(b,0) \times (0,1)]$$

Exercise A. Verify the above identities.

We then set up the following relationship between the two notations:

	(a,b) **notation**	$a + bi$ **notation**
Real numbers	$(a,0)$	a
Unit imaginary	$(0,1)$	i

Using the identities above, we then derive the correspondences:

Pure imaginaries	$(0,b)$	bi
Complex numbers	(a,b)	$a + bi$

From these we show that the rules for the equality, addition, and multiplication of complex numbers in the $a + bi$ notation, which were stated as definitions at the beginning of this section, are in agreement with the corresponding definitions in the (a,b) notation.

Finally, we observe that with these definitions the complex numbers form a field. The details of the proof of this are included in the problems.

Problems 2.7

In Probs. 1 to 14 find the sum or difference of the given complex numbers.

1. $(9 + 8i) + (4 - 7i)$.
2. $(-19 + 5i) + (7 + 2i)$.
3. $(15 - 7i) - (20 + 9i)$.
4. $(-13 + 9i) - (21 + 5i)$.
5. $(10 + 6i) + (-13 - 18i)$.
6. $(6 - 5i) + (-4 + 2i)$.
7. $-(5 - 3i) + (7 + 4i)$.
8. $-(33 + 5i) + (13 - 7i)$.
9. $-(3 + i) - (-8 + 6i)$.
10. $-(-8 + 6i) - (6 - 4i)$.
11. $6 - (4 - 7i)$.
12. $(8 - 9i) + 6$.
13. $(5 + 13i) - 11i$.
14. $(21 - 16i) + 29i$.

In Probs. 15 to 30 find the product of the given complex numbers.

15. $(3 + 7i)(-5 + 2i)$.
16. $(5 - 9i)(4 + 3i)$.
17. $(7 + 3i)(5 - 2i)$.
18. $(6 + 2i)(5 + 9i)$.
19. $(\sqrt{3} + i)(\sqrt{3} - i)$.
20. $(\sqrt{5} - 3i)(\sqrt{5} + 3i)$.
21. $(8 + 5i)(8 - 5i)$.
22. $(3 - 7i)(3 + 7i)$.
23. $3(7 - 5i)$.
24. $8(-3 + 6i)$.
25. $(9i)(1 + 5i)$.
26. $(-4i)(-13 + 7i)$.
27. $(3i)(7i)$.
28. $(4i)(-6i)$.
29. i^5.
30. i^6.

In Probs. 31 to 40 find the quotient of the given complex numbers.

31. $(5 + 4i)/(5 + 6i)$.
32. $(1 + 2i)/(3 + 5i)$.
33. $(8 + 7i)/(4 - 3i)$.
34. $(5 + 9i)/(3 - i)$.
35. $(-7 + 4i)/(-2 + 3i)$.
36. $(-8 + 4i)/(-5 - 6i)$.
37. $5i/(6 + 9i)$.
38. $7i/(6 - 3i)$.
39. $(3 + 8i)/(6i)$.
40. $(4 - i)/(3i)$.

In Probs. 41 to 46 solve for x and y by equating real and imaginary parts.

41. $(x + iy)(1 - 5i) = 2 + 5i$.
42. $(x + iy)(3 - 7i) = 2 + 4i$.
43. $(x + iy)(-1 + 4i) = -9 + 5i$.
44. $(x + iy)(2 + i) = 1 + 3i$.
45. $(x + iy)(6i) = 9$.
46. $(x + iy)(-3i) = 13i$.

In Probs. 47 to 52 show that the given complex number satisfies the given equation

47. $3 + 3i$; $x^2 - 6x + 18 = 0$.
48. $3 - 3i$; $x^2 - 6x + 18 = 0$.
49. $3 - i$; $x^2 - (1 + 4i)x + (-1 + 17i) = 0$.
50. $-2 + 5i$; $x^2 - (1 + 4i)x + (-1 + 17i) = 0$.
51. $3 - 5i$; $x^2 + (-8 + 3i)x + (25 - 19i) = 0$.
52. $5 + 2i$; $x^2 + (-8 + 3i)x + (25 - 19i) = 0$.
53. Verify that R1 to R11 are satisfied when $a = 1 + 2i$, $b = 2 - i$, and $c = 2 + 3i$.
54. What is the additive inverse of $a + bi$? The multiplicative inverse?

In Probs. 55 to 60 carry out the indicated operations on complex numbers in the (a,b) notation.

55. $(3,-4) + (2,3)$.
56. $(1,2) - (-4,1)$.
57. $(2,3) \times (1,4)$.
58. $(-1,2) \times (2,-3)$.
59. $(4,1)/(3,-2)$.
60. $(-2,3)/(3,4)$.

2.8 Finite Fields

The fields considered so far contain an infinite number of elements. In this section we shall define certain fields which have only a finite set of elements.

In order to describe these, let us consider some number systems which are closely related to the integers. These arise in many practical situations, one of which is the behavior of the hour hand of a clock. This runs from 12 (or 0) to 1 to 2 to . . . to 11 to 12 and then starts over. In Europe and in the United States armed forces time is measured on a 24-hr basis, with 0 to 12 representing A.M. and 12 to 24 representing P.M.

Suppose that we are using an ordinary 12-hr clock and that it is now 10 o'clock. What time will it be in 4 hr? Certainly, 2 o'clock, but how do we get this answer? If we add $10 + 4$, we get 14, and $14 - 12 = 2$. A harder problem is: If it is now 7 o'clock, what time will it be 32 hr from now? *Answer:* $7 + 32 = 39$; $39 - 3(12) = 3$, or 3 o'clock.

This suggests a new kind of addition. The "sum of two integers" now is their ordinary sum reduced by a suitable multiple of 12.

Exercise A. Show that in this system:

(a) $4 + 15 = 7$. (b) $9 + 6 = 3$. (c) $10 + 11 = 9$.

Exercise B. Find the values of the following in this system:

(a) $8 + 9$. (b) $2 + 11$. (c) $7 + 25$.

There is nothing essential here about the number 12; any other positive integer would serve as well. A positive integer used for this purpose is called a *modulus* and will be represented by the symbol m. Furthermore, these ideas may equally well be applied to multiplication as illustrated below.

Illustration 1. Using 9 as a modulus, compute 7×5.

Solution:

$$\begin{aligned} 7 \times 5 &= 35 - 3(9) \\ &= 8 \end{aligned}$$

If we were to continue for long in this free and easy style, we would run into trouble. For it looks rather strange to write

$$9 + 6 = 3 \qquad \text{and} \qquad 7 \times 5 = 8$$

For this reason we introduce the following conventions.

Definition: If the difference of two integers a and b is divisible by an integer m, a and b are said to be *congruent* with respect to the modulus m.
We write this (congruence) relation

$$a \equiv b, \bmod m$$

where $\equiv$ is read *congruent to* and *mod* is read *modulo*. From the definition, this relation is equivalent to the statement:

There exists an integer k such that $a = b + km$.

In this notation we write the above equations

$$9 + 6 = 3 \qquad \text{and} \qquad 7 \times 5 = 8$$

in the preferred form:

$$9 + 6 \equiv 3, \bmod 12 \qquad \text{and} \qquad 7 \times 5 \equiv 8, \bmod 9$$

Let us now, for example, examine the set of integers which are congruent to 3, mod 5. This is the set

$$\{\ldots, -12, -7, -2, 3, 8, 13, 18, \ldots\}$$

We observe that the elements of this set are congruent, mod 5, not only to 3 but also to each other. This suggests the following theorem to us.

Theorem 9. The relation $a \equiv b$, mod m, is an equivalence relation.

Proof: The definition of an equivalence relation is given in Sec. 1.10. We must check that:

(1) $a \equiv a$, mod m.
(2) If $a \equiv b$, mod m, then $b \equiv a$, mod m.
(3) If $a \equiv b$, mod m, and $b \equiv c$, mod m, then $a \equiv c$, mod m.

Conditions (1) and (2) are self-evident from the definition. In order to prove (3), we note that

$$\begin{aligned} & a \equiv b, \text{ mod } m && \text{implies that } a = b + k_1 m \\ & b \equiv c, \text{ mod } m && \text{implies that } b = c + k_2 m \\ \text{Hence} \quad & a = (c + k_2 m) + k_1 m \\ & = c + (k_1 + k_2)m \\ \text{Therefore} \quad & a \equiv c, \text{ mod } m \end{aligned}$$

A set of mutually equivalent elements is called an *equivalence class.* In the present case an equivalence class of integers mutually congruent, mod m, is called a *residue class,* mod m, on the ground that the elements of this class all have the same remainder, or *residue,* when they are divided by m. We can specify a residue class, mod m, by mentioning any one of its elements, but it is customary in doing this to use the least nonnegative element of the class. Thus we shall call the set

$$\{\ldots, -12, -7, -2, 3, 8, 13, 18, \ldots\}$$

the residue class 3, mod 5. In general, we follow the definition:

Definition: The *residue class a,* mod m, is the set of integers congruent to a, mod m.

Exercise C. Prove that the residue class a, mod m, is equal to the residue class b, mod m, if and only if $a \equiv b$, mod m.

Illustration 2. The set of even integers is the residue class 0, mod 2.
The set of odd integers is the residue class 1, mod 2.

Exercise D. Show that there are precisely m distinct residue classes, mod m.

We can now construct an arithmetic of residue classes by defining their addition and multiplication.

Definition: The *sum* of the residue class a, mod m, and the class b, mod m, is the class $a + b$, mod m.

The *product* of these two classes is the class ab, mod m.

Illustration 3. The sum of the residue classes 4, mod 7, and 5, mod 7, is the class 9, mod 7. According to our convention this is usually written as the residue class 2, mod 7.

The product of the two given classes is the class 20, mod 7, or the class 6, mod 7.

A convenient way to illustrate the procedures for multiplication and addition of residue classes is to write tables such as:

Residue classes, mod 5

Addition

+	0	1	2	3	4
0	0	1	2	3	4
1	1	2	3	4	0
2	2	3	4	0	1
3	3	4	0	1	2
4	4	0	1	2	3

Multiplication

×	0	1	2	3	4
0	0	0	0	0	0
1	0	1	2	3	4
2	0	2	4	1	3
3	0	3	1	4	2
4	0	4	3	2	1

An element in the table represents the residue class of integers congruent to it, mod 5.

Now we ask: With the above definitions of addition and multiplication, do the residue classes, mod 5, have properties R1 to R11; i.e., do they form a field?

The laws of closure, associativity, and commutativity and the distributive law are immediate consequences of the properties of integers and the above definitions of + and ×. Similarly, 0 and 1 are the additive and multiplicative identities. The delicate question concerns the existence of the additive and multiplicative inverses. From the tables we observe the following:

Element	Additive inverse	Multiplicative inverse
0	0	....
1	4	1
2	3	3
3	2	2
4	1	4

Since the required inverses exist for each element, the properties R1 to R11 are all fulfilled, and we conclude that the *residue classes, mod 5, form a field.* Notice that 0 does not have a multiplicative inverse.

On the other hand, the residue classes, mod 4, do not form a field. To see this, construct the tables:

Residue classes, mod 4

Addition

+	0	1	2	3
0	0	1	2	3
1	1	2	3	0
2	2	3	0	1
3	3	0	1	2

Multiplication

×	0	1	2	3
0	0	0	0	0
1	0	1	2	3
2	0	2	0	2
3	0	3	2	1

There is no multiplicative inverse for 2, since in the table for multiplication the column headed 2 does not contain a 1.

From these two examples we see that, for some moduli, the residue classes form a field, and for others, they do not. The following theorem, which we give without proof, settles this matter.

Theorem 10. The residue classes, mod m, form a field if and only if m is prime.

Problems 2.8

In Probs. 1 to 4 write the addition and multiplication tables for the residue classes with the given modulus.

1. mod 2. **2.** mod 3.
3. mod 6. **4.** mod. 7.

5. Find five integers congruent to 3, mod 8.
6. Find five integers congruent to -4, mod 11.
7. Find five integers congruent to -2, mod 7.
8. Find five integers congruent to 5, mod 3.
9. Prove that if $a \equiv b$, mod m, and $c \equiv d$, mod m, then $a + c \equiv b + d$, mod m.
10. Prove that the sum of the residues class a, mod$m$, and b, mod m, is equal to the sum of the residue classes $a + k_1m$, mod m, and $b + k_2m$, mod m.
11. Prove that the product of the residue classes a, mod m, and b, mod m, is equal to the product of the classes $a + k_1m$, mod m, and $b + k_2m$, mod m.
12. Why must we exclude zero as a possible modulus? Can m be negative?
13*. Let m be a composite number, so that $m = ab$ where $a \neq \pm 1$ and $b \neq \pm 1$. Let us try to find the multiplicative inverse of the residue class a, mod m, where $a \neq 0$. We must find an integer x such that

$$ax \equiv 1 \text{ mod } ab$$

In other words, we must find integers x and k such that

$$ax = 1 + kab$$

or

$$a(x - kb) = 1$$

Why is this impossible?

2.9 Definition of a Field

Now that we are familiar with a number of examples of fields, we wish to change our point of view. We have observed that, although the reals, the rationals, the complex numbers, and the finite fields of residue classes, modulo a prime, each has its own individual characteristics, they have much in common: i.e., they satisfy R1 to R11. This leads us to distill these common properties from our examples and to consider the abstract concept of a field. Our purpose in doing so is to construct an axiomatic system, such as is described in Chap. 1, which has all our examples as concrete representatives. We shall prove theorems in the abstract system and conclude that these are true in each concrete system which satisfies the axioms. The great advantage of this procedure is that we prove these theorems just once (in the abstract system) rather than having to construct separate proofs in each of our examples.

Definition: A *field* is an abstract mathematical system given by the following.

Undefined Terms

ELEMENTS: $a, b, c, \ldots$ of a set F. We assume that at least two elements exist.

OPERATIONS: $+$; $\times$. The "product" $a \times b$ will frequently be written ab, or $(a)(b)$. The symbols $+$ and $\times$ do not necessarily represent ordinary addition and multiplication, but are undefined operations. We could just as well have used any two other symbols such as $\circ$ and $*$, but $+$ and $\times$ are usually employed.

Axioms

R1. The sum $a + b$ of each pair of elements of F is a unique element c of F. [Closure.]

R2. For any triple of elements of F,

$$(a + b) + c = a + (b + c)$$

[Associative law.]

R3. There exists a unique element of F, called zero, such that for every element a of F

$$a + 0 = 0 + a = a$$

The element zero is called the additive identity of F. [Existence of zero.]

R4. Corresponding to each a of F there is a unique element $-a$ in F such that

$$a + (-a) = (-a) + a = 0$$

The element $(-a)$ is called the additive inverse of a. [Existence of additive inverse.]

R5. For every pair of elements of F,

$$a + b = b + a$$

[Commutative law.]

R6. The product $a \times b$ of each pair of elements of F is a unique element c of F. [Closure.]

R7. For any triple of elements of F,

$$(a \times b) \times c = a \times (b \times c)$$

[Associative law.]

R8. There exists a unique element of F, called the unit element and written 1, such that for every element a of F,

$$a \times 1 = 1 \times a = a$$

The unit element is called the multiplicative identity of F. [Existence of the unit element.]

R9. Corresponding to each a of F (*except zero*) there is a unique element $1/a$ in F such that

$$a \times \frac{1}{a} = \frac{1}{a} \times a = 1$$

The element $1/a$ is called the multiplicative inverse of a. [Existence of multiplicative inverse.]

R10. For every pair of elements of F,

$$a \times b = b \times a$$

[Commutative law]

R11. For every triple of elements of F,

$$a \times (b + c) = (a \times b) + (a \times c)$$

[Distributive law]

These axioms differ in their wording in some places from the Properties R1 to R11 of the reals. Their content is so similar, however, that we use the same designations for these two sets of statements.

Remark. The symbols 0, 1, $-$, and / are not to be interpreted in their usual way. Zero is the additive identity and not the number *zero*. Similarly, 1 is the multiplicative identity and not the number *one*. The element $-a$ is the additive inverse of a, and $-$ has no connotation of *negative*. Indeed, *positive* and *negative* are meaningless terms. Finally, $1/a$ is the multiplicative inverse and has nothing to do with a fraction in the usual sense. Frequently the symbol a^{-1} is used to represent this inverse.

2.10 Theorems on a Field

We begin our discussion of theorems on a field with an elementary result which depends only on the definition of equality (Sec 1.10).

Theorem 11. If a, b, c, d are any elements of a field, then:

$$[(a = b) \wedge (c = d)] \rightarrow (a + c = b + d)$$
$$[(a = b) \wedge (c = d)] \rightarrow (ac = bd)$$

The proof is trivial, since the two sides of the equalities on the right are just different names for the same element (Sec. 1.10, Illustration 2).

Theorem 12. $[a + b = c] \rightarrow [b = (-a) + c]$.

$$[(ab = c) \wedge (a \neq 0)] \rightarrow \left[b = \left(\frac{1}{a}\right) \times c\right].$$

Proof: To prove the first of these we write:

(1) $a + b = c$	[Given]
(2) $(-a) + (a + b) = (-a) + c$	[Theorem 11]
(3) $[(-a) + a] + b = (-a) + c$	[R2]
(4) $0 + b = (-a) + c$	[R4]
(5) $b = (-a) + c$	[R3]

The second follows in the same way.

Exercise A. Write out the details of the proof of the second statement in Theorem 12.

Theorem 13. $[(a + c) = (b + c)] \to (a = b)$.
$[(ac = bc) \wedge (c \neq 0)] \to (a = b)$.
In the first, add $(-c)$ to both sides; in the second, multiply both sides by $(1/c)$.

Exercise B. Why must we include the hypothesis $c \neq 0$ in the second part of Theorem 13?

Definitions: The symbol $a - b$ is defined to mean $a + (-b)$. We call $a - b$ the *difference* of a and b and call this process *subtraction*.

The symbol a/b, where $b \neq 0$, is defined to mean $a(1/b)$. We call a/b the *quotient* of a by b and call this process *division*. This symbol is not defined if $b = 0$.

Theorem 14. $[(a = b) \wedge (c = d)] \to (a - c = b - d)$.
$[(a = b) \wedge (c = d) \wedge (c \neq 0)] \to (a/c = b/d)$.
This results immediately from the definition of equality (Sec. 1.10, Illustration 2).

Theorem 15. $-(-a) = a; 1(1/a) = a$ if $a \neq 0$.

Proof: (1) $a + (-a) = 0$ [R4]
(2) $-(-a) + (-a) = 0$ [R4]
(3) $a = -(-a)$ [since by R4 $(-a)$ has a unique additive inverse]

Exercise C. Prove the second part of Theorem 15.

Problems 2.10

In Probs. 1 to 18 prove the given theorems, assuming the axioms of a field.

1. $-(a + b) = (-a) + (-b)$.
2. $1/(ab) = (1/a)(1/b)$ if $a \neq 0, b \neq 0$.
3. $-(a - b) = -a + b$.
4. $1/(a/b) = b/a$ if $a \neq 0, b \neq 0$.
5. $-(0) = 0$.
6. $1/1 = 1$.
7. $a - 0 = a$.
8. $a/1 = a$.

9. $a \times 0 = 0$. HINT: See Theorem 2.
10. $a \times (-b) = -(a \times b)$. HINT: Use the method of Theorem 4.
11. $(-a) \times (-b) = (a \times b)$. HINT: Use the result of Prob. 10 and the method of Theorem 4.
12. $(a \times b = 0) \leftrightarrow [(a = 0) \vee b = 0]$. HINT: See Theorem 3.
13. $0/a = 0$ if $a \neq 0$.
14. $[(a/b = 0) \wedge (b \neq 0)] \to (a = 0)$.
15. $[(a/b = c) \wedge (b \neq 0)] \leftrightarrow [(a = bc) \wedge (b \neq 0)]$.

16. $\frac{a}{b} + \frac{c}{d} = \frac{ad + bc}{bd}$ if $b \neq 0, d \neq 0$. HINT: See Theorem 7.

17. $\left(\frac{a}{b}\right) \times \left(\frac{c}{d}\right) = \frac{ac}{bd}$ if $b \neq 0, d \neq 0$. HINT: See Theorem 8.

18. $\frac{a/b}{c/d} = \left(\frac{a}{b}\right)\left(\frac{d}{c}\right)$ if $b \neq 0, c \neq 0, d \neq 0$. HINT: Use Prob. 4.

2.11 Ordered Fields

Certain (but not all) fields satisfy additional axioms of order and are called ordered fields. Since the prototype of an ordered field is the field of real numbers, let us first develop our intuition by looking at the order properties of real numbers.

We have already noted that the set of real numbers is the union of the subsets:

(1) The positive real numbers
(2) The negative real numbers
(3) Zero

Moreover, each real number belongs to just one of the subsets; the sum of two positive numbers is positive, and the product of two positive numbers is positive. In view of these remarks, we define inequality of real numbers as follows:

Definition: For two real numbers a and b, a is *greater* than b, written $a > b$, if and only if $a - b$ is positive. Moreover, a is *less* than b, written $a < b$, if and only if $a - b$ is negative.

The symbol $a \geq b$, read *a is greater than or equal to b*, means $(a > b) \vee (a = b)$.* Similarly, $a \leq b$, read *a is less than or equal to b*, means $(a < b) \vee (a = b)$.

We next observe that we can prove that the real numbers have the following properties:

R12. For each pair of real numbers a and b, one and only one of the following relations is true:

$$a < b \qquad a = b \qquad a > b$$

This follows from the fact that $a - b$ is negative, zero, or positive.

* Since $a > b$ and $a = b$ cannot both be true, we could have used the *exclusive or* $\underline{\vee}$ here. However, since $\vee$ is simpler and causes no confusion, we shall use it in cases of this kind.

Now consider the inequalities

$$a > b \qquad b > c$$

From these we can conclude that $a > c$. To see this we note that

$$a - b \text{ is positive} \qquad b - c \text{ is positive}$$

Hence the sum $(a - b) + (b - c) = a - c$ is positive, which shows that $a > c$. We call this the *transitive law* for inequalities and write it as property R13.

R13. If a, b, and c are real numbers, and if $a > b$ and $b > c$, then $a > c$.

Illustration 1. Since $2 > -3$ and $8 > 2$, it follows that $8 > -3$.

Next let us suppose that $a > b$, so that $a - b$ is positive. Then it follows that $a + c > b + c$, for $(a + c) - (b + c) = a - b$, which is positive. We write this as property R14.

R14. If a, b, and c are real numbers and if $a > b$, then

$$a + c > b + c$$

Illustration 2. Since $5 > 3$, it follows that $5 + 2 > 3 + 2$, or $7 > 5$.

Finally, we again suppose that $a > b$ and that c is positive. Then $ac - bc = (a - b)c$ is positive, for each factor is positive. Therefore $ac > bc$. We write this as property R15.

R15. If a, b, and c are real numbers, and if $a > b$, and $c > 0$, then $ac > bc$.

Illustration 3. Since $6 > 2$, it follows that $6 \times 3 > 2 \times 3$, or that $18 > 6$.

Exercise A. The conclusion of R15 is false if c is negative. Find a counterexample which illustrates this fact.

Exercise B. Do the rational numbers have properties R12 to R15?

Exercise C. Show that $a > 0$ if and only if a is positive.

Exercise D. Show that $a < b$ if and only if $b > a$.

Let us now adopt a different point of view and define an ordered field as an abstract system modeled on the real numbers.

Definition: A field F is called an *ordered field* if and only if we add to its previous definition the following statements:

Undefined Relation. $a > b$, read: *a is greater than b.*

Defined Relation. $a < b$, read: *a is less than b.* Its definition is given by the equivalence $(a < b) \leftrightarrow (b > a)$.

Axioms

R12. For each pair of elements of F exactly one of the following relations is true:

$$a < b \qquad a = b \qquad a > b$$

R13. For each triple of elements of F

$$[(a > b) \wedge (b > c)] \rightarrow (a > c)$$

R14. For each triple of elements of F

$$(a > b) \rightarrow (a + c > b + c)$$

R15. For each triple of elements of F

$$[(a > b) \wedge (c > 0)] \rightarrow (ac > bc)$$

2.12 Theorems on an Ordered Field

There are a number of theorems which follow from the axioms of an ordered field. We prove these briefly below.

Theorem 16. $[(a > b) \wedge (c > d)] \rightarrow [(a + c) > (b + d)]$.

Proof: (1) $a + c > b + c$ [R14]
(2) $b + c > b + d$ [R14]
(3) $a + c > b + d$ [R13]

Illustration 1. From $6 > -3$ and $8 > 4$ we conclude from Theorem 16 that $6 + 8 > -3 + 4$, or $14 > 1$. Note that Theorem 16 says nothing about adding two inequalities, one of which contains a *less than* ($<$) and the other a *greater than* ($>$).

Exercise A. Prove that $[(a < b) \wedge (c < d)] \rightarrow [(a + c) < (b + d)]$.

Theorem 17. $(a > 0) \leftrightarrow (-a < 0)$.

Proof (Left to right):

(1) $a + (-a) > 0 + (-a)$ [R14]
(2) $0 > (-a)$ [R4 and R3]
(3) $-a < 0$ [Definition]

Exercise B. Complete the proof (right to left) of Theorem 17.

Exercise C. Prove: $(a < 0) \leftrightarrow (-a > 0)$.

Theorem 18. $(a > b) \leftrightarrow (-a < -b)$.

Proof (Left to right):

(1) $[(-a) + (-b)] + a > [(-a) + (-b)] + b$ [R14]
(2) $(-b) > (-a)$ [R2, R3, and R4]
(3) $(-a) < (-b)$ [Definition]

Exercise D. Complete the proof (right to left) of Theorem 18.

Exercise E. Prove: $(a < b) \leftrightarrow [(-a) > (-b)]$.

Illustration 2

From $7 > 2$, we conclude that $-7 < -2$.
From $10 > -3$, we conclude that $-10 < 3$.
From $-4 > -8$, we conclude that $4 < 8$.

For real numbers this theorem is sometimes stated in the form:

If we change the signs of both sides of an inequality, we change its sense. By changing the sense of an inequality we mean that we have replaced $>$ by $<$ or $<$ by $>$.

Theorem 19. $[(a > b) \wedge (c < 0)] \rightarrow (ac < bc)$.

Proof:

(1) Let $c = -d$, where $d > 0$.
(2) $ad > bd$ [R15]
(3) $-ad < -bd$ [Theorem 18]
(4) $a(-d) < b(-d)$ [Sec. 2.10, Prob. 10]
(5) $ac < bc$ [Substitution from (1)]

For real numbers this means that if we multiply both sides of an inequality by a negative number, we change its sense.

Problems 2.12

In Probs. 1 to 8 prove the stated theorems from the axioms of an ordered field.

1. $(a \neq 0) \leftrightarrow (a^2 > 0)$.
2. $(a > 0) \leftrightarrow (1/a > 0)$. HINT: Use the result of Prob. 1.
3. $[(a > b) \wedge (c > 0)] \rightarrow (a/c > b/c)$.
4. $[(a > b) \wedge (c < 0)] \rightarrow (a/c < b/c)$.
5. $[(a > b) \wedge (ab > 0)] \rightarrow (1/a < 1/b)$.
6. $(ab > 0) \rightarrow [(a > 0) \wedge b > 0] \vee [(a < 0) \wedge (b < 0)]$. HINT: Use indirect proof. Consider the other alternatives for a and b and show that these lead to contradictions.
7. $(a > b) \rightarrow (a^2 > b^2)$.
8. $[(a > 0) \wedge (b > 0) \wedge (a^2 > b^2)] \rightarrow (a > b)$.
9. For what hypotheses on a and b is the implication $(a^2 > b^2) \rightarrow (a > b)$ false?

In Probs. 10 to 22 the letters a, b, etc., represent real numbers.

10. From the Pythagorean relation $a^2 + b^2 = c^2$, prove that the hypotenuse of a right triangle is longer than either leg.
11. Apply the result of Prob. 10 to the general triangle given in Fig. 2.1 to show that $a + b > c + d$. Hence the sum of two sides of a triangle is greater than the third side.

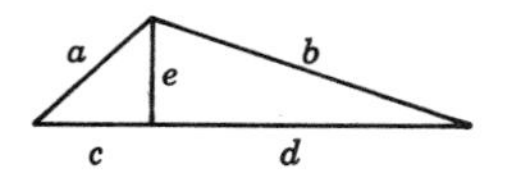

Figure 2.1

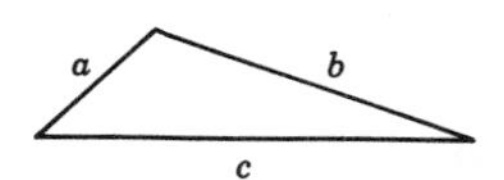

Figure 2.2

12. In Fig. 2.2 assume that $c > a$. Use the result of Prob. 11 to show that $c - a < b$. Hence any side of a triangle is greater than the absolute value of the difference of the other two sides.
13. Prove that $|a + b| \leq |a| + |b|$. HINT: Since $|a|^2 = a^2$, etc., let us consider

$$\begin{aligned}(a + b)^2 &= a^2 + 2ab + b^2 \\ &\leq a^2 + 2|a|\,|b| + b^2 \\ &\leq [|a| + |b|]^2\end{aligned}$$

When does equality hold in the above relation?

14. Using the result of Prob. 13, prove that

$$|a - b| \geq |a| - |b|$$

15. Prove that $\sqrt{ab} \leq \dfrac{a + b}{2}$ if a and b are positive. HINT: Show that this is equivalent to

$$0 \leq a^2 - 2ab + b^2$$

$\sqrt{ab}$ is called the "geometric mean" of a and b; $(a + b)/2$ is called the "arithmetic mean".

The above result is a special case of the general theorem that, for a set of positive quantities $a_1, \ldots, a_n$, the geometric mean $\sqrt[n]{a_1a_2 \cdots a_n}$ is less than or equal to the arithmetic mean $\frac{1}{n}(a_1 + \cdots + a_n)$.

When does equality hold in the above relation?

16. Show that $(|x| < a) \leftrightarrow (-a < x < a)$ where $(-a < x < a)$ is a brief notation for $[(-a < x) \wedge (x < a)]$.

17. Show that $(|x - b| < a) \leftrightarrow (-a + b < x < a + b)$.

18. Show that $(|x| > a) \leftrightarrow [(x > a) \vee (x < -a)]$.

19*. Show that $ax^2 + bx + c = a\left(x + \frac{b}{2a}\right)^2 - a\left(\frac{b^2 - 4ac}{4a^2}\right)$, where $a \neq 0$, and a, b, c, and x are real numbers. Use this identity to demonstrate that for $a > 0$, $ax^2 + bx + c \geq 0$ for all real x if and only if $b^2 - 4ac \leq 0$.

20*. Prove that for any real numbers a_1, a_2, b_1, and b_2,

$$(a_1b_1 + a_2b_2)^2 \leq (a_1^2 + a_2^2)(b_1^2 + b_2^2)$$

This is known as "Cauchy's inequality." HINT: Consider

$$(a_1x + b_1)^2 = a_1^2x^2 + 2a_1b_1x + b_1^2 \geq 0$$
$$(a_2x + b_2)^2 = a_2^2x^2 + 2a_2b_2x + b_2^2 \geq 0$$

Adding, we find that

$$(a_1^2 + a_2^2)x^2 + 2(a_1b_1 + a_2b_2)x + (b_1^2 + b_2^2) \geq 0$$

for all x.

Now apply the result of Prob. 19. When does equality hold in the above relation?

21*. Generalize the result of Prob. 20 to two sets of numbers: $a_1, \ldots, a_n$ and $b_1, \ldots, b_n$. The formula so obtained is of great importance in statistics and higher geometry.

22*. Use Prob. 21 to prove that

$$-1 \leq \frac{a_1b_1 + a_2b_2 + \cdots + a_nb_n}{\sqrt{a_1^2 + \cdots + a_n^2}\sqrt{b_1^2 + \cdots + b_n^2}} \leq 1$$

23*. Show that the complex numbers do not form an ordered field by completing the following argument. Suppose there is such an ordering of the complex numbers. Then $i^2 > 0$, or $-1 > 0$ (Prob. 1). Similarly, $1^2 > 0$ or $1 > 0$. This is a contradiction (Theorem 17).

24*. Show that the residue classes, mod 5, do not form an ordered field.

2.13 Complete Ordered Fields

Both the rational numbers and the real numbers are ordered fields, but the real numbers have an additional property known as *completeness*. Hence they are said to form a complete, ordered field. In order to explain what this means, let us examine decimal expansions of real numbers in some detail.

By carrying out the ordinary process of division, any rational number can be represented as a decimal. Some representations "terminate" after a finite number of steps; i.e., all later terms in the expansion are zero. For example,

$$\tfrac{1}{2} = 0.5000 \ldots$$
$$\tfrac{1}{4} = 0.2500 \ldots$$

But other expansions never terminate, such as

$$\tfrac{1}{3} = 0.3333 \ldots$$
$$1\tfrac{1}{7} = 1.142857142857 \ldots$$

By experimenting you may assure yourself that in each expansion the digits after a certain point repeat themselves in certain groups like (0), (3), and (142857) above. This is always true for rational numbers.

It is awkward to express numbers in this form since we cannot be sure what the . . . at the end really mean. To clear up this ambiguity, we place a bar over the set of numbers which is to be repeated indefinitely. In this notation we write

$$\tfrac{1}{2} = 0.5\overline{0}$$
$$\tfrac{1}{4} = 0.25\overline{0}$$
$$\tfrac{1}{3} = 0.\overline{3}$$
$$1\tfrac{1}{7} = 1.\overline{142857}$$

It is also true that any repeating decimal expansion of this type represents a rational number. We state this as Theorem 20.

Theorem 20. Every repeating decimal expansion is a rational number.

Before giving the general proof, we give several illustrations.

Illustration 1. Prove that $a = 3.\overline{3}$ is a rational number.

Solution: If we multiply by 10, we merely shift the decimal point; thus

$$10a = 33.\overline{3} = 30 + a$$

Hence

$$9a = 30$$
$$a = \tfrac{30}{9} = 3\tfrac{1}{3}$$

Illustration 2. Now consider the harder case where $b = 25.\overline{12}$.

Solution:

$$b = 25.\overline{12}$$
$$100b = 2512.\overline{12}$$

Subtracting, we find

$$99b = 2{,}487$$

$$b = \frac{2{,}487}{99} = \frac{829}{33} = 25\frac{4}{33}$$

Illustration 3. Finally consider $c = 2.3\overline{12}$.

Solution: The change here is that the repeating part begins one place to the right of the decimal point. We can correct this easily by writing

$$10c = 23.\overline{12}$$

and then proceeding as in Illustration 2.

To prove the general theorem, suppose that

$$c = a_k \, . \, a_1 \cdots \overline{a_k b_1 \cdots b_p}$$

where the a's and b's represent digits in the expansion of c. Using the idea of Illustration 3, we write

$$d = 10^k c = a_0 a_1 \cdots a_k \, . \, \overline{b_1 \cdots b_p}$$

Then

$$10^p d = a_0 a_1 \cdots a_k b_1 \cdots b_p \, . \, \overline{b_1 \cdots b_p}$$

$$(10^p d) - d = (a_0 a_1 \cdots a_k b_1 \cdots b_p - a_0 a_1 \cdots a_k)$$

Solving, we find that d is rational and that therefore c is rational.

We have thus seen that the set of rational numbers is identical with the set of repeating decimals. However, we may perfectly well conceive of a nonrepeating decimal. This might be constructed in infinite time by a man throwing a 10-sided die who records the results of his throws in sequence. Thus he might obtain the number (nonrepeating)

$$0.352917843926025 \ldots$$

Such a number is not rational but is included among the reals; thus the reals include irrationals as well as rationals.

Perhaps this example is far-fetched, and therefore we consider the very practical question of solving the equation $x^2 = 2$. The value of x is equal to the length of the hypotenuse of a right triangle whose legs are each 1. We now wish to show that $x = \sqrt{2}$ is not rational. We prove this by a sequence of theorems, in which a is assumed to be an integer.

Theorem 21. If a^2 is divisible by 2, then a is divisible by 2.

Proof: Every integer a can be written in one of the two forms:

$$a = \begin{cases} 2n & \text{where } n \text{ is an integer} \\ 2n + 1 \end{cases}$$

Hence
$$a^2 = \begin{cases} 4n^2 \\ 4n^2 + 4n + 1 \end{cases}$$

Since a^2 is divisible by 2, according to the hypothesis, a^2 must equal $4n^2$. Hence $a = 2n$, and a is divisible by 2.

Theorem 22. $\sqrt{2}$ is not a rational number.

Proof: (By contradiction.) Suppose p/q is a rational number in lowest terms; that is, p and q have no common factor. Suppose also that $p^2/q^2 = 2$, or that $p^2 = 2q^2$.

Then p^2 is divisible by 2, and thus p is divisible by 2 (Theorem 21). Write $p = 2r$, where r is an integer. Then $4r^2 = 2q^2$, or $2r^2 = q^2$. Hence q^2 is divisible by 2, and thus q is divisible by 2 (Theorem 21). Hence p and q have a common factor contrary to our assumption. This proves the theorem.

2.14 Complete Ordered Fields (Continued)

We have just shown that $\sqrt{2}$ is not a rational number; but what exactly is it? Let us try to approximate it with rational numbers. Surely it is between 1 and 2, for

$$1^2 = 1 < 2 < 2^2 = 4$$

We can do successively better by choosing

$$\begin{aligned} (1.4)^2 &= 1.96 < 2 < (1.5)^2 = 2.25 \\ (1.41)^2 &= 1.9881 < 2 < (1.42)^2 = 2.0264 \\ (1.414)^2 &= 1.999396 < 2 < (1.415)^2 = 2.002225 \\ (1.4142)^2 &= 1.99996164 < 2 < (1.4143)^2 = 2.0002449 \end{aligned}$$

etc. We suppose this process to continue indefinitely. Then the rational numbers on the left side generate an infinite decimal expansion for $\sqrt{2}$.

The numbers 1.4, 1.41, 1.414, 1.4142, etc., are members of the set:

$$S = \{x \mid x^2 < 2\}$$

This set has two important properties which will permit us to state our final property of the real numbers.

(1) The set S has an upper bound. This means that there exists a real number b such that every element s of S satisfies the inequality: $s \leq b$. We see that $\sqrt{2}$ is such a bound, but that 1.5, 2, 10, etc., are also upper bounds.
(2) The set S has a least upper bound. This means that there is an upper bound which is less than every other upper bound. For our set S this least upper bound is $\sqrt{2}$.

Since our special set S is typical of other subsets of the real numbers which have upper bounds, we are led to state the following property of the real numbers:

R16. Every nonempty subset of the real numbers which has an upper bound has a least upper bound.

Notice that R16 is false for the rationals, for there is no rational number which serves as the least upper bound of the set of rationals whose squares are less than $\sqrt{2}$. In this sense there are numbers missing from the set of rationals, and so the system of rational numbers is incomplete. Because of the existence of such a least upper bound, we call the real number system *complete*.

Illustration 1

(a) The set $\{x \mid x^2 < 5\}$ is bounded and has the least upper bound $\sqrt{5}$.
(b) The set $\{x \mid x^2 \leq 4\}$ is bounded and has the least upper bound 2. Note that this least upper bound is also an element of the given set.
(c) The finite set $\{-3, 1, 6, 15\}$ is bounded and has the least upper bound 15.
(d) The set of odd numbers is not bounded and does not have a least upper bound.
(e) The set whose elements are the lengths of the perimeters of closed polygons inscribed in a circle of radius 1 is bounded and has a least upper bound, which we call 2π.

In order to give a formal definition of a complete ordered field, let us first define some terms.

Definitions: Let A be a nonempty subset of the elements of an ordered field F. Then an element b of F is called *an upper bound* of A if and only if every element a of A satisfies the inequality $a \leq b$.

An upper bound of A is called *the least upper bound* of A if and only if it is less than all other upper bounds of A.

Now we can rephrase R16 in general terms.

R16. [Completeness] Every nonempty subset of the ordered field F which has an upper bound has a least upper bound.

Definition: A complete ordered field is an abstract system which satisfies Axioms R1 to R16.

Thus we see that the real numbers form a complete ordered field. You may wonder if there are other such fields. The answer is that, apart from changes in nomenclature, the real numbers are the only such field.

Problems 2.14

In Probs. 1 to 6 find decimal expansions for the given rational numbers.

1. $\frac{4}{7}$.
2. $\frac{5}{6}$.
3. $\frac{1}{9}$.
4. $\frac{3}{8}$.
5. $3\frac{7}{9}$.
6. $6\frac{2}{11}$.

In Probs. 7 to 12 find expressions of the form a/b for the given decimal expansions.

7. $0.\overline{5}$.
8. $3.\overline{6}$.
9. $15.\overline{23}$.
10. $5.3\overline{74}$.
11. $2.7\overline{354}$.
12. $1.3\overline{123}$.
13. Prove that the decimal expansion of any rational number is repeating. HINT: Try dividing, and see what happens.
14. When a/b is expressed as a repeating decimal, what is the maximum length of the period? HINT: Try dividing, and see what happens.
15. State and prove the converse of Theorem 21.
16. Prove that $\sqrt{3}$ is irrational. HINT: First prove the analog of Theorem 21: "If a^2 is divisible by 3, then a is divisible by 3." To do so, note that every integer can be written in one of the forms:

$$a = \begin{cases} 3n \\ 3n + 1 \\ 3n + 2 \end{cases} \quad \text{where } n \text{ is an integer}$$

Hence
$$a^2 = \begin{cases} 9n^2 \\ 9n^2 + 6n + 1 \\ 9n^2 + 12n + 4 \end{cases}$$

Since a^2 is divisible by 3, according to the hypothesis, a^2 must equal $9n^2$, etc.

17. Prove that $\sqrt{5}$ is irrational.
18. Where does the method of Probs. 16 and 17 fail when we try to prove $\sqrt{4}$ to be irrational?
19. Prove that $1 + \sqrt{2}$ is irrational. HINT: Suppose that $1 + \sqrt{2} = a/b$. Then $\sqrt{2} = a/b - 1 = (a - b)/b$. Why is this impossible?
20. Prove that $2 - \sqrt{3}$ is irrational.
21. Find the least upper bound of the set $\{\frac{1}{2}, \frac{3}{4}, \frac{7}{8}, \frac{15}{16}, \ldots\}$.
22. Find the least upper bound of the set $\{0.3, 0.33, 0.333, 0.3333, \ldots\}$.
23. Find a set which has $\sqrt{7}$ as its least upper bound.
24. Find a set which has $-\sqrt{3}$ as its least upper bound.

References

Birkhoff, Garrett, and Saunders MacLane: "A Survey of Modern Algebra", chaps. 1–3, Macmillan, New York (1953).

Courant, Richard, and Herbert Robbins: "What Is Mathematics?" chaps. 1 and 2, Oxford, New York (1941).
Dantzig, Tobias: "Number, the Language of Science", Macmillan, New York (1930).
Lieber, L. R.: "Infinity", pp. 87–203, Rinehart, New York (1953).
McCoy, Neal H.: "Introduction to Modern Algebra", chaps. 5–7, Allyn and Bacon, Boston (1960).
Weiss, M. J.: "Higher Algebra for the Undergraduate", chaps. 1 and 2, Wiley, New York (1962).
Whitehead, A. N., "An Introduction to Mathematics", chaps. 5–8, Holt, New York (1911).

Also consult the following articles in the *American Mathematical Monthly:*

Allen, E. S.: Definitions of Imaginary and Complex Numbers, vol. 29, p. 301 (1922)
Boyer, C. B.: An Early Reference to Division by Zero, vol. 50, p. 487 (1943).
Cajori, Florian: Historical Note on the Graphic Representation of Imaginaries before the Time of Wessel, vol. 19, p. 167 (1912).

The Integers | 3

3.1 Introduction

Because of their frequent use in mathematics and in daily life, the integers are a most important subset of the real numbers, and hence deserve special attention. Later in this chapter we shall consider them as a formal system, but before doing so we wish to introduce you to two of their most inviting properties, namely, *divisibility* and *induction*. For the moment we shall merely define divisibility, and then shall return to it after we have discussed induction.

Definition: An integer b ($\neq 0$) is called a *divisor* of an integer a if and only if there is an integer x such that $a = bx$. In this case a is said to be divisible by b.

3.2 Mathematical Induction

Let us suppose that we are faced with the proof of theorems such as the following:

For all integers $n \geq 2$, the product of n odd integers is odd; i.e., the product of any number of odd integers is odd.
For any set of positive integers x, y, n, where $x > y$, $x^n - y^n$ is divisible by $x - y$.
For all integers $n \geq 1$, $2^n \geq 1 + n$.
For all integers $n \geq 1$, $1 + 2 + \cdots + n = \dfrac{n(n+1)}{2}$.
For all integers $n \geq 2$, and all real numbers $a, b_1, \ldots, b_n$:

$$a(b_1 + \cdots + b_n) = ab_1 + \cdots + ab_n$$

For all integers $n \geq 1$, $n(n+1)(n+2)$ is divisible by 3; i.e., the product of every three consecutive positive integers is divisible by 3.
For all integers $n \geq -2$, $2n^3 + 3n^2 + n + 6 \geq 0$.

All these theorems have the common feature: their statement contains an open sentence p_n with variable n (Sec. 1.5), which is to be proved true for all n that are elements of a given infinite subset of the integers. Since we cannot prove such a theorem by verification in any finite number of cases, we must devise a general proof by some other means.

We shall illustrate the ideas behind such a proof in several informal examples.

Illustration 1. Prove: For all integers $n > 2$, the product of n odd integers is odd.

Let $\{a_1, a_2, \ldots, a_n\}$ be a set of n odd integers. We must show that $a_1 \times a_2 \times \cdots \times a_n$ is odd. First we note that we have proved earlier that the product of two odd integers is odd. Therefore $a_1 \times a_2$ is odd. Now $a_1 \times a_2 \times a_3 = (a_1 \times a_2) \times a_3$, which is the product of the odd integers $(a_1 \times a_2)$ and a_3. Since the product of two odd integers is odd, it follows that $a_1 \times a_2 \times a_3$ is odd. This argument can be repeated successively to show, in turn, that $a_1 \times a_2 \times a_3 \times a_4$ is odd, etc. Hence we are convinced of the truth of the theorem for all $n \geq 2$.

Illustration 2. Prove: For all integers $n \geq 2$ and all real numbers $a, b_1, \ldots, b_n$, it is true that $a(b_1 + \cdots + b_n) = ab_1 + \cdots + ab_n$.

First we recall (Secs. 2.2 and 2.4) that

$$\begin{aligned} a(b_1 + b_2) &= ab_1 + ab_2 && \text{[Distributive Law]} \\ b_1 + b_2 + b_3 &= (b_1 + b_2) + b_3 && \text{[Definition]} \end{aligned}$$

Hence

$$\begin{aligned} a(b_1 + b_2 + b_3) &= a[(b_1 + b_2) + b_3] \\ &= a(b_1 + b_2) + ab_3 && \text{[Distributive Law]} \\ &= (ab_1 + ab_2) + ab_3 && \text{[Distributive Law]} \\ &= ab_1 + ab_2 + ab_3 && \text{[Definition]} \end{aligned}$$

Continuing in this fashion, we have:

$$\begin{aligned} a(b_1 + b_2 + b_3 + b_4) &= a(b_1 + b_2 + b_3) + ab_4 \\ &= (ab_1 + ab_2 + ab_3) + ab_4 \\ &= ab_1 + ab_2 + ab_3 + ab_4 \end{aligned}$$

Since the process can continue indefinitely, we have again convinced ourselves of the truth of the theorem.

The trouble with these informal arguments is that in the end we must say: "this continues indefinitely", "etc.", "and so on", or something of the kind. How do we really know that we shall not be blocked at some stage far in the future? Since there is no logical means of giving such an assurance, we must formulate an axiom which effectively says that no such blocks will appear. In order to formulate this axiom, let us analyze the informal arguments above. In each case we began with a true fact for $n = 2$, namely, in Illustration 1, that the product of two odd integers is odd and, in Illustration 2, that $a(b_1 + b_2) = ab_1 + ab_2$. We used this fact to derive a true statement for $n = 3$, $n = 4$, $n = 5$, etc., in turn, and concluded that the process would never stop. By analogy we can think of the positive integers as the rungs of an infinitely tall ladder based on the ground and reaching to the sky. The bottom rung is 1, the next 2, and so on. We wish to climb this ladder to any desired rung. To do so, there are two essential steps:

(I) We must first get our foot on a low rung (the second rung in the above illustrations).
(II) We must be able to climb from any rung to the next rung. Clearly, if we can do these two things, we can climb up as far as we please.

To formalize this idea, we now state our axiom.

Axiom of Mathematical Induction. Let a be an integer (positive, negative, or zero), and let A be the set of integers which are greater than or equal to a; that is, $A = \{n \mid n \geq a\}$.
If S is a subset of A with the two properties:

(I) S contains a
(II) For all integers k in A: if k belongs to S, then $k + 1$ belongs to S

then the set S is equal to the set A.

In many applications of this axiom we have $a = 1$, so that A is the set of all positive integers.

When we use this axiom to prove theorems of the type under consideration, the set A and the open sentence p_n are given to us in the statement of the theorem. We choose S to be the subset of A consisting of those integers for which p_n is true. With these interpretations, we can reformulate the axiom as an operational procedure, which we call the Principle of Mathematical Induction.

Principle of Mathematical Induction. Let us be given a set of integers $A = \{n \mid n \geq a\}$ and a proposition of the form: For all n in A, p_n. We can prove the truth of this proposition by establishing the following:

(I) p_a is true. (We use the symbol p_a to denote the proposition obtained from the open sentence p_n by substituting a for n.)
(II) For all k in A, the implication $p_k \rightarrow p_{k+1}$ is true.

Let us now illustrate this method.

Illustration 3. (See Illustration 1.) Prove: For all integers $n \geq 2$, the product of n odd integers is odd.

We have: $a = 2$; $A = \{n \mid n \geq 2\}$.

p_n: the product of n odd integers is odd.

Following the Principle of Mathematical Induction, we establish the two necessary facts:

(I) p_2 is true, for the product of two odd integers is odd.
(II) For all integers $k \geq 2$: if the product of k odd integers is odd, then the product of $k + 1$ odd integers is odd.

Proof of (II): Let $a_1, a_2, a_3, \ldots, a_k, a_{k+1}$ be odd integers. By hypothesis, $a_1 \times a_2 \times \cdots \times a_k$ is odd. Moreover, $a_1 \times \cdots \times a_k \times a_{k+1} = (a_1 \times \cdots \times a_k) \times a_{k+1}$ by definition. The expression in parentheses is odd by hypothesis, and so this is the product of two odd integers. Since such a product is odd, we conclude that $a_1 \times \cdots \times a_k \times a_{k+1}$ is odd.

From the Principle of Mathematical Induction we now conclude that the given proposition is true.

Illustration 4. (See Illustration 2). Prove: For all integers $n \geq 2$ and all real numbers $a, b_1, \ldots, b_n$, it is true that

$$a(b_1 + \cdots + b_n) = ab_1 + \cdots + ab_n$$

We have: $a = 2$; $A = \{n \mid n \geq 2\}$.
p_n: $a(b_1 + \cdots + b_n) = ab_1 + \cdots + ab_n$.

(I) p_2 is true, for $a(b_1 + b_2) = ab_1 + ab_2$ by the distributive law.
(II) For all integers $k \geq 2$: if $a(b_1 + \cdots + b_k) = ab_1 + \cdots + ab_k$, then $a(b_1 + \cdots + b_{k+1}) = ab_1 + \cdots + ab_{k+1}$.

Proof: By definition, $b_1 + \cdots + b_{k+1} = (b_1 + \cdots + b_k) + b_{k+1}$. Hence

$$\begin{aligned} a(b_1 + \cdots + b_{k+1}) &= a[(b_1 + \cdots + b_k) + b_{k+1}] \\ &= a(b_1 + \cdots + ab_k) + ab_{k+1} \end{aligned}$$

by the distributive law. By hypothesis, $a(b_1 + \cdots + b_k) = ab_1 + \cdots + ab_k$. Hence

$$\begin{aligned} a(b_1 + \cdots + b_{k+1}) &= (ab_1 + \cdots + ab_k) + ab_{k+1} \\ &= ab_1 + \cdots + ab_k + ab_{k+1} \end{aligned}$$

The stated theorem is now proved.

Illustration 5. Prove: For all integers $n \geq 1$, $2^n \geq 1 + n$.
We have: $a = 1$; $A = \{n \mid n \geq 1\}$.
p_n: $2^n \geq 1 + n$.

(I) p_1 is true, for $2 \geq 1 + 1$.
(II) For all integers $k \geq 1$: if $2^k > 1 + k$, then $2^{k+1} \geq 1 + (k + 1) = 2 + k$.

Proof: By hypothesis, $2^k \geq 1 + k$. Multiplying both sides of this inequality by 2, we obtain

$$2^{k+1} \geq 2 + 2k$$

Since $2k > k$ where k is positive, we have

$$2^{k+1} \geq 2 + 2k > 2 + k$$

which is the desired result.

Illustration 6. Prove: For all integers ≥ 1, $1 + 2 + \cdots + n = \frac{n(n+1)}{2}$.

We have: $A = \{n \mid n \geq 1\}$.

$$p_n: 1 + 2 + \cdots + n = \frac{n(n+1)}{2}.$$

(I) p_1 is true, for $1 = \frac{1(2)}{2}$.

(II) For all integers $k \geq 1$: if $1 + 2 + \cdots + k = \frac{k(k+1)}{2}$, then

$$1 + 2 + \cdots + k + k + 1 = \frac{(k+1)(k+2)}{2}$$

Proof: By hypothesis, $1 + 2 + \cdots + k = \frac{k(k+1)}{2}$. Adding $k + 1$ to each side of the equality, we have

$$\begin{aligned} 1 + 2 + \cdots + k + k + 1 &= \frac{k(k+1)}{2} + k + 1 \\ &= \frac{k^2 + k + 2k + 2}{2} \\ &= \frac{k^2 + 3k + 2}{2} \\ &= \frac{(k+1)(k+2)}{2} \end{aligned}$$

Illustration 7. Prove: For all integers $n \geq 1$, $3^{2n} - 1$ is divisible by 8.

We have: $a = 1$; $A = \{n \mid n \geq 1\}$.

p_n: $3^{2n} - 1$ is divisible by 8.

(I) p_1 is true, for $3^2 - 1 = 8$ is divisible by 8.

(II) For all integers $k \geq 1$, if $3^{2k} - 1$ is divisible by 8, then $3^{2k+2} - 1$ is divisible by 8.

Proof: By hypothesis, $3^{2k} - 1$ is divisible by 8, so that there is an integer x such that $3^{2k} - 1 = 8x$, or $3^{2k} = 1 + 8x$. Multiplying both sides of this equality by $3^2 = 9$, we get

$$3^{2k+2} = 9 + 8(9x)$$

or

$$\begin{aligned} 3^{2k+2} &= 1 + [8 + 8(9x)] \\ &= 1 + 8(1 + 9x) \end{aligned}$$

So $3^{2k+2} - 1 = 8(1 + 9x)$, or $3^{2k+2} - 1$ is divisible by 8.

Illustration 8. Prove: For all integers $n \geq 4$, $n! > 2^n$. (By definition $n!$ is the product $1 \times 2 \times \cdots \times n$. The symbol $n!$ is read *n factorial*. See Sec. 12.11.)

We have: $a = 4$; $A = \{n \mid n \geq 4\}$.

p_n: $n! > 2^n$.

(I) p_4 is true, for $4! = 24$, $2^4 = 16$, and $24 > 16$.
(II) For all integers $k \geq 4$: if $k! > 2^k$, then $(k+1)! > 2^{k+1}$.

Proof: By hypothesis, $k! > 2^k$. Moreover, $k \geq 4$, so that $k + 1 > 4$. Combining these two equalities we have

$$\begin{aligned} k! \times (k+1) &> 2^k \times 4 \\ &> 2^k \times 2 \end{aligned}$$

Hence $(k+1)! > 2^{k+1}$.

Illustration 9. Prove: For all integers $n \geq -2$, $2n^3 + 3n^2 + n + 6 \geq 0$.
We have: $a = -2$; $A = \{n \mid n \geq -2\}$.
p_n: $2n^3 + 3n^2 + n + 6 \geq 0$.

(I) p_{-2} is true, since $2(-2)^3 + 3(-2)^2 + (-2) + 6 \geq 0$.
(II) For all integers, $k \geq -2$: if $2k^3 + 3k^2 + k + 6 \geq 0$, then $2(k+1)^3 + 3(k+1)^2 + (k+1) + 6 \geq 0$.

Proof: We write:

$$\begin{aligned} 2(k+1)^3 + 3(k+1)^2 + (k+1) + 6 &= 2k^3 + 9k^2 + 13k + 12 \\ &= (2k^3 + 3k^2 + k + 6) + (6k^2 + 12k + 6) \\ &= (2k^3 + 3k^2 + k + 6) + 6(k+1)^2 \end{aligned}$$

In the above equation $(2k^3 + 3k^2 + k + 6) \geq 0$ by hypothesis, and $(k+1)^2 \geq 0$ from Sec. 2.11, Prob. 1. Hence the sum of these is greater than or equal to zero.

Notice that there is no standard way of proving the implication in (II). You must exercise your ingenuity.

Problems 3.2

In Probs. 1 to 18 use mathematical induction to show the truth of the stated proposition for all integers $n \geq 1$.

1. $1 + 3 + 5 + \cdots + (2n - 1) = n^2$.
2. $1 + 4 + 7 + \cdots + (3n - 2) = \dfrac{n(3n-1)}{2}$.
3. $n^3 + 2n$ is divisible by 3.
4. $n^2 + n$ is divisible by 2.
5. $5^n \geq 1 + 4n$.
6. $3^n \geq 1 + 2n$.
7. $1^2 + 2^2 + \cdots + n^2 = \dfrac{n(n+1)(2n+1)}{6}$.
8. $1^3 + 2^3 + \cdots + n^3 = \dfrac{n^2(n+1)^2}{4}$.
9. $n^4 + 2n^3 + n^2$ is divisible by 4.
10. $4^n - 1$ is divisible by 3.
11. $\dfrac{1}{1 \cdot 2} + \dfrac{1}{2 \cdot 3} + \cdots + \dfrac{1}{n(n+1)} = \dfrac{n}{n+1}$.

12. $2 + 5 + 13 + \cdots + (2^{n-1} + 3^{n-1}) = 2^n - 1 + \frac{1}{2}(3^n - 1)$.
13. For all integers $x \neq 1$, $x^n - 1$ is divisible by $x - 1$.
HINT: Use the identity $x^{n+1} - 1 = x(x^n - 1) + (x - 1)$.
14. For all integers $x \neq -1$, $x^{2n} - 1$ is divisible by $x + 1$.
15. $a + ar + \cdots + ar^{n-1} = \dfrac{a(1 - r^n)}{1 - r}$ for $r \neq 1$.
16. $a + (a + d) + \cdots + [a + (n - 1)d] = (n/2)[2a + (n - 1)d]$.
17. $n(n + 1)(n + 2)$ is divisible by 3.
18. For all positive integers, x and y where $x > y$: $x^n - y^n$ is divisible by $x - y$.
19. If a is a real number, we define $a^1 = a$ and $a^{n+1} = a^n \cdot a$. Prove that, for $n \geq 1$, $1^n = 1$.
20. For all real numbers a and b and all integers $n \geq 1$, $(ab)^n = a^n b^n$.
21*. For all integers $n \geq 5$, $2^n > n^2$.
22*. For all integers $n \geq 5$, $4^n > n^4$.
23*. For all integers $n \geq -1$, $2n^3 - 9n^2 + 13n + 25 > 0$.
24*. For all integers $n \geq -2$, $2n^3 + 9n^2 + 13n + 7 > 0$.
25*. For all integers $n \geq 2$, $|x_1 + x_2 + \cdots + x_n| \leq |x_1| + |x_2| + \cdots + |x_n|$.
HINT: See Sec. 2.11, Prob. 13.
26*. For all integers $n \geq 1$ and for all real numbers $a > -1$: $(1 + a)^n \geq 1 + na + \dfrac{n(n - 1)}{2} a^2$.
27*. Define the symbol $\binom{n}{r}$ to be equal to $\dfrac{n!}{r!(n - r)!}$, where n and r are integers and $r < n$. Prove that, for fixed r and for $n \geq 2$,

$$\binom{n}{r - 1} + \binom{n}{r} = \binom{n + 1}{r}$$

28*. Consider the set of $n \geq 2$ lines in the plane such that no two lines are parallel and no three lines pass through the same point. Prove that the number of points of intersection is $\frac{1}{2}(n^2 - n)$.

In Probs. 29 to 38 try to establish the indicated theorem by mathematical induction. Point out why the method fails.

29. For $n \geq 1$, $3 + 6 + 9 + \cdots + 3n = \dfrac{3n(n + 1)}{2} + 1$.
30. For $n \geq 1$, $3 + 5 + 7 + \cdots + (2n + 1) = n^2 + 2$.
31. For $n \geq 1$, $2^n > n^2$.
32. For $n \geq 1$, $4^n > n^4$.
33. For $n \geq 2$, $2^n > n^2$.
34. For $n \geq 2$, $4^n > n^4$.
35. For $n \geq 3$, $2^n > n^2$.
36. For $n \geq 3$, $4^n > n^4$.
37. For $n \geq 4$, $2^n > n^2$.
38. For $n \geq 4$, $4^n > n^4$.

39*. Prove that there is no integer between 0 and 1. HINT: Suppose there is an integer a such that $0 < a < 1$. Consider the sets: $A = \{n \mid n \geq a\}$, $S = \{a, a + 1, a + 2, \ldots\}$. Then $S \subset A$, and S satisfies the Axiom of Induction. So $S = A$, which is a contradiction.
40*. Let a be an integer. Then there is no integer between a and $a + 1$. HINT: Use Prob. 39.
41*. Let b be an element of the set $A = \{n \mid n \geq a\}$. Then there are only a finite number of elements of A less than b.
42*. Let a be any integer, $A = \{n \mid n \geq a\}$, and S any nonempty subset of A. Prove that S has a least element. This theorem is called the *Well-ordering Principle.*
43*. Assume the Well-ordering Principle as an axiom and prove the Axiom of Induction.

3.3 Divisibility

In our discussion of divisibility we shall assume for simplicity that a and b are positive and shall leave details of other cases to you if you are interested.

In our first theorem we prove that the usual process of division is legitimate.

Theorem 1. *Division Algorithm.* If a and b are positive integers, there exist positive integers q and r such that

$$a = qb + r \qquad \text{and} \qquad 0 \leq r < b$$

The integer q is called the *quotient,* and r the *remainder.*

Proof: The first three cases below are trivial:

Case 1. If $b > a$, then $q = 0$, $r = a$.

Case 2. If $b = a$, then $q = 1$, $r = 0$.

Case 3. If $b < a$, and b divides a, then $a = qb$, $r = 0$.

The serious proof begins with:

Case 4. $b < a$, and b not a divisor of a. Consider the set of integers: $\{0, b, 2b, 3b, 4b, \ldots, ab\}$. It is evident that $0 < a \leq ab$ and that no member of the set is equal to a. Let the largest member of this set which is less than a be qb. That is, let

$$qb < a < (q + 1)b = qb + b$$

Subtracting qb from each term of this inequality, we get:

$$0 < a - qb < b$$

Now define $r = a - qb$, so that $a = qb + r$. Then r satisfies the required inequalities. Note that in this case $r \neq 0$.

3.4 Factorization

We now turn to the question of the factorization of a positive integer. For this purpose we divide the set of positive integers into three subsets: (1) the *unit* 1, (2) the prime integers, (3) the composite integers.

Definition: A positive integer $\neq 1$ is called a *prime* if and only if its only positive divisors are itself and 1.

A positive integer which is neither the unit nor a prime is called *composite.*

Exercise A. List the first 12 prime integers.

You are undoubtedly aware that an integer such as 60 can be written as the product of other integers (its *factors*) in many ways. For example,

$$60 = 10 \times 6 = 12 \times 5 = 15 \times 4$$

Often these factors can themselves be factored, and the process continues until only prime factors appear. Thus

$$60 = 10 \times 6 = (2 \times 5) \times (2 \times 3)$$

Also

$$60 = 12 \times 5 = (6 \times 2) \times 5 = (3 \times 2 \times 2) \times 5$$

We observe that, in these two cases, the prime factors are the same except for their order. This illustrates an important result (Theorem 5) which we shall prove later in this section. To prove this theorem we shall need several preliminary results.

Definition: Two positive integers a and b are *relatively prime* if and only if there is no positive integer except 1 which is a divisor of each of them.

Theorem 2. If a and b are two relatively prime positive integers, there exist integers x and y such that

$$ax + by = 1$$

Proof: Since a cannot equal b, suppose that $b < a$. Using Theorem 1, write

$$a = q_1 b + r_1 \qquad (0 < r_1 < b)$$

If $r_1 = 1$, we stop, since the theorem is proved with $x = 1$, $y = -q_1$. If $r_1 \neq 1$, we observe that b and r_1 are relatively prime, for if they had a common divisor, it would also divide a. Then a and b would not be relatively prime. Now divide b by r_1 and write

$$b = q_2 r_1 + r_2 \qquad (0 < r_2 < r_1)$$

For the same reason r_1 and r_2 are relatively prime. Unless $r_2 = 1$, we continue by dividing r_1 by r_2 and keep going until a remainder is reached which is equal to 1.

We write this sequence of equations in the form:

$$(1)\qquad a = q_1 b + r_1$$
$$(2)\qquad b = q_2 r_1 + r_2$$
$$(3)\qquad r_1 = q_3 r_2 + r_3$$
$$\cdots$$
$$(4)\qquad r_{k-2} = q_k r_{k-1} + r_k$$
$$(5)\qquad r_{k-1} = q_{k+1} r_k + r_{k+1}$$

A remainder $r_{k+1} = 1$ must be reached since the r's are all positive integers and each r is less than the previous one. The above sequence of Eqs. (1) to (5) is called the *Euclidean Algorithm.*

Before proceeding with the proof of the theorem, consider the example:

Illustration 1.

$$a = 23,\ b = 17$$
$$23 = 1 \times 17 + 6$$
$$17 = 2 \times 6 + 5$$
$$6 = 1 \times 5 + 1$$

Exercise B. Write out the Euclidean Algorithm for $a = 31$, $b = 13$.

Exercise C. What goes wrong with this procedure when $a = 25$, $b = 15$?

Let us now continue the proof of Theorem 2. Equation (1) expresses r_1 in the form

$$r_1 = ax + by$$

where $x = 1$, $y = -q_1$. Let us now substitute

$$(6)\qquad r_1 = a - q_1 b \qquad \text{[from (1)]}$$

into (2). We get

$$b = q_2(a - q_1 b) + r_2$$
$$(7)\qquad r_2 = b(1 + q_1 q_2) - q_2 a$$

which is of the form

$$r_2 = ax + by$$

with $x = -q_2$ and $y = 1 + q_1 q_2$. If we substitute (6) and (7) into (3), we get

$$a - q_1 b = q_3[b(1 + q_1 q_2) - q_2 a] + r_3$$

From this we can obtain r_3 in the form

$$r_3 = ax + by \tag{8}$$

Exercise D. Find x and y in (8).

Continuing this process and using induction, we can write

$$1 = r_{k+1} = ax + by$$

where x and y are integers. This completes the proof.

Illustration 2. Let us carry out this process for the example of Illustration 1, where $a = 23$, $b = 17$. The equations numbered (6′), (7′), etc., are special cases of (6), (7), etc.

$$\begin{aligned}
&(6') & 6 &= 23 - (1 \times 17) = 23 - 17 \\
& & 17 &= 2 \times (23 - 17) + 5 \\
&(7') & 5 &= 17(1 + 2) - 2 \times 23 \\
& & &= 17 \times 3 - 2 \times 23 \\
& & 23 - 17 &= 1 \times (17 \times 3 - 2 \times 23) + 1 \\
&(8') & 1 &= 23(1 + 2) + 17(-1 - 3) \\
& & &= 23 \times 3 + 17(-4).
\end{aligned}$$

So, when $a = 23$, $b = 17$, the x and y which satisfy

$$ax + by = 1$$

are $x = 3$, $y = -4$.

Exercise E. Find suitable x and y when $a = 31$, $b = 13$.

Exercise F. Show that if $ax + by = 1$, then $a(x + kb) + b(y - ka) = 1$. Thus the integers x and y of Theorem 2 are not unique.

Theorem 3. If the product of two positive integers a and b is divisible by a prime number p, at least one of the integers a or b is divisible by c.

Proof: By hypothesis, $ab = cp$. Assume a not divisible by p. Then a and p are relatively prime. Hence, by Theorem 2, there exist x and y, which satisfy

$$ax + py = 1$$

Multiplying this equation by b, we have

$$bax + bpy = b$$

Since $ab = cp$, this gives $cpx + bpy = b$, or $(cx + by)p = b$. Since $cx + by$ is an integer, this shows that b is divisible by p.

Corollary: If the product of any number of positive integers is divisible by a prime number p, at least one of the integers is divisible by p.

Exercise G. Prove this corollary by induction.

The proof of Theorem 3 also establishes:

Theorem 4. If the product of two positive integers a and b is divisible by a number d, which is relatively prime to a, then d divides b.

We are now ready to prove the *Unique Factorization Theorem*, which is often called the *Fundamental Theorem of Arithmetic.*

Theorem 5. The factorization of any positive integer into primes is unique, apart from the order of the factors.

Proof: Suppose that we have two factorizations for a positive integer a:

$$a = p_1p_2 \cdots p_r = q_1q_2 \cdots q_s$$

where the p's and q's are primes, not necessarily distinct, and r is not necessarily equal to s. Now divide out any primes that are common to both factorizations so that we have a new equality

$$p_1p_2 \cdots p_t = q_1q_2 \cdots q_u$$

where no prime on the left occurs on the right, and vice versa. The product $q_1 \cdots q_u$ is certainly divisible by p_1, and so, from the corollary to Theorem 3, p_1 must divide one of the q's, say, q_1. Since p_1 and q_1 are both primes, the only possibility is that $p_1 = q_1$. This, however, was ruled out earlier.

Problems 3.4

In Probs. 1 to 10 find integers x and y such that $ax + by = 1$.

1. $a = 5, b = 2$.
2. $a = 13, b = 3$.
3. $a = 25, b = 13$.
4. $a = 17, b = 5$.
5. $a = 31, b = 29$.
6. $a = 6, b = 7$.
7. $a = 9, b = 4$.
8. $a = 12, b = 5$.
9. $a = 15, b = 14$.
10. $a = 25, b = 16$.

11*. At the end of the proof of Theorem 2 we said, "Continuing this process and using induction we can write . . . ". Write out the details of this process.

12*. Write out the details of the proof of Theorem 4.

13*. Prove (see Sec. 2.8): The residue classes, mod m, form a field if and only if m is a prime. HINT: Examine the existence of the multiplicative inverse.

3.5 Formal Properties of the Integers

The integers can be treated as an abstract system satisfying a set of axioms. Most of these are very familiar to us, for they are the same as the corresponding axioms of a field. For brevity, we shall refer to these only by name.

Axioms for the Integers

Addition. The addition of integers is closed, associative, and commutative and has the identity 0, and each element has an additive inverse.

Multiplication. The multiplication of integers is closed, associative, and commutative and has the identity 1.

The existence of a multiplicative inverse, however, is omitted, and this axiom is replaced by the following one:

Cancellation Law: $[(ac = bc) \wedge (c \neq 0)] \rightarrow (a = b)$

Distributive Law. The usual distributive law is to hold.

Order. The four axioms of inequality R12 to R15 are to hold.

Induction. The axiom of mathematical induction.

3.6 Integral Domains

From the integers we can abstract two mathematical systems which have a number of uses in higher mathematics.

Definition: An *integral domain* consists of an undefined set of at least two elements and two undefined binary operations $+$ and $\times$ which satisfy the axioms of the integers for addition, multiplication, and distributivity.

Definition: An *ordered integral domain* is an integral domain in which there is the further undefined operation $>$ which satisfies the axioms of order for the integers.

Problems 3.6

In Probs. 1 to 8 prove the stated theorems using the axioms of an integral domain.

1. $(ab = 0) \rightarrow [(a = 0) \vee (b = 0)]$.
2. $-(a + b) = (-a) + (-b)$.
3. $-(a - b) = -a + b$.
4. $-0 = 0$.
5. $a - 0 = a$.
6. $a \times 0 = 0$.
7. $a \times (-b) = -(a \times b)$.
8. $(-a) \times (-b) = (a \times b)$.

9*. Assume the implication of Prob. 1 as an axiom and prove the Cancellation Law, using any of the other axioms of an integral domain.

10*. Prove that in an ordered integral domain the Cancellation Law follows from the other axioms.

In Probs. 11 to 16 state whether the given set is an integral domain.

11. The even integers.

12. The odd integers.

13. The positive integers.

14. The complex numbers $a + bi$, where a and b are integers.

15. The numbers of the form $a + b\sqrt{3}$, where a and b are integers.

16. The residue classes, mod 4.

17*. Let D be an integral domain which is the union of three subsets $\{0\}$, {positive elements}, {negative elements} such that:

(1) The sum of two positive elements is positive.
(2) The product of two positive elements is positive.
(3) For a given element a, one and only one of the following alternatives holds: $a = 0$, a is positive, a is negative.

Then prove that D is an ordered integral domain if $>$ is properly defined.

3.7 Rings

We have previously seen (Sec. 2.8) that the residue classes, mod m, form a field if and only if m is a prime. Now let us examine their properties when m is composite. An easy but tedious verification shows us that all the properties of a field are satisfied except for the existence of a multiplicative inverse for certain elements. In the same way, these classes have all the properties of an integral domain, except that the Cancellation Law does not hold.

Illustration 1. In the residue classes, mod 4, we have $2 \times 3 = 2$ and $2 \times 1 = 2$. Hence $2 \times 3 = 2 \times 1$, but $3 \neq 1$.

The existence of such classes leads us to define our final abstract algebraic system.

Definition: A *commutative ring* consists of an undefined set of at least two elements and two undefined binary operations $+$ and $\times$ such that:

(1) Addition and multiplication are closed, associative, and commutative.
(2) The distributive laws: $a(b + c) = ab + ac$ and $(b + c)a = ba + ca$ hold.
(3) The additive identity and additive inverses exist.

If, further, there is a multiplicative identity, we speak of a *commutative ring with unit element.* In a general ring (noncommutative) we do not assume that multiplication is commutative.

The following familiar systems are examples of rings with unit element: the real numbers; the rational numbers; the integers; the residue classes, mod m; any field; any integral domain.

The set of even integers is an example of a commutative ring *without* unit element.

3.8 2×2 Matrices

Although the subject of matrices is a large and important branch of mathematics, we present only a brief introduction here. Our chief purpose is to show you an example of a noncommutative ring.

Definition: A 2×2 *matrix* is an array of the form $\begin{pmatrix} a & b \\ c & d \end{pmatrix}$ For simplicity in presentation, we assume that a, b, c, and d are real numbers.

Our object is to develop the algebra of 2×2 matrices so that in some sense they may be treated as numbers. By analogy with our definitions of equality and addition of complex numbers, we define these concepts for matrices.

Definition: $\begin{pmatrix} a & b \\ c & d \end{pmatrix}$ is equal to $\begin{pmatrix} p & q \\ r & s \end{pmatrix}$ if and only if $a = p$, $b = q$, $c = r$, and $d = s$.

Definition: The *sum* of two 2×2 matrices is the 2×2 matrix given by the formula:

$$\begin{pmatrix} a & b \\ c & d \end{pmatrix} + \begin{pmatrix} p & q \\ r & s \end{pmatrix} = \begin{pmatrix} a + p & b + q \\ c + r & d + s \end{pmatrix}$$

Exercise A. Prove that the addition of 2×2 matrices is closed, associative, and commutative.

The definition of the product of two matrices seems arbitrary, but there are good reasons for it which we cannot present here.

Definition: The *product* of two 2×2 matrices is the 2×2 matrix given by the formula:

$$\begin{pmatrix} a & b \\ c & d \end{pmatrix} \times \begin{pmatrix} p & q \\ r & s \end{pmatrix} = \begin{pmatrix} ap + br & aq + bs \\ cp + dr & cq + ds \end{pmatrix}$$

Exercise B. Prove that the multiplication of 2×2 matrices is closed and associative.

Exercise C. Prove that the distributive law holds for the addition and multiplication of 2×2 matrices.

The salient fact here is that the multiplication of matrices is *not commutative.*

Illustration 1

$$\begin{pmatrix} 1 & 2 \\ -1 & 4 \end{pmatrix} \times \begin{pmatrix} 2 & 3 \\ 1 & -2 \end{pmatrix} = \begin{pmatrix} 4 & -1 \\ 2 & -11 \end{pmatrix}$$

$$\begin{pmatrix} 2 & 3 \\ 1 & -2 \end{pmatrix} \times \begin{pmatrix} 1 & 2 \\ -1 & 4 \end{pmatrix} = \begin{pmatrix} -1 & 16 \\ 3 & -6 \end{pmatrix}$$

Now let us examine the sensitive question of identity and inverse elements. It is immediate that $\begin{pmatrix} 0 & 0 \\ 0 & 0 \end{pmatrix}$ is the additive identity, for

$$\begin{pmatrix} a & b \\ c & d \end{pmatrix} + \begin{pmatrix} 0 & 0 \\ 0 & 0 \end{pmatrix} = \begin{pmatrix} 0 & 0 \\ 0 & 0 \end{pmatrix} + \begin{pmatrix} a & b \\ c & d \end{pmatrix} = \begin{pmatrix} a & b \\ c & d \end{pmatrix}$$

Also, the additive inverse of $\begin{pmatrix} a & b \\ c & d \end{pmatrix}$ is $\begin{pmatrix} -a & -b \\ -c & -d \end{pmatrix}$, for

$$\begin{pmatrix} a & b \\ c & d \end{pmatrix} + \begin{pmatrix} -a & -b \\ -c & -d \end{pmatrix} = \begin{pmatrix} -a & -b \\ -c & -d \end{pmatrix} + \begin{pmatrix} a & b \\ c & d \end{pmatrix} = \begin{pmatrix} 0 & 0 \\ 0 & 0 \end{pmatrix}$$

There is also a multiplicative identity, namely, $\begin{pmatrix} 1 & 0 \\ 0 & 1 \end{pmatrix}$, for

$$\begin{pmatrix} a & b \\ c & d \end{pmatrix} \times \begin{pmatrix} 1 & 0 \\ 0 & 1 \end{pmatrix} = \begin{pmatrix} 1 & 0 \\ 0 & 1 \end{pmatrix} \times \begin{pmatrix} a & b \\ c & d \end{pmatrix} = \begin{pmatrix} a & b \\ c & d \end{pmatrix}$$

In summary of the above observations, we see that the 2×2 *matrices form a noncommutative ring with unit element.*

The final open question is the existence of a multiplicative inverse. In other words, we ask whether, given $\begin{pmatrix} a & b \\ c & d \end{pmatrix}$, there is a matrix $\begin{pmatrix} w & x \\ y & z \end{pmatrix}$ such that $\begin{pmatrix} a & b \\ c & d \end{pmatrix} \times \begin{pmatrix} w & x \\ y & z \end{pmatrix} = \begin{pmatrix} 1 & 0 \\ 0 & 1 \end{pmatrix}$. If there is such a matrix, we must have that

$$\begin{pmatrix} aw + by & ax + bz \\ cw + dy & cx + dz \end{pmatrix} = \begin{pmatrix} 1 & 0 \\ 0 & 1 \end{pmatrix}$$

From the definition of the equality of two matrices, this gives us the simultaneous system:

$$\begin{aligned} aw + by &= 1 \\ ax + bz &= 0 \\ cw + dy &= 0 \\ cx + dz &= 1 \end{aligned}$$

Writing $\Delta = ad - bc$ and supposing that this is not zero, we find that the solution is:

$$w = \frac{d}{\Delta} \qquad x = -\frac{b}{\Delta} \qquad y = -\frac{c}{\Delta} \qquad z = \frac{a}{\Delta}$$

Therefore

$$\begin{pmatrix} a & b \\ c & d \end{pmatrix} \times \begin{pmatrix} \frac{d}{\Delta} & -\frac{b}{\Delta} \\ -\frac{c}{\Delta} & \frac{a}{\Delta} \end{pmatrix} = \begin{pmatrix} 1 & 0 \\ 0 & 1 \end{pmatrix}$$

Since it also follows that

$$\begin{pmatrix} \frac{d}{\Delta} & -\frac{b}{\Delta} \\ -\frac{c}{\Delta} & \frac{a}{\Delta} \end{pmatrix} \times \begin{pmatrix} a & b \\ c & d \end{pmatrix} = \begin{pmatrix} 1 & 0 \\ 0 & 1 \end{pmatrix}$$

we conclude that the multiplicative inverse of

$$\begin{pmatrix} a & b \\ c & d \end{pmatrix} \text{ is } \begin{pmatrix} \frac{d}{\Delta} & -\frac{b}{\Delta} \\ -\frac{c}{\Delta} & \frac{a}{\Delta} \end{pmatrix}$$

On the other hand, suppose that $\Delta = ad - bc = 0$. Then we shall show that there is no multiplicative inverse. Using indirect proof, we assume that this inverse exists, so that the simultaneous system above has a solution (w,x,y,z). Then from these equations we have that:

$$(aw + by)(cx + dz) - (ax + bz)(cw + dy) = 1$$

From simple algebra we can prove the identity:

$$(aw + by)(cx + dz) - (ax + bz)(cw + dy) = (ad - bc)(wz - xy)$$

so that

$$(ad - bc)(wz - xy) = 1$$

Since, however, $ad - bc = 0$ by hypothesis, this equality is impossible. This gives the needed contradiction.

Exercise D. Prove the identity above.

Definition: A 2×2 matrix for which $\Delta = ad - bc \neq 0$ is *nonsingular;* if $ad - bc = 0$, it is *singular.*

Thus we have proved the theorem.

Theorem 6. A 2×2 matrix has a multiplicative inverse if and only if it is nonsingular.

Since a 2×2 matrix other than the additive identity $\begin{pmatrix} 0 & 0 \\ 0 & 0 \end{pmatrix}$ may well be singular, we cannot assert that these matrices form a field. Moreover, they do not form an integral domain, for the cancellation law does not hold in some cases.

Illustration 2. The cancellation law fails, for example, in the following case:

$$\begin{pmatrix} 2 & -1 \\ 3 & 1 \end{pmatrix} \times \begin{pmatrix} 1 & 2 \\ 1 & 2 \end{pmatrix} = \begin{pmatrix} 1 & 2 \\ 4 & 8 \end{pmatrix}$$

$$\begin{pmatrix} 3 & -2 \\ 2 & 2 \end{pmatrix} \times \begin{pmatrix} 1 & 2 \\ 1 & 2 \end{pmatrix} = \begin{pmatrix} 1 & 2 \\ 4 & 8 \end{pmatrix}$$

Hence

$$\begin{pmatrix} 2 & -1 \\ 3 & 1 \end{pmatrix} \times \begin{pmatrix} 1 & 2 \\ 1 & 2 \end{pmatrix} = \begin{pmatrix} 3 & -2 \\ 2 & 2 \end{pmatrix} \times \begin{pmatrix} 1 & 2 \\ 1 & 2 \end{pmatrix}$$

But

$$\begin{pmatrix} 2 & -1 \\ 3 & 1 \end{pmatrix} \neq \begin{pmatrix} 3 & -2 \\ 2 & 2 \end{pmatrix} \quad \text{and} \quad \begin{pmatrix} 1 & 2 \\ 1 & 2 \end{pmatrix} \neq \begin{pmatrix} 0 & 0 \\ 0 & 0 \end{pmatrix}$$

Problems 3.8

In Probs. 1 to 8, prove the following, using the axioms of a commutative ring.

1. $-(a + b) = (-a) + (-b)$.
2. $-(a - b) = -a + b$.
3. $-0 = 0$.
4. $a - 0 = a$.
5. $a \times 0 = 0$.
6. $a \times (-b) = -(a \times b)$.
7. $(-a) \times (-b) = (a \times b)$.
8. $a(b - c) = ab - ac$.

In Probs. 9 to 18, carry out the indicated process.

9. $\begin{pmatrix} 1 & -2 \\ 2 & 0 \end{pmatrix} + \begin{pmatrix} 3 & -1 \\ 1 & 4 \end{pmatrix}$.

10. $\begin{pmatrix} 4 & -3 \\ 1 & 2 \end{pmatrix} + \begin{pmatrix} 1 & 7 \\ 2 & 6 \end{pmatrix}$.

11. $\begin{pmatrix} 3 & 4 \\ 1 & 2 \end{pmatrix} \times \begin{pmatrix} 2 & -7 \\ 1 & 3 \end{pmatrix}$.

12. $\begin{pmatrix} 4 & 5 \\ -6 & 2 \end{pmatrix} \times \begin{pmatrix} 3 & -2 \\ 4 & 1 \end{pmatrix}$.

13. $\begin{pmatrix} 2 & -7 \\ 1 & 3 \end{pmatrix} \times \begin{pmatrix} 3 & 4 \\ 1 & 2 \end{pmatrix}$.

14. $\begin{pmatrix} 3 & -2 \\ 4 & 1 \end{pmatrix} \times \begin{pmatrix} 4 & 5 \\ -6 & 2 \end{pmatrix}$.

15. $\begin{pmatrix} 1 & -1 \\ -1 & 1 \end{pmatrix} \times \begin{pmatrix} 1 & 1 \\ 1 & 1 \end{pmatrix}$.

16. $\begin{pmatrix} 1 & 3 \\ 2 & 6 \end{pmatrix} \times \begin{pmatrix} 3 & 3 \\ -1 & -1 \end{pmatrix}$.

17. $\begin{pmatrix} 5 & 1 \\ 1 & 3 \end{pmatrix} \times \begin{pmatrix} -2 & 4 \\ 3 & 6 \end{pmatrix} + \begin{pmatrix} -1 & -4 \\ 2 & 3 \end{pmatrix}$.

18. $\begin{pmatrix} 1 & 4 \\ 3 & -1 \end{pmatrix} + \begin{pmatrix} 0 & 5 \\ 1 & 6 \end{pmatrix} \times \begin{pmatrix} 1 & 1 \\ 2 & 6 \end{pmatrix}$.

In Probs. 19 to 24 find the multiplicative inverse (if any) of the given matrix.

19. $\begin{pmatrix} 4 & 1 \\ 1 & 2 \end{pmatrix}$.

20. $\begin{pmatrix} 3 & -1 \\ -1 & 3 \end{pmatrix}$.

21. $\begin{pmatrix} 1 & 5 \\ 2 & 6 \end{pmatrix}$.

22. $\begin{pmatrix} 4 & -1 \\ 7 & 3 \end{pmatrix}$.

23. $\begin{pmatrix} 2 & 4 \\ 4 & 8 \end{pmatrix}$.

24. $\begin{pmatrix} 1 & -2 \\ -3 & 6 \end{pmatrix}$.

25. Define $k\begin{pmatrix} a & b \\ c & d \end{pmatrix} = \begin{pmatrix} ka & kb \\ kc & kd \end{pmatrix}$, $I = \begin{pmatrix} 1 & 0 \\ 0 & 1 \end{pmatrix}$, and $0 = \begin{pmatrix} 0 & 0 \\ 0 & 0 \end{pmatrix}$. Then show that the matrix $A = \begin{pmatrix} 3 & 1 \\ 2 & 2 \end{pmatrix}$ satisfies the equation $A^2 - 5A + 4I = 0$.

26. Show that the matrix $A = \begin{pmatrix} 1 & 2 \\ 3 & 4 \end{pmatrix}$ satisfies the équation $A^2 - 5A - 2I = 0$.

27*. Prove that the product of two nonsingular matrices is nonsingular.

28*. Prove that the product of a nonsingular matrix and a singular matrix is singular.

References

Birkhoff, Garrett, and Saunders MacLane: "A Survery of Modern Algebra", chaps. 1, 2, 4, 8, Macmillan, New York (1953).

Henkin, Leon: "Mathematical Induction", Mathematical Association of America, Buffalo, N.Y. (1961).

Hohn, Franz E.: "Elementary Matrix Algebra", Macmillan, New York (1958).

McCoy, Neal H.: "Introduction to Modern Algebra", chaps. 2–4, Allyn and Bacon, Boston (1960).

Niven, Ivan, and Herbert S. Zuckerman: "An Introduction to the Theory of Numbers", Wiley, New York (1960).

Sorninskii, I. S.: "The Method of Mathematical Induction", Blaisdell, New York (1961).

4 | Groups

4.1 Introduction

A common feature of the fields, integral domains, and rings, discussed in Chaps. 2 and 3, was the existence of binary operations $+$ and $\times$. In these chapters we were interested in the question of whether these operations were closed, associative, and commutative, whether there was an identity element, and whether there was an inverse element.

We found that in all these systems, addition had all the above properties and that multiplication was closed and associative. In a field, multiplication was also commutative, there was a multiplicative identity, and every element except the additive identity had a multiplicative inverse. In an integral domain, the existence of a multiplicative inverse was replaced by the Cancellation Law. Multiplication in a general ring, however, was not commutative, and there was no multiplicative identity or inverse.

Apparently there is something fundamental about the five properties:

(1) Closure
(2) Associativity
(3) Existence of an identity
(4) Existence of an inverse
(5) Commutativity

We should therefore examine these in detail. First, let us discard all irrelevant matters. It is unimportant to know what the symbols a, b, . . . mean. Let us call them *undefined*. It is also unimportant whether the operation is $+$ or $\times$ or something else. Let us use an undefined operation "$\circ$". Then we can derive an abstract mathematical system from these ideas, which we call a *commutative group*. The detailed discussion of groups will be given in the remainder of this chapter.

Before beginning this, we should answer two questions which are probably in your mind. First, you may ask, What are we going to gain from studying groups? To answer this, we shall show you a fair number of specific mathematical theories whose elements satisfy the axioms of a group. In each of these theories we naturally desire to find a set of significant theorems and to learn how to calculate within each theory. If we do this for each theory separately, we shall have a large quantity of information to remember. We can save ourselves work by proving the theorems just once for the abstract system called a *group*. The results thus proved can then be applied to each concrete case without serious mental labor. The advantage, then, is that we have one inclusive theory to remember instead of a large number of separate ones.

The second question is, Why do we select axioms based upon the five properties above? Why not choose other properties as motivation for our axioms? The answer here is harder to give. One reason for our choice is that the chosen axioms involve just one operation $\circ$, instead of a pair of operations such as $+$ and $\times$. An abstract system involving a single operation is likely to be simpler than one involving a pair of operations, and we want to keep things simple here. But why do we abstract from all five of the properties above; why not take (1), (2), and (3) alone, or possibly some other combination? As a matter of fact, all such combinations have been investigated, and the *group* turns out to have more important applications than any of the others. With this preamble we now turn to the theory of groups themselves.

4.2 Definition of a Group

The idea of a group is due to the Frenchman E. Galois (1811–1832), who died at the age of twenty in a duel, but is still remembered among the mathematical great. Its structure is given by the following:

Undefined elements: $a, b, c, \ldots$ belonging to a nonempty set G.

Undefined operation: $\circ$, used to pair two elements: $a \circ b$.

Axioms

G1. For each ordered pair, a and b in G, the combination $a \circ b$ is a unique element c of G. [Axiom of closure]

G2. For each triple, a, b, and c in G:

$$(a \circ b) \circ c = a \circ (b \circ c)$$

[Associative axiom]

G3. There exists a unique element e of G having the property that for every a in G:

$$a \circ e = e \circ a = a$$

The element e is called the identity. [Identity axiom]

G4. Corresponding to each a in G, there is a unique element a' having the property that

$$a \circ a' = a' \circ a = e$$

The element a' is called the inverse of a. [Inverse axiom]

Often it is desirable to add a fifth axiom:

G5. For every a and b in G,

$$a \circ b = b \circ a$$

[Commutative axiom]

A group which also satisfies axiom G5 is called a *commutative group*. A group will not be assumed to be commutative unless we specifically say that it is.

Remark. The operation $\circ$ is often called *multiplication* even though it is completely different from the multiplication of numbers. When we use this terminology we shall write "multiplication", in quotation marks.

4.3 Examples of Groups

(I) By comparing axioms G1 to G5 with the axioms of Sec. 3.5, we see at once that the integers are a commutative group with respect to the operation $+$. The identity e is zero. The inverse of an integer a is the integer $-a$.

(II) By comparing axioms G1 to G5 with properties R1 to R5, we see that the real numbers also form a commutative group with respect to addition. The identity is zero, and the inverse of a is $-a$.

(III) By comparing axioms G1 to G5 with properties R6 to R10, we see that the set of real numbers excluding zero forms a commutative group with respect to multiplication. The identity is 1, and the inverse of a is $1/a$.

Exercise A. Does the set of all real numbers form a commutative group with respect to multiplication? Why?

Exercise B. Recall that the rational numbers and the complex numbers also satisfy R1 to R11. Describe how these form groups with respect to addition and multiplication. State the identity in each group you describe, and tell how to find the inverse of each element in the groups.

(IV) Consider the residue classes, mod m. We have already seen (Sec. 2.7) how to add these and have examined their properties under addition. We see immediately that these form a commutative group with respect to addition. For example, the addition table for the residue classes, mod 4, is:

+	0	1	2	3
0	0	1	2	3
1	1	2	3	0
2	2	3	0	1
3	3	0	1	2

(V) Now let us take the set of residue classes, mod m, but ignore the class 0 mod m. With the remaining classes we can form a multiplication table such as the following for mod 4:

×	1	2	3
1	1	2	3
2	2	0	2
3	3	2	1

This is not a group, for it is not closed (the product $2 \times 2 = 0$, which is not one of the elements: 1, 2, 3), and, moreover, 2 has no inverse.

On the other hand, the nonzero residue classes, mod 3, do form a commutative group with respect to multiplication. The multiplication table is:

×	1	2
1	1	2
2	2	1

(VI) The 2×2 matrices form a commutative group with respect to addition. The identity is $\begin{pmatrix} 0 & 0 \\ 0 & 0 \end{pmatrix}$, and the inverse of $\begin{pmatrix} a & b \\ c & d \end{pmatrix}$ is $\begin{pmatrix} -a & -b \\ -c & -d \end{pmatrix}$.

(VII) The nonsingular 2×2 matrices form a (noncommutative) group with respect to multiplication. The closure axiom holds since the product of two nonsingular matrices is nonsingular (Sec. 3.8, Prob. 21). The identity is $\begin{pmatrix} 1 & 0 \\ 0 & 1 \end{pmatrix}$, and the inverse of $\begin{pmatrix} a & b \\ c & d \end{pmatrix}$ is

$$\begin{pmatrix} \dfrac{d}{\Delta} & -\dfrac{b}{\Delta} \\ \dfrac{-c}{\Delta} & \dfrac{a}{\Delta} \end{pmatrix} \qquad \text{where } \Delta = ad - bc \neq 0$$

Problems 4.3

1. Show why the following are not groups.

(a) The real numbers with respect to subtraction.
(b) The nonzero integers with respect to division.

2. Show why the following are not groups.

(a) The residue classes, mod 6, with respect to multiplication.
(b) The 2×2 matrices with respect to multiplication.

3. In group (IV) above (the residue classes, mod 4, with respect to addition):

(a) Why is the closure axiom satisfied?
(b) Verify the associative law when $a = 1, b = 2, c = 3$.
(c) What is the identity?
(d) What are the inverses $0', 1', 2', 3'$?
(e) Verify the commutative law when $a = 1, b = 2$, and when $a = 0, b = 3$.

4. In group (V) above (the nonzero residue classes, mod 3, with respect to multiplication):

(a) Why is the closure axiom satisfied?
(b) Verify the associative law when $a = 1, b = 1, c = 2$.
(c) What is the identity?
(d) What are the inverses $1', 2'$?
(e) Verify the commutative law for all possible cases.

5. (a) Form the addition table for the residue classes, mod 5.
(b) Verify the associative law when $a = 1, b = 3, c = 2$.
(c) Is there an identity? What is it?
(d) What are the inverses $0', 1', 2', 3', 4'$?
(e) Verify the commutative law when $a = 2, b = 3$.
(f) Is this a group?

6. (a) Form the multiplication table for the residue classes, mod 5, excluding zero.
(b) Verify the associative law when $a = 2, b = 4, c = 1$.
(c) Is there an identity? What is it?
(d) What are the inverses $1', 2', 3', 4'$?
(e) Verify the commutative law when $a = 3, b = 4$.
(f) Is this a group?

7*. Determine those moduli m such that the nonzero residue classes, mod m, form a group with respect to multiplication.

8. Form the multiplication table for the set of matrices below, and show that these form a group under multiplication:

$$a = \begin{pmatrix} 1 & 0 \\ 0 & 1 \end{pmatrix} \quad b = \begin{pmatrix} 0 & 1 \\ 1 & 0 \end{pmatrix} \quad c = \begin{pmatrix} 0 & -1 \\ -1 & 0 \end{pmatrix} \quad d = \begin{pmatrix} -1 & 0 \\ 0 & -1 \end{pmatrix}$$

9. Form the multiplication table for the set of matrices below, and show that these form a group under multiplication. What is the identity?

$$a = \begin{pmatrix} 0 & 0 \\ -1 & 1 \end{pmatrix} \quad b = \begin{pmatrix} 0 & 0 \\ 1 & -1 \end{pmatrix}$$

10. Do the nonsingular 2×2 matrices form a group under addition? Why?

4.4 Further Examples of Groups

(VIII) We assume that you are acquainted with the representation of a point in the plane by a pair of real numbers (x,y), called its coordinates. Suppose that we move every point in the plane a units horizontally and b units vertically. Then a point (x,y) has the new position $(x + a, y + b)$. Such a motion is called a *translation* of the plane. We shall write it as $[a,b]$, using brackets to distinguish translations from points. Now, if we perform the translation $[a,b]$ and follow it by the translation $[c,d]$, the point (x,y) first goes into $(x + a, y + b)$ and then into $(x + a + c, y + b + d)$. In other words, the result of these two translations is the single translation $[a + c, b + d]$. Using $\circ$ to mean "followed by", we can write this symbolically:

$$[a,b] \circ [c,d] = [a + c, b + d]$$

If G consists of the set of all translations with this definition of the operation $\circ$, we can show that G is a commutative group, the "Group of Translations".

This proof is based upon known properties of the numbers a, b, c, d, etc., that are used to express translations. For example, to prove

$$[a,b] \circ [c,d] = [c,d] \circ [a,b]$$

we note that

$$[a,b] \circ [c,d] = [a + c, b + d]$$

and

$$[c,d] \circ [a,b] = [c + a, d + b]$$

The result follows from the known fact that, for real numbers,

$$a + c = c + a$$

and

$$b + d = d + b$$

(IX) In physics and other applied subjects, one often meets the idea of a *vector*. A vector is represented in the plane as a directed distance, $\overrightarrow{PQ}$, and is said to have magnitude and direction. The magnitude is represented by the length of the line segment in the plane, and the direction is given by the angle which this line makes with the horizontal and by the sense in which the arrow points. A common example of a vector is *velocity*, which is described by a phrase such as "the train is going 50 mi/hr northwest". Force and acceleration are two other elementary physical notions which are best treated as vectors.

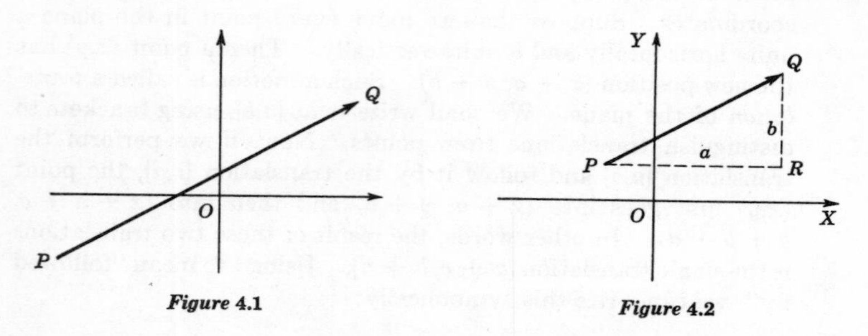

Figure 4.1 *Figure 4.2*

Corresponding to the vector $\overrightarrow{PQ}$ we may draw a right triangle PQR with PR horizontal and QR vertical. The length of PR is the *x-component*, a of $\overrightarrow{PQ}$; a is positive if $\overrightarrow{PQ}$ points to the right and is negative if $\overrightarrow{PQ}$ points to the left. Similarly, RQ is the *y-component*, b of $\overrightarrow{PQ}$; b is positive if $\overrightarrow{PQ}$ points up and negative if it points down. Clearly, these components are known if the vector is known, and conversely, a pair of components determine a vector if we know its initial point P. To simplify the discussion, we shall suppose that all our vectors have the origin O as their initial point, so that the coordinates of their end points are equal to the components of the vectors. Then any vector is determined by its components (a,b).

We now introduce the concept of the *sum* of two vectors. Since vectors are not numbers, this is a new idea, and ordinary addition will not be meaningful. By definition, the sum (designated by $\oplus$) of (a,b) and (c,d) is given by

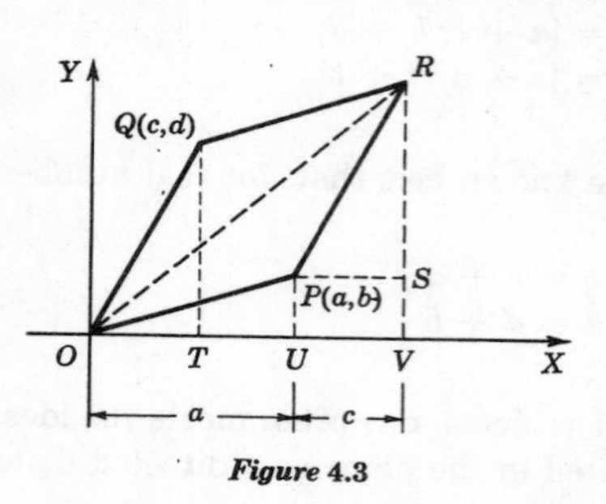

Figure 4.3

$$(a,b) \oplus (c,d) = (a + c, b + d)$$

The geometric interpretation of this sum is given in Fig. 4.3. The point R is obtained by completing the parallelogram with O, P, and Q as vertices. Then triangle PRS is congruent to triangle OQT so that $PS = c$ and $RS = d$. Now $OU = a$ and OV is the

x-component of OR.

From the figure

$$\begin{aligned} OV &= OU + UV \\ &= OU + PS \\ &= a + c \end{aligned}$$

Similarly, $RV = b + d$. Thus $\overrightarrow{OR}$ is $\overrightarrow{OP} \oplus \overrightarrow{OQ}$.

With these definitions, let G be the set of vectors in the plane whose initial points are at the origin. Let $\oplus$ represent vector addition. Then G is a commutative group.

(X) Consider a wheel, and let G be the set of rotations of this wheel through angles equal to $n \times 60°$, where n is an integer (positive, negative, or zero). Let $\circ$ represent "followed by"; that is, $a \circ b$ means "the rotation a followed by the rotation b". This is clearly a rotation through an angle equal to the sum of the angles of a and b. (We do not reduce mod 360°.) These rotations also form a commutative group.

(XI) Let G be the set of elements e, a, b, c, with the rule of combination given below.

$\circ$	e	a	b	c
e	e	a	b	c
a	a	e	c	b
b	b	c	e	a
c	c	b	a	e

This is a commutative group called the Klein "Four Group".

(XII) Consider three chairs in a row, and call their positions, respectively, 1, 2, and 3. Let the symbol $\begin{pmatrix} 1 & 2 & 3 \\ 2 & 3 & 1 \end{pmatrix}$ mean that the chair in position 1 is moved to position 2, that in position 2 is moved to position 3, and that in position 3 is moved to position 1. This rearrangement is called a *permutation*. There are six such permutations:

$$e = \begin{pmatrix} 1 & 2 & 3 \\ 1 & 2 & 3 \end{pmatrix} \qquad p = \begin{pmatrix} 1 & 2 & 3 \\ 3 & 1 & 2 \end{pmatrix} \qquad q = \begin{pmatrix} 1 & 2 & 3 \\ 2 & 3 & 1 \end{pmatrix}$$

$$r = \begin{pmatrix} 1 & 2 & 3 \\ 2 & 1 & 3 \end{pmatrix} \qquad s = \begin{pmatrix} 1 & 2 & 3 \\ 3 & 2 & 1 \end{pmatrix} \qquad t = \begin{pmatrix} 1 & 2 & 3 \\ 1 & 3 & 2 \end{pmatrix}$$

These form the set G.

Let $p \circ q$ mean that we carry out first p and then q. We see that

$$p \circ q = \begin{pmatrix} 1 & 2 & 3 \\ 3 & 1 & 2 \end{pmatrix} \circ \begin{pmatrix} 1 & 2 & 3 \\ 2 & 3 & 1 \end{pmatrix} = \begin{pmatrix} 1 & 2 & 3 \\ 1 & 2 & 3 \end{pmatrix} = e$$

for p replaces 1 by 3 and q replaces 3 by 1. Hence, 1 goes finally into 1, etc. With this understanding, G is a group. However, it is not commutative, for in particular, $r \circ p = t$ and $p \circ r = s$. To see this, examine

$$r \circ p = \begin{pmatrix} 1 & 2 & 3 \\ 2 & 1 & 3 \end{pmatrix} \circ \begin{pmatrix} 1 & 2 & 3 \\ 3 & 1 & 2 \end{pmatrix} = \begin{pmatrix} 1 & 2 & 3 \\ 1 & 3 & 2 \end{pmatrix} = t$$

Note that r takes 1 into 2; p takes 2 into 1; therefore $r \circ p$ takes 1 into 1. Further, r takes 2 into 1; p takes 1 into 3; therefore $r \circ p$ takes 2 into 3. Finally, r takes 3 into 3; p takes 3 into 2; therefore $r \circ p$ takes 3 into 2. The net result is

$$r \circ p = \begin{pmatrix} 1 & 2 & 3 \\ 1 & 3 & 2 \end{pmatrix} = t$$

Similarly,

$$p \circ r = \begin{pmatrix} 1 & 2 & 3 \\ 3 & 1 & 2 \end{pmatrix} \circ \begin{pmatrix} 1 & 2 & 3 \\ 2 & 1 & 3 \end{pmatrix} = \begin{pmatrix} 1 & 2 & 3 \\ 3 & 2 & 1 \end{pmatrix} = s$$

Problems 4.4

1. Prove that the set of translations [(VIII) above] forms a commutative group. In proving each step, use the method outlined in the above discussion of this group. Be sure to state what is the identity and what is the inverse of $[a, b]$.
2. Prove that the set of vectors whose initial points are at the origin [(IX) above] forms a commutative group with respect to addition. Use the method suggested in Prob. 1.
3. How is the group of translations related to the additive group of vectors?
4. How is vector addition related to the addition of complex numbers?
5. For the group of rotations [(X) above], find:

   (a) The identity.
   (b) The inverse of a rotation through $n \times 60°$.

6. For the Klein Four Group [(XI) above]:

   (a) Prove that the closure axiom is satisfied.
   (b) Verify the associative law for the case

$$(a \circ b) \circ c = a \circ (b \circ c)$$

(c) What is the identity?
(d) What are the inverses a', b', c'?
(e) Verify the commutative law for the cases

$$a \circ b = b \circ a$$
$$b \circ c = c \circ b$$

7. For the group of permutations [(XII) above], show that

$$p \circ p = q \qquad p \circ s = t \qquad p \circ t = r$$

8. For the group (XII), show that

$$q \circ p = e \qquad q \circ q = p \qquad q \circ r = t \qquad q \circ s = r \qquad q \circ t = s$$

9. For the group (XII), show that

$$r \circ q = s \qquad r \circ r = e \qquad r \circ s = q \qquad r \circ t = p$$

10. For the group (XII), show that

$$s \circ p = r \qquad s \circ q = t \qquad s \circ r = p \qquad s \circ s = e \qquad s \circ t = q$$

11. For the group (XII), show that

$$t \circ p = s \qquad t \circ q = r \qquad t \circ r = q \qquad t \circ s = p \qquad t \circ t = e$$

12. Using results in the text and Probs. 7 to 11, construct a "multiplication" table for the group of permutations (XII).
13. In the group (XII), what is the identity? What are the elements p', q', r', s', t'?
14. Consider the equilateral triangle in Fig. 4.4 where a, b, and c are altitudes meeting in O. As elements of a group we take the following rotations of this

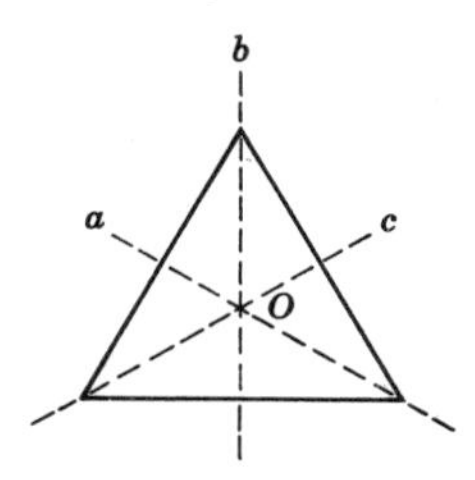

***Figure* 4.4**

triangle into itself and agree that two rotations are equal if and only if they transform the triangle into the same position. The lines a, b, and c are fixed in the plane and do not rotate with the triangle.

e = identity, no rotation
p = counterclockwise rotation about O through 120°
q = counterclockwise rotation about O through 240°
r = rotation about a through 180°
s = rotation about b through 180°
t = rotation about c through 180°

The operation here is "followed by"; that is, $p \circ r$ is "rotation p followed by rotation r".

(**a**) Write out the "multiplication" table for this group. A convenient way of computing these products is to cut a triangle out of cardboard and label the vertices A, B, C, or 1, 2, 3, or red, green, blue, on both sides. Then physically carry out the required rotations and note the result.

(**b**) Compare this table with the table for the permutation group (Prob. 12 above). Draw conclusions.

15. Consider the square in Fig. 4.5 with center O. The lines h and v are the horizontal and vertical lines, respectively, through O, and a and b are diagonals. These

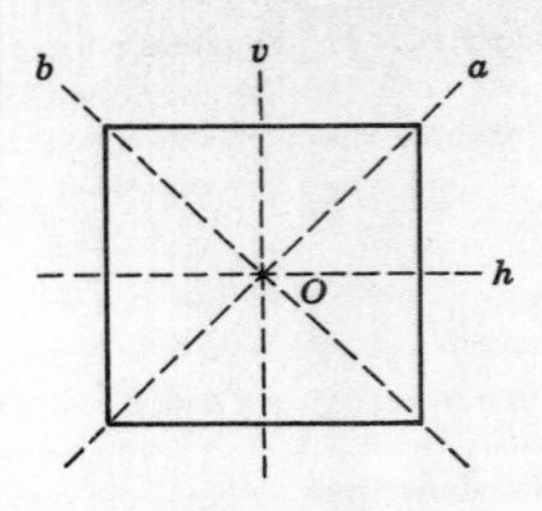

***Figure* 4.5**

are fixed lines in the plane and do not rotate with the square. As elements of a group, we take the following rotations of this square into itself, agreeing that two rotations are equal if and only if they transform the square into the same position. The group operation is "followed by".

e = identity, no rotation
m = counterclockwise rotation about O through 90°
n = counterclockwise rotation about O through 180°
p = counterclockwise rotation about O through 270°
q = rotation about h through 180°
r = rotation about v through 180°
t = rotation about a through 180°
u = rotation about b through 180°

(**a**) Write out the "multiplication" table for this group.
(**b**) Is it commutative?
(**c**) What are the inverses: m', n', p', q', r', s', t', and u'?

16. Show that the complex numbers 1, -1, i, $-i$ form a commutative group under multiplication.

17. Write a "multiplication" table for a group with elements e, a, b. Can you find two distinct tables of this kind?

18. In a group G, define a^n (n a positive integer) by $a^n = a \circ a \circ \cdots \circ a$ (n factors). If there is a least positive integer m such that $a^m = e$, we say that *a is of order m in G*. If there is no such integer m, a is said to be of *infinite order in G*.

The group G is *cyclic* if and only if it contains some element a whose powers exhaust G; this element is said to *generate* G.

Show that, for an additive group of residue classes, mod m:

(a) The group is cyclic.
(b) The group is generated by the class 1, mod m.
(c) The order of 1, mod m, is m.

19. Show that the multiplicative group of nonzero residue classes, mod 5, is cyclic and that it is generated by the class 2, mod 5, which is of order 4.

20. Show that the group (X) of rotations is cyclic. What is the order of a rotation through 60°?

21. Show that the Four Group (XI) is not cyclic. In this group what is the order of a, of b, of c?

22. Show that the group whose elements are $1, -1, i, -i$ is cyclic. Find a generator, and determine its order.

23. A system of abstract elements with a binary "product" designated by $\circ$ is called a *semigroup* if and only if "multiplication" is closed and associative.

(a) Show that the following are semigroups.
(b) Which are groups?

A

$\circ$	a	b
a	a	b
b	b	a

B

$\circ$	a	b
b	a	b
a	a	b

C

$\circ$	a	b
a	a	a
b	a	a

D

$\circ$	a	b
a	a	b
b	b	b

24. Show that the following are semigroups. Which are groups?

A

$\circ$	a	b	c
a	a	b	c
b	b	c	a
c	c	a	b

B

$\circ$	a	b	c
a	a	b	c
b	a	b	c
c	a	b	c

C

$\circ$	a	b	c
a	a	b	c
b	c	b	c
c	c	b	c

25*. Find a semigroup with three elements which is distinct from those in Prob. 24.

4.5 Elementary Theorems about Groups

There are some elementary consequences of the axioms of a group which we shall prove next. You have met these before in some of our examples of groups, but we shall repeat them here for the general case.

Theorem 1. There exists a unique element x such that $a \circ x = b$.

Proof: First, $x = a' \circ b$ satisfies this equation, for

$$\begin{aligned} a \circ (a' \circ b) &= (a \circ a') \circ b && \text{[Axiom G2]} \\ &= e \circ b && \text{[Axiom G4]} \\ &= b && \text{[Axiom G3]} \end{aligned}$$

This shows that there is a solution. To prove it unique, suppose there is another solution y such that

$$a \circ y = b$$

Then $$a \circ x = a \circ y$$

"Multiply" both sides of this equation on the left by a', and get

$$a' \circ (a \circ x) = a' \circ (a \circ y) \qquad \text{[Sec. 1.10, Illustration 2]}$$

That is, $$(a' \circ a) \circ x = (a' \circ a) \circ y \qquad \text{[Axiom G2]}$$

or $$e \circ x = e \circ y \qquad \text{[Axiom G4]}$$

or $$x = y \qquad \text{[Axiom G3]}$$

Therefore the solution is unique.

Theorem 2. If $a \circ c = b \circ c$, then $a = b$.

Proof: "Multiply" both sides of (3) on the right by c', and get

$$(a \circ c) \circ c' = (b \circ c) \circ c'$$

Then $$a \circ (c \circ c') = b \circ (c \circ c') \qquad \text{[Axiom G2]}$$

$$a \circ e = b \circ e \qquad \text{[Axiom G4]}$$

$$a = b \qquad \text{[Axiom G3]}$$

Theorem 3. $(a')' = a$.

Proof: By definition, $(a')' \circ a' = e$

Also $$a \circ a' = e \qquad \text{[Axiom G4]}$$

Therefore $$(a')' = a \qquad \text{[Theorem 2]}$$

Problems 4.5

1. Prove $(a \circ b)' = b' \circ a'$. HINT: By definition $(a \circ b)' \circ (a \circ b) = e$. "Multiply" on the right by b', and obtain $(a \circ b)' \circ a = b'$. Continue.

2. Prove $(a \circ b')' = b \circ a'$.

3. Prove $(a \circ b) \circ (c \circ d) = [(a \circ b) \circ c] \circ d$.

4. Prove if $a \circ x = a$, then $x = e$.

5. Prove $e' = e$.

6. Prove that if $a \circ a = a$ in a group, then $a = e$.

7*. Let us replace Axioms G3 and G4 by the following:

G3*. There exists a unique element e of G having the property that, for every a in G,

$$e \circ a = a$$

This is the *left-identity* axiom. Note that we do not assume that $a \circ e = a$.

G4*. Corresponding to each a in G, there is a unique element a' having the property that

$$a' \circ a = e$$

This is the *left-inverse* axiom. Note that we do not assume that $a \circ a' = e$.

Prove that a system satisfying Axioms G1, G2, G3*, and G4* is a group.

8*. Let G be a nonempty set with associative multiplication such that all equations $xa = b$ and $ay = b$ have unique solutions x and y in G. Prove that G is a group.

9*. In Axiom G3* omit the word *unique*. Then prove that there is a unique identity.

10*. In Axiom G4* omit the word *unique*. Then prove that each a has a unique inverse a'.

4.6 Isomorphism

Let us recall the "multiplication" tables of two groups which we have already considered:

(1) The residue classes, mod 3, whose table under addition is:

+	0	1	2
0	0	1	2
1	1	2	0
2	2	0	1

(2) The permutations

$$e = \begin{pmatrix} 1 & 2 & 3 \\ 1 & 2 & 3 \end{pmatrix} \qquad p = \begin{pmatrix} 1 & 2 & 3 \\ 3 & 1 & 2 \end{pmatrix} \qquad q = \begin{pmatrix} 1 & 2 & 3 \\ 2 & 3 & 1 \end{pmatrix}$$

whose table under "followed by" is:

$\circ$	e	p	q
e	e	p	q
p	p	q	e
q	q	e	p

Although these groups involve quite different types of elements and operations, they are essentially the same in a sense to be defined. If we establish the 1 to 1 correspondence:

$$\begin{matrix} 0 & 1 & 2 \\ e & p & q \end{matrix}$$

we see that it carries corresponding sums of residue classes, mod 3, into corresponding "products" of permutations. For example, $1 + 2 = 0$ and $p \circ q = e$. Here we chose 1 and p to correspond and 2 and q to correspond and observe that, in the result, 0 and e correspond. Similarly, $2 + 2 = 1$ and $q \circ q = p$. Correspondences having this property are called isomorphisms.

Definition: Let G be a group with elements a, b, etc., and operation $\circ$; also let G^* be a group with elements a^*, b^*, etc., and operation $\oplus$. Then a 1 to 1 correspondence between G and G^* defined by $a \leftrightarrow a^*$, $b \leftrightarrow b^*$, etc., is an *isomorphism* if and only if, for every a and b,

$$a \circ b \leftrightarrow a^* \oplus b^*$$

The groups G and G^* are called *isomorphic*.

Exercise A. Show that isomorphism is an equivalence relation.

Illustration 1. The groups of Probs. 12 and 14, Sec. 4.4, are isomorphic.

Illustration 2. The additive group of residue classes, mod 4, is isomorphic to the group whose elements are $1, -1, i, -i$ (Sec. 4.4, Prob. 16). Their "multiplication" tables are, respectively;

+	0	1	2	3
0	0	1	2	3
1	1	2	3	0
2	2	3	0	1
3	3	0	1	2

$\times$	1	$-i$	-1	i
1	1	$-i$	-1	i
$-i$	$-i$	-1	i	1
-1	-1	i	1	$-i$
i	i	1	$-i$	-1

The 1 to 1 correspondence is given by

$$\begin{array}{cccc} 0 & 1 & 2 & 3 \\ 1 & -i & -1 & i \end{array}$$

Problems 4.6

In Probs. 1 to 8 prove that the two given groups are isomorphic.

1. The additive group of residue classes, mod 4; the multiplicative group of nonzero residue classes, mod 5.
2. The additive group of residue classes, mod 4; the group of rotations of the square consisting of the elements e, m, n, p of Sec. 4.4, Prob. 15.
3. The group of translations (VIII) and the additive group of vectors (IX) of Sec. 4.4.
4. The group of matrices in Sec. 4.3, Prob. 8, and the Four Group (XI) of Sec. 4.4.
5. The multiplicative group of positive real numbers; the additive group of all real numbers.
6. The set of elements 1, 3, 5, 7 under multiplication, mod 8; the Four Group (XI) of Sec. 4.4.

7*. A cyclic group of order m; the additive group of residue classes, mod m.
8*. The multiplicative group of nonzero complex numbers $a + bi$; the multiplicative group of matrices $\begin{pmatrix} a & -b \\ b & a \end{pmatrix}$ where $a^2 + b^2 \neq 0$.

4.7 Subgroups

In some of the groups we have studied, a subset of the elements of the given group also forms a group.

Illustration 1

(**a**) The elements e, p, q of the permutation group (XII) form a group.
(**b**) The elements e, a of the Four Group (XI) form a group.
(**c**) The elements e, m, n, p of the group of rotations of the square (Sec 4.4, Prob 15) form a group.

These illustrations lead us to the following definition of a subgroup:

Definition: A subset of a group G is a *subgroup* of G if and only if its elements form a group relative to the operation $\circ$ of G.

It is evident that a subgroup S of G must contain the identity e of G and that, if a is in S, its inverse a' in S is the same as its inverse in G.

Exercise A. Prove the truth of the above remark.

There is much to say about subgroups, but for brevity we shall restrict ourselves to a few ideas concerning the subgroups of a group with a finite number of elements, i.e., a finite group. For convenience we write ab instead of $a \circ b$.

Consider a finite group G with a subgroup S with the distinct elements: $\{e, a_1, a_2, \ldots, a_r\}$. Let b be any element of G. Then consider the set $bS = \{be, ba_1, ba_2, \ldots, ba_r\}$, which is called a *left-coset* of S. Similarly, the set $Sb = \{eb, a_1b, \ldots, a_rb\}$ is a *right-coset* of S. Each of these cosets has precisely $r + 1$ distinct elements. Suppose, to the contrary, that $ba_1 = ba_2$; then $a_1 = a_2$, which is ruled out by hypothesis.

Next consider two left-cosets of S, say, bS and cS. Either these are identical (as sets) or they have no element in common. Suppose there is an element in common so that $ba_p = ca_q$ for some p and q. Then an arbitrary element in bS, say, ba, can be written:

$$ba = ba_pa'_pa = ca_qa'_pa = c(a_qa'_pa)$$

and hence is an element of cS. Similarly, every element in cS is an element of bS, so that $bS = cS$.

Moreover, every element G is in some left-coset of S, for the element g

of G is in the coset gS. Hence the set of left-cosets exhausts the group G. The left-cosets therefore decompose G into a collection of nonoverlapping sets, each having the same number of elements $r + 1$.

Illustration 2. Let G be the permutation group (XII) and S be the subgroup with elements $\{e, p, q\}$. Then the left-cosets are:

$$eS = \{e, p, q\}$$
$$rS = \{r, t, s\}$$

Note that $pS = eS$, $qS = eS$, $tS = rS$, $sS = rS$, so that there are only two distinct cosets.

Illustration 3. Let G be the Four Group (XI) and S be the subgroup with elements $\{e, a\}$. Then the left-cosets are:

$$eS = \{e, a\}$$
$$bS = \{b, c\}$$

Note that $aS = eS$, $cS = bS$, so there are only two distinct cosets.

The important theorem to which this argument leads is due to Lagrange. In order to state it we make the definition:

Definition: The *order* of a group or a subgroup is the number of its elements.

Theorem 4. The order of a finite group G is an integral multiple of the order of each of its subgroups.

Problems 4.7

1. Find the right-cosets in Illustration 2.
2. Find the right-cosets in Illustration 3.
3. Find the left-cosets of the subgroup $\{e, m, n, p\}$ of the group of Prob. 15, Sec. 4.4.
4. Find the right-cosets of the subgroup $\{e, m, n, p\}$ of the group of Prob. 15, Sec. 4.4.

5*. (a) Let $a \neq e$ be an element of a finite group G. Show that the cyclic group generated by a is a subgroup of G.
(b) Show that the order of a in G is a divisor of the order of G.

6*. Show that every group G of prime order is cyclic.

7*. Show that the number of left-cosets defined by a subgroup S of a finite group G is equal to the number of right-cosets.

8*. Show that every group of order 6 is isomorphic to either the cyclic group of order 6 or the permutation group XII.

References

Birkhoff, Garrett, and Saunders MacLane: "A Survey of Modern Algebra", chap. 6, Macmillan, New York (1953).

McCoy, Neal H.: "Introduction to Modern Algebra", chap. 9, Allyn and Bacon, Boston (1960).
Weiss, M. J.: "Higher Algebra for the Undergraduate", chap. 3, Wiley, New York (1962).

Also consult the following papers in the *American Mathematical Monthly:*

Burns, J. E.: The Foundation Period in the History of Group Theory, vol. 20, p. 141 (1913).
Miller, G. A.: Examples of a Few Elementary Groups, vol. 7, p. 9 (1900).
Miller, G. A.: On the Concepts of Number and Group, vol. 8, p. 137 (1901).
Miller, G. A.: Appreciative Remarks on the Theory of Groups, vol. 10, p. 87 (1903).
Miller, G. A.: On the Groups of the Figures of Elementary Geometry, vol. 10, p. 215 (1903).
Miller, G. A.: Groups of Elementary Trigonometry, vol. 11, p. 225 (1904).

5 | Equations and Inequalities

5.1 Solutions of Equations and Inequalities

One of the most common problems in mathematics is that of *solving an equation.* In this chapter we shall tackle a number of problems of this kind, but first we must know what is meant by the *solution of an equation.* In order to avoid unnecessary complications we shall begin by considering the simple case of polynomial equations.

Definition: A *polynomial* $P(x)$ is an expression of the form:

$$P(x) = a_0x^n + a_1x^{n-1} + \cdots + a_{n-1}x + a_n$$

where n is an integer which is positive or zero, the coefficients a_0, a_1, . . . , a_n are elements of a given field F, and x is a variable whose universal set is F. We say that $P(x)$ is a polynomial *over the field F.*

Definition: If $a_0 \neq 0$, the polynomial

$$P(x) = a_0x^n + a_1x^{n-1} + \cdots + a_{n-1}x + a_n$$

is said to be of *degree n.*

Remarks. It is usual to take F to be the field of real numbers or the field of complex numbers, but we may occasionally use other fields. In any case, the field involved must be specified in advance.

Illustration 1. The following expressions are polynomials over the field of complex numbers:

(a) 2
(b) $\sqrt{3}\, x^5 - \pi x^2 - 1$
(c) $(2 + i)x^2 - 3x + (-3 + i)$

Illustration 2. The following expressions are not polynomials over the field of complex numbers:

(**a**) $1/x - 3i$
(**b**) 2^x
(**c**) $\sqrt{x}$
(**d**) $|x|$

We can now define a polynomial equation and its solution as follows:

Definition: *A polynomial equation* is an open sentence of the form $P(x) = Q(x)$, where $P(x)$ and $Q(x)$ are polynomials.

Definition: *The solution set* of the polynomial equation $P(x) = Q(x)$ is the set $\{x \mid P(x) = Q(x)$; i.e., it is the truth set of the given open sentence. Any element of this solution set is called a *solution* of $P(x) = Q(x)$.

Illustration 3

(**a**) The solution set of $2x - 6 = 0$ is $\{3\}$.
(**b**) The solution set of $x^2 - 5x + 6 = 0$ is $\{2, 3\}$. The solutions are the integers 2 and 3.
(**c**) The solution set of $(x - 1)^2 = 0$ is $\{1\}$.

If $P(x)$ and $Q(x)$ are polynomials over an *ordered* field F, we can give a similar definition for a polynomial inequality. Since the field of real numbers is the most important ordered field with which we are familiar, we shall automatically assume that all polynomial inequalities refer to polynomials over the real numbers.

Definition: A *polynomial inequality* is an open sentence of the form $P(x) > Q(x)$, $P(x) < Q(x)$, $P(x) \geq Q(x)$, or $P(x) \leq Q(x)$, where $P(x)$ and $Q(x)$ are polynomials over the field of real numbers. Its *solution set* is the truth set of this open sentence.

Illustration 4

(**a**) The solution set of $5x - 10 > 0$ is the set $\{x \mid x > 2\}$.
(**b**) The solution set of $x^2 - 5x + 6 \leq 0$ is the set $\{x \mid 2 \leq x \leq 3\}$.

Polynomials may involve more than one variable. For example, the expressions

$$x^2 - 2xy + y^2 \qquad 3x + 4y - 6 \qquad x^3y + x^2y^2 + 3x + 7y$$

are polynomials in two variables.

Definition: A *polynomial* $P(x,y)$ in two variables is an expression of the form

$$P(x,y) = P_0(y)x^n + P_1(y)x^{n-1} + \cdot \cdot \cdot + P_{n-1}(y)x + P_n(y)$$

where n is an integer which is positive or zero, $P_0(y), \ldots, P_n(y)$ are polynomials over a field F, and x is a variable whose universal set is F.

Definition: The *solution sets* of the polynomial equations and inequalities $P(x,y) = Q(x,y)$, $P(x,y) > Q(x,y)$, $P(x,y) < Q(x,y)$, $P(x,y) \geq Q(x,y)$, and $P(x,y) \leq Q(x,y)$ are the truth sets of the corresponding open sentences.

Since, in general, the solution sets of polynomials in two variables and of inequalities in either one or two variables have an infinite number of elements, we cannot describe them by listing their elements as we did in Illustration 3 for equations in one variable. When F is the field of real numbers, the best procedure is to plot their graphs. So that you may fully understand what is involved in the process of graphing, we interrupt our discussion of solutions of equations and inequalities to tell you something about coordinate systems and graphs.

5.2 Geometric Representation of Real Numbers

The basic idea of the graph of a set of numbers is the representation of this set by a set of points on a line. You are probably familiar with the representation of the real numbers as points on a line (Fig. 5.1). In order to obtain this, we start with the points 0 and 1 chosen at random, except that 0 is to the left of 1. The segment [0, 1] is said to have length 1 by definition of *length*. It is now assumed that this segment can be slid along the line without altering its length. Doing this step by step, we locate the other integers, so that the length of the segment between any two successive integers is still equal to 1. The location of $1/b$ (where b is a positive integer) is found by dividing [0, 1] into b equal parts by the usual geometric construction. Then by sliding the segment $[0, 1/b]$ along the line we locate the points a/b. The location of the irrational numbers is more complicated, and we pass over this point. Their approximate positions, however, can be obtained from the first few decimals in their decimal expansions.

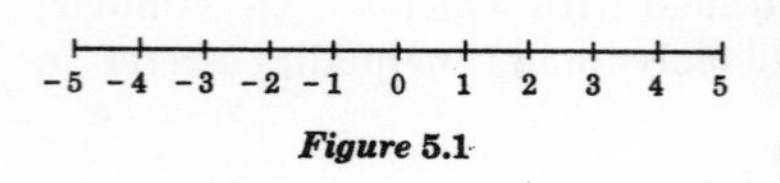

Figure 5.1

The most important fact about this representation is *that every point corresponds to one and only one real number and that every real number corresponds to one and only one point.* We cannot prove this fact here and consequently must take it as an assumption.

This representation has another property; namely, it preserves order.

We have already defined order for the real numbers, but need to introduce it on the line by means of the notion *beyond*. First of all we place an arrow on the line and thus define a *positive direction* on the line as the direction determined by the arrow. We now call this line a *directed line*. It is customary to direct horizontal lines to the *right* and vertical lines *upward*. Then we define *beyond* as follows.

Definition: A point P is *beyond* a point Q on a directed line if the segment (or vector) from Q to P points in the given positive direction. If P is beyond Q, we write $P \succ Q$.

Let us now return to our assumption about real numbers and points on the line and describe how this preserves order. Let P_a be the point corresponding to the real number a and P_b to b. Then our correspondence is such that:

$$P_a \succ P_b \qquad \textit{if and only if } a > b$$

In summary, we have defined a correspondence between real numbers and points on a line which is 1 to 1 and which preserves *order*. The number associated with a point is called its *coordinate*, and we can use coordinates to identify points. Thus the point whose coordinate is the number a will henceforth be written as the point a, and not as P_a, as was done above. This use of coordinates is the foundation of the application of real numbers to geometry and to geometrical representations of nature.

By an extension of this idea we can use points in the plane to represent ordered pairs of real numbers. By such an *ordered pair* we mean a pair of real numbers (x,y) in which x is the first element and y is the second element. Because of the ordering (x,y) is to be distinguished from (y,x).

Definition: The set $X \times Y$, called the *Cartesian Product* of the real line with itself, is the set of ordered pairs of real numbers (x,y).

In order to establish the correspondence between ordered pairs of real numbers and points in the plane, we first construct two perpendicular lines in the plane (Fig. 5.2) which we call the X-axis and the Y-axis. Their point of intersection is called the origin O. We put the X-axis into an exact correspondence with the real numbers by placing zero at O, the positive reals to the right of O and the negative reals to its left. We do the same for the Y-axis, putting the positive reals above O and the negative reals below O. We remind ourselves of these conventions by putting arrows on the right end of the X-axis and the upper end of the Y-axis. These lines divide the plane into four regions called *quadrants*, which are numbered I, II, III, and IV in Fig. 5.2.

Using this scheme, we can now associate an ordered pair of real numbers (x,y) with each point P of the plane. Let P be a point on the X-axis. It corresponds to a real number x on this axis. We associate with P the ordered pair $(x,0)$. Now let P be a point on the Y-axis. Similarly, we associate the ordered pair $(0,y)$ with it. When P is not on either axis, draw PQ perpendicular to the X-axis and PR perpendicular to the Y-axis

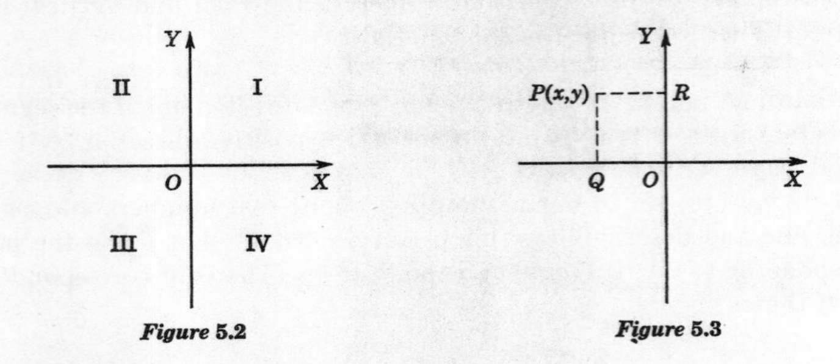

Figure 5.2 *Figure 5.3*

(Fig. 5.3). Suppose that Q corresponds to the real number x on the X-axis and that R corresponds to the real number y on the Y-axis. Then we associate the ordered pair (x,y) with P.

By this process we find an ordered pair (x,y) which corresponds to each P in the plane. It is also evident that every pair (x,y) determines a point in the plane, for suppose x and y are given. These locate points Q and R (Fig. 5.3). Draw PQ and PR as perpendiculars to the X-axis and the Y-axis at Q and R, respectively. These lines intersect at P, which is the desired point.

Thus we have established a 1 *to* 1 *correspondence between the points of the plane and the ordered pairs* (x,y).

Definition: The real numbers x and y in the ordered pair (x,y) are called the *coordinates* of the point P. Sometimes x is called the *x-coordinate*, or the *abscissa*, and y is called the *y-coordinate*, or the *ordinate*.

We often identify the point P with its pair of coordinates and speak of the *point* (x,y). By using this identification, we can convert geometric statements about points into algebraic statements about numbers and can convert geometric reasoning into algebraic manipulation. The methods of algebra are usually simpler than those of geometry, and therefore the algebraic approach is now the common one. The detailed elaboration of this method is called *analytic geometry*, which is discussed in Chap. 9.

5.3 Lengths of Segments; Units on the Axes

There is now a simple method of defining the lengths of line segments which are parallel to one of the axes. For the purposes of this chapter we shall not need to discuss the lengths of other segments.

Definition: Let P_1 and P_2 have coordinates (x_1,a) and (x_2,a), so that P_1P_2 is parallel to the X-axis. Then the *length* of the segment P_1P_2 is defined to be the real number $|x_2 - x_1|$.

Similarly, the *length* of Q_1Q_2, where Q_1 is (b,y_1) and Q_2 is (b,y_2), is defined to be $|y_2 - y_1|$.

We have said nothing about the relation of distance on the X-axis to that on the Y-axis, and we prefer not to make any rigid requirements about this at present. Indeed, it is often useful to use different scales of measurement on the two axes. Unequal scales are used for a variety of reasons, of which the following are the most common:

(1) The range of values to be plotted on the Y-axis is much greater (or smaller) than the range to be plotted on the X-axis. In this case we must contract (or expand) the scale on the Y-axis in order to get a graph on a reasonably shaped piece of paper.

Illustration 1. Suppose that we are plotting $y = x^{10}$ for x in the range 0 to 2. Then y lies in the range 0 to 1,024. In this case it would be extremely awkward to use equal scales on the two axes.

(2) In applications to science the physical significance of the numbers on the two axes may be very different. In such cases the physical units of measurement (such as time, distance, velocity, etc.) are not comparable, and suitable scales on the two axes should be chosen independently.

Illustration 2. In order to illustrate the motion of a particle, it is customary to plot the distance traveled on the vertical axis and the corresponding time on the horizontal axis. The units of measurement are *feet* and *seconds*, respectively, and it would be absurd to equate feet and seconds. Hence separate, convenient scales are used on the two axes.

In geometry, however, it is necessary to plot distance on each of the axes and to use the same scale on each. When we do this, it is meaningful to speak of the lengths of segments on slanting lines, and we shall develop a formula for this in Sec. 8.2. The notion of slant distance, however, is quite meaningless in cases (1) and (2) above, and we shall avoid mention of it until we begin our study of geometry.

Problems 5.3

1. Plot the points whose coordinates are $(2,6)$, $(-3,2)$, $(-4,-6)$, $(5,-1)$, $(0,0)$.
2. Plot the points whose coordinates are $(1,5)$, $(-2,4)$, $(-3,-2)$, $(2,-3)$, $(0,0)$.
3. What signs do the coordinates of points in quadrant I have; quadrant III?
4. What signs do the coordinates of points in quadrant II have; quadrant IV?
5. State the quadrant in which each of the following points lies: $(2,-4)$, $(3,7)$, $(-5,-6)$, $(-9,11)$, $(-4,7)$.
6. State the quadrant in which each of the following points lies: $(-8,-6)$, $(19,-4)$, $(-6,9)$, $(2,2)$, $(8,-4)$.
7. Find the lengths of the segments joining the following pairs of points: $(2,3)$ and $(5,3)$; $(-3,-5)$ and $(-3,-7)$; $(-4,2)$ and $(7,2)$; $(5,5)$ and $(5,-6)$; $(0,0)$ and $(0,3)$.

8. Find the lengths of the segments joining the following pairs of points: (4,7) and (4, −10); (3, −5) and (7, −5); (6,6) and (13,6); (4,0) and (0,0); (−6, −9) and (−6, −11).

9. If P_1P_2 is parallel to the X-axis, show that the length of P_1P_2 is equal to the length of P_2P_1.

10. If Q_1Q_2 is parallel to the Y-axis, show that the length of Q_1Q_2 is equal to the length of Q_2Q_1.

11. If P_1P_2 is parallel to the X-axis, show that the square of its length is $(x_2 - x_1)^2 = (x_1 - x_2)^2$.

12. If Q_1Q_2 is parallel to the Y-axis, show that the square of its length is $(y_2 - y_1)^2 = (y_1 - y_2)^2$.

Transformation of Coordinates. If we are given a coordinate system on a line in terms of numbers x, we can define a new coordinate system x' by giving a relationship between x and x'. This relabels the points with new numbers and is called a *transformation of coordinates*. The following problems give some important illustrations of these.

In Probs. 13 to 16, we take $x' = a + x$. This transformation is called a *translation*.

13. Prove that a translation leaves the lengths of segments unchanged. HINT: Prove that $|x_2' - x_1'| = |x_2 - x_1|$.

14. Prove: If the coordinate of any one point is left unchanged by a translation, then the coordinates of all points are unchanged. HINT: Prove that $a = 0$.

15. Express as a translation the relationship between absolute temperature K (degrees Kelvin) and centigrade temperature C (degrees centigrade).

16. Express as a translation the relationship between the distance s of a rocket from the center of the earth and its height h above the surface of the earth.

In Probs. 17 to 20, we take $x' = ax$, where $a \neq 0$. This transformation is called a *dilatation*.

17. Prove that a dilatation multiplies the lengths of segments by $|a|$.

18. Prove that, if the coordinate of any *one* point (other than $x = x' = 0$) is left unchanged by a dilatation, then the coordinates of all points are unchanged.

19. Express the relationship between feet F and inches I as a dilatation.

20. Express the relationship between seconds S and hours H as a dilatation.

In Probs. 21 to 26 we take $x' = ax + b$, where $a \neq 0$. This transformation is called a *linear transformation*.

21. Prove that a linear transformation multiplies the lengths of segments by $|a|$.

22. Prove that a linear transformation with $a \neq 1$ leaves the coordinate of just one point unchanged. Find this point.

23. Express the relationship between degrees Fahrenheit F and degrees centigrade C as a linear transformation. What temperature is the same in both systems?

24. Express the relationship between degrees Fahrenheit F and degrees Kelvin K as a linear transformation. What temperature is the same in both systems?

25. Prove that the linear transformation $x' = ax + b$ with $a > 0$ preserves the order relationship; i.e., if $x_1 > x_2$, then $x_1' > x_2'$.

26. Prove that the linear transformation $x' = ax + b$ with $a < 0$ reverses the order relationship; i.e., if $x_1 > x_2$, then $x_1' < x_2'$.

27. Let A represent the linear transformation $x' = ax + b$, and let A' represent the linear transformation $x'' = cx' + d$. Define the "product" $A' \circ A$ to be the linear transformation $x'' = c(ax + b) + d = cax + (cb + d)$. Prove that the set of linear transformations with this product forms a group.

28. Show that the translations form a subgroup of the group of linear transformations.
29. Show that the dilatations form a subgroup of the group of linear transformations.
30. Are the groups of Probs. 28 and 29 isomorphic?

5.4 Graphs

These geometric representations of real numbers and ordered pairs of real numbers enable us to define the graphs of subsets of X and $X \times Y$, respectively.

Definition: Let S be a subset of X. Then the *graph* of S is the subset of points on the line whose coordinates are the elements of S.

Illustration 1

(a) The graph of $\{2, 4, 6\}$ is given in Fig. 5.4.

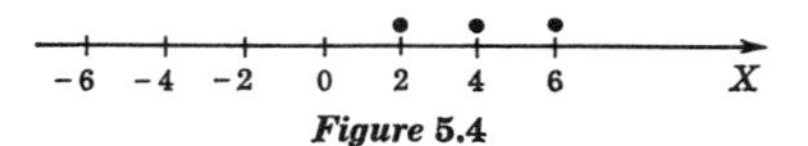

Figure 5.4

(b) The graph of $\{x \mid x > 2\}$ is given in Fig. 5.5. Here we indicate that 2 is not in the graph by placing an open circle above it.

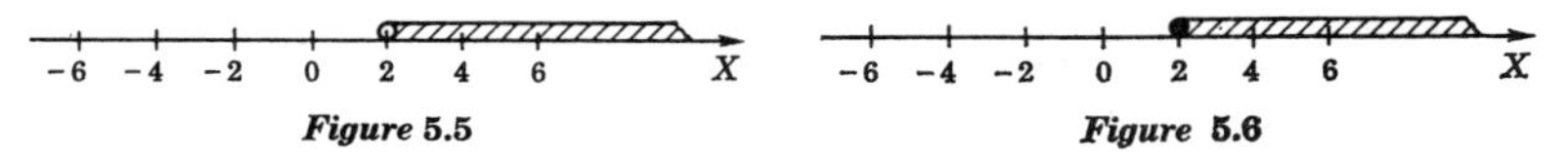

Figure 5.5 *Figure* 5.6

(c) The graph of $\{x \mid x \geq 2\}$ is given in Fig. 5.6. Since 2 is a point of this graph, we fill in the circle above it.

Definition: Let S be a subset of $X \times Y$. Then the *graph* of S is the subset of points in the plane whose ordered pairs of coordinates are the elements of S.

Illustration 2

(a) The graph of $\{(1, 2), (-1, 3)\}$ is given in Fig. 5.7.

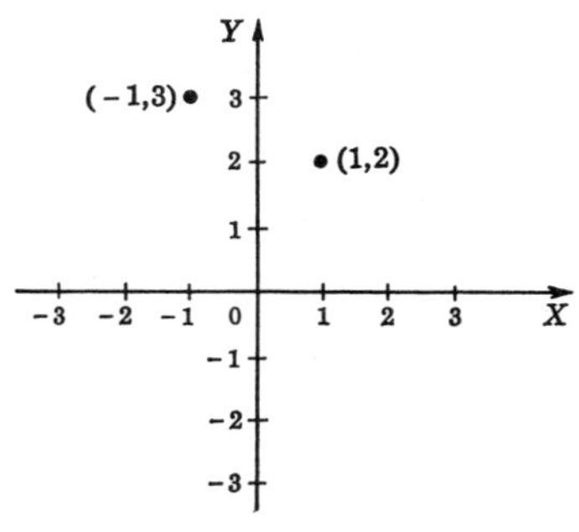

Figure 5.7

(b) The graph of $\{(x, y) \mid x > 1\}$ is given in Fig. 5.8. Here we shade the right half plane, leaving the line $x = 1$ light to indicate that it is not included in the graph.

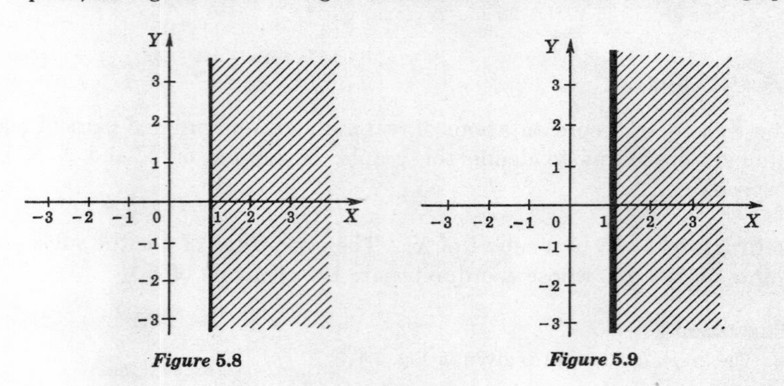

Figure 5.8 *Figure* 5.9

(c) The graph of $\{(x, y) \mid x \geq 1\}$ if given in Fig. 5.9. Here we darken the line $x = 1$ to indicate that it is included in the graph.

Definition: The *graph* of a polynomial equation or inequality over the reals is the graph of its solution set.

Our task now is to develop methods for finding the solution sets of equations and inequalities and for plotting their graphs.

5.5 Equivalent Equations and Inequalities

Unless the solution set of an equation or an inequality is obvious at sight, our general procedure is to simplify it step by step until it is reduced to an equation or inequality which we can solve by inspection. At each step of this process we replace one equation or inequality with another one. We must be certain, however, that the solution set of our new equation is the same as that of the old one. This leads us to the concept of equivalent equations or inequalities.

Definition: Two equations (or inequalities) are *equivalent* if and only if they have the same solution sets.

Remark. This definition is merely a rewording of our definition in Sec. 1.11 of the equivalence of two open sentences.

The problem of solving equations or inequalities thus requires us to answer the question: What algebraic operations on an equation or an inequality transform it into an equivalent equation or inequality, and what operations do not? Operations of the first type may be used safely at any time; those of the second kind are dangerous to use and may lead to wrong answers.

Let us examine the most important kinds of operations to see how they behave.

First of all, let us add (or subtract) the same polynomial from both sides of an equation or inequality. For purposes of exposition we restrict ourselves to an equation in one variable and leave the other cases to you.

Theorem 1. Let $P(x) = Q(x)$ be a polynomial equation over a field F, and let $G(x)$ be another polynomial over F. Then the equations

$$P(x) = Q(x)$$

and

$$P(x) + G(x) = Q(x) + G(x)$$

are equivalent; i.e.,

$$\{x \mid P(x) = Q(x)\} = \{x \mid P(x) + G(x) = Q(x) + G(x)\}$$

Proof: Let a be any element of F for which $P(a) = Q(a)$. Then $P(a) + G(a) = Q(a) + G(a)$, since the two sides are just different names of the same element of F. Therefore:

$$\{x \mid P(x) = Q(x)\} \subseteq \{x \mid P(x) + G(x) = Q(x) + G(x)\}$$

Conversely, if a is such that $P(a) + G(a) = Q(a) + G(a)$, it follows by subtraction of $G(a)$ from both sides that $P(a) = Q(a)$. Therefore

$$\{x \mid P(x) + G(x) = Q(x) + G(x)\} \subseteq \{x \mid P(x) = Q(x)\}$$

Hence the two sets are identical and the equations are equivalent.

Remark. By virtue of this theorem we can rewrite any polynomial equation $P(x) = Q(x)$ in the *standard* form $R(x) = 0$, where

$$R(x) = P(x) - Q(x)$$

In later sections of this chapter we shall assume that polynomial equations are written in this form.

Exercise A. State and prove a theorem parallel to Theorem 1 for the inequalities $P(x) > Q(x)$ and $P(x) + G(x) > Q(x) + G(x)$.

Exercise B. State and prove a theorem parallel to Theorem 1 for the equations: $P(x,y) = Q(x,y)$ and $P(x,y) + G(x,y) = Q(x,y) + G(x,y)$.

Now let us turn to multiplication of both sides by the same expression.

Theorem 2. Let $P(x) = Q(x)$ be a polynomial equation over F, and let a be any nonzero element of F. Then the equations

$$P(x) = Q(x)$$

and

$$aP(x) = aQ(x)$$

are equivalent; i.e.,

$$\{x \mid P(x) = Q(x)\} = \{x \mid aP(x) = aQ(x)\}$$

The proof is immediate and is included in the problems.

Remarks

(1) Since a may be $1/b$, Theorem 2 includes division by a nonzero element of F.

(2) If $a = 0$, the two equations in Theorem 2 are not necessarily equivalent; since any x in F is a solution of $0 \cdot P(x) = 0 \cdot Q(x)$. Thus we must exclude $x = 0$ in the statement of the theorem.

Exercise C. Prove that if $a > 0$, then $P(x) > Q(x)$ and $aP(x) > aQ(x)$ are equivalent.

Exercise D. Prove that if $a \leq 0$, then $P(x) > Q(x)$ and $aP(x) < aQ(x)$ are not in general equivalent.

When we try to extend Theorem 2 to multiplication by a polynomial, we run into trouble.

Theorem 3. Let $P(x) = Q(x)$ be a polynomial equation over F, and let $G(x)$ be a polynomial of degree ≥ 1. Then the equations

$$P(x) = Q(x)$$

and

$$G(x) \cdot P(x) = G(x) \cdot Q(x)$$

may not be equivalent. Indeed,

$$\{x \mid P(x) = Q(x)\} \subseteq \{x \mid G(x) \cdot P(x) = G(x) \cdot Q(x)\}$$

Proof: The set on the right contains all the elements of the set on the left, and in addition it contains those elements a for which $G(a) = 0$. If there is an a such that $G(a) = 0$ and $P(a) \neq Q(a)$, the equations are not equivalent. If, however, every a with $G(a) = 0$ also satisfies $P(a) = Q(a)$, the equations are equivalent.

Illustration 1

(**a**) The equations $2x + 1 = 3$ and $(x - 1)(2x + 1) = 3(x - 1)$ are equivalent.
(**b**) The equations $2x + 1 = 3$ and $x(2x + 1) = 3x$ are not equivalent.

Exercise E. Under what circumstances are the inequalities $P(x) > Q(x)$ and $G(x) \cdot Q(x) > G(x) \cdot P(x)$ not equivalent?

Illustration 2

(**a**) The inequalities $4x + 1 > 2x + 3$ and $(x - 2)^2(4x + 1) > (x - 2)^2(2x + 3)$ are not equivalent.
(**b**) The inequalities $4x + 1 > 2x + 3$ and $x^2(4x + 1) > x^2(2x + 3)$ are equivalent.
(**c**) The inequalities $4x + 1 > 2x + 3$ and $(3x - 2)(4x + 1) > (3x - 2)(2x + 3)$ are not equivalent.

Theorem 4. Let $P(x) = A(x) \cdot G(x)$ and $Q(x) = B(x) \cdot G(x)$, where $A(x)$, $B(x)$, and $G(x)$ are polynomials. Then the equations

$$P(x) = Q(x) \qquad \text{and} \qquad A(x) = B(x)$$

may not be equivalent. Indeed,

$$\{x \mid A(x) = B(x)\} \subseteq \{x \mid P(x) = Q(x)\}$$

This is in fact just a restatement of Theorem 3.

Illustration 3. The solution of $(x + 4)(x + 1) = 0$ is the set $\{-1, -4\}$. But the solution of $x + 4 = 0$ is the set $\{-4\}$. By dividing by $(x + 1)$ we have lost the element -1.

Finally, we shall need the following theorem of a somewhat different type.

Theorem 5. Let $P(x)$ be a polynomial which factors into the product: $P(x) = P_1(x) \cdot P_2(x) \cdots P_r(x)$. Then:

$$\{x \mid P(x) = 0\} = \{x \mid P_1(x) = 0\} \cup \{x \mid P_2(x) = 0\} \cup \cdots \cup \{x \mid P_r(x) = 0\}$$

Proof: If a satisfies any one of the equations $P_1(a) = 0, P_2(a) = 0, \ldots, P_r(a) = 0$, it is clear that $P(a) = 0$. Hence the right-hand set is a subset of the left-hand side. Conversely, suppose that a satisfies $P(a) = 0$. Then the product

$$P_1(a) \cdot P_2(a) \cdots P_r(a) = 0$$

and at least one of the factors must be zero. Hence the left-hand set is a subset of the right-hand set. Combining these results, we get a proof of the theorem.

As we have noted earlier (Sec. 2.4), this theorem is extremely useful in the solution of polynomial equations. We factor the given polynomial and then set each factor equal to zero. The union of the solution sets of the equations so obtained is the solution set of the given equation.

Illustration 4. Solve $x^2 + 12x + 32 = 0$.
Since $x^2 + 12x + 32 = (x + 8)(x + 4)$, we consider

$$x + 8 = 0 \quad \text{and} \quad x + 4 = 0$$

Since their solution sets are $\{-8\}$ and $\{-4\}$, respectively, the solution set of the given equation is

$$\{-8\} \cup \{-4\} = \{-8, -4\}$$

Problems 5.5

In each of Probs. 1 to 10 one of the following relations is true: $A \subset B$; $A = B$; $A \supset B$. Write the correct relation in each case. All polynomials are defined over the field of real numbers.

	A	B
1.	$\{x \mid 3x - 2 = x + 4\}$	$\{x \mid 4x + 3 = 2x + 9\}$
2.	$\{x \mid x^2 + 4 = -2x + 3\}$	$\{x \mid x^2 + 2x = -1\}$
3.	$\{x \mid (x + 2)(x - 3) = 0\}$	$\{x \mid x + 2 = 0\}$
4.	$\{x \mid (x + 1)(x - 4) = 0\}$	$\{x \mid x + 1 = 0\} \cup \{x \mid x - 4 = 0\}$
5.	$\{x \mid (x + 3)(x + 5) = 0\}$	$\{x \mid x + 3 = 0\} \cup \{x \mid x + 5 = 0\}$
6.	$\{x \mid x + 3 = 7\}$	$\{x \mid (x + 3)^2 = 49\}$
7.	$\{x \mid \sqrt{x + 2} = -3\}$	$\{x \mid x + 2 = 9\}$
8.	$\{x \mid 2x + 5 = 9\}$	$\{x \mid \sqrt{2x + 5} = -3\}$
9.	$\{x \mid 3x(x - 2) = 4(x - 2)\}$	$\{x \mid x = 2\}$
10.	$\{x \mid 6x - 3 = 9x + 12\}$	$\{x \mid 2x - 1 = 3x + 4\}$

11. Show that the equations $P(x) = Q(x)$ and $[P(x)]^2 = [Q(x)]^2$ may not be equivalent.

12. Show that the inequalities $P(x) > 0$ and $[P(x)]^2 > 0$ may not be equivalent.

13. Prove Theorem 2.

14. Prove Theorem 4.

5.6 Linear Equations and Inequalities in One Variable

Our wish to solve the linear equation provided the chief motivation for the extension of our number system beyond the elementary system of counting numbers, or positive integers. We saw in Chap. 2 that if we restricted ourselves to the positive integers, we could not solve equations like $x + 3 = 2$. This led us to invent zero and the negative integers.

We were still limited, however, in that, in terms of the integers, we could not solve equations such as $2x + 1 = 4$. Hence we invented the rational numbers. We have seen that, in the field of rational numbers, every equation of the form $ax + b = 0$ with $a \neq 0$ has a unique solution. This theorem is indeed true in any field and is the most characteristic property of a field.

Theorem 6. If a, b, and c are elements of a field F and $a \neq 0$, then the equation $ax + b = 0$ has a unique solution.

Proof: From Theorem 1, $ax + b = 0$ is equivalent to $ax = -b$. From Theorem 2, this is equivalent to $x = -(b/a)$. Therefore $-(b/a)$ is the one and only solution of $ax + b = 0$.

The familiar process of solving linear equations with real or complex coefficients can thus be extended to other fields as illustrated below.

Illustration 1. Solve the congruence $3x + 2 \equiv 4$, mod 5. This is equivalent to the solution of the equation $3x + 2 = 4$, or $3x = 2$, where the coefficients represent residue classes, mod 5. Since these form a field, we multiply both sides by the multiplicative inverse of 3, which is 2. Doing this, we obtain

$$2 \cdot 3x = 2 \cdot 2$$

or

$$x = 4$$

The solution of the congruence is then

$$x \equiv 4, \text{ mod } 5$$

In the same way we can solve the linear inequalities $ax + b > 0$ or $ax + b \geq 0$ with $a \neq 0$ over an ordered field. The solution is given by the simpler inequality $x > -(b/a)$ or $x \geq -(b/a)$. The solution set here has infinitely many elements and cannot be expressed in a simpler form. When the ordered field involved is the real number system, the usual representation of this solution set is given by its graph on a line as described in Sec. 5.3.

5.7 Linear Equations and Inequalities in Two Variables

The linear equation in two variables has the form:

$$ax + by + c = 0$$

where we assume that at least one of a and b is not zero. Its solution set is a set of ordered pairs (x,y) containing more than one element. When the basic field F is the real numbers, the best method of describing this set is to plot its graph, which you already know is a straight line. (For a proof see Sec. 9.8.) The customary procedure is to find two pairs (x,y) which satisfy the equation, to plot them, and then to draw the line through them.

In the case of the linear inequality

$$ax + by + c > 0 \qquad (a \neq 0) \vee (b \neq 0)$$

we first plot the graph of the line

$$ax + by + c = 0$$

This line divides the plane into two half planes A and B, neither of which contains the line. These half planes are each *connected*, in the sense that, if points P_1 and P_2 lie in one of them, then there is a smooth curve joining P_1 and P_2 which lies entirely in that half plane (Fig. 5.10). In particular, the curve does not have a point in common with the given line. Our procedure for graphing the solution set of $ax + by + c > 0$ depends strongly on the next theorem.

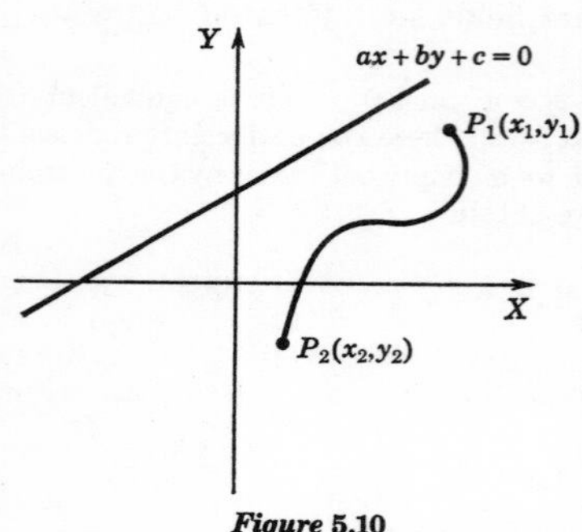

Figure 5.10

Theorem 7. Let P lie in one of the two half planes into which the graph of $ax + by + c = 0$ divides the plane. If $ax + by + c > 0$ at P, then $ax + by + c > 0$ at every point of the half plane in which P lies.

Proof: Assume that P_1 and P_2 lie in the same half plane, that $ax + by + c > 0$ at P_1, and that $ax + by + c \leq 0$ at P_2.

We shall prove that this is absurd (indirect proof).

Let P_1 have coordinates (x_1,y_1) and P_2 have coordinates (x_2,y_2).

In the first place, we cannot have $ax_2 + by_2 + c = 0$, for in this case P_2 is on the given line, which is contrary to its assumed location. Hence our hypotheses reduce to: $ax_1 + by_1 + c > 0$; $ax_2 + by_2 + c < 0$; P_1 and P_2 in the same half plane.

Draw the curve P_1P_2, which lies entirely in the given half plane. Let $\bar{P}(\bar{x},\bar{y})$ move along this curve from P_1 to P_2. At P_1, $a\bar{x} + b\bar{y} + c > 0$, and at P_2, $a\bar{x} + b\bar{y} + c < 0$. Hence there must be a point on the curve P_1P_2 at which $a\bar{x} + b\bar{y} + c = 0$. This, however, is impossible, for such a point would have to be on the given line, and P_1P_2 has no points in common with this line.*

Exercise A. State similar theorems for the inequalities: $ax + by + c < 0$, $ax + by + c \geq 0$, and $ax + by + c \leq 0$.

As a result of this theorem, we have the following procedure for plotting the graph of the solution set of a linear inequality. (1) Plot the graph of the line $ax + by + c = 0$. (2) For a single point in each half plane determined by this line, test the truth of the given inequality. (3) Shade the half plane or half planes corresponding to points at which the given inequality is true.

Remark. You will observe that if $ax + by + c > 0$ in one half plane, then $ax + by + c < 0$ in the other half plane. Nevertheless, it is wise to check both half planes. There are situations in which an inequality is true in both half planes. For instance, this is the case for the inequality

$$(ax + by + c)^2 > 0$$

Illustration 1. Plot the graph of the solution set of $3x + 2y - 6 > 0$. The points (0,3) and (2,0) satisfy the equation $3x + 2y - 6 = 0$, so we plot the line through them (Fig. 5.11). The point (0,0) is in the lower half plane, and $0 + 0 - 6 > 0$ is false. On the other hand, (3,0) is in the upper half plane and $9 + 0 - 6 > 0$ is true. Hence we shade the upper half plane. The line is drawn lightly, since it is not part of the desired graph.

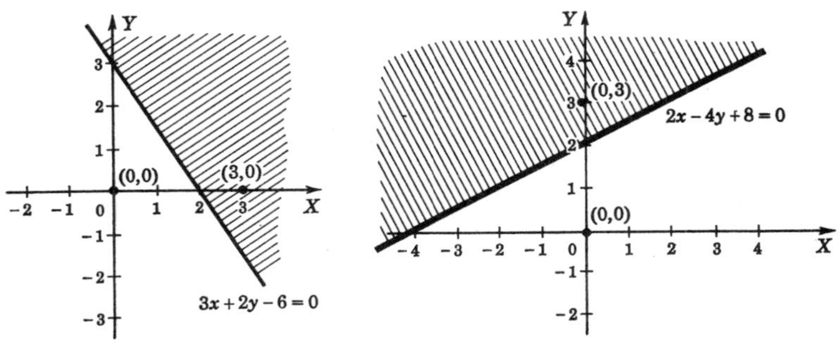

Figure 5.11

Figure 5.12

Illustration 2. Plot the graph of the solution set of $2x - 4y + 8 \leq 0$.

We draw the line through (−4,0) and (0,2) (Fig. 5.12). The point (0,0) does not satisfy the inequality. The point (0,3) does satisfy the inequality, so we shade the upper half plane, drawing the line heavily, since it is part of the graph.

* This proof assumes properties of continuity which you have not studied so far. Although precise details must be omitted here, the idea should be clear.

Problems 5.7

In Probs. 1 to 6 solve the given congruences.

1. $2x \equiv 3$, mod 7.
2. $4x - 1 \equiv 2$, mod 7.
3. $4x - 2 \equiv 2x + 1$, mod 5.
4. $x + 3 \equiv 2x - 2$, mod 5.
5. $7x \equiv 5$, mod 11
6. $5x \equiv 9$, mod 11.

In Probs. 7 to 10 prove that the given congruences do not have a solution.

7. $2x \equiv 3$, mod 6.
8. $3x \equiv 5$, mod 6.
9. $5x \equiv 7$, mod 10.
10. $2x \equiv 9$, mod 10

In Probs. 11 to 32 plot the graph of the solution set of the given inequality.

11. $3x + 7 > 6$.
12. $2x - 2 > 4$.
13. $2x - 5 \leq 8$.
14. $-5x + 2 \leq 3$.

15. $|x| > 1$. HINT: This is equivalent to $(x > 1) \vee (x < -1)$.
16. $|4x| > 8$.
17. $|2x - 4| > 3$. HINT: This is equivalent to $[(2x - 4) > 3] \vee [-(2x - 4) > 3]$.
18. $|3x + 6| > 2$.
19. $|2x| < 4$. HINT: This is equivalent to $(2x < 4) \wedge (2x > -4)$, or to $-4 < 2x < 4$.
20. $|3x| < 9$.
21. $|2x - 4| < 5$. HINT: This is equivalent to $[(2x - 4) < 5] \wedge [(2x - 4) > -5]$, or to $-5 < 2x - 4 < 5$, or to $-1 < 2x < 9$.
22. $|3x - 5| < 2$.
23. $|x - 5| < \frac{1}{10}$.
24. $|x - 3| < \frac{1}{100}$.
25. $3x - 4y + 12 > 0$.
26. $-2x + 5y + 10 > 0$.
27. $4x - y - 8 \leq 0$.
28. $5x + 3y - 15 \leq 0$.
29. $x + 7y < 0$.
30. $2x - 5y \geq 0$.
31. $(3x + 2y - 6)^2 > 0$.
32. $(3x + 2y - 6)^2 < 0$.

In Probs. 33 to 38 plot the graph of the given set.

33. $\{x \mid 0 \leq x \leq 1\}$.
34. $\{(x, y) \mid x - |x| = 0\}$.
35. $\{x \mid -1 + |x| + |x - 1| = 0\}$.
36. $\{(x, y) \mid y + |y| = 0\}$.

37. $\{(x, y) \mid -1 + |x| + |x - 1| + |y| = 0\}$.
38. $\{(x, y) \mid y - x + |y - x| = 0\}$.

5.8 Simultaneous Linear Equations in Two Variables

Here we complicate the situation a little by considering a pair of simultaneous linear equations in two variables. The general expression for such a system is

$$(1) \qquad \begin{cases} a_1x + b_1y + c_1 = 0 & (a_1 \neq 0) \vee (b_1 \neq 0) \\ a_2x + b_2y + c_2 = 0 & (a_2 \neq 0) \vee (b_2 \neq 0) \end{cases}$$

where the coefficients and variables are elements of a field F. By a solution of (1) we mean an ordered pair (x,y) which satisfies both equations.

In set language, the solution set of (1) is the set

$$\{(x,y) \mid a_1x + b_1y + c_1 = 0\} \cap \{(x,y) \mid a_2x + b_2y + c_2 = 0\}$$

When the coefficients and variables are real numbers, we are looking for the points which lie on the graphs of both the two given lines.

The method of solution is to transform (1) into an equivalent system whose solution is obvious. You are undoubtedly familiar with this procedure as the method of elimination by *addition and subtraction.* Let us state it formally.

Theorem 8. The linear systems

$$(1) \qquad \begin{cases} a_1x + b_1y + c_1 = 0 & (a_1 \neq 0) \vee (b_1 \neq 0) \\ a_2x + b_2y + c_2 = 0 & (a_2 \neq 0) \vee (b_2 \neq 0) \end{cases}$$

and

$$(2) \qquad \begin{cases} a_1x + b_1y + c_1 = 0 \\ k_1(a_1x + b_1y + c) + k_2(a_2x + b_2y + c_2) = 0 \end{cases}$$

with $k_2 \neq 0$ are equivalent.

Proof: It is clear that every solution of (1) satisfies (2). Conversely, if $(\bar{x},\bar{y})$ satisfies (2), it follows that $a_1\bar{x} + b_1\bar{y} + c_1 = 0$ and

$$k_2(a_2\bar{x} + b_2\bar{y} + c_2) = 0$$

Since $k_2 \neq 0$, $(\bar{x},\bar{y})$ satisfies (1).

Using this theorem, we can eliminate x and y in turn by choosing suitable values of k_1 and k_2 and thus solve the system. Let us do this systematically. By hypothesis, either $a_1 \neq 0$ or $b_1 \neq 0$. Let us suppose that $a_1 \neq 0$. Then, choosing $k_1 = -a_2$ and $k_2 = a_1$ in the system (2), we replace (1) by:

$$(3) \qquad \begin{cases} ax_1 \qquad + b_1y \qquad + c_1 = 0 \\ \qquad (a_1b_2 - a_2b_1)y + (a_1c_2 - a_2c_1) = 0 \end{cases}$$

If $a_1b_2 - a_2b_1 \neq 0$, we solve the second equation of (3) for y and obtain

$$y = -\frac{a_1c_2 - a_2c_1}{a_1b_2 - a_2b_1}$$

We may substitute this value of y in the first equation of (3) and solve for x:

$$x = -\frac{b_2c_1 - b_1c_2}{a_1b_2 - a_2b_1}$$

If we suppose that $b_1 \neq 0$, we choose $k_1 = b_2$ and $k_2 = -b_1$ and replace (1) by

$$(4) \qquad \begin{cases} a_1x + b_1y + c_1 = 0 \\ (a_1b_2 - a_2b_1)x + (b_2c_1 - b_1c_2) = 0 \end{cases}$$

Solving as before, we obtain the values for x and y derived above.

If, on the other hand, $a_1b_2 - a_2b_1 = 0$, there are two possibilities:

(i) $a_1c_2 - a_2c_1 = 0$, and $b_2c_1 - b_1c_2 = 0$. Then the system is equivalent to the single equation $a_1x + b_1y + c_1 = 0$, which has more than one solution.

Exercise A. Show that the number of pairs (x,y) which satisfy $a_1x + b_1y + c_1 = 0$ is equal to the number of elements in the field F. Thus, if F is the real numbers, there are infinitely many solutions.

(ii) At least one of $(a_1c_2 - a_2c_1)$ and $(b_2c_1 - b_1c_2)$ is not zero. Then (3) or (4) or both contains a contradiction and there is no solution.

Exercise B. Show that if $a_1b_2 - a_2b_1 = 0$ and $a_1c_2 - a_2c_1 = 0$, then $b_2c_1 - b_1c_2 = 0$.

We summarize these results in the theorem:

Theorem 9. The simultaneous equations

$$\begin{aligned} a_1x + b_1y + c_1 &= 0 \qquad (a_1 \neq 0) \vee (b_1 \neq 0) \\ a_2x + b_2y + c_2 &= 0 \qquad (a_2 \neq 0) \vee (b_2 \neq 0) \end{aligned}$$

with coefficients in a field F:

(a) Have a unique solution if $a_1b_2 - a_2b_1 \neq 0$.
(b) Have no solution if $a_1b_2 - a_2b_1 = 0$ and at least one of $(a_1c_2 - a_2c_1)$ and $(b_2c_1 - b_1c_2)$ is not zero.
(c) Have more than one solution if $a_1b_2 - a_2b_1 = 0$, $a_1c_2 - a_2c_1 = 0$, and $b_2c_1 - b_2c_2 = 0$.

Remark. When the coefficients are real, case (*a*) of Theorem 9 corresponds to two intersecting lines; case (*b*), to two parallel lines; and case (*c*), to a single line.

Illustration 1. Since the coefficients may be in any field, let us apply this method to solve the simultaneous congruences:

$$\left.\begin{aligned} x + 3y &\equiv 4 \\ 3x - y &\equiv 3 \end{aligned}\right\} \quad \text{mod } 7$$

These congruences are equivalent to the following equations in the field of residue classes, mod 7:

$$\begin{aligned} x + 3y - 4 &= 0 \\ 3x - y - 3 &= 0 \end{aligned}$$

To eliminate y, multiply the second equation by 3 and add it to the first. The result is

$$10x - 13 = 0 \qquad \text{or } 3x - 6 = 0$$

Multiplying by 5, which is the multiplicative inverse of 3 in this field, we obtain

$$x - 2 = 0 \qquad \text{or } x = 2$$

Similarly, we eliminate x by multiplying the first equation by -3 and adding it to the second. The result is:

$$-10y + 9 = 0 \qquad \text{or } 4y + 2 = 0$$

Multiplying by 2, which is the multiplicative inverse of 4 in this field, we obtain

$$y + 4 = 0 \qquad \text{or } y = -4 \qquad \text{or } y = 3$$

The solution is then the pair (2,3).

Illustration 2. In the field of real numbers solve:

$$\begin{aligned} 2x - 5y - 19 &= 0 \\ 3x + 4y + 6 &= 0 \end{aligned}$$

To eliminate x, we multiply the first equation by 3 and the second by -2 and add. The result is

$$-23y - 69 = 0 \qquad \text{or } y = -3$$

To eliminate y, we multiply the first equation by 4 and the second by 5 and add. The result is

$$23x - 46 = 0 \qquad \text{or } x = 2$$

The solution is then the pair $(2,-3)$.

Illustration 3. In the field of real numbers solve:

$$\begin{aligned} 3x + 2y + 5 &= 0 \\ 6x + 4y - 4 &= 0 \end{aligned}$$

To eliminate x, we multiply the first equation by -2 and add it to the second equation. The result is:

$$0y - 14 = 0$$

Since this is impossible, there is no solution. The graph of the two equations is a pair of parallel lines.

Illustration 4. In the field of real numbers solve:

$$\begin{aligned} 4x - y + 3 &= 0 \\ 8x - 2y + 6 &= 0 \end{aligned}$$

Elimination of x gives the result:

$$0y = 0$$

Therefore the system is equivalent to a single equation. The solution set contains infinitely many elements, and the graphs of the two equations are coincident.

5.9 Simultaneous Linear Inequalities in Two Variables

A typical system of two linear inequalities in two variables is:

$$\begin{aligned} a_1x + b_1y + c_1 > 0 \qquad & (a_1 \neq 0) \lor (b_1 \neq 0) \\ a_2x + b_2y + c_2 > 0 \qquad & (a_2 \neq 0) \lor (b_2 \neq 0) \end{aligned}$$

where we assume that the coefficients are real numbers. The solution set of each inequality is a half plane, and the solution set of the pair of inequalities is the intersection of these two half planes. The best method of procedure is graphical.

Illustration 1. Graph the set determined by:

$$\begin{aligned} 2x + y - 3 &> 0 \\ x - 2y + 1 &< 0 \end{aligned}$$

First draw the two lines, which intersect at (1,1). We find that $2x + y - 3 > 0$ determines its right half plane and that $x - 2y + 1 < 0$ determines its left half plane. The region common to the two is shaded in Fig. 5.13. It is the interior of an angle whose vertex is (1,1).

This illustration is typical for two inequalities. But we can consider three or more simultaneous inequalities. The ideas and procedure are the same.

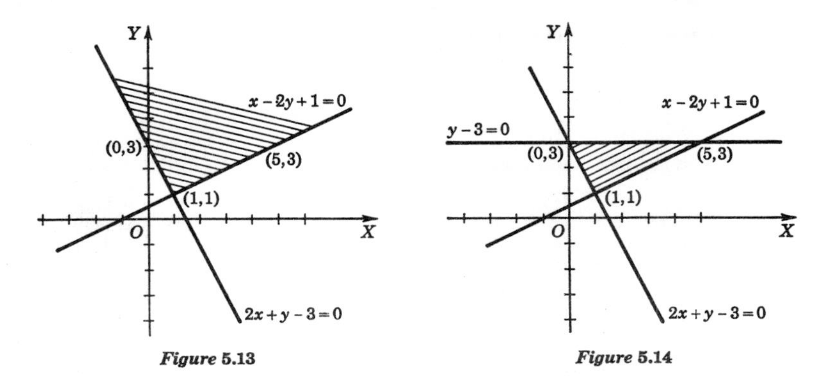

Figure 5.13 *Figure* 5.14

Illustration 2. Graph the set determined by

$$\begin{aligned} 2x + y - 3 &> 0 \\ x - 2y + 1 &< 0 \\ y - 3 &< 0 \end{aligned}$$

The first two inequalities are the same as in Illustration 1, so we merely add the third line to Fig. 5.13. This gives a triangle with vertices (1,1), (0,3), and (5,3). This line divides the shaded region of Fig. 5.13 into two parts. We see that $y - 3 < 0$ determines its lower half plane. Hence the desired set is the interior of the triangle shaded in Fig. 5.14.

Illustration 3. Let us add one more inequality to our picture, and consider the system:

$$\begin{aligned} 2x + y - 3 &> 0 \\ x - 2y + 1 &< 0 \\ y - 3 &< 0 \\ x + y - 5 &< 0 \end{aligned}$$

Since $x + y + 5 < 0$ determines its left half plane, the desired set is the quadrilateral shaded in Fig. 5.15.

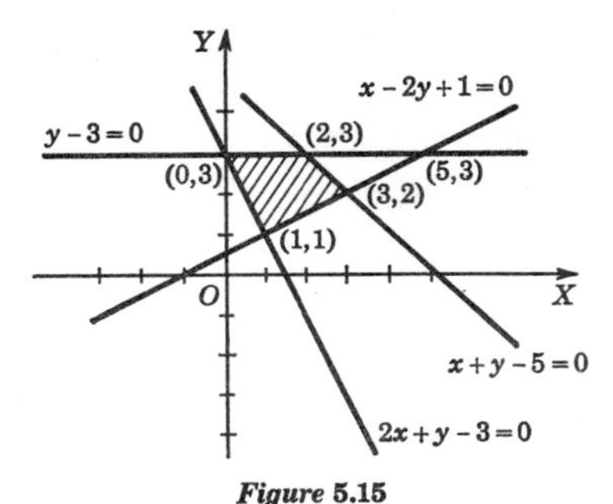

Figure 5.15

Illustration 4. Instead of the system of Illustration 3, consider:

$$\begin{aligned} 2x + y - 3 &> 0 \\ x - 2y + 1 &< 0 \\ y - 3 &< 0 \\ x + y + 2 &< 0 \end{aligned}$$

The inequality $x + y + 2 < 0$ determines its left half plane, which has no points in common with the triangle of Fig. 5.14. Hence the desired set is empty, and nothing is shaded in Fig. 5.16.

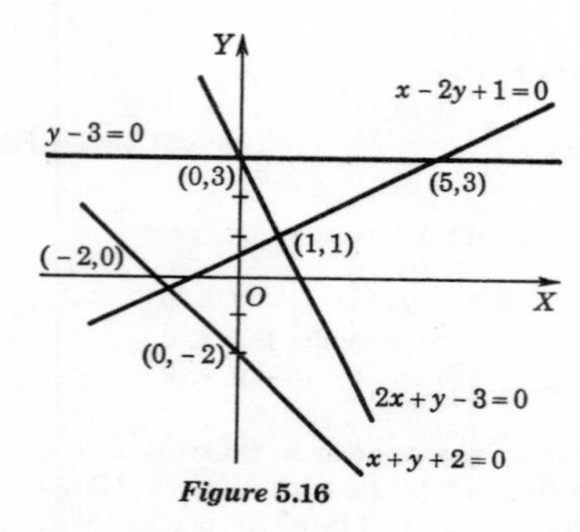

Figure **5.16**

We see in this way that the graph of a system of linear inequalities can take many forms. In the usual cases it is the interior of a convex polygon, which may be bounded as in Figs. 5.14 and 5.15 or unbounded as in Fig. 5.13. In other cases it may be the empty set. When the inequalities include $\geq$ or $\leq$, the possibilities are even more numerous.

Simultaneous inequalities can arise in unexpected places in practical situations. Let us consider two of these.

Illustration 5. The situation in a simplified version of a football game is as follows. There are just two plays, a running play and a pass play. We assume these facts to be true:

Play	Distance gained, yd	Time required, sec
Running	3	30
Pass	6	10

Also suppose that there are 60 yd to go for a touchdown and that 150 sec remain in the game. Ignore the requirement of having to make 10 yd in four downs and other considerations of score and strategy.

Problem. What combinations of running and pass plays will secure a touchdown in the allotted time?

Solution: Let r and p, respectively, represent the numbers of running and pass plays. Then the conditions of the problem may be written:

$$\begin{aligned} 3r + 6p &\geq 60 \\ 30r + 10p &\leq 150 \end{aligned}$$

These may be simplified to:

$$\begin{aligned} r + 2p &\geq 20 \\ 3r + p &\leq 15 \end{aligned}$$

Plot the graph of this simultaneous system as in Fig. 5.17. The solution set is the shaded triangle. Since r and p must be integers, the solution consists of the following combinations:

r	0	0	...	0	1	1	...	1	2
p	10	11	...	15	10	11	...	12	9

The quarterback can then select his pattern of plays from this array according to his best judgment regarding strategy or other matters.

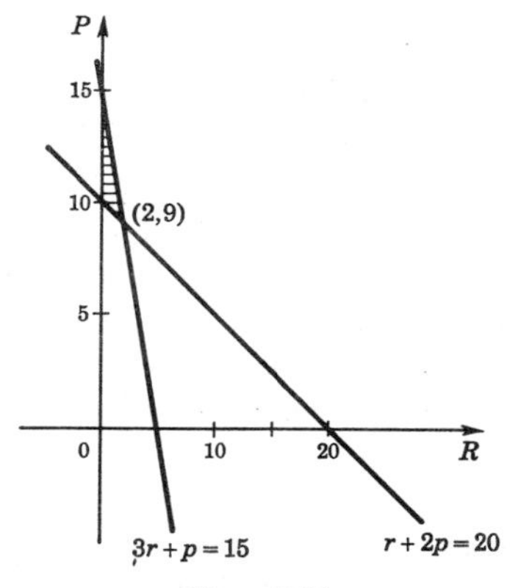

Figure **5.17**

Illustration 6. The Minneapolis and Seattle Lumber Company can convert logs into either lumber or plywood. In a given week the mill can turn out 400 units of production, of which 100 units of lumber and 150 units of plywood are required by regular customers. The profit on a unit of lumber is \$20 and on a unit of plywood is \$30.

Problem. How many units of lumber and plywood should the mill produce (totaling 400) in order to maximize the total profit?

Solution: Let L and P, respectively, represent the number of units of lumber and plywood. Then the conditions of the problem are:

$$L + P = 400 \qquad L \geq 100 \qquad P \geq 150$$
$$\text{Total profit} = 20L + 30P$$

Let us graph these in Fig. 5.18.

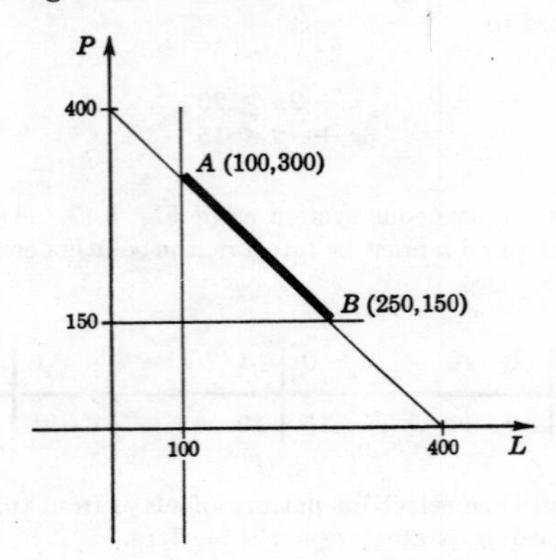

Figure 5.18

The possible solutions must lie on the line $L + P = 400$, on or above $P = 150$, and on or to the right of $L = 100$. Hence they lie on the segment AB. The total profit at A is \$11,000 and at B is \$9,500. At other points of AB the profit lies between these values. Hence the mill should produce 100 units of lumber and 300 units of plywood.

Perhaps this solution is obvious to you without all this analysis, and it should be. The point of the illustration is that methods of this kind are extremely useful in more complicated problems where the answers are far from obvious. The mathematical subject which deals with such problems is called *Linear Programing.*

Problems 5.9

In Probs. 1 to 10 solve the given pair of equations algebraically. Then plot the graph of the two lines, and check your solution graphically.

1. $3x - 4y + 2 = 0.$
$x + 2y - 6 = 0.$

2. $3x - 2y - 11 = 0.$
$5x + 2y + 3 = 0.$

3. $4x + y - 3 = 0.$
$8x - y + 3 = 0.$

4. $6x + y + 10 = 0.$
$3x - 2y + 10 = 0.$

5. $x + 3y - 6 = 0.$
$-2x + 4y + 2 = 0.$

6. $5x + 2y + 2 = 0.$
$x - y - 8 = 0.$

7. $x + y - 2 = 0.$
$2x + 2y + 1 = 0.$

8. $3x + 5y + 15 = 0.$
$6x + 10y - 2 = 0.$

9. $-2x + 7y + 19 = 0.$
$4x - 14y - 38 = 0.$

10. $3x - 2y + 2 = 0.$
$9x - 6y + 6 = 0.$

In Probs. 11 to 18 solve the given pair of congruences.

11. $\left.\begin{aligned} 2x + 3y &\equiv 4 \\ x + y &\equiv 0 \end{aligned}\right\}$ mod 5

12. $\left.\begin{aligned} x + 3y &\equiv 3 \\ 2x - y &\equiv 2 \end{aligned}\right\}$ mod 5

13. $\left.\begin{aligned} 3x - 4y &\equiv 5 \\ 2x + 5y &\equiv 4 \end{aligned}\right\}$ mod 7

14. $\left.\begin{aligned} 5x + 2y &\equiv 4 \\ 3x - 5y &\equiv 0 \end{aligned}\right\}$ mod 7

15. $\left.\begin{aligned} 2x + 3y &\equiv 4 \\ x - y &\equiv 2 \end{aligned}\right\}$ mod 5

16. $\left.\begin{aligned} 3x + y &\equiv 3 \\ x + 2y &\equiv 1 \end{aligned}\right\}$ mod 5

17. $\left.\begin{aligned} x - y &\equiv 1 \\ 2x + 5y &\equiv 3 \end{aligned}\right\}$ mod 7

18. $\left.\begin{aligned} 4x + y &\equiv 5 \\ x + 2y &\equiv 5 \end{aligned}\right\}$ mod 7

In Probs. 19 to 34 plot the graph of the given inequality or system of inequalities.

19. $3x + y - 2 > 0.$

20. $x + 2y + 3 > 0.$

21. $-2x + 4y + 8 \leq 0.$

22. $4x - y + 12 \leq 0.$

23. $5x - y \geq 0.$

24. $4x + 7y \geq 0.$

25. $\{(x, y) \mid 3x - 7 > 0\}.$

26. $\{(x, y) \mid -2y + 9 \leq 0\}.$

27. $(3x - 2y + 6)^2 > 0.$

28. $(-x + 2y - 4)^2 \leq 0.$

29. $3x + y - 4 > 0.$
$x - 2y + 1 < 0.$
$2x + 3y - 19 < 0.$

30. $4x - 5y + 25 > 0.$
$5x + 3y + 22 > 0.$
$9x - 2y + 10 > 0.$

31. $3x + y - 4 > 0.$
$x - 2y + 1 < 0.$
$2x + 3y - 19 < 0.$
$3x + y - 11 < 0.$

32. $4x - 5y + 25 > 0.$
$5x + 3y + 22 > 0.$
$2x - 7y - 24 < 0.$

33. $3x + y - 4 > 0.$
$x - 2y + 1 < 0.$
$2x + 3y - 19 < 0.$
$3x + y + 4 < 0.$

34. $4x - 5y + 25 > 0.$
$5x + 3y + 22 > 0.$
$2x - 7y - 24 < 0.$
$7x + 5y - 25 < 0.$

35. In the football problem (Illustration 5 above) what combinations of plays will meet the required conditions if there are 180 sec left to play? If there are 90 sec left to play?

36. How should the Minneapolis and Seattle Lumber Company (Illustration 6 above) arrange its production if the profit on a unit of plywood is \$10 and the profit on a unit of lumber is \$15?

Problems 37 to 40 refer to linear transformations on the line. (See Sec. 5.3, Probs. 21 to 28.)

37. Find the linear transformation which relabels the point $x = 1$ with $x' = 5$ and $x = 2$ with $x' = 7$.

38. Given that 0°C corresponds to 32°F, and 100°C corresponds to 212°F, derive the linear transformation which expresses F in terms of C.

39. In one grading system 60 is passing and 100 perfect. In a second grading system 70 is passing and 100 is perfect. Find a linear transformation between these two grading systems which takes passing into passing and perfect into perfect. What grade remains unchanged?

40. Let x_1, x_2' be the corresponding labels for two points in the x and x' coordinate systems, respectively. Assume $x_2 > x_1$ and $x_2' > x_1$. Then find a linear transformation between these coordinate systems. What point has the same coordinates in both systems?

Transformation of Coordinates in the Plane. If we are given a coordinate system in the plane in terms of pairs (x,y), we can define a new coordinate system (x',y') by means of the *linear* transformation:

$$x' = a_1x + b_1y + c_1$$
$$y' = a_2x + b_2y + c_2$$

where we assume $a_1b_2 - a_2b_1 \neq 0$. The new pairs (x',y') serve as new labels for the points in the plane. The point O', where $x' = 0$, $y' = 0$, is the new origin, the X'-axis is the line $y' = 0$, and the Y'-axis is the line $x' = 0$. These axes, however, need not be at right angles.

A point P is called a *fixed point* if its coordinates in both systems are equal, i.e., if $x' = x$, $y' = y$ at P. If every point is a fixed point, the transformation has the equations $x' = x$, $y' = y$ and is called the *identity transformation*.

In working the problems below you will need to anticipate the following result, which will be proved in Sec. 8.2. Let s be the distance between $P_1(x_1,y_1)$ and $P_2(x_2,y_2)$. Then:

$$s^2 = (x_2 - x_1)^2 + (y_2 - y_1)^2$$

This is really nothing but the Pythagorean theorem.

In Probs. 41 to 44 we take $x' = x + a$, $y' = y + b$. This transformation is called a *translation*.

41. Prove that a translation leaves the lengths of segments unchanged.
42. Prove: if a translation has a fixed point, then it is the identity transformation.
43. Show that the correspondence $(x,y) \leftrightarrow (x',y')$ defined by a translation is 1 to 1.
44. For the translation $x' = x - 2$, $y' = y - 5$, find the new origin and sketch the new axes.

In Probs. 45 to 48 we take $x' = ax$, $y' = ay$, where $a \neq 0$. This transformation is called a *dilatation*.

45. Prove that a dilatation multiplies lengths of segments by $|a|$ and areas of rectangles by a^2. Hence show that a triangle is transformed into a similar triangle.
46. Prove that the origin is a fixed point under a dilatation.
47. Prove that, if a dilatation has a fixed point in addition to the origin, then it is the identity transformation.
48. Show that the correspondence $(x,y) \leftrightarrow (x',y')$ defined by a dilatation is 1 to 1.

In Probs. 49 to 51 we take $x' = -x$; $y' = y$. This transformation is called a *reflection* in the Y-axis.

49. Prove that a reflection leaves the lengths of segments unchanged.
50. Find the fixed points for the above reflection.
51. Prove that the correspondence $(x,y) \leftrightarrow (x',y')$ defined by a reflection is 1 to 1.

In Probs. 52 to 56 we take

$$\left.\begin{aligned} x' &= ax + by \\ y' &= -bx + ay \end{aligned}\right\} \quad \text{where } a^2 + b^2 = 1$$

This transformation is called a *rotation*.

52. Sketch the new axes when

$$x' = \frac{1}{\sqrt{2}}x + \frac{1}{\sqrt{2}}y$$
$$y' = -\frac{1}{\sqrt{2}}x + \frac{1}{\sqrt{2}}y$$

53. Prove that a rotation leaves the lengths of segments unchanged.

54. Prove that the origin is a fixed point under a rotation.

55. Prove that, if a rotation has a fixed point other than the origin, then it is the identity transformation. HINT: Solve

$$\left.\begin{aligned} x &= ax + by \\ y &= -bx + ay \end{aligned}\right\} \quad \text{for } a \text{ and } b$$

assuming $(x,y) \neq (0,0)$.

56. Prove that the correspondence $(x,y) \leftrightarrow (x',y')$ defined by a rotation is 1 to 1.

In Probs. 57 to 62 we consider the *centered* linear transformation

$$\left.\begin{aligned} x' &= a_1x + b_1y \\ y' &= a_2x + b_2y \end{aligned}\right\} \quad \text{where } a_1b_2 - a_2b_1 \neq 0$$

57. Prove that every point on the line $x + y = 0$ is a fixed point for the transformation

$$\begin{aligned} x' &= 2x + y \\ y' &= x + 2y \end{aligned}$$

HINT: Solve

$$\left.\begin{aligned} x &= 2x + y \\ y &= x + 2y \end{aligned}\right\} \quad \text{for } x \text{ and } y$$

58. Find the fixed points of the transformation

$$\begin{aligned} x' &= 3x - y \\ y' &= 2x \end{aligned}$$

59. Prove that the origin is the only fixed point of the general centered linear transformation unless $a_1b_2 - a_2b_1 - b_2 - a_1 + 1 = 0$.

60. Solve the equations of the general centered linear transformation for (x,y) in terms of (x',y'). This is the *inverse* transformation.

61. Prove that the correspondence $(x,y) \leftrightarrow (x',y')$ defined by a centered linear transformation is 1 to 1.

62. Consider the pair of transformations:

$$\begin{aligned} x' &= ax + by \qquad & x'' &= px' + qy' \\ y' &= cx + dy \qquad & y'' &= rx' + sy' \end{aligned}$$

Find (x'',y'') in terms of (x,y). Compare the matrix of coefficients so obtained with the product

$$\begin{pmatrix} p & q \\ r & s \end{pmatrix}\begin{pmatrix} a & b \\ c & d \end{pmatrix}$$

This result motivates our definition of the product of two matrices.

5.10 Quadratic Equations in One Variable

Earlier we saw that the quadratic equation $x^2 = 2$ could not be solved in the field of rational numbers and that $x^2 = -1$ could not be solved in the field of real numbers. Thus it is clear that the axioms of a field are not strong enough to assure the solution of every quadratic equation. It was for this reason that we invented the field of complex numbers, in which every quadratic equation has a solution.

In order to prove this, let us first consider the equation

$$(1) \qquad ax^2 + bx + c = 0$$

where a, b, and c are real and $a \neq 0$. We proceed to solve this by a method known as *completing the square*. This depends upon the fact that

$$(2) \qquad x^2 + 2\,dx + d^2 = (x + d)^2$$

Since $a \neq 0$, let us write Eq. (3), which is equivalent to (1):

$$(3) \qquad x^2 + \frac{b}{a}x + \frac{c}{a} = 0 \quad \text{[Theorem 2, Sec. 5.4]}$$

If we put $d = b/2a$, the first two terms on the left side of (2) are equal to the corresponding terms in (3). In general, however, $d^2 \neq c/a$. Therefore we write (4), which is equivalent to (3):

$$(4) \qquad x^2 + \frac{b}{a}x + \left(\frac{b}{2a}\right)^2 = \left(\frac{b}{2a}\right)^2 - \frac{c}{a}$$

Now the left-hand side of (4) is of the same form as the left-hand side of (2). Thus:

$$(5) \qquad \left(x + \frac{b}{2a}\right)^2 = \frac{b^2 - 4ac}{4a^2}$$

We can solve (5) by extracting the square root of both sides. We use the fact that, if $y^2 = r$, then $y = +\sqrt{r}$ or $y = -\sqrt{r}$. Hence from (5) we find that:

$$x + \frac{b}{2a} = \frac{\sqrt{b^2 - 4ac}}{2a} \quad \text{or} \quad x + \frac{b}{2a} = -\frac{\sqrt{b^2 - 4ac}}{2a}$$

Therefore

$$x = \frac{-b + \sqrt{b^2 - 4ac}}{2a} \quad \text{or} \quad x = \frac{-b - \sqrt{b^2 - 4ac}}{2a}$$

This proves the following theorem:

Theorem 10. The quadratic equation

$$ax^2 + bx + c = 0 \qquad a \neq 0$$

where, a, b, and c are real numbers, has two solutions (which may not be real numbers), namely:

$$x = \frac{-b \pm \sqrt{b^2 - 4ac}}{2a}$$

As a matter of fact, this same method will work even if a, b, and c are complex numbers. Everything goes as above until we reach the expression $\sqrt{b^2 - 4ac}$. The question now is, Is there a complex number whose square is the complex number $b^2 - 4ac$? The best proof that the answer is "yes" is given in Sec. 8.13, but we shall give another proof here.

Theorem 11. Every complex number has two square roots which are complex numbers.

Proof: In effect we are asked to solve the equation

$$(x + yi)^2 = c + di$$

or

$$(x^2 - y^2) + 2xyi = c + di$$

This separates into the two equations

$$x^2 - y^2 = c \qquad 2xy = d$$

From the second we have $y = d/2x$, so from the first

$$x^2 - \frac{d^2}{4x^2} = c$$

or

$$4x^4 - 4cx^2 - d^2 = 0$$

Since all coefficients are real, we apply Theorem 10 and obtain:

$$x^2 = \frac{c \pm \sqrt{c^2 + d^2}}{2}$$

Since x is to be real, x^2 is positive and we can use only the positive sign in our expression on the right. Hence

$$x = \pm \sqrt{\frac{c + \sqrt{c^2 + d^2}}{2}}$$

which is real, and consequently

$$y = \frac{d}{2x} = \pm \frac{d}{|d|} \sqrt{\frac{-c + \sqrt{c^2 + d^2}}{2}}$$

Therefore the square roots of $c + di$ are

$$\pm \left[\sqrt{\frac{c + \sqrt{c^2 + d^2}}{2}} + i \frac{d}{|d|} \sqrt{\frac{-c + \sqrt{c^2 + d^2}}{2}} \right]$$

Using the method of Theorem 10 and the result of Theorem 11, we have now proved Theorem 12.

Theorem 12. The quadratic equation

$$ax^2 + bx + c = 0 \qquad (a \neq 0)$$

where a, b, and c are complex numbers, has the two solutions

$$x = \frac{-b \pm \sqrt{b^2 - 4ac}}{2a}$$

in the field of complex numbers.

Remarks

(a) If a, b, and c are real, we may easily deduce the following properties of the solutions:

When $b^2 - 4ac$ is positive, the two solutions are real and unequal.

When $b^2 - 4ac$ is zero, the two solutions are real and equal.

When $b^2 - 4ac$ is negative, the two solutions are unequal and neither of them is real.

(**b**) Let r_1 and r_2 be the two solutions of $ax^2 + bx + c = 0$; that is,

$$(6) \qquad r_1 = \frac{-b + \sqrt{b^2 - 4ac}}{2a} \qquad r_2 = \frac{-b - \sqrt{b^2 - 4ac}}{2a}$$

Then:

$$r_1 + r_2 = -\frac{b}{a} \qquad r_1 r_2 = \frac{c}{a}$$

Exercise A. By direct calculation verify the above statements.

(**c**) The above formulas permit us to give a general expression for the factors of $ax^2 + bx + c$, namely:

$$ax^2 + bx + c = a(x - r_1)(x - r_2)$$

where r_1 and r_2 are defined in (6). To prove this correct we note that

$$\begin{aligned} a(x - r_1)(x - r_2) &= ax^2 - a(r_1 + r_2)x + ar_1r_2 \\ &= ax^2 - a\left(-\frac{b}{a}\right)x + a\left(\frac{c}{a}\right) \\ &= ax^2 + bx + c \end{aligned}$$

5.11 Quadratic Inequalities in One Variable

The properties of quadratic inequalities follow easily from those of quadratic equations. We are concerned with the inequalities:

$$\begin{aligned} ax^2 + bx + c &> 0 \\ ax^2 + bx + c &< 0 \\ ax^2 + bx + c &\geq 0 \\ ax^2 + bx + c &\leq 0 \end{aligned}$$

where $a > 0$, and a, b, and c are real. Only real values of x will be permitted.

Let us suppose, first, that $b^2 - 4ac > 0$, so that the solutions of $ax^2 + bx + c = 0$ are real and unequal. Then we can write

$$ax^2 + bx + c = a(x - r_1)(x - r_2)$$

where r_1, r_2 are real and $r_1 \neq r_2$. Suppose $r_1 < r_2$. Let us imagine x moving from left to right along the real line, and consider how the sign of $ax^2 + bx + c$ varies in the process. At the extreme left (where $x < r_1 < r_2$), $(x - r_1)$ is negative and $(x - r_2)$ is negative. Hence, in this region, $a(x - r_1)(x - r_2)$ is positive. When x reaches r_1, $a(x - r_1)(x - r_2)$ becomes zero. Between r_1 and r_2, $(x - r_1)$ is negative and $(x - r_2)$ is positive. So, in this region, $a(x - r_1)(x - r_2)$ is negative. At r_2, $a(x - r_1)(x - r_2)$ is zero. To the right of r_2, $(x - r_1)$ and $(x - r_2)$ are both positive, so that $a(x - r_1)(x - r_2)$ is positive. These results are summarized in Fig. 5.19. We state them formally in the theorem.

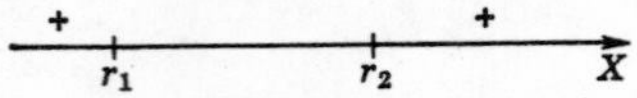

Figure 5.19

Theorem 13. If $ax^2 + bx + c = 0$ $(a > 0)$ has distinct real solutions r_1 and r_2 with $r_1 < r_2$, then when:

$$\begin{aligned} x < r_1,&\ ax^2 + bx + c > 0 \\ x = r_1,&\ ax^2 + bx + c = 0 \\ r_1 < x < r_2,&\ ax^2 + bx + c < 0 \\ x = r_2,&\ ax^2 + bx + c = 0 \\ x > r_2,&\ ax^2 + bx + c > 0 \end{aligned}$$

A similar method gives the next result.

Theorem 14. If $ax^2 + bx + c = 0$ $(a > 0)$ has equal real solutions $r_1 = r_2 = r$, then when:

$$\begin{aligned} x < r,&\ ax^2 + bx + c > 0 \\ x = r,&\ ax^2 + bx + c = 0 \\ x > r,&\ ax^2 + bx + c > 0 \end{aligned}$$

The final situation is that in which $ax^2 + bx + c = 0$ has nonreal solutions; that is, $b^2 - 4ac < 0$. In Sec. 5.10 we showed that

$$ax^2 + bx + c = a\left[\left(x + \frac{b}{2a}\right)^2 - \frac{b^2 - 4ac}{4a^2}\right]$$

In this expression we have $a > 0$, $\left(x + \dfrac{b}{2a}\right)^2 \geq 0$, $\left(-\dfrac{b^2 - 4ac}{4a^2}\right) > 0$. Hence the sum is >0. This proves the theorem:

Theorem 15. If $ax^2 + bx + c = 0$ $(a > 0)$ has nonreal solutions, then for all x, $ax^2 + bx + c > 0$.

Exercise A. Prove the converse of Theorem 15.

Illustration 1

(**a**) Solve $3(x - 1)(x - 4) > 0$. Answer: $\{x \mid (x < 1) \vee (x > 4)\}$. (See Fig. 5.20*a*.)
(**b**) Solve $2(x + 1)(x - 2) < 0$. Answer: $\{x \mid -1 < x < 2\}$. (See Fig. 5.20*b*.)

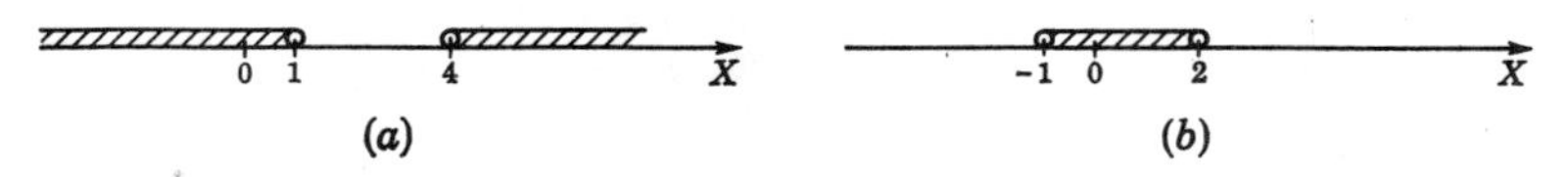

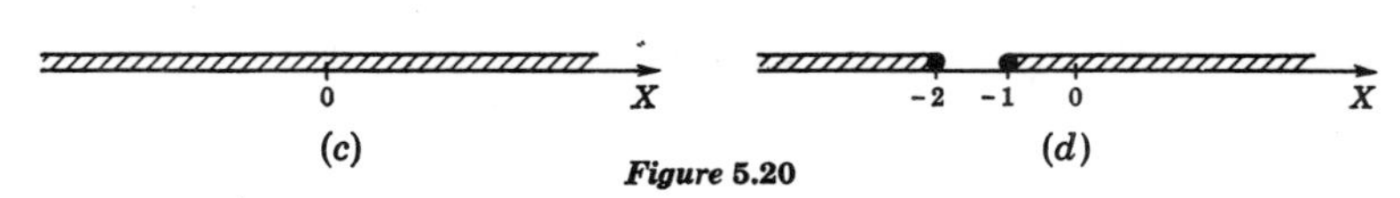

Figure 5.20

(**c**) Solve $4(x - 3)^2 \geq 0$. Answer: All real x. (See Fig. 5.20*c*.)
(**d**) Solve $(x + 1)(x + 2) \geq 0$. Answer: $\{x \mid (x \leq -2) \vee (x \geq -1)\}$. (See Fig. 5.20*d*.)
(**e**) Solve $x^2 + x + 1 > 0$. Answer: All real x. (See Fig. 5.20*c*.)

The method of this section is easily extended to inequalities of the form:

$$a(x - r_1)(x - r_2)(x - r_3) \cdots (x - r_n) > 0, <0$$

etc., where $a > 0$ and $r_1, \ldots, r_n$ are real.

Illustration 2. Solve: $2(x - 1)(x - 2)(x - 3) > 0$.

When $x < 1$, each factor is negative, so the product is negative.

When $1 < x < 2$, $(x - 1)$ is positive, and $(x - 2)$ and $(x - 3)$ are negative. Hence the product is positive.

Continuing in this way we find the solution set: $\{x \mid (1 < x < 2) \vee (3 < x)\}$, whose graph is given in Fig. 5.21.

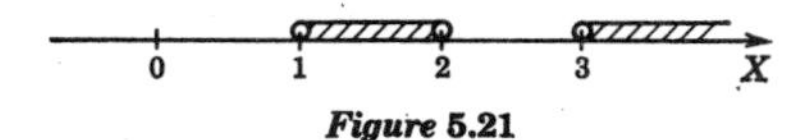

Figure 5.21

Problems 5.11

In Probs. 1 to 10 find the solutions of the quadratic equations.

1. $x^2 + 4x + 5 = 0$.
2. $x^2 + 2x + 2 = 0$.
3. $9x^2 + 6x + 1 = 0$.
4. $x^2 + 10x + 25 = 0$.
5. $36x^2 + 17x - 35 = 0$.
6. $18x^2 + 27x - 56 = 0$.
7. $x^2 - (3 + 2i)x + (5 + 5i) = 0$.
8. $x^2 - (3 + 2i)x + (1 + 3i) = 0$.
9. $x^2 - (5 + 2i)x + (5 + 5i) = 0$.
10. $x^2 + (4 + 3i)x + (7 + i) = 0$.

In Probs. 11 to 16 add terms to both sides so that the left-hand side becomes a perfect square.

11. $x^2 + 5x + 1 = 0$.
12. $x^2 - 2x - 3 = 0$.
13. $x^2 - 8x + 2 = 0$.
14. $x^2 + 7x - 2 = 0$.
15. $4x^2 - 10x + 3 = 0$.
16. $3x^2 + 12x + 1 = 0$.

In Probs. 17 to 22 find the sum and product of the solutions without solving the equation.

17. $x^2 - 3x + 10 = 0$.
18. $x^2 + 4x - 7 = 0$.
19. $4x^2 - 7x + 12 = 0$.
20. $5x^2 + 3x + 6 = 0$.
21. $(2 - i)x^2 + (4 + 3i)x + (-1 + i) = 0$.
22. $(3 + 2i)x^2 + (1 - 2i)x + (3 + i) = 0$.

In Probs. 23 to 28 find the value of k for which the solutions of the given equation are equal.

23. $x^2 + 3x + k = 0$.
24. $x^2 - 5x + k = 0$.
25. $2x^2 + 7x + k = 0$.
26. $3x^2 - x + k = 0$.
27. $x^2 + 3kx + 4 = 0$.
28. $x^2 - 4kx + 5 = 0$.

In Probs. 29 to 34 find a quadratic equation the sum and product of whose solutions have the given values.

29. Sum 3, product 4.
30. Sum 0, product 5.
31. Sum $\frac{1}{3}$, product $\frac{2}{3}$.
32. Sum $\frac{2}{5}$, product $-\frac{4}{5}$.
33. Sum $3 - i$, product $2 + 5i$.
34. Sum $1 + i$, product $2 - i$.

In Probs. 35 to 50 solve the given inequality (see Illustrations 1 and 2 above) and plot the graph.

35. $(x + 2)(x - 1) > 0$.
36. $(x - 3)(x + 4) > 0$.
37. $5(x - 2)^2 > 0$.
38. $-3(x - 1)^2 \leq 0$.
39. $2x^2 + 3x - 2 \leq 0$.
40. $3x^2 + 10x + 3 \geq 0$.
41. $4x^2 + 3x - 1 < 0$.
42. $2x^2 - 3x - 2 > 0$.
43. $x^2 + 3x + 3 > 0$.
44. $-x^2 + 2x - 4 \leq 0$.
45. $-x^2 - x - 1 \geq 0$.
46. $3x^2 + 2x + 6 < 0$.
47. $4(x - 3)(x + 2)(x + 5) > 0$.
48. $3(x + 1)(x + 2)(x - 3) < 0$.
49. $-5(x + 2)^2(x - 3) \geq 0$.
50. $-2(2x - 1)(x + 4)(x) < 0$.

5.12 Quadratic Equations and Inequalities in Two Variables

Because of the great variety of quadratic equations in two variables, we shall limit ourselves here to those of the form

$$y = ax^2 + bx + c \qquad (a \neq 0)$$

and to the related inequalities

$$y > ax^2 + bx + c \qquad y < ax^2 + bx + c$$

etc., where a, b, and c are real.

The solution set of an equation of this type contains an infinite number of pairs (x,y), and so our discussion centers about its graph.

The usual procedure for plotting the graph of such an equation is to compute a small number of points on it and then to connect these with a smooth curve.

Illustration 1. Plot the graph of

$$y = x^2 - 6x + 5$$

We write the table of pairs:

x	-1	0	1	2	3	4	5	6	7
y	12	5	0	-3	-4	-3	0	5	12

and then draw Fig. 5.22.

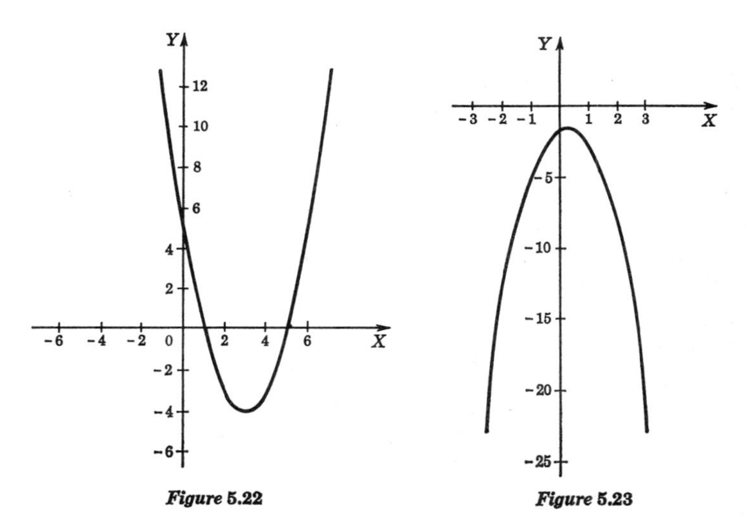

Figure **5.22**

Figure **5.23**

Illustration 2. Plot the graph of

$$y = -2x^2 + x - 2$$

The table of pairs is

x	-2	-1	0	1	2	3
y	-12	-5	-2	-3	-8	-23

and we draw Fig. 5.23

The graphs of these equations have the following important properties:

(1) If $a > 0$, the graph opens upward.
(2) If $a < 0$, the graph opens downward.
(3) If $ax^2 + bx + c = 0$ has the real solutions r_1 and r_2, where $r_1 \neq r_2$, the graph crosses the X-axis at $x = r_1$ and $x = r_2$.
(4) If $ax^2 + bx + c = 0$ has two equal real solutions r, the graph is tangent to the X-axis at $x = r$.
(5) If $ax^2 + bx + c = 0$ has nonreal solutions, the graph does not cross the X-axis.
(6) The graph is symmetric about the line

$$x = -\frac{b}{2a}$$

(7) The highest (or lowest) point on the graph is the point

$$\left(-\frac{b}{2a}, \frac{-b^2 + 4ac}{4a}\right)$$

Exercise A. Verify as many as possible of these properties by examining the graphs in Illustrations 1 and 2 above.

Definition: The graph of $y = ax^2 + bx + c$ $(a \neq 0)$ is called a *parabola*. Its highest (or lowest) point is called its *vertex*.

From our illustrations we see that a parabola divides the plane into two regions, one above it and one below it. Each of these regions is *connected* in the sense that if P_1 and P_2 lie in the same region, there is a smooth curve lying entirely in that region which connects them. By the reasoning similar to that used in the proof of Theorem 7, we can establish the following result here.

Theorem 16. Let P lie in one of the regions into which a parabola divides the plane. If $y > ax^2 + bx + c$ at P, then $y > ax^2 + bx + c$ at every point of this region.

As a result of this theorem we can plot the solution set of inequalities like $y > ax^2 + bx + c$ by (1) drawing the graph of $y = ax^2 + bx + c$, (2) testing the truth of the given inequality at a point in each of the two regions into which this parabola divides the plane, (3) shading the region or regions in which the above tests are affirmative.

Illustration 3. Plot the graph of

$$y < x^2 - 5x + 6$$

First draw (Fig. 5.24) the graph of $y = x^2 - 5x + 6$. Then (0,0) lies below the graph and $0 < 0 - 0 + 6$ is true. On the other hand, (2,1) lies above the graph and $1 < 4 - 10 + 6$ is false. Hence we shade the region below the graph.

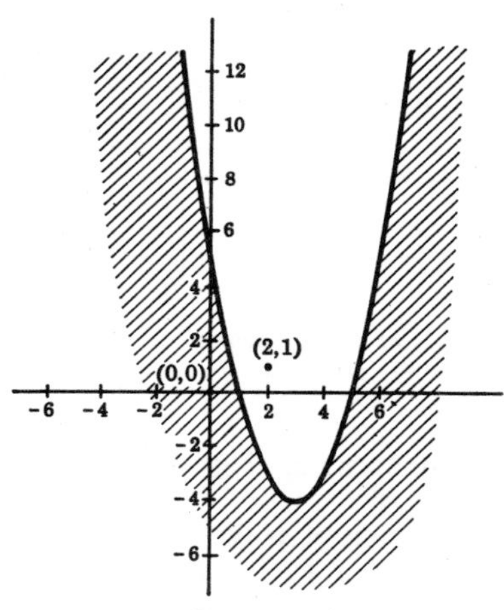

Figure **5.24**

Problems 5.12

In Probs. 1 to 10 plot the graph of the given equations.

1. $y = x^2 + 9x + 8$.
2. $y = x^2 - 8x + 12$.
3. $y = -x^2 + 9x - 18$.
4. $y = -x^2 + 10x - 24$.
5. $y = 4x^2 + 4x + 1$.
6. $y = 9x^2 + 6x + 1$.
7. $y = 2x^2 - x + 8$.
8. $y = -3x^2 + 2x - 2$.
9. $y = -x^2 - 2x - 1$.
10. $y = -16x^2 - 8x - 1$.

In Probs. 11 to 20 plot the graph of the given inequality. (See Probs. 1 to 10.)

11. $y > x^2 + 9x + 8$.
12. $y > x^2 - 8x + 12$.
13. $y \leq -x^2 + 9x - 18$.
14. $y \geq -x^2 + 10x - 24$.
15. $y < 4x^2 + 4x + 1$.
16. $y \leq 9x^2 + 6x + 1$.
17. $y - 2x^2 + x - 8 \geq 0$.
18. $y + 3x^2 - 2x + 2 < 0$.
19. $(y + x^2 + 2x + 1)^2 \geq 0$.
20. $(y + 16x^2 + 8x + 1)^2 \leq 0$.

In Probs. 21 to 24 graph the common solution set of the two inequalities.

21. $y > x^2 - 10x + 24$.
$x + y - 6 < 0$.
22. $y > x^2 + 6x + 8$.
$x - y + 8 < 0$.
23. $y > x^2 + x + 1$.
$y < -1$.
24. $y \geq x^2 + 2x + 2$.
$y \leq 1$.

Problems 25 to 33 refer to properties 1 to 7 of parabolas, which are stated just after Illustration 2 above.

25. Why is property (1) true?

26. Why is property (2) true?

27. Why is property (3) true?

28. Why is property (4) true?

29. Why is property (5) true?

30. Why is property (6) true? HINT: Write $y = ax^2 + bx + c$ in the form

$$y + \frac{b^2 - 4ac}{4a} = a\left(x + \frac{b}{2a}\right)^2$$

31. Why is property (7) true? HINT: Write $y = ax^2 + bx + c$ in the form

$$y = \frac{-b^2 + 4ac}{4a} + a\left(x + \frac{b}{2a}\right)^2$$

32. From properties (1) and (5) prove Theorem 15.

33. From properties (1) and (7) prove Theorem 15.

5.13 Polynomial Equations of Higher Degree

We now turn to the solution of polynomial equations $P(x) = 0$, where $P(x)$ is of arbitrary degree $n(\geq 1)$. It would be reasonable to search for a formula like the quadratic formula, which would give us the desired solutions automatically, and mathematicians looked long and hard for this. Formulas were, indeed, discovered for polynomial equations of degrees three and four, but these are much too complicated for us to present here. No one was able to find a similar formula for an equation of degree five. Finally, the Norwegian mathematician Abel (1802–1829) proved that it is impossible to find a formula for the solution of such an equation in terms of the usual algebraic operations. Indeed, there is no such formula for polynomial equations of any degree greater than four.

This may well cause us to wonder whether there are, in fact, solutions of polynomial equations of higher degree. Since we had to invent the complex numbers to solve the general quadratic equation, perhaps we must go to some hypercomplex number system to solve equations of higher degree. Fortunately, our fears are groundless, for we have the following theorem, whose proof is too difficult for us to include here.

Theorem 17. Fundamental theorem of algebra. Every polynomial equation $P(x) = 0$ of degree ≥ 1, over the field of complex numbers, has at least one solution.

Now that we are assured of the existence of solutions, let us learn as much about their properties as we can. First let us discuss the *Division Algorithm* for polynomials.

Theorem 18. Division algorithm. If $P(x)$ is a polynomial of degree $n(\geq 1)$ over a field F and b is an element of F, there exists a polynomial $Q(x)$ over F of degree $n - 1$ called the *quotient* and an element R of F (called the *remainder*) such that

$$P(x) = (x - b)Q(x) + R \tag{1}$$

for all x in F.

For the proof see Prob. 44, Sec. 5.13.

Since (1) is true for all x in F, let us substitute b for x in this equation. The result is

$$P(b) = 0 + R \qquad \text{or} \qquad P(b) = R$$

This result is the theorem:

Theorem 19. Remainder Theorem. If $P(x)$ is divided by $(x - b)$ so that a remainder R is obtained, then $P(b) = R$.

Illustration 1. Let $P(x) = x^3 + 2x^2 - 3$. Find $P(x)$ by the Remainder Theorem. By division, we have

$$x^3 + 2x^2 - 3 = (x - 2)(x^2 + 4x + 8) + 13$$

Hence $R = 13$ and $P(2) = 13$. This can be checked by noting that

$$P(2) = 8 + 8 - 3 = 13$$

Theorem 20. Factor Theorem. If r is a solution of $P(x) = 0$, then $(x - r)$ is a factor of $P(x)$.

Proof: The statement "r is a solution of $P(x) = 0$" is equivalent to the statement "$P(r) = 0$". Then divide $P(x)$ by $(x - r)$ as in Eq. (1) and obtain

$$P(x) = (x - r)Q(x) + R$$

By the Remainder Theorem, $R = P(r) = 0$. Hence $(x - r)$ is a factor of $P(x)$.

Exercise A. Prove the converse of the Factor Theorem: If $(x - r)$ is a factor of $P(x)$, then r is a solution of $P(x) = 0$.

Illustration 2. We may use the Factor Theorem to find a polynomial equation with given solutions. Suppose we are given $r_1 = 1$, $r_2 = -2$, $r_3 = 0$ and asked to find an equation with these solutions. From the Factor Theorem, $x - 1$, $x + 2$, and x are factors. Hence an equation with the desired property is

$$(x - 1)(x + 2)x = x^3 + x^2 - 2x = 0$$

Illustration 3. We use the converse of the Factor Theorem to help us solve polynomial equations which we can factor. Consider the problem: Solve

$$(x + 2)(x - 1)(x^2 + x = 1) = 0$$

Solution: Since $x + 2$ and $x - 1$ are factors, we know that two solutions are $r_1 = -2$, $r_2 = 1$. The other solutions are solutions of

$$x^2 + x + 1 = 0$$

or

$$x = \frac{-1 \pm i\sqrt{3}}{2}$$

Theorem 21. Number-of-factors theorem. A polynomial $P(x)$ of degree n (≥ 1) over the complex numbers has exactly n factors of first degree.

Proof: From Theorem 17 there is at least one solution r_1 and therefore one factor $(x - r_1)$ because of Theorem 20. Hence

$$P(x) = (x - r_1)Q(x)$$

where $Q(x)$ is a polynomial of degree $n - 1$. Unless $Q(x)$ is of degree zero, the equation $Q(x) = 0$ also has a solution r_2; thus $Q(x) = (x - r_2)S(x)$, where $S(x)$ is a polynomial of degree $n - 2$. Thus

$$P(x) = (x - r_1)(x - r_2)S(x)$$

Continue this process as long as possible. It must stop when n factors have been obtained, for the product of more factors would have a degree higher than n. Hence we have

$$P(x) = (x - r_1)(x - r_2) \cdot \cdot \cdot (x - r_n)a_0$$

Each of the numbers $r_1, \ldots, r_n$ is a solution of $P(x) = 0$, and so we are tempted to say that $P(x) = 0$ *has n solutions.* This is somewhat misleading, for some of the r's may be equal. We say that r is a solution of $P(x) = 0$ of multiplicity k if $(x - r)^k$ is a factor and $(x - r)^{k+1}$ is not a factor of $P(x)$. With the understanding that we count solutions with their proper multiplicities, it is true that $P(x) = 0$ *has precisely n solutions.*

Illustration 4. In the equation

$$(x - 2)(x - 2)(x + 1)(x + 1)(x + 1) = 0$$

$(x - 2)$ is a factor of multiplicity 2 and $(x + 1)$ is of multiplicity 3. We then say that 2 is a solution of multiplicity 2 and that -1 is a solution of multiplicity 3. The solution set, however, is $\{-1, 2\}$.

Problems 5.13

In Probs. 1 to 10 find a polynomial equation of lowest degree which has the following solutions:

1. $2, -2$.
2. $0, 2$.
3. $1, 2, 3$.
4. $-1, 0, 1$.
5. $2 + 3i$.
6. $2 + 3i, 2 - 3i$.
7. $1, 2, 1 + 2i, 1 - 2i$.
8. $0, i, 1$.
9. $1, -\frac{1}{2} + \frac{i\sqrt{3}}{2}, -\frac{1}{2} - \frac{i\sqrt{3}}{2}$.
10. $1, i, -1, -i$.

In Probs. 11 to 20 find a polynomial equation of third degree which has the following solutions:

11. $-3, \sqrt{3}, -\sqrt{3}$.
12. $2, 2\sqrt{2}, -2\sqrt{2}$.
13. $1, 3 + 2i, 3 - 2i$.
14. $0, 2 + 2i, 2 - 2i$.
15. $1, a + ib, a - ib$.
16. $-1, i, i^3$.
17. 1.
18. $1, 1$.
19. $1, 1, 1$.
20. $1, 1, 1, 1$. (BT)

In Probs. 21 to 28 use the Remainder Theorem to find:

21. $P(1)$, when $P(x) = x^2 + x + 3$.
22. $P(-2)$, when $P(x) = 2x^2 - 3x - 4$.
23. $P(0)$, when $P(x) = 3x^3 + x^2 - 5x$.
24. $P(-1)$, when $P(x) = 2x^3 + x - 1$.
25. $P(2)$, when $P(x) = 3x^4 - x^2 + 15$.
26. $P(-2)$, when $P(x) = x^4 - x^3 + x^2 - x + 1$.
27. $P(i)$, when $P(x) = x^2 + x$.
28. $P(-i)$, when $P(x) = x^2 - x + 1$.

In Probs. 29 to 32, by using the converse of the Factor Theorem, find all the solutions of:

29. $x^3 - 8 = 0$.
30. $x^4 - 81 = 0$.
31. $x^4 - 2x^2 + 1 = 0$.
32*. $x^5 + x^3 + x = 0$.

33. How many solutions does the equation $x^3 - 1 = 0$ have? Hence, how many numbers are cube roots of 1? Find them.
34. Find all fourth roots of 1.

In Probs. 35 to 40 find the number of solutions.

35. $x^2 + 3x^4 - x + 5 = 0$.
36. $x^5 - x^2 + 1 = 0$.

37. $(3 + i)x^2 + (4 - i)x^3 + (1 + i) = 0.$
38. $i + 2ix + 3x^2 + x^4 = 0.$
39. $\sqrt{x^2 - 1} = 4.$ (BT)
40. $x^2 - 2^x - 1 = 0.$ (BT)
41*. Show that a polynomial equation of degree n cannot have more than $n/2$ solutions of multiplicity 2.
42*. As a consequence of Theorem 21 show that if

$$P(x) = a_0x^n + a_1x^{n-1} + \cdots + a_{n-1}x + a_n = 0$$

has $n + 1$ solutions, then each coefficient a_i is zero.
43*. Prove that $P(x) = 0$ and $Q(x) = 0$ have all their solutions equal if and only if there is some constant c such that, for all x, $P(x) - cQ(x) = 0$.
44*. Prove the Division Algorithm by completing the following outline. Let

$$P(x) = a_0x^n + a_1x^{n-1} + \cdots + a_{n-1}x + a_n$$
$$Q(x) = c_0x^{n-1} + c_1x^{n-2} + \cdots + c_{n-2}x + c_{n-1}$$

Then, for $b \neq 0$,

$$(x - b)Q(x) = c_0x^n + (c_1 - bc_0)x^{n-1} + (c_2 - bc_1)x^{n-2} + \cdots + (c_{n-1} - bc_{n-2})x + (-bc_{n-1})$$

Hence

$$a_0 = c_0 \qquad a_1 = c_1 - bc_0 \qquad \cdots \qquad a_{n-1} = c_{n-1} - bc_{n-2} \qquad a_n = R - bc_{n-1}$$

These can be solved in turn to find $c_0 \ldots c_{n-1}$ and R. Complete the proof for $b = 0$.

5.14 Synthetic Division

The Remainder Theorem gives us a convenient short cut for finding the value $P(b)$, say, for it tells us that $P(b) = R$ and R is easy to compute. To perform the division called for in the Remainder Theorem, we use a short method, called synthetic division. To illustrate the method, we consider the case of the general cubic (complex) polynomial

$$P(x) = a_0x^3 + a_1x^2 + a_2x + a_3$$

which is to be divided by $x - b$. The work is exhibited in all detail below, where R is the remainder.

$$
\begin{array}{l|l|l}
x - b & a_0x^3 + a_1x^2 + a_2x + a_3 & a_0x^2 + (a_0b + a_1)x + (a_0b^2 + a_1b + a_2) \\
& a_0x^3 - a_0bx^2 & \\
& \quad (a_0b + a_1)x^2 + a_2x & \\
& \quad (a_0b + a_1)x^2 - (a_0b^2 + a_1b)x & \\
& \qquad (a_0b^2 + a_1b + a_2)x + a_3 & \\
& \qquad (a_0b^2 + a_1b + a_2)x - (a_0b^3 + a_1b^2 + a_2b) & \\
& \qquad\qquad a_0b^3 + a_1b^2 + a_2b + a_3 = R &
\end{array}
$$

But, surely, we have written down more detail than we actually need; the following, where we have suppressed every x, is quite clear:

$$\begin{array}{l} -b \,|\, a_0 + a_1 + a_2 + a_3 \,|\, a_0 + (a_0b + a_1) + (a_0b^2 + a_1b + a_2) \\ \quad - a_0b \\ \quad (a_0b + a_1) + a_2 \\ \qquad - (a_0b^2 + a_1b) \\ \qquad (a_0b^2 + a_1b + a_2) + a^3 \\ \qquad\qquad - (a_0b^3 + a_1b^2 + a_2b) \\ \qquad\qquad a_0b^3 + a_1b^2 + a_2b + a_3 = R \end{array}$$

We have also omitted the second writing of a_0, $a_0b + a_1$, and $a_0b^2 + a_1b + a_2$ inasmuch as they are going to cancel by subtraction anyway. We shall further simplify the process by changing the sign of $-b$ to $+b$ in the divisor, and hence the subtractive process to an additive one. Also, there is no need of writing the quotient Q in the little box to the right since every term there is to be found in the work below, which is finally written on just three lines:

$$\begin{array}{c|lllll} b & a_0 & a_1 & a_2 & a_3 & \\ & & a_0b & a_0b^2 + a_1b & a_0b^3 + a_1b^2 + a_2b & \\ \hline & a_0 & a_0b + a_1 & a_0b^2 + a_1b + a_2 & a_0b^3 + a_1b^2 + a_2b + a_3 & = R \end{array}$$

with no

Although we have skeletonized the work, the details can still be extracted: We are dividing $a_0x^3 + a_1x^2 + a_2x + a_3$ by $x - b$, and we get a quotient of $a_0x^2 + (a_0b + a_1)x + (a_0b^2 + a_1b + a_2)$ and a remainder of $a_0b^3 + a_1b^2 + a_2b + a_3$. Note that the remainder is $P(b)$, as it should be by the Remainder Theorem.

Illustration 1. Divide $x^4 - 3x^3 + x + 3$ by $x - 2$ synthetically.

Solution: Form the array, noting that the coefficient of x^2 is zero. (We normally place the "2" associated with the divisor $x - 2$ on the right.)

$$\begin{array}{rrrrrl} 1 & -3 & 0 & 1 & 3 & |\underline{2} \\ & 2 & -2 & -4 & -6 & \\ \hline 1 & -1 & -2 & -3 & -3 & = R \end{array}$$

The quotient $Q(x) = x^3 - x^2 - 2x - 3$, and the remainder $R = -3$. By direct computation we also find that $P(2) = -3$.

Illustration 2. Given $P(x) = 3x^4 - 4x^3 - 2x^2 + 1$, compute $P(-1)$, $P(0)$, $P(1)$, $P(2)$, $P(3)$, and $P(-0.2)$, and sketch $y = P(x)$.

Solution: Directly from $P(x)$ we compute $P(0) = 1$, $P(1) = -2$, and $P(-1) = 6$. In the slightly more complicated cases of $P(2)$, $P(3)$, and $P(-0.3)$, we use synthetic division:

3	−4	−2	0	1	|2
	6	4	4	8	
3	2	2	4	9 = $P(2)$	

3	−4	−2	0	1	|3
	9	15	39	117	
3	5	13	39	118 = $P(3)$	

3	−4	−2	0	1	|−0.3
	−0.9	1.47	0.159	−0.0477	
3	−4.9	−0.53	0.159	0.9523 = $P(-0.3)$	

The preceding table is self-explanatory. Note especially the value $P(-0.3) = 0.9523$

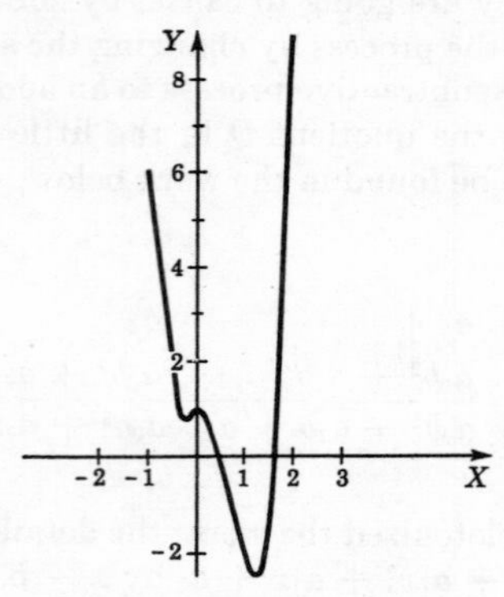

Figure 5.25

and the corresponding dip in the graph (Fig. 5.25). This kind of variation cannot be discovered in general without the methods of the calculus (Chap. 11).

Exercise A. Compute $P(-1)$ by synthetic division, and note the alternating signs in the last line of your work. Explain why this tells us there are no real solutions of $P(x) = 0$ to the left of $x = -1$. Generalize for the case where $P(x)$ is a real polynomial.

Exercise B. Examine the line where $P(2) = 9$, and state what follows about the real solutions of $P(x) = 0$ to the right of $x = 2$. Generalize for the case where $P(x)$ is a real polynomial.

Problems 5.14

In Probs. 1 to 10 compute by synthetic division:

1. $P(2)$, when $P(x) = x^3 - 2x^2 + x + 1$.
2. $P(-2)$, when $P(x) = 2x^3 + 3x - 4$.
3. $P(-3)$, when $P(x) = x^4 + x^2 - x - 2$.
4. $P(3)$, when $P(x) = x^4 - x^3 + 2$.

5. $P(1.1)$, when $P(x) = x^2 + x + 1$.
6. $P(-1.1)$, when $P(x) = 2x^2 + x - 1$.
7. $P(\frac{1}{3})$, when $P(x) = x^3 + 2x + 1$.
8. $P(\frac{2}{3})$, when $P(x) = x^3 + 2x + 1$.

9. $P(\frac{1}{4})$, when $P(x) = 8x^4 - 14x^3 - 9x^2 + 11x - 2$.
10. $P(\frac{1}{2})$, when $P(x) = 8x^4 - 14x^3 - 9x^2 + 11x - 2$.

In Probs. 11 to 14 find the interval of shortest length with integral end points which contains the real solutions of $P(x) = 0$ and which can be determined by the methods of Exercises A and B above.

11. $P(x) = x^3 - 12x + 4$.
12. $P(x) = 8x^3 - 4x^2 - 18x + 9$.
13. $P(x) = x^4 - 3x^3 + 5x - 6$.
14. $P(x) = x^4 - 8x^3 - x^2 + 1$.

15. Divide $ax^2 + bx + c$ by $x - r$ by long division and by synthetic division. Compare the results.
16. Divide $ax^2 + bx + c$ by $x + r$ by long division and by synthetic division. Compare the results.
17. For what value of k does $x^2 + kx + 4$ yield the same remainder when divided by either $(x - 1)$ or $(x + 1)$?
18. For what value of k is the remainder 7 when $x^2 - 3x + 2k$ is divided by $(x + 2)$?
19. If the polynomial $P(x) = x^4 + Ax^3 + Ax + 4$ is such that $P(2) = 6$, find $P(-2)$.
20. If the solutions of the polynomial equation $x^2 + 2Ax + B = 0$ are equal, find B in terms of A. Find the (equal) solutions.

5.15 Roots of Polynomial Equations

Because of common practice, it is desirable here to introduce some new terms.

Definition: A *root* of a polynomial equation $P(x) = 0$ is a solution of this equation. In this context *root* and *solution* are synonyms. A *zero* of a polynomial $P(x)$ which is defined over a field F is an element $a \in F$ such that $P(a) = 0$. Thus a zero of $P(x)$ is a root of $P(x) = 0$, and vice versa.

Because of its practical importance, much effort has been spent on the question of how to calculate the roots of a polynomial equation. We have mentioned that formulas for the roots exist for $n = 1, 2, 3,$ and 4; but there is no simple method of handling equations of higher degree. The general procedure consists of two steps:

(I) Find all roots which can be obtained by elementary means; then use the factor theorem or other methods to factor the polynomial into polynomials of lower degree.
(II) Find the zeros of the factors by known formulas or by approximate methods.

When the coefficients of $P(x)$ are general complex numbers, there is little that can be said here which will help you in these steps, for the known methods are too complicated to be treated in this book. We can make progress, however, if we consider only polynomials whose coefficients are real numbers. In this case we can prove the theorem:

Theorem 22. A polynomial $P(x)$, with real coefficients, can always be represented as a product of factors each of which is either of the form $ax + b$ or $cx^2 + dx + e$, where a, b, c, d, and e are real numbers.

Proof: We know that the roots of $P(x) = 0$ are complex numbers, but some of them may actually be real. Corresponding to each real root r, the Factor Theorem tells us that there is a factor $(x - r)$. Therefore we can write

$$P(x) = (x - r_1)(x - r_2) \cdots (x - r_s)Q(x) = 0$$

where $r_1, r_2, \ldots, r_s$ are its real roots and $Q(x)$ is a polynomial of degree $n - s$ which has no real zeros. We must show that $Q(x)$ can be factored into quadratic factors of the form $cx^2 + dx + e$.

Suppose that $\alpha + i\beta$ with $\beta \neq 0$ is a root of $Q(x) = 0$. Construct the quadratic polynomial:

$$(x - \alpha - i\beta)(x - \alpha + i\beta) = (x - \alpha)^2 + \beta^2 = S(x)$$

Note that $S(\alpha + i\beta) = 0$ and $S(\alpha - i\beta) = 0$. Now divide $Q(x)$ by $S(x)$, and obtain

$$Q(x) = S(x) \cdot R(x) + px + q$$

Substitute $x = \alpha + i\beta$ into this equation. Since $Q(\alpha + i\beta) = 0$ and $S(\alpha + i\beta) = 0$, we get

$$p(\alpha + i\beta) + q = 0$$

or

$$p\alpha + q = 0 \qquad p\beta = 0$$

Since $\beta \neq 0$, this shows that $p = 0$ and $q = 0$. Therefore $S(x)$ is a factor of $Q(x)$ and hence of $P(x)$. The same process can now be applied to $R(x)$, and we continue until we get

$$P(x) = (x - r_1) \cdots (x - r_s)S_1(x) \cdots S_t(x)a_0$$

where $s + 2t = n$. This is of the required form.

Corollary. If $\alpha + i\beta$ is a root of a real polynomial equation, then $\alpha - i\beta$ is also a root of this equation.

Exercise A. Construct an example which shows this corollary false when the coefficients of $P(x)$ are no longer real.

Exercise B. Show that the degree of $Q(x)$ must be even.

This theorem tells us a lot about the nature of the roots of $P(x) = 0$, but it does not help us to find them. Special methods for finding the roots of certain simple types of equations are given in the next two sections.

Problems 5.15

Solve by factoring:

1. $2x^2 + 5x - 3 = 0$.
2. $3x^2 - 5x - 2 = 0$.
3. $(2x + 1)^2 + 3(2x + 1) - 4 = 0$.
4. $(3x - 1)^2 - 4(3x - 1) - 12 = 0$.
5. $x^3 + 3x^2 - x - 3 = 0$.
6. $x^5 + x^4 - x - 1 = 0$.
7. $1/x^2 - 4 = 0$.
8. $x^4 - 16 = 0$.
9. $x^3 - x = 0$.
10. $4/x^2 - x^2 = 0$.
11. $y^5 - y = 0$.
12. $z^5 + 5z^4 - 6z^3 = 0$.
13. $(x^2 - x - 2)^2 - 14(x^2 - x - 2) + 40 = 0$.
14. $(x^2 + 5x + 4)^2 - 22(x^2 + 5x + 4) + 72 = 0$.

5.16 Rational Roots of Rational Polynomial Equations

We now restrict ourselves to rational polynomial equations, i.e., to polynomial equations of the form $P(x) = 0$, where the coefficients in $P(x)$ are rational numbers.

Exercise A. Show that a rational polynomial can be written in the form $A \cdot P(x)$, where A is a rational number and where $P(x)$ has integer coefficients. Hence show that a given rational polynomial equation has the same roots as a certain polynomial equation in which the coefficients are integers. [Multiplying both sides of an equation by a constant ($\neq 0$) does not change the roots.]

There is a simple method in this case for obtaining quickly all those roots of $P(x) = 0$ which happen to be rational numbers. Of course, there is no necessity that any of these roots be rational; therefore this method may not produce any of the roots at all since it exhibits only the rational roots.

Theorem 23. Rational-root theorem. If

$$P(x) = a_0x^n + a_1x^{n-1} + \cdot \cdot \cdot + a_{n-1}x + a_n$$

has integers for coefficients, and if $r = p/q$ is a rational root (in lowest terms) of $P(x) = 0$, then p is a factor of a_n and q is a factor of a_0.

Proof: We are given that

$$a_0 \frac{p^n}{q^n} + a_1 \frac{p^{n-1}}{q^{n-1}} + \cdots + a_{n-1} \frac{p}{q} + a_n = 0$$

Multiply through by q^n; the result is

$$a_0 p^n + a_1 p^{n-1} q + \cdots + a_{n-1} p q^{n-1} + a_n q^n = 0$$

This may be written

$$p(a_0 p^{n-1} + a_1 p^{n-2} q + \cdots + a_{n-1} q^{n-1}) = -a_n q^n$$

Now p is a factor of the left-hand side of this equation, and therefore p is a factor of the right-hand side, $-a_n q^n$. Since p/q is in lowest terms, p and q^n are relatively prime, and since p is a factor of $a_n q^n$, it follows (from Sec. 3.4, Theorem 4) that *p is a factor of a_n.*

The second equation above can also be written

$$a_0 p^n = -q(a_1 p^{n-1} + \cdots + a_{n-1} p q^{n-2} + a_n q^{n-1})$$

By a similar argument *q is a factor of a_0.*

Illustration 1. Solve the equation $2x^4 + 5x^3 - x^2 + 5x - 3 = 0$.

Solution: The possible rational roots are $\pm 1, \pm 3, \pm \frac{1}{2}, \pm \frac{3}{2}$. Using synthetic division, we find that -3 is a root, for

$$\begin{array}{rrrrr|l} 2 & 5 & -1 & 5 & -3 & -3 \\ & -6 & 3 & -6 & 3 & \\ \hline 2 & -1 & 2 & -1 & 0 & \end{array}$$

Therefore

$$2x^4 + 5x^3 - x^2 + 5x - 3 = (x + 3)(2x^3 - x^2 + 2x - 1)$$

The remaining roots of the given equation are thus roots of the "reduced" equation

$$2x^3 - x^2 + 2x - 1 = 0$$

Its possible rational roots are $\pm 1, \pm \frac{1}{2}$. Using synthetic division, we find that $\frac{1}{2}$ is a root, for

$$\begin{array}{rrrr|l} 2 & -1 & 2 & -1 & \frac{1}{2} \\ & 1 & 0 & 1 & \\ \hline 2 & 0 & 2 & 0 & \end{array}$$

The new reduced equation is

$$2x^2 + 2 = 0 \qquad \text{or } x^2 + 1 = 0$$

This is solved by the usual methods for quadratic equations and yields $x = \pm i$.

The roots of the original equation are therefore $\frac{1}{2}$, -3, i, $-i$. In this case each real root is a rational number.

Problems 5.16

In the following equations find the rational roots and, where possible, solve completely.

1. $x^3 + 3x^2 + 3x + 2 = 0$.
2. $2x^3 - 7x^2 + x + 1 = 0$.
3. $3x^3 + x^2 + 3x + 1 = 0$.
4. $3x^3 - 2x^2 - 3x + 2 = 0$.
5. $2x^4 - x^3 + 2x - 1 = 0$.
6. $2x^4 - x^3 - 2x + 1 = 0$.
7. $x^4 - x^3 - 2x + 1 = 0$.
8. $x^4 + 3x^3 + 2x^2 + x + 2 = 0$.
9. $x^2 - 2 = 0$. How does this prove that $\sqrt{2}$ is irrational?
10. $x^2 - 3 = 0$. How does this prove that $\sqrt{3}$ is irrational?

5.17 Real Roots of Real Polynomial Equations

In Sec. 5.16 we discussed the general method of obtaining the roots of rational polynomial equations when those roots are rational numbers. There is no simple general way in which a root can be determined exactly when it is not rational and when the degree of the polynomial exceeds four. Indeed, about the only method available to us is an approximation method, which is best described as a graphical one. This method will yield those roots of $P(x) = 0$ which are real, but gives no information concerning other roots. The method applies equally to other types of equations, provided that the graphs of these equations are continuous (see Sec. 10.10).

A real root of $P(x) = 0$ corresponds to a value of x at which the graph of $y = P(x)$ crosses or touches the X-axis. Hence the procedure is to construct an accurate graph from which the zeros may be read off (approximately).

Most graphs will be accurate enough only to locate the desired zero between successive integers, and a refined technique is needed to obtain more decimal places. To be definite, suppose that we have located a single (nonmultiple) root between 2 and 3, so that $P(2)$ and $P(3)$ have opposite signs. We may calculate $P(2)$, $P(2.1)$, $P(2.2)$, . . . , $P(2.9)$, $P(3)$, in turn and thus locate the root between the adjacent pair of these which have opposite signs. Since this process is tedious, we try to speed it up graphically by a procedure which suggests which of these tenths to try first.

Suppose the situation is as shown in Fig. 5.26. Draw a straight line between the points [2, $P(2)$] and [3, $P(3)$], and observe where this crosses the axis. Now try tenths in the neighborhood of this crossing. When the root is located between successive tenths, the process may be repeated for hundredths, etc., as far as desired. Usually, however, the graphic method is abandoned after the tenths have been obtained, and refined numerical techniques (beyond the scope of this book) are employed.

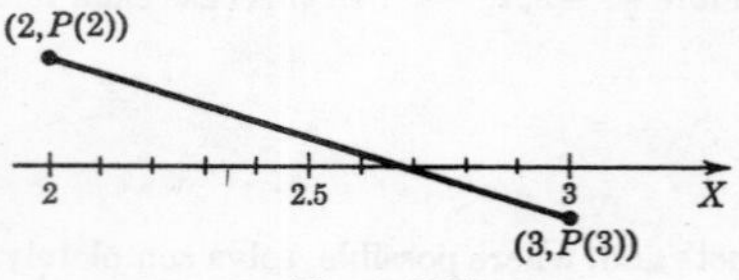

Figure 5.26

We should say a final word about the use of a straight line with which to approximate a (continuous) curve. Our remarks must necessarily be somewhat vague since we have not presented the mathematical background necessary to a full understanding of the problem. (A thorough knowledge of Chap. 11 is a necessary condition for such an understanding.) Consider a continuous curve in a very small interval from $x = a$ to $x = b$, say. It can be proved that if $|b - a|$ is sufficiently small, then $|P(b) - P(a)|$ is small. In effect, this says that a small portion of a decently behaving graph is somewhat like a straight line. This is the basis on which (linear) interpolation is made in various tables (such as a table of logarithms).

Illustration 1. Find the real roots of the equation:

$$y = x^3 - 2x^2 + x - 3$$

Solution: We prepare the table of values

x	-1	1	0	2	3
y	-7	-3	-3	-1	9

and plot the graph as in Fig. 5.27. We see that there is a root between 2 and 3. We

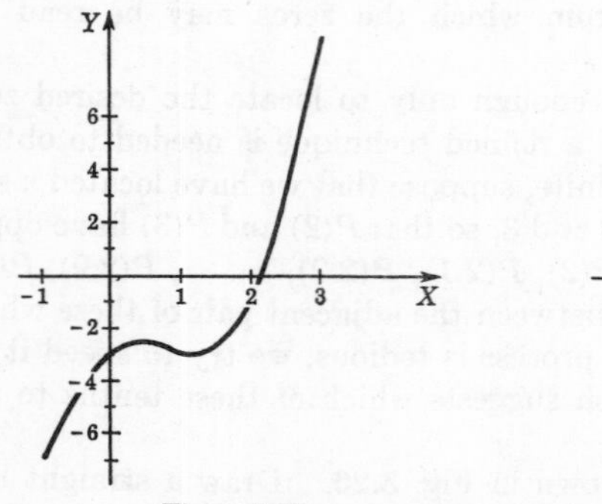

Figure 5.27

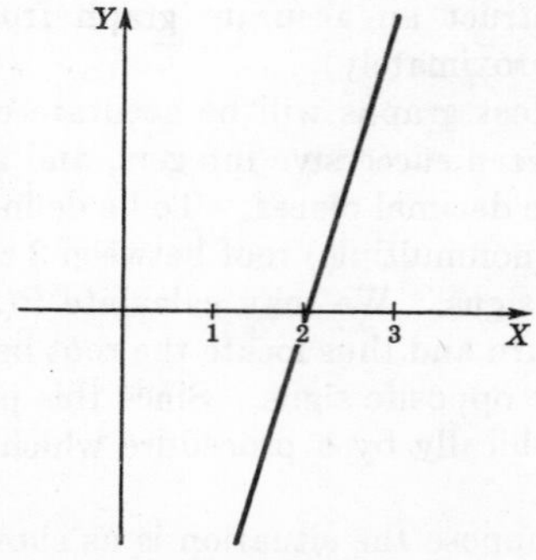

Figure 5.28

cannot prove it with our present knowledge, but this is the only real root of this equation. We plot Fig. 5.28. The line crosses the axis at exactly 2.1; therefore we calculate the following table.

x	2.0	2.1	2.2
y	-1	-0.459	$+0.168$

Thus the root is between 2.1 and 2.2.

To obtain the next decimal place, we plot Fig. 5.29. The line crosses the axis

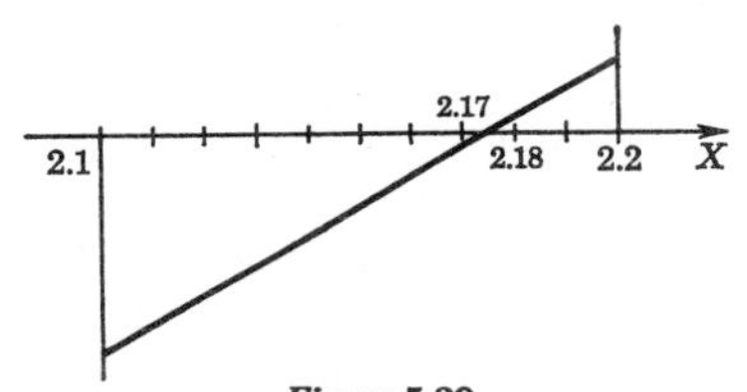

Figure 5.29

between 2.17 and 2.18; therefore we calculate the following table.

x	2.17	2.18
y	-0.03	0.04

Hence the root is 2.17+.

Repeated, this process will determine the decimal expansion of the root in question. But note that, to obtain the best approximation to, say, two decimal places, we should compute the expansion to three places and then round off to two places.

Problems 5.17

Find the first decimal place of the indicated real root x_0 of the equations:

1. $y = x^3 - 3x + 1$, $1 < x_0 < 2$.
2. $y = x^3 + x^2 - 4x - 2$, $0 < x_0$.
3. $y = x^4 - 6x^3 + x^2 + 12x - 6$, $x_0 < 0$.
4. $y = x^4 - 6x^3 + x^2 + 12x - 6$, $2 < x_0$.
5. $y = x^3 + 3x^2 - 4x + 1$, the largest x_0.
6. $y = x^3 + 3x^2 - 4x + 1$, the smallest x_0.
7. $y = x^4 + 10x^3 - 17x^2 + 8x - 1$, the largest x_0.
8. $y = x^4 + 10x^3 - 17x^2 + 8x - 1$, the smallest x_0.
9. $y = x^2 - 2$, $0 < x_0$. Check by solving.
10. $y = x^3 - 2$, $0 < x_0$. Check by solving.
11. $y = x^4 - 2$, $0 < x_0$. Check by solving.
12. $y = x^3 - 3x^2 - 4x + 13$, $x_0 < 2$.

5.18 Equations Containing Fractions

In the next two sections we consider two types of equations which are closely associated with polynomial equations. It is not uncommon to meet equations like

$$\frac{1}{x} + \frac{2}{x+1} = 3 \qquad \text{or} \qquad \frac{x-1}{x+4} - \frac{x+2}{x-3} = 5$$

in which fractions appear whose numerator and denominator are polynomials. The solution of these depends upon one important fact:

Basic principle. Let a/c and b/c be two fractions with equal denominators, $c \neq 0$. Then $a/c = b/c$ if and only if $a = b$. (See Sec. 2.6.)

We apply this principle in the following way: First we express all the given fractions in terms of a common denominator. Then we write the equation obtained by equating the numerators of both sides and solve this equation. This gives us a tentative set of solutions, but there is still the possibility that some or all of these will make one of the denominators of the given fractions equal to zero. Such values must be discarded. It is therefore wise to check all solutions in the original equation before announcing the final answer.

Illustration 1. Solve $\frac{1}{x} + \frac{6}{x+4} = 1$.

In terms of a common denominator this becomes:

$$\frac{(x+4) + (6x)}{x(x+4)} = \frac{x(x+4)}{x(x+4)}$$

Putting the numerators equal, we obtain:

$$7x + 4 = x^2 + 4x$$

or

$$x^2 - 3x - 4 = 0$$

Hence $x = 4, -1$.

Both of these satisfy the given equation.

Illustration 2. Solve $\frac{7}{x-1} - \frac{6}{x^2-1} = 5$.

$$\frac{7(x^2-1) - 6(x-1)}{(x-1)(x^2-1)} = \frac{5(x-1)(x^2-1)}{(x-1)(x^2-1)}$$

$$7(x^2-1) - 6(x-1) = 5(x-1)(x^2-1)$$

or

$$(x-1)(x-2)(5x+3) = 0$$

So $x = 1, 2, -\frac{3}{5}$.

If we put $x = 1$ in the original equation, we obtain $\frac{7}{0} - \frac{6}{0} = 5$, which is certainly false. However, $x = 2$ and $x = -\frac{3}{5}$ do satisfy the original equation. The correct solution set is therefore $\{2, -\frac{3}{5}\}$.

The above method of solution is not as elegant as it might have been, for we did not use the LCD. If we had observed that $x^2 - 1$ serves as the LCD, we should have written:

$$\frac{7(x+1) - 6}{x^2 - 1} = \frac{5(x^2 - 1)}{x^2 - 1}$$
$$(7x + 7) - 6 = 5x^2 - 5$$
$$5x^2 - 7x - 6 = 0$$
$$(x - 2)(5x + 3) = 0$$
$$x = 2, -\tfrac{3}{5}$$

In this case the incorrect solution $x = 1$ does not appear. Hence it is advisable to use the LCD whenever you can find it. But even the consistent use of the LCD will not excuse you from testing every tentative solution. See the next illustration.

Illustration 3

$$\frac{x}{x+2} - \frac{4}{x+1} = \frac{-2}{x+2}$$

Using the LCD $(x + 2)(x + 1)$, we get:

$$\frac{x(x+1) - 4(x+2)}{(x+2)(x+1)} = \frac{-2(x+1)}{(x+2)(x+1)}$$

Equating numerators, we find:

$$x^2 - x - 6 = 0$$

So
$$x = 3, -2$$

Testing, we observe that $x = 3$ satisfies the given equation but that $x = -2$ makes two denominators zero and hence is not a solution. The correct solution set is therefore $\{3\}$.

5.19 Equations Containing Radicals

In this section we are interested in equations like:

$$\sqrt{x + 13} - \sqrt{7 - x} = 2 \qquad \text{or} \qquad 2\sqrt{x + 4} - x = 1$$

in which x appears under a radical. For simplicity we shall consider square roots only. The only possible method of procedure involves squaring both sides and hence is subject to the cautions expressed in Prob. 11, Sec. 5.5.

When there is only one radical in the given equation, write the equivalent equation in which the radical is on one side and all the other terms on the other side. Then squaring both sides removes the radical and leaves an equation without radicals to be solved. Since this equation is not equivalent to the given equation, all solutions must be checked in the given equation.

Illustration 1. Solve: $2\sqrt{x+4} - x = 1$.

$$2\sqrt{x+4} = x + 1$$
$$4(x+4) = x^2 + 2x + 1$$
$$x^2 - 2x - 15 = 0$$
$$(x-5)(x+3) = 0$$
$$x = 5, -3$$

Checking $x = 5$, we have $2\sqrt{9} - 5 = 1$, or $6 - 5 = 1$, which is true
Checking $x = -3$, we have $2\sqrt{1} + 3 = 1$, or $2 + 3 = 1$, which is false.
The solution set of the given equation is therefore $\{5\}$.

When there are two radicals, the method is similar, but two squarings are required. Proceed as in the illustration below.

Illustration 2. Solve: $\sqrt{x+13} - \sqrt{7-x} = 2$.

$$\sqrt{x+13} = 2 + \sqrt{7-x}$$
$$x + 13 = 4 + 4\sqrt{7-x} + 7 - x$$
$$2x + 2 = 4\sqrt{7-x}$$
$$x + 1 = 2\sqrt{7-x}$$
$$x^2 + 2x + 1 = 28 - 4x$$
$$x^2 + 6x - 27 = 0$$
$$(x-3)(x+9) = 0$$
$$x = 3, -9$$

Checking $x = 3$, we have $\sqrt{16} - \sqrt{4} = 2$, or $4 - 2 = 2$, which is true.
Checking $x = -9$, we have $\sqrt{4} - \sqrt{16} = 2$, or $2 - 4 = 2$, which is false.
Hence the solution set is $\{3\}$.

Illustration 3. Solve: $\sqrt{x+1} - \sqrt{x+6} = 1$.

$$\sqrt{x+1} = 1 + \sqrt{x+6}$$
$$x + 1 = 1 + 2\sqrt{x+6} + x + 6$$
$$-6 = 2\sqrt{x+6}$$
$$36 = 4(x+6)$$
$$4x = 12$$
$$x = 3$$

Testing, we find

$$\sqrt{3+1} - \sqrt{3+6} \neq 1$$

Therefore the equation has no solution and the solution set is the null set $\emptyset$.

Problems 5.19

Solve:

1. $\dfrac{6}{x+2} + \dfrac{7}{x} = 9$.

2. $\dfrac{8}{x-2} - \dfrac{4}{x} = 3$.

3. $\dfrac{6}{x-1} + \dfrac{16}{x^2-1} = 5$.

4. $\dfrac{4}{x-3} - \dfrac{16}{x^2-9} = 1$.

5. $\frac{6}{x} - \frac{1}{x-1} - \frac{10}{x+3} = 0.$

6. $-\frac{6}{x} - \frac{1}{x+1} + \frac{20}{x-3} = 0.$

7. $\frac{-4}{x+1} + \frac{5x-5}{x^2+1} = 0.$

8. $\frac{-2}{x+2} + \frac{3x-6}{x^2+2} = 0.$

9. $\frac{2x}{x+2} - \frac{2}{x-2} = \frac{-4}{x+2}.$

10. $\frac{-3x}{x+1} + \frac{6}{x} = \frac{3}{x+1}.$

11. $\sqrt{x+7} + 5x - 13 = 0.$

12. $\sqrt{x-2} - 3x + 16 = 0.$

13. $\sqrt{x+3} + \sqrt{x+8} = 0.$

14. $\sqrt{x-11} + \sqrt{x+4} = 0.$

15. $\sqrt{x+21} - \sqrt{x+14} = 1.$

16. $\sqrt{x+3} + \sqrt{x+15} = 6.$

17. $\sqrt{2x+9} + \sqrt{3x+16} = 7.$

18. $\sqrt{5x-1} - \sqrt{2x-3} = 2.$

19. $\frac{1}{\sqrt{x}} - \frac{2}{\sqrt{x+27}} = 0.$

20. $\frac{x}{\sqrt{x+1}} + \frac{2x}{\sqrt{x+3}} = 0.$

5.20 General Methods of Graphing

In your later work you will find it necessary to plot the graphs of equations which are more complicated than those considered so far in this chapter. In order to do so efficiently you will need some methods which are more sophisticated than those we have been using.

Up to the present, our basic method for plotting a graph has been to find a reasonable number of points (x,y) whose coordinates satisfy the equation. Then we have joined these points by a smooth curve, but in doing so we have run some major risks. How do we know that the graph is actually a smooth curve? Can we be sure that between the plotted points there is no strange behavior of the graph which we have overlooked? Are we sure that the graph does not contain isolated points? Questions like these cannot be answered with precision until you have studied calculus and, in particular, the concept of *continuity*, but we can still give you some hints that will speed your graphing and help you to avoid major blunders.

(1) *Intercepts.* The x-intercepts are the x-coordinates of the points at which the graph crosses (or meets) the X-axis, and the y-intercepts are the y-coordinates of the corresponding points on the Y-axis. To find the x-intercepts, put $y = 0$ in the equation, and solve for x. To find the y-intercepts, put $x = 0$, and solve for y.

Illustration 1. Find the intercepts of the graph of $y = x^2 - 3x + 2$.

Solution: Setting $y = 0$, we find that the solution set of $x^2 - 3x + 2 = 0$ is $\{1, 2\}$. The x-intercepts are 1 and 2. By setting $x = 0$, we find $y = 2$, which is the y-intercept. We now plot the points (1,0), (2,0), (0,2) (Fig. 5.30).

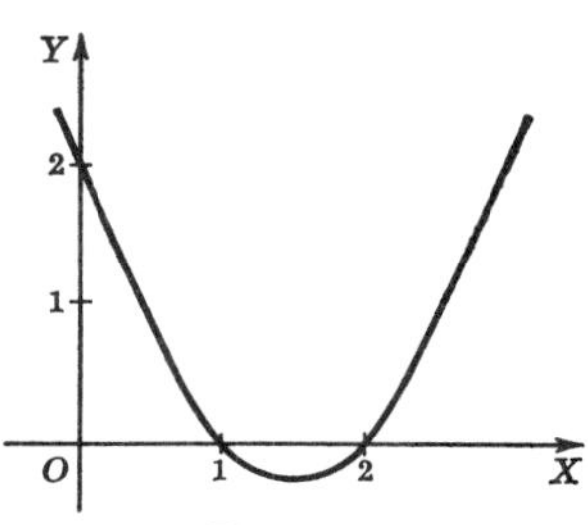

Figure 5.30

In general, if the given equation is of the form $y = P(x)$, the x-intercepts are the zeros of $P(x)$.

(2) *Domain and Range.* In our search for pairs (x,y) which satisfy our equation, we can save time if we know that certain values of x or y are impossible, so that no pairs can contain them. The remaining values which need to be considered form sets called the domain and range.

Definition: The *domain* of the graph of an equation is the subset of X: $\{x \mid x$ is the first element of at least one pair (x,y) which satisfies the equation$\}$. The range is the subset of Y: $\{y \mid y$ is the second element of at least one pair (x,y) which satisfies the equation$\}$.

The domain can be determined fairly easily if we assume that we can solve the given equation for y and express it in the form: $y = F(x)$, where $F(x)$ is an algebraic expression. We then look for values of x to be excluded from the domain. The chief reasons for such an exclusion are the following:

(i) We should be suspicious of $x = a$ if the substitution of a for x makes any denominator in $F(x)$ equal to zero. Usually such values of x must be excluded from the domain, but as in Illustration 2*b*, there are other situations which may arise.

Illustration 2

(a) In the equation

$$y = \frac{x}{x-1}$$

We must exclude $x = 1$ from the domain.

(b) In the equation

$$xy - 4y = \sqrt{x^2 - 7} - 3$$

we solve for y and obtain

$$y = \frac{\sqrt{x^2 - 7} - 3}{x - 4}$$

At $x = 4$, both numerator and denominator of the right side are zero. In such a case we put $x = 4$ in the original equation and attempt to solve for y. This yields

$$0y = 0$$

Any value of y satisfies this equation, so all the points $(4,y)$ are on the graph and we do not exclude $x = 4$.

(ii) We must also exclude from the domain values of x whose substitution into $y = F(x)$ gives us nonreal values for y. This can happen, for example, if $F(x)$ contains a term like $\sqrt{g(x)}$, where $g(x)$ is negative for certain values of x.

Illustration 3. Discuss the domain of $x^2 + y^2 = 4$.

Solution: First we solve for y and obtain

$$y = \pm \sqrt{4 - x^2}$$

We see that y is real if and only if $4 - x^2 \geq 0$, or $-2 \leq x \leq 2$. Therefore we exclude other values of x and state that the domain is $-2 \leq x \leq 2$.

Illustration 4. Discuss the domain of

$$y^2 = (x - 1)(x + 3)$$

Solution: Solving for y, we have

$$y = \pm \sqrt{(x - 1)(x + 3)}$$

We see that y is real if and only if $(x - 1)(x + 3) \geq 0$. Hence the domain is $\{x \mid (x \leq -3) \vee (x \geq 1)\}$.

Illustration 5. Discuss the domain of $y^2 - 2xy + 2x^2 + x + 1 = 0$.

Solution: Solving for y by the quadratic formula, we obtain

$$y = x \pm \sqrt{-x^2 - x - 1}$$

Since $-x^2 - x - 1 < 0$ for all x, no value of x is acceptable. The domain is the empty set, and there is no graph.

To find the range, we solve for x (if possible) and proceed to exclude values of y by the methods just discussed.

In summary, to find the domain, solve the given equation for y and look for trouble. To find the range, solve for x and look for trouble. If you are unable to solve for x or y as directed, these methods no longer are helpful.

Illustration 6. Find the range of the equation (see Illustration 2a)

$$y = \frac{x}{x - 1}$$

Solution: Solving for x, we obtain

$$x = \frac{y}{y - 1}$$

Hence $y = 1$ must be excluded from the range.

Illustration 7. Find the range of $y^2 = (x - 1)(x + 3)$ (see Illustration 4).

Solution: Solving for x, we obtain

$$x = -1 \pm \sqrt{4 + y^2}$$

Since $4 + y^2 > 0$ for all y, there is no restriction on y, and the range is Y.

(3) *Symmetry.* The points (x,y) and $(x,-y)$ are symmetric with respect to the X-axis, the one being the mirror image of the other. Either point is called a *reflection* of the other about the X-axis. The graph will be symmetric about the X-axis if for every point (x,y) on the graph the corresponding point $(x,-y)$ also lies on it. To test for symmetry about the X-axis, we therefore replace y in the equation by $-y$. If the resulting equation is the same as the given one, the graph is symmetric about the X-axis. In particular, the graph is symmetric about the X-axis when y appears in the given equation to an *even* power only, for $y^{2k} = (-y)^{2k}$.

In a similar manner, a graph is symmetric about the Y-axis when replacement of x by $-x$ leaves the equation unchanged, e.g., when x occurs to an even power only.

Further, since a line joining (x,y) and $(-x,-y)$ passes through the origin and the distance from (x,y) to the origin is the same as the distance from $(-x,-y)$ to the origin, the graph will be symmetric about the origin if replacement of (x,y) with $(-x,-y)$ leaves the given equation unchanged.

Exercise A. Examine $|y| - x = 0$, $y - |x| = 0$, $|x| + |y| - 1 = 0$ for symmetry.

Exercise B. Show that, if there is symmetry with respect to both axes, there is necessarily symmetry with respect to the origin, but not conversely.

Exercise C. Show that the graph of $y = (x - a)^4 + 3(x - a)^2 + 5$ is symmetric about the line $x = a$.

Illustration 8

(a) The graph of $x^2 - x + y^4 - 2y^2 - 6 = 0$ is symmetric about the X-axis, but not about the Y-axis or the origin.
(b) The graph of $x^2 - x^4 + y - 5 = 0$ is symmetric about the Y-axis, but not about the X-axis or the origin.
(c) The graph of $x^4 + 2x^2y^2 + y^4 - 10 = 0$ is symmetric about both axes and the origin.
(d) The graph of $xy = 1$ is symmetric about the origin, but not about either axis.

(4) *Asymptotes.* When we solve the given equation for x or y, we may get an expression which contains a variable in the denominator. For example, we may have

$$y = \frac{x}{x - 1}$$

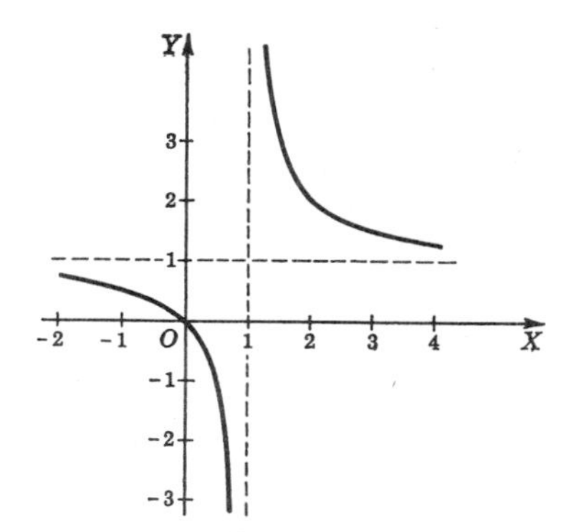

Figure 5.31

We have seen before that we cannot substitute $x = 1$ on the right, for this would make the denominator zero. We can, however, let x take values nearer and nearer to 1 and see how the graph behaves. Construct the table of values:

x	1.1	1.01	1.001	1.0001
y	11	101	1,001	10,001

It is clear that, as x approaches 1 from the right, y is becoming very large in the positive direction. Similarly, as x approaches 1 from the left, y becomes very large in the negative direction (Fig. 5.31).* The line $x = 1$ is called a *vertical asymptote.*

If we solve the above equation for x, we obtain

$$x = \frac{y}{y - 1}$$

The same argument can now be applied to show that $y = 1$ is a horizontal asymptote.

To find asymptotes, the procedure is as follows: Solve for y and x if possible. Examine values of x or y which make the corresponding denominator zero to see whether or not there is an asymptote there. The behavior of the graph near an asymptote must be determined by examining points near it, as was done above.

There is a more general definition of asymptote which applies to lines in other directions, but we shall not give it here.

Illustration 9. Find the horizontal and vertical asymptotes, if any, of

$$x = \frac{y(y - 1)}{y + 2}$$

Since the denominator is zero for $y = -2$, there is a horizontal asymptote at $y = -2$. Solving for y, we find:

$$y = \frac{1 + x + \sqrt{x^2 + 10x + 1}}{2}$$

Since x does not appear in the denominator, there are no vertical asymptotes.

* The language here is very imprecise, but is the best that can be presented to you at this stage. Later (Chap. 10) we shall write $\lim_{x \to 1^+} \frac{x}{x - 1} = +\infty$ and $\lim_{x \to 1^-} \frac{x}{x - 1} = -\infty$ and shall define these terms more precisely.

In the illustrations below we shall use these methods as needed to plot several graphs.

Illustration 10. Plot the graph of $x^2 + y^2 = 4$.

The x-intercepts are ± 2; the y-intercepts are ± 2. The domain is $-2 \leq x \leq 2$; the range is $-2 \leq y \leq +2$. There is symmetry with respect to both axes. There are no asymptotes.

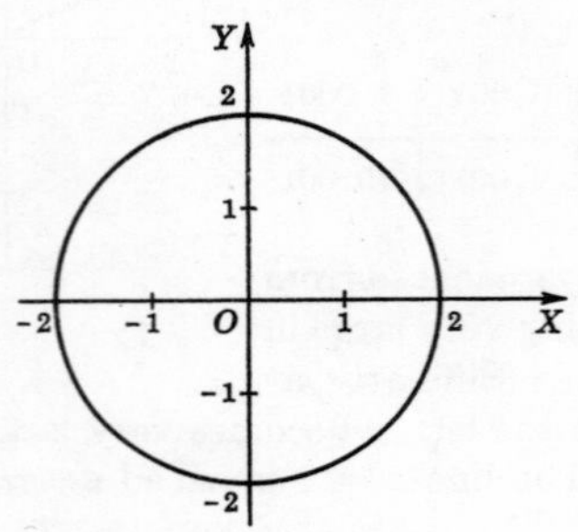

Figure 5.32

A short table of values is

x	0	1	2
y	± 2	$\pm\sqrt{3}$	0

The graph is plotted in Fig. 5.32.

Illustration 11. Plot the graph of $y^2 = (x - 1)(x + 3)$.

The x-intercepts are 1 and -3; there is no y-intercept. The domain is the set

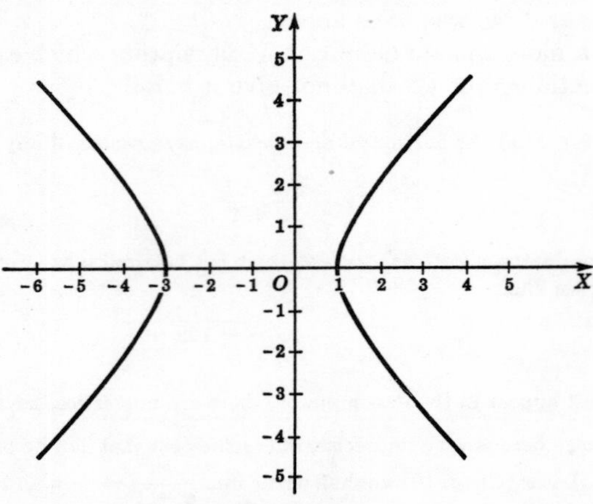

Figure 5.33

$\{x \mid (x \leq -3) \vee (x \geq 1)\}$. The range is Y. There is symmetry with respect to the X-axis. There are no asymptotes.

A short table of values is

x	2	3	4	-4	-5	-6
y	± 2.2	± 3.5	± 4.6	± 2.2	± 3.5	± 4.6

The graph is plotted in Fig. 5.33.

Illustration 12. Plot the graph of

$$y^2 = \frac{x + 3}{(x + 2)(x - 1)}$$

The x-intercept is -3; there is no y-intercept.

The domain is $\{x \mid (-3 \leq x < -2) \vee (1 < x)\}$. The range is Y. There is symmetry with respect to the X-axis. There are vertical asymptotes at $x = 1$ and $x = -2$. There is a horizontal asymptote at $y = 0$.

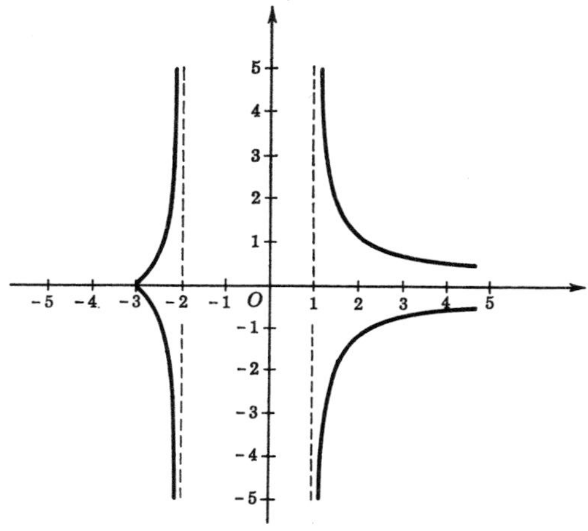

Figure **5.34**

We construct the table of values:

x	2	3	4	-2.5
y	± 1.1	± 0.8	± 0.6	± 0.5

The graph is plotted in Fig. 5.34.

Problems 5.20

In Probs. 1 to 28 find the intercepts, domain, and range (if possible), symmetry (if it exists), and asymptotes (if any). Sketch.

1. $9x^2 + 4y^2 = 36$.
2. $25x^2 + 9y^2 = 225$.
3. $y = 2/(x + 3)$.
4. $y = 4/(x - 1)$.
5. $y = x - x^3$.
6. $y = x^3 - 1$.
7. $y = x^3 + 2x^2 - x - 2$.
8. $y = x^3 + x^2 - 4x - 4$.
9. $y = x^4 - x^2$.
10. $y = x^4 + x^2$.
11. $y^2 = x - x^3$.
12. $y^2 = x^3 - 1$.
13. $y^2 = x^4 - x^2$.
14. $y^2 = x^4 + x^2$.
15. $y^2 = x^2/(1 - x)$.
16. $y^2 = x^2/(1 - x^2)$.
17. $y = 8/(4 + x^2)$ [The Witch of Agnesi].
18. $y^2 = x^3/(2 - x)$ [Cissoid of Diocles].
19. $\sqrt{x} + \sqrt{y} = 1$ [Portion of a parabola].
20. $\sqrt[3]{x} + \sqrt[3]{y} = 1$.
21. $x^{\frac{2}{3}} + y^{\frac{2}{3}} = 1$ [Hypocycloid].
22. $y^2 + 4x^2 = 0$.
23. $y = \dfrac{x+1}{x-2}$.
24. $y = \dfrac{x-3}{x+2}$.
25. $y^2 = \dfrac{x+1}{x-2}$.
26. $y^2 = \dfrac{x-3}{x+2}$.
27. $y = \dfrac{x(x-1)}{x+2}$.
28. $y^2 = \dfrac{x+3}{(x+2)(x-1)}$.

29. Sketch the graph defined jointly by $y = 4/(x^2 + 3x)$ and the condition $y > 1$.
30. Discuss the graph of $y = x^n$, n an even integer, again for n odd.

In Probs. 31 to 40 plot the graph of the given inequality.

31. $x^2 + y^2 < 9$.
32. $x^2 + y^2 > 9$.
33. $9x^2 + 4y^2 \geq 36$. (See Prob. 1.)
34. $25x^2 + 9y^2 \leq 225$. (See Prob. 2.)
35. $y \neq 2/x + 3$.
36. $y \neq 4(x - 1)$.
37. $x^{\frac{2}{3}} + y^{\frac{2}{3}} > 1$. (See Prob. 21.)
38. $y^2 > x^4 + x^2$. (See Prob. 14.)
39. $4x^2 + 9y^2 \leq 0$.
40. $x^2 + y^2 < 0$.

References

Fine, Henry B: "College Algebra", chaps. 29, 30, Dover, New York (1961)
Kemeny, John G., J. Laurie Snell, and Gerald L. Thompson: "Finite Mathematics", chap. 6, Prentice-Hall, Englewood Cliffs, N.J. (1956).
Weisner, Louis: "Theory of Equations", chaps. 4, 5, 7, 8, Macmillan, New York (1938).

Functions | 6

6.1 Examples and Definitions

So far we have been considering pairs of numbers (x,y) which were related to each other by means of an equation. Now we turn to a more general relationship between two sets of numbers, which we call a *function*. In doing so we are breaking away from the subject of algebra and are beginning a new mathematical subject called *analysis*. Let us first illustrate a few functions before we come to a formal definition.

Illustration 1. The rate of postage on first-class mail is 5 cents per ounce or fraction thereof. Consider the sets:

$$\begin{aligned} W &= \{w \mid w \text{ is weight of a letter in ounces}\} \\ &= \text{set of positive real numbers, } w \\ P &= \{p \mid p \text{ is a possible amount of postage in cents on a letter}\} \\ &= \{5, 10, 15, \ldots\}. \end{aligned}$$

To each element of W there corresponds a unique element of P, and we say that "weight determines postage" and write:

$$W \to P \qquad \text{or} \qquad \text{weight} \to \text{postage}$$

The precise relationship is given if we write down the set of corresponding pairs (w,p), a few of which are: $(\frac{1}{2},5)$, $(1,5)$, $(6,30)$, $(6\frac{3}{4},35)$, etc.

The arrow used here does not have the same meaning as the one used for "implication". However, there is some similarity: "weight → postage" can be read, "If a weight is given, *then* the postage is determined".

Illustration 2. Consider the relation given by the following table, which pairs a value r with the corresponding value l:

r	10	25	50	75	100
l	1.0000	1.3979	1.6990	1.8751	2.0000

(Those of you who have studied logarithms may think the table looks familiar.) Let us suppose for the moment that interpolation in this table is meaningful, and consider the two sets: $R = \{r \mid 10 \leq r \leq 100\}$, where r is real, and $L = \{l \mid 1 \leq l \leq 2\}$, where l is real. Then we say that "to a given value r of R there corresponds a unique value of L" and write

$$R \rightarrow L$$

We note that the table thus provides us with a set of ordered pairs (r,l).

Illustration 3. Let A be the set of all integers no one of whose squares exceeds 15, and let B be the set of the cubes of every such integer. There is no pairing of the elements of these two sets until a rule is stated which associates a given element a of A with a unique element b of B. Let a rule to do this pairing be: the b corresponding to a given a shall be the cube of a. We write

$$b = a^3 \qquad \text{for } a = -3,\ -2,\ -1,\ 0,\ 1,\ 2,\ 3$$

which is an algebraic way of describing the relationship given above, in words. We say that "b depends upon a" and write

$$A \rightarrow B$$

Exercise A. Write down the seven pairs (a,b) for Illustration 3.

Illustration. 4. Consider $x \in X$, $y \in Y$, where X and Y represent the set of real numbers. Suppose that, for a given x, we always choose $y = x^2 + 4$. Then x determines y, which is the real number $x^2 + 4$. We write $X \rightarrow Y$ and say that "to every real number x of the set X there corresponds a real number y of the set Y, namely, $y = x^2 + 4$". Some pairs are $(-4,20)$, $(0,4)$, $(5,29)$, $(\pi, \pi^2 + 4)$, $(\sqrt{2},6)$.

In the above illustrations we have dealt with numbers. However, the notion of function, which is to be defined later, is not restricted to sets of numbers. It may apply to any pair of sets whatsoever, as in Illustration 5.

Illustration 5. Consider T the set of towns in Pennsylvania and P the set of locations marked "towns" on a detailed map of Pennsylvania. Since the map position of a town is determined by its actual geographical position, we write

$$T \rightarrow P \qquad \text{or} \qquad \text{geographical position} \rightarrow \text{map position}$$

and say that "To a geographical position there corresponds a map position". We have pairs of the form (geographical position, map position). Because of examples like this one, the relationships we are illustrating are sometimes called "mappings"

Let us examine simultaneously the above five illustrations in order to determine what abstract principles they hold in common. In the first place, in each case there are two initial sets involved. Let us designate these by X and Y.† Then some rule is given which determines a unique

† These do not necessarily refer to sets of real numbers.

element y of Y whenever an element x of X is selected. These rules in the illustrations given are (1) "postage is 5 cents per ounce or fraction", (2) a "table" (we do not need to know for the moment how the table was made), (3) "$b = a^3$, for $a = -3, -2, -1, 0, 1, 2, 3$", (4) "$y = x^2 + 4$", and (5) "geographical position determines map position", respectively. A rule thus generates a third set, namely, the set of ordered pairs (x,y).

Exercise B. Write down some of the ordered pairs in each of the above five illustrations.

Putting these ideas together we have our first definition of a function.

Definition: A *function* f is a relationship between two sets: (1) a set X called the *domain of definition* and (2) a set Y called the *range*, or *set of values*, which is defined by (3) a rule that assigns to each element of X a unique element of Y.

This definition may be more compactly stated as follows:

Definition: A *function* f is a set of ordered pairs (x,y) where (1) x is an element of a set X, (2) y is an element of a set Y, and (3) no two pairs in f have the same first element.

For illustrations 1 to 3 above we summarize these notions in the following table:

	1	2	3
Function f	Set of all ordered pairs (w,p) following:	Set of all ordered pairs (r,l) given by the table; also all pairs that could be obtained from:	Seven ordered pairs $(0,0)$, $(2,8)$, $(-3,-27)$, $(1,1)$, $(3,27)$, $(-2,-8)$, $(-1,-1)$ (order in which elements of f are listed, not important)
Rule determining f	Rule: postage p is 5¢ per oz weight w or fraction thereof	Mathematical formula from which this particular table was originally prepared	$b = a^3$, for a given a
Domain of f	All real positive numbers w	All real numbers $10 \leq r \leq 100$	Numbers (a) 1, 2, 3, 0, −1, −2, −3 (order not important)
Range of f	All numbers p of form $p = 5n$, where n is a positive integer	All real numbers $1 \leq l \leq 2$	Numbers (b) 8, −1, 0, 27, −27, −8, 1 (order not important)

Exercise C. Make out a similar table for Illustrations 4 and 5.

We recall from Sec. 1.5 that a *variable* is a symbol for which any element of a given set may be substituted. In the case of a function, we say that x (which refers to the domain of definition) is the *independent variable* and that y (which refers to the range) is the *dependent variable*. The dependent variable y is also called the *value of the function* f at x and is represented by $f(x)$, which is read "f of x", or better, "f at x".

We should pause to clarify the case where the rule is a table. Reconsider Illustration 2. Actually, what we had given there was a portion of a table of common logarithms with a characteristic supplied. A complicated mathematical formula was used to compute the entries in the table. We should therefore consider the formula as the rule rather than the table. However, we might be given a table such as the following:

x	-3	-2	-1	0	1	2	3
$f(x)$	0	27	-8	1	-27	8	-1

where we are told that the function is completely defined by the table. If this is the case, then the function f is the set $(-3,0)$, $(-2,27)$, $(-1,-8)$, $(0,1)$, $(1,-27)$, $(2,8)$, $(3,-1)$. (Compare this function with that of Illustration 3.) Here $f(\frac{3}{2})$ has no significance; neither has $f(4)$. In other words, interpolation and extrapolation are meaningless; the complete domain of definition is the set $\{0, 1, 2, 3, -1, -2, -3\}$, the order of listing being unimportant, and the range is the set $\{-1, -27, 0, 8, -8, 27\}$, again the order of listing being unimportant. The pairing which establishes an element of f is important, but not the order in which the pairs are written down.

However, interpolated and extrapolated values would have meaning if we were told that the above table was not complete, that it constituted but a few of the possible entries, that the domain of the function f was actually all real numbers between -3 and $+3$, inclusive, and that $f(x)$ represented, really, the number of feet above the 1,000-ft level of a blimp x min after 12 o'clock. The negative values of $f(x)$ would represent the number of feet below the 1,000-ft mark, and the negative values of x would represent minutes before 12 o'clock. Presumably the blimp moves up and down in some curious way—buffeted by gusts of wind, perhaps—but in a manner which we loosely describe as "continuous". Here we cannot hope to know the function completely. We take some readings and interpolate and extrapolate for others. We say that the table is an approximation to some function f.

Exercise D. Given that a function f is completely defined by the following table:

x	0	2	3
$f(x)$	-2	4	$\sqrt{10}$

Write down the function. State the rule that determines this function. What is the domain of definition, and what is the range? What is the meaning of $f(0)$? What is the meaning of $f(1)$?

A particularly simple function is the set of ordered pairs (x,c), where X is the set of real numbers and Y consists of the single real number c. Such a function is called a constant (function). A more general definition is the following one:

Definition: The function $f:X \to Y$ is called a constant function if Y consists of a single element.

6.2 Notations for a Function

The study of functions is complicated by the variety of notations which are in common use. Let us discuss some of these here.

A most natural notation is $f:X \to Y$, read "the function f which maps X onto Y". When we wish to consider several functions having the same domain X and range Y, we use symbols such as $g:X \to Y$, $F:X \to Y$, $\phi:X \to Y$, etc. This notation, while not in common use at an elementary level, is suggestive, however, of a simple geometric interpretation and leads directly to some very modern phraseology. We have a set X, the elements of which are represented by points on a vertical line (Fig. 6.1) and similarly for Y; we have a rule which assigns a unique y to a given x. We write $f:X \to Y$ and say, "The set X is mapped, by means of f, onto the set Y". It is implied that, for an element (x_1,y_1) of the function f, the first element x_1 is mapped onto the second element y_1, etc. Thus the rule which determines the ordered pairs of the function may be thought of as a "mapping", or a "transformation", which carries a given first element into the corresponding second element. The language "The set X is mapped, by means of f, onto the set Y" sounds almost as if the function f were being identified with the rule. We do not mean this to be the case: the function f is the set of ordered pairs, the pairing being established by a rule. However, the more standard notations used in most books and research papers do almost make such an identification.

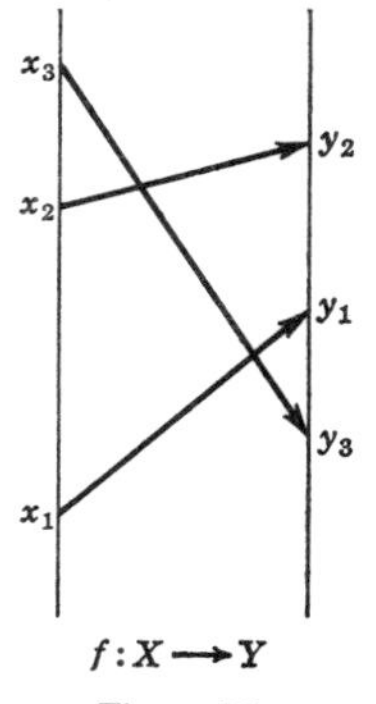

$f:X \to Y$

Figure **6.1**

Another appropriate notation for a function is $f:(x,y)$, read "the function f whose ordered pairs are (x,y)". Similarly, we can write $g:(A,r)$, or $\phi:(u,v)$, etc., to represent other functions.

Often, however, we need a notation which tells us precisely what function we are dealing with. This will vary with the circumstances, but there is one situation which is common enough to deserve mention here. Let us consider a function f whose domain and range are subsets of the real numbers and whose rule is given by an equation such as $y = 2x^2 - 5x + 1$. We may write this function in any of the following ways:

(**a**) $y = 2x^2 - 5x + 1$.
(**b**) $f(x) = 2x^2 - 5x + 1$.
(**c**) $f:X \to Y$ is the function whose values are given by $f(x) = 2x^2 - 5x + 1$.
(**d**) $f:(x,y)$ is the function whose ordered pairs are $(x, 2x^2 - 5x + 1)$.
(**e**) $\{(x,y) \mid y = 2x^2 - 5x + 1\}$.

Since the last three, though complete, are long to state, mathematicians generally abbreviate and use (*a*) or (*b*). Note carefully that, when we write

$$f(x) = 2x^2 - 5x + 1$$

and say that "$f(x)$ is the function $2x^2 - 5x + 1$", we are using the symbol $f(x)$ in two senses: First, it stands for the function itself. Second, it stands for the value of f corresponding to a particular value of x. This dual usage is thoroughly established in mathematical literature, but the ambiguity rarely causes trouble. (We shall be guilty of using it on occasion.) You should bear this in mind, however, as you progress with your mathematical studies.

You may be interested to know that, before the day of the Prussian mathematician Dirichlet (1805–1859), a function had to be expressible in terms of some mathematical formula using $+$, $-$, $\times$, $\div$, etc. Dirichlet himself dismissed this requirement and gave the first modern definition. It was: "$g(x)$ is a real function of a real variable x if to every real number x there corresponds a real number $g(x)$." Note the dual usage of the symbol $g(x)$; it has stood for more than a hundred years.

Exercise A. Analyze carefully Dirichlet's definition, comparing it with ours.

When we use this method of defining a function we leave open the question of its precise domain and range. We shall understand that these are the entire X- and Y-axes, with the exception of those points which are excluded for reasons given in Sec. 5.20. We can then find the domain by using the techniques of that section. If we can solve for x, we can find the range in a similar fashion; otherwise the range may be difficult to determine.

Problems 6.2

1. If $f(x) = 2x^3 - 3x^2 + 1$, find $f(-1)$, $f(3)$, $f(0)$.
2. If $f(x) = x^4 - 2x - 3$, find $f(-1)$, $f(0)$, $f(4)$.
3. The following table completely defines a function:

x	1	3	4
y	0	2	3

Write down the elements of this function.
4. Write down the elements of f where f is completely determined by the table

x	2	3	7	9
y	0	-1	0	-1

5. Which of the following tables define a function $f:(x,y)$?

(a)

x	2	2
y	4	1

(b)

x	1	3
y	-1	-1

(c)

x	1	1
y	1	2

6. Which of the following tables define a function $f:(x,y)$? **(BT)**

(a)

x	0	0
y	0	1

(b)

x	0	1
y	0	0

(c)

x	1	1
y	1	1

7. Does the following table define a function? If so, what is the independent variable?

s	0	0
t	1	0

8. Does the following table define a function? If so, what is the dependent variable?

w	1	2
z	1	1

9. For f such that $f(x) = (\sqrt{x} - 2x)/3$, compute the values of the function for $x = 0, 1, 2, 3, -1, -2$.
10. Given $f(x) = x^2 + 1$, find:

(a) $f(2)$, $f(6)$.

(b) $f(1 + \sqrt{2})$, $f(t)$.

11. Given $f(x) = 1 - x^2$, find:

(**a**) $f(a + h) - f(a - h)$. (**b**) $3f(1) + f(2) - f(3)$.

12. Construct a table showing the average number (assumed) of home accidents in the United States per day of the week. Will this table constitute a function? Explain.

13. Why must $x = 3$ be excluded from the domain of the function f defined by $f(x) = 2x/(x - 3)$?

14. A bookstore reduces the price of a certain book for quick sale. In the first 5 days of the sale, the numbers sold are, respectively, 3,000, 5,000, 1,000, 700, and 500. Write this in functional notation. State domain and range.

15. A packing house processes 200 animals each hour of the day. Choose variables, units, etc., and write in some functional notation.

16. If f is a function defined by $f(x) = (x - 1)/(x - 1)$, should $x = 1$ be included in the domain?

17. What is the range of the function defined by $f(x) = x^2 - 6$?

18. What is the range of the function defined by $f(x) = -\sqrt{9 - x^2}$?

19. Find the value of $\dfrac{f(2 + h) - f(2)}{h}$ for the functions defined by:

(**a**) $f(x) = x^2$. (**b**) $f(x) = 4x^2 + 2x - 3$.
(**c**) $y = x$. (**d**) $y = k$, k constant.

20. Find the value of $\dfrac{f(a + h) - f(a)}{h}$ for the functions defined by:

(**a**) $f(x) = x^3$. (**b**) $f(x) = 1/x$.
(**c**) $f(x) = x$. (**d**) $f(x) = k$, k constant.

6.3 Special Functions

In the course of your studies you will meet some of the especially important functions, such as the polynomial, logarithmic, exponential, and trigonometric functions. Now we should like to introduce you to two rather unusual but very useful functions. The domain of each of these functions is the set X of real numbers.

The first special function is the absolute-value function f whose values are given by $f(x) = |x|$, read "$f(x)$ is the absolute value of x", defined by

$$y = |x| = \begin{cases} x & x > 0 \\ 0 & x = 0 \\ -x & x < 0 \end{cases}$$

Some of the elements of the set f are (3,3), (−3,3), (−8,8); regardless of the sign of x in the ordered pair (x,y), the corresponding value y is nonnegative. Sometimes "the absolute value" is called "the numerical value" for this reason.

Exercise A. Show that the two functions f and g whose values, respectively, are given by $f(x) = |x|$ and $g(x) = \sqrt{x^2}$ are identical.

Exercise B. Compute $|x|$ for $x = 3, -1, \pi, 0, -\sqrt{2}, \frac{1}{3}$.

The second special function is the greatest integer function f whose values are given by $f(x) = [\![x]\!]$, read "$f(x)$ is the greatest integer not greater than x". You can easily compute elements of this function which are of the form $(x, [\![x]\!])$. Thus (2,2), (2.1, 2), (2.99, 2), (3,3) are some ordered pairs of this function.

The notations $y = |x|$ and $y = [\![x]\!]$ are probably new to you as they are here used. Study them until you are thoroughly familiar with them. Do not confuse $[\![\quad]\!]$ with the usual parenthetical brackets [ ].

Exercise C. Compute $[\![x]\!]$ for $x = 1.2$, $x = 2$, $x = \pi$, $x = -3$, $x = -2.4$, $x = \sqrt{2}$, $x = |-6.999|$.

Exercise D. Write an equation for the postage function of Illustration 1, Sec. 6.1. HINT: Use $|\quad|$ *and* $[\![\quad]\!]$.

6.4 Relations

So far we have associated a unique element of Y with each element of X. When this is true it has been customary to call the function "single-valued". We shall speak of function in no other case; that is, as in our definitions, a function is defined if and only if to every element of X there is assigned a specific and unique element of Y. But other possibilities do exist when two arbitrary sets are related. There may be two, three, or even infinitely many values of Y for each element of X. In this case, throughout classical mathematical literature, the function is called "multiple-valued". However, many modern writers prefer to say that this correspondence between X and Y is a "relation". We adopt this point of view. For example, consider Illustration 1.

Illustration 1. Let X be the set of all football squads in the country, and let Y be the set of all football players. Then, with each squad (element of X), there is associated something like fifty players (elements of Y). We say that the two sets X and Y are related.

Definition: A relation is a set of ordered pairs. The domain of definition of a relation is the set of all first elements, and the range of a relation is the set of all second elements of the ordered pairs. Note how a relation differs from a function which is a special case of a relation.

Exercise A. Indicate how you could modify Illustration 1 in order to obtain a function from the sets X and Y.

Illustration 2. Consider the equation $y^2 - x = 0$. This equation defines a relation three of whose ordered pairs are (0,0), (1,1), (1, −1). Note that (1,1) and (1, −1) could not be the elements of a function. As a matter of fact, there are infinitely many functions $f,g,h,$ . . . , such as the following, which are special cases of the above relation.

f is the function whose values are given by $f(x) = \sqrt{x}$, $0 \leq x$.
g is the function whose values are given by $g(x) = -\sqrt{x}$, $0 \leq x$.
h is the function whose values are given by

$$h(x) = \begin{cases} \sqrt{x} & 0 \leq x \text{ rational} \\ -\sqrt{x} & 0 < x \text{ irrational} \end{cases}$$

Of all such functions derivable from $y^2 - x = 0$, f and g are of special importance.

Exercise B. Write down three more functions which are special cases of the relation given by $y^2 - x = 0$.

Problems 6.4

1. Which of the following sets of ordered pairs are functions? Which are not?

(**a**) (0,1), (0,2), (1,1). (**b**) (0,1), (1,1), (2,1).
(**c**) (0,2), (0,1), (0, −1). (**d**) (−1,2), (−2,1), (0,0).

2. Which of the following tables define functions? Write down the ordered pairs of the function defined.

(**a**)

t	0	1	2	4	4
s	0	−1	2	−3	4

(**b**)

p	1	0	1
w	1	0	−1

(**c**)

m	−1	0	1
n	1	2	1

3. Let X and Y be sets of real numbers. Which of the following define functions? State the domain, range, and rule of each.

(**a**) $y = x - 1$. (**b**) $y^2 = x - 1$. (**c**) $y^3 = x - 1$.
(**d**) $y = x^2 + 1$. (**e**) $y = x^3 + 1$. (**f**) $y^2 = x^2 + 1$.

4. Show that $y > x$ defines a relation.
5. Given $f(x) = |x|$. What is the domain? The range? State the rule in words.
6. Given $f(x) = [\![x]\!]$. What is the domain? The range? State the rule in words.
7. For $y = f(x) = x - |x| - [\![x]\!]$, compute $f(0)$, $f(1)$, $f(2.5)$, $f(-\sqrt{3})$, $f(-2)$, $f(-4.2)$.
8. The equation $x^2 + y^2 = 16$ defines a relation. What is the relation? Write down three functions which are special cases of this relation. State the domain, range, and rule of each. Assume x and y to be real.
9. Show that a relation is defined by each of the following, where x and y are real numbers:

(**a**) $x - y < 2$.
(**b**) $|x - y| > 1$.
(**c**) $[\![x]\!] + |y| < 1$.

10. Let X be the set "parents" and Y the set "children". Show how this can be thought of as defining a relation.
11. From your own experience construct a nonnumerical relation. State domain, range, and rule. From this relation extract several functions, stating domain, range, and rule for each.

6.5 Rule, Domain, and Range

In the definition of a function in a preceding section we stated that a rule had to be given which would assign a unique y to a particular x. A rule may be given in any one of several ways; for example, the rule may (a) be stated in words, (b) be given by a mathematical formula or an equation, (c) consist of a table, or (d) be presented in the form of a graph.

We have already considered illustrations of (a), (b), and (c). An example of (d), a function given by a graph, is the curve drawn by a continuously recording thermometer which suggests that temperature depends upon the time of day. We shall say little about case (d), although much mathematics has been written on functions defined by graphs.

When the defining rule is stated in words, it will be desirable generally to translate those words into an equation or formula if possible. If the function is given by a table, it is essential to know whether the table is the complete rule or whether it is only an approximation wherein interpolation and extrapolation are permitted. A similar remark applies to case (d). The important thing here is the following: regardless of the form a particular rule takes, it is quite necessary to know in detail the domain of definition of the function and the range of the function. Yet in many cases no mention is made of either domain or range, it being tacitly assumed that they can be determined with relative ease or that they are obvious. For instance, when we say that the area of a square of side x is given by the function: $A = x^2$, we certainly mean to assume that $x > 0$. Similarly, when we state that the price P of n items is given by $P = 10(n - 1) + 15$, we surely assume that n is an integer and $n \geq 1$. Although restrictions of this kind are often implicit, they should be explicitly stated in any practical problem if you wish to avoid silly answers.

In order to simplify matters in this book we shall make the following agreement concerning functions $f\colon(x,y)$, where x and y are numbers: *The function will be defined by its rule, and (unless otherwise explicitly stated) the domain and range are the largest subsets of the real numbers for which this rule is meaningful.*

We now give a number of illustrations of such functions and discuss the domain and range of each.

Illustration 1. Given f, the function defined by $f(x) = x^4 + x^2$. It is assumed that the domain of definition is the set X of real numbers, in which case the range is the set Y of nonnegative reals since only even powers of x occur.

Illustration 2. Given $f:(x,y)$, the function defined by $y = 2/x(x - 1)$.

Solution: The domain is the set of reals with 0 and 1 excluded. To determine the range, we proceed as follows: Write $x(x - 1)y = 2$, or $yx^2 - yx - 2 = 0$, which is a quadratic in x. Solving for x, we get

$$x = \frac{y \pm \sqrt{y^2 + 8y}}{2y}$$

Since x must be real, we require that $y^2 + 8y \geq 0$. This is true when y is in the set $\{y \mid (y \leq -8) \vee (y \geq 0)\}$. Since no value of x corresponds to $y = 0$, $y = 0$ must be excluded, and the range is the set $\{y \mid (y \leq -8) \vee (y > 0)\}$.

Illustration 3. $y = \sqrt{16 - x^2}$.

We must again stress that we are dealing here with real numbers. Hence the domain is the set $\{x \mid -4 \leq x \leq 4\}$, for otherwise the range would not be made up wholly of real numbers. The range Y is the set $\{y \mid 0 \leq y \leq 4\}$.

Illustration 4. $y = (x - 4)/(x - 4)$.

Solution: The value of the function at $x = 4$, namely, $f(4)$, is not defined; $f(4) = 0/0$, which does not mean anything. But $f(a) = (a - 4)/(a - 4) = 1$, if $a \neq 4$. Hence the domain is the set of reals (omitting 4) and the range consists of the single element 1. We say, "$f(x) = (x - 4)/(x - 4)$ is a constant, namely, 1, over the domain".

Illustration 5

$$y = \begin{cases} 1 & -2\pi \leq x \leq -\pi \\ -1 & x = 0 \\ 0 & 0 < x \leq \pi \\ \frac{1}{2} & \pi < x \\ \text{otherwise undefined} \end{cases}$$

You can readily see that the domain is the set $\{x \mid (-2\pi \leq x \leq -\pi) \vee (0 \leq x)\}$. The range is the set $\{1, 0, -1, \frac{1}{2}\}$, and the rule is as is shown. We often refer to such a function as being "piecewise-constant".

Illustration 6

x	1	2	3	4	5
y	2	7	24	55	201

The table associates an x and a y. But we know nothing about intermediate or other values. If y is the temperature of an object x sec after it is placed in a furnace, then, presumably, there would be sense to the question, What is a good estimate for the value of the temperature at the end of $3\frac{1}{2}$ sec? (A reasonable answer might be 37.6.) Or again, we might say, "Estimate the temperature at $5\frac{2}{3}$ sec"; the answer might not turn out to be an integer. Such data are to be associated with continuous portions of the real number system.

On the other hand, if y represents the number of rabbits in a hutch on the first day of successive months, then intermediate values are of no significance and the data are discrete and are associated with only some of the integers. The domain is the set $\{1, 2, 3, 4, 5\}$; the range is the set $\{2, 7, 24, 55, 201\}$.

Illustration 7. $f:X \to Y$, where X is the set of positive integers >1, and Y is the set of primes and the rule associating x and y is: given x, then y is the least prime not less than x.

For this function the domain is the set X of all positive integers with 1 omitted and the range is the set Y of all primes. (For special reasons 1 is usually omitted from the list of primes.) To a given x there is assigned a unique y; some pairs (x,y) are listed in the following table (which is made up from the verbally given rule).

x	2	3	4	5	6	7	8	9	10	11	12	13	...
y	2	3	5	5	7	7	11	11	11	11	13	13	...

Illustration 8. $f:(x,y)$, where X is the set of all men in New Orleans of age twenty-seven to twenty-nine years inclusive with "Pierre" as the first name and where Y is the set of three elements consisting of the tallest twenty-seven-year-old Pierre, the tallest twenty-eight-year-old Pierre, and the tallest twenty-nine-year-old Pierre. The rule is such that it assigns to a given x the unique y of that age.

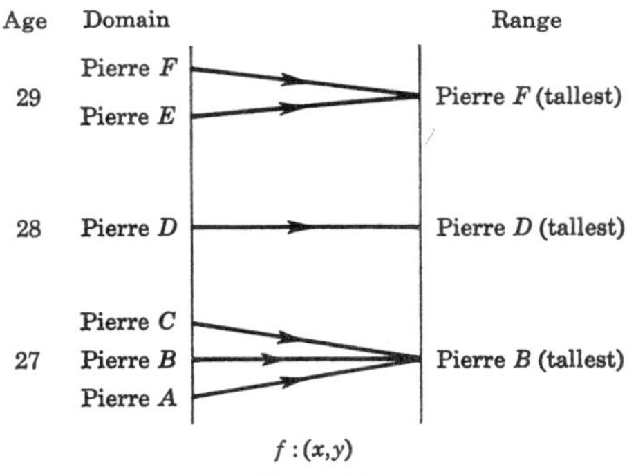

Figure 6.2

Some of the details are not given since we do not know how many "Pierres" there are in New Orleans of a given age. Also, we have assumed that there *is* a tallest (perhaps there is no Pierre of age twenty-seven or perhaps there are exactly five such all the same height; in this latter case we have a relation, not a function). But functional relationships of this general type are often useful in sociological studies. A graph of this mapping might look something like that in Fig. 6.2.

Problems 6.5

In Probs. 1 to 6 state whether the given equation defines a unique function, and if so find its domain and range.

1. $y = 3x + 4$.
2. $y = 3x + 2$.
3. $y^2 = 5x + 5$.
4. $y = (1/x) + 3$.
5. $y^2 = 7\sqrt{x}$.
6. $y = \sqrt{4 + x}$.

7. Find the range and domain of $y = \frac{1}{x(x-1)}$.

8. Find the range and domain of $y = \frac{1}{(x-2)(x+3)}$.

In Probs. 9 to 12 state the range.

9. $y = \frac{x}{(x-1)(x+2)}$.

10. $y = \frac{x+1}{x^2+1}$.

11. $y = \frac{x+1}{x^2-1}$.

12. $y = x^3 + x + 3$.

13. For $y = (x-1)(x+2)$, state the range if the domain is restricted to

$$\{x \mid 1 \leq x \leq 2\}$$

14. For $y = (x-1)(x+2)$, state the range if the domain is restricted to

$$\{x \mid x \leq 2\} \cup \{x \mid 1 \leq x\}$$

15. How does a merchant's price list define a function? State the rule, domain, and range.

16. Let x be the set of books in the Library of Congress and Y be the set of authors. State a rule that would define a function. State a rule that would define a relation that is not a function.

17. Given the function $f(x) = [\![x]\!]$. State its domain and range.

18. Given the function $f(x) = \sqrt{[\![x]\!]}$. State its domain and range.

19. Given the function $f(x) = [\![\sqrt{|x|}]\!]$. State its domain and range.

20. Given the function $f(x) = [\![|x|]\!]$. State its domain and range.

6.6 Algebra of Functions

We have studied the four elementary operations of arithmetic $+, -, \times, \div$ in connection with numbers (Chap. 2). These ideas can also be applied to functions according to the following definitions.

Definitions: Consider the functions $f:(x,y)$ and $g:(x,z)$ whose domains are, respectively, indicated by d_f and d_g. The sum $f + g$, the difference $f - g$, the product fg, and the quotient f/g are defined as follows:

(1) $f:(x,y) + g:(x,z) = (f+g):(x, y+z)$.

(2) $f:(x,y) - g:(x,z) = (f-g):(x, y-z)$.

(3) $f:(x,y) \times g:(x,z) = (fg):(x,yz)$.

(4) $\frac{f:(x,y)}{g:(x,z)} = \left(\frac{f}{g}\right):\left(x, \frac{y}{z}\right)$, $(z \neq 0)$.

(1′) In the addition $f + g$ of two functions, the functional values are added.

(2′) In the subtraction $f - g$, the functional values are subtracted (in the proper order).

(3′) In the multiplication fg, the functional values are multiplied.

(**4′**) In the division f/g, the functional values are divided (in the proper order).

The functional values of f and g are y and z, respectively. Thus, when f and g are defined by $y = f(x)$ and $z = g(x)$, the above definitions become:

(**1″**) $y + z = f(x) + g(x)$.
(**2″**) $y - z = f(x) - g(x)$.
(**3″**) $yz = f(x) \times g(x)$.
(**4″**) $\dfrac{y}{z} = \dfrac{f(x)}{g(x)}$.

The domain of each of $f + g$, $f - g$, and fg is the set of all elements x common to the domains of f and g; that is, it is the intersection of the sets d_f and d_g. Thus, in symbols $d_{f+g} = d_f \cap d_g$, $d_{f-g} = d_f \cap d_g$, and $d_{fg} = d_f \cap d_g$. The domain $d_{f/g} = d_f \cap d_g$ except for those x's for which $g(x) = 0$. (Division by zero is impossible.)

Illustration 1. Given the two functions f and g, defined by $y = x^2$ and $z = x^3$; the domain of each is the set of real numbers.

Solution: Then

$$\begin{aligned} y + z &= x^2 + x^3 \\ y - z &= x^2 - x^3 \\ yz &= x^2 \times x^3 = x^5 \\ \frac{y}{z} &= \frac{x^2}{x^3} = \frac{1}{x} \qquad x \neq 0 \end{aligned}$$

Note that $x = 0$ is not in the domain of y/z since $g(0) = 0$.

Illustration 2. Let f and g be defined by $y = 1 + 1/x$, $z = \sqrt{1 - x^2}$. The domain d_f is the set of all real numbers excluding 0; the domain d_g is the set of all real numbers between -1 and 1, inclusive.

Solution: We have:

$$\begin{aligned} f + g &= 1 + \frac{1}{x} + \sqrt{1 - x^2} && d_{f+g} \text{ is } -1 \le x \le 1,\ x \neq 0 \\ f - g &= 1 + \frac{1}{x} - \sqrt{1 - x^3} && d_{f-g} \text{ is } -1 \le x \le 1,\ x \neq 0 \\ fg &= \left(1 + \frac{1}{x}\right)\sqrt{1 - x^2} && d_{fg} \text{ is } -1 \le x \le 1,\ x \neq 0 \\ \frac{f}{g} &= \frac{1 + (1/x)}{\sqrt{1 - x^2}} && d_{f/g} \text{ is } -1 < x < 1,\ x \neq 0 \end{aligned}$$

In f/g we must exclude $x = \pm 1$, since $g(-1) = g(1) = 0$.

One further operation in the algebra of functions is of great importance; it is the operation of forming the *composite* of two functions.

The method of representing graphically the function $f:X \to Y$ used in Figs. 6.1 and 6.2 suggests a generalization in which more than two initial sets are involved. For instance, consider three sets X, Y, and Z and the two correspondences $f:X \to Y$ and $g:Y \to Z$. Figure 6.3 shows that, first, there is a mapping of the set X onto the set Y by means of the f function and that, second, Y is mapped onto Z by g. Obviously more sets could be used. In terms of classical mathematical language, this situation is described by saying "z is a function of a function"; for z is a function of y and, in turn, y is a function of x. We write

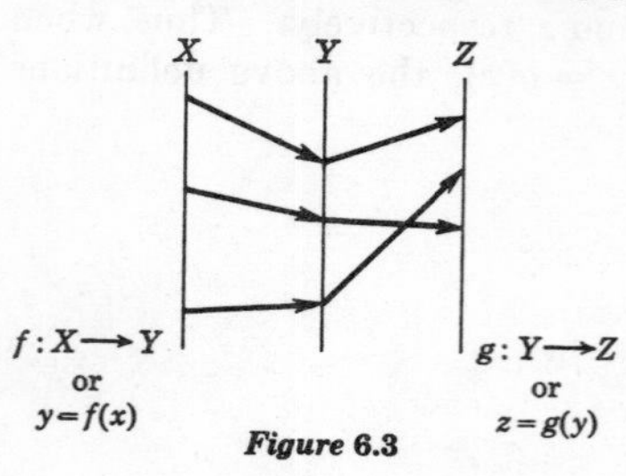

Figure 6.3

$$z = g(y) \qquad y = f(x)$$

that is,

$$z = g(f(x))$$

This language, "z is a function g of the function f", although inaccurate, could be used with the understanding that what we mean is the following: First, a set X is made to correspond to a set Y under the function f; then a correspondence is set up between the set Y and the set Z by means of the function g; hence X is mapped, by compounding the operations, onto the set Z. However, we prefer to use the phrase "$g(f):X \to Z$ is the composite of g and f" to describe the way in which z depends upon x.

Definition: When

$$f:X \to Y \qquad g:Y \to Z$$

we say that $g(f):X \to Z$ is the composite of g and f. Sometimes $g(f)$ is called a composite function of x.

Another notation is common. For two given functions $f:(x,y)$ and $g:(x,z)$ the composite of g and f is denoted by $g \circ f:(x,z)$, and its value at x by $(g \circ f)(x)$.

The domain $d_{g \circ f}$ is the set of all x's for which $f(x)$ is contained in the domain d_g. Since there may be no such x's, the composite function may not be defined at all. We can also speak of the composite of f and g, namely, $f \circ g$.

Illustration 3. Let $z = g(y) = 3y^2 - 2y + 1$ and $y = f(x) = 4x + 7$. The composite $g \circ f$ is given by

$$z = g(f(x)) = 3(4x + 7)^2 - 2(4x + 7) + 1$$

This can be simplified to yield

$$z = 48x^2 + 160x + 134$$

Illustration 4. A stone is dropped into a liquid, forming circles which increase in radius with time according to the formula $r = 4t$. How does the area of a given circle depend upon time?

Solution: The area A of a circle is $A = \pi r^2$, and we are given that $r = 4t$; hence A is a composite function of t given by

$$A = \pi r^2 = \pi(4t)^2 \qquad \text{or reduced} \qquad A = 16\pi t^2$$

Here we are not interested in negative values of t, although, mathematically, the maximal domain is the set of all real numbers. For the physical problem, a subset such as $0 \leq t < t_1$, where t_1 is sufficiently large, would suffice.

Since the letters used to represent the independent and dependent variables of a function can be replaced by other letters without change of meaning, we can speak of the composite of the functions $y = g(x)$ and $y = f(x)$. For we can write the first function as $z = g(y)$ and thus obtain $z = g(f(x))$.

Illustration 5. Given f and g defined by

$$f(x) = x^2 + 2 \qquad \text{and} \qquad g(x) = 1 - \frac{1}{x}$$

form the composites $g \circ f$ and $f \circ g$.

Solution: We first write $y = x^2 + 2$; $z = 1 - (1/y)$. Then the composite of g and f is $z = g(f(x)) = 1 - 1/(x^2 + 2)$. Similarly, the composite of f and g is given by $z = f(g(x)) = [1 - (1/x)]^2 + 2$.

Illustration 6. Find the composite $g(f)$ when

$$g(x) = |x| \qquad \text{and} \qquad f(x) = x^2 - 3x + 1$$

Solution: Rewrite in the forms

$$z = g(y) = |y| \qquad \text{and} \qquad y = f(x) = x^2 - 3x + 1$$

Then $z = g(f(x)) = |x^2 - 3x + 1|$. To evaluate $|x^2 - 3x + 1|$ for a given x, say, $x = 1$, we first find $x^2 - 3x + 1$, which equals -1. Then we take its absolute value which is $+1$.

Problems 6.6

In Probs. 1 to 7 and for the functions indicated, find $f + g$, $f - g$, $f \times g$, and f/g. State domain.

1. $f(x) = \dfrac{1}{x} + \dfrac{1}{x+1}$, $g(x) = x$.

2. $f(x) = \sqrt{x+1}$, $g(x) = \sqrt{x-2}$.

3. $f(x) = \dfrac{1}{x+1}$, $g(x) = x - 1$.

4. $f(x) = \dfrac{1}{x+1}$, $g(x) = (x+1)^2$.

5. $f(x) = x + 1$, $g(x) = \dfrac{1}{(x+1)^2}$.

6. $f(x) = x + 1$, $g(x) = \dfrac{1}{x+1}$.

7. $f(x) = \begin{cases} \dfrac{1}{(x-1)}, & x \neq 1, \\ 4, & x = 1; \end{cases} \qquad g(x) = \begin{cases} \dfrac{2}{(x-1)(x+4)}, & x \neq 1, -4, \\ \sqrt{7}, & x = 1, \\ \text{undefined}, & x = -4. \end{cases}$

In Probs. 8 to 10 form the composite $(g \circ f)(x)$. State domain.

8. $z = g(y) = y^4 + 2y + 1$, $y = f(x) = 2x + x^{-2}$.
9. $z = g(y) = 1/(1 + y)$, $y = f(x) = 1/(2 - x)$.
10. $z = g(y) = 3y^2 + 2y$, $y = f(x) = c$.

In Probs. 11 to 14 form $(g \circ f)(x)$ and also $(f \circ g)(x)$. State domain.

11. $f(x) = 5 + (1/x) + |x|$, $g(x) = x^2 + 4$.
12. $f(x) = |x|$, $g(x) = |x| - 1$.
13. $f(x) = |x|$, $g(x) = [\![x]\!]$.
14. $f(x) = \sqrt{x}$, $g(x) = |x|$.

15. Given $f(x) = |x|$, show that $f(f(x)) = f(x)$.
16. Given $f(x) = [\![x]\!]$, find $(f \circ f)(x)$.
17. Evaluate $|3x^2 - x - 5|$ for $x = -2, -1, 0$.
18. Simplify:

(a) $\sqrt{x^2 + 2x + 1}$.

(b) $\sqrt{5/x^2}$.

HINT: Use absolute values.

6.7 Graph of a Function

For a function whose domain and range are subsets of the reals, we can now plot the graph. As in Chap. 5, the graph consists of those points in the plane whose coordinates (x,y) represent the ordered pairs of the given function. The graph of such a function may look like Fig. 6.4. In this figure the domain is indicated by the heavy portions of the X-axis. These may be intervals, isolated points, or other sets of points such as the set of all rationals. If we draw a vertical line through a point $P(x,0)$ of the domain, it intersects the graph in exactly one point $Q(x,y)$. If we draw a horizontal line through this intersection, it intersects the Y-axis in a point $R(0,y)$ such that y is the value of the given function at x. The broken line PQR shows how we can determine y from x.

The graph of a function differs from that of a general relation in that a vertical line can intersect the graph of a function in only a single point,

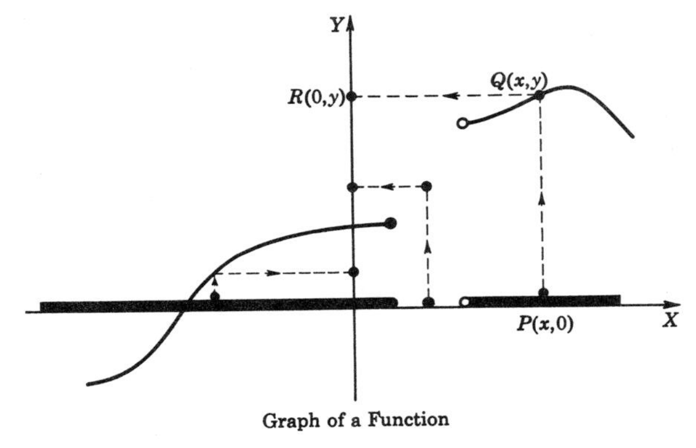

Graph of a Function

Figure 6.4

whereas such a vertical line can intersect the graph of a relation in many points (Fig. 6.5).

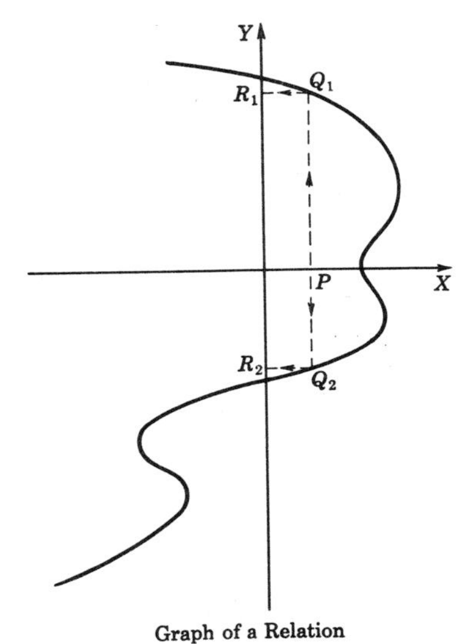

Graph of a Relation

Figure 6.5

The technique of plotting the graphs of functions defined by equations is precisely that discussed in Chap. 5. Let us consider the graphs of some other types of functions.

Illustration 1. Plot the graph of $y = |x|$.

Solution: For $x \geq 0$, this is equivalent to $y = x$, and for $x \leq 0$, to $y = -x$. Hence we plot the graphs of these two equations, restricting them to their appropriate domains (Fig. 6.6).

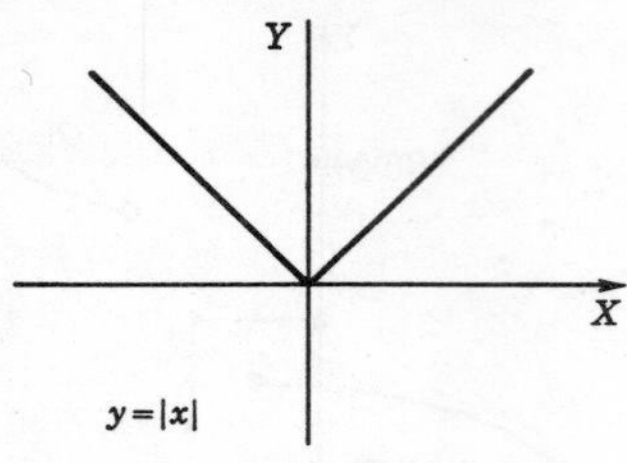

Figure 6.6

Illustration 2. Plot the graph of $y = [\![x]\!]$.

Solution: We observe, for instance, that $y = 1$ for $1 \leq x < 2$; $y = 2$ for $2 \leq x < 3$; etc. Therefore the graph is a sequence of horizontal segments each of which includes its left end point but not its right end point (Fig. 6.7).

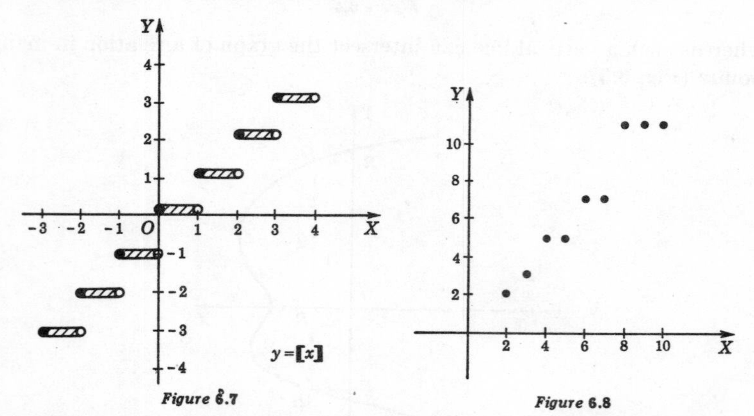

Figure 6.7 *Figure 6.8*

Illustration 3. Plot the graph of $f:(x,y)$, where X is the set of positive integers > 1 and Y is the set of primes. The rule is: given x, then y is the least prime not less than x (see Sec. 6.5, Illustration 7). This graph is a set of isolated points (Fig. 6.8).

Problems 6.7

In Probs. 1 to 19 plot the graphs of the given functions.

1. $y = \begin{cases} 2 + x, & x > 0, \\ x, & x \leq 0. \end{cases}$

2. $y = \begin{cases} 3, & x \text{ an integer}, \\ -2, & \text{otherwise}. \end{cases}$

3. $y = \begin{cases} 0, & x \text{ an even integer}, \\ -1, & x \text{ an odd integer}, \\ \text{otherwise undefined}. \end{cases}$

4. $y = \begin{cases} -1, & x \text{ rational}, \\ 1, & x \text{ irrational}. \end{cases}$

5. $y = \begin{cases} x^2, & x < 0, \\ 1, & x = 0, \\ 0, & x > 0. \end{cases}$

6. $y = \begin{cases} 1, & x \equiv 0, \bmod 3, \\ -1, & x \equiv 1, \bmod 3, \\ \text{otherwise undefined}. \end{cases}$

7. $y = \begin{cases} 0, x \equiv 0, \text{ mod } 3, \\ 1, x \equiv 1, \text{ mod } 3, \\ 2, x \equiv 2, \text{ mod } 3, \\ 3, \text{ otherwise.} \end{cases}$

8. $y = \begin{cases} 2x, x < 1, \\ 4, x = 1, \\ x + 5, x > 1. \end{cases}$

9. $y = x - |x|$.
10. $y = |x - 1|$.
11. $y = -|x|$.
12. $y = |x| - x$.
13. $y = |2x - 1|$.
14. $y = (|x|)^2$.
15. $y = [\![\sqrt{x}]\!]$.
16. $y = \sqrt{[\![x]\!]}$.
17. $y = x + [\![x]\!]$.
18. $y = 2x - [\![x]\!]$.
19. The *postage function* of Illustration 1, Sec. 6.1.
20. Can the graph of a function be symmetric with respect to the X-axis? To the Y-axis?

In Probs. 21 to 26 plot the graphs of the given relations.

21. $|y| = 2x$.
22. $|y| = [\![x]\!]$.
23. $|x| + |y| = 1$.
24. $|x| + |y| \geq 1$.
25. $[\![y]\!] = |x|$. Note a restriction on the domain.
26. $[\![y]\!] = 2 + x$. Note a restriction on the domain.

6.8 Inverse Function

A function $f: (x,y)$ is a set of ordered pairs such that no two of the ordered pairs have the same first element x. Several ordered pairs could have the same second element y, however. If a function f is of such character that no two pairs have the same second element, then there exists a function f^{-1} called the *inverse function* of f defined below.

Definition: Given the function f such that no two of its ordered pairs have the same second element, the inverse function f^{-1} is the set of ordered pairs obtained from f by interchanging in each ordered pair the first and second elements.

Thus the function f has elements (ordered pairs) of the form (x_1,y_1), (x_2,y_2), . . . , while the inverse function has elements (ordered pairs) of the form (y_1,x_1), (y_2,x_2), ... We may write $f: X \to Y$ and $f^{-1}: Y \to X$. The range of f is the domain of f^{-1}, and the domain of f is the range of f^{-1}.

This notion is well illustrated by Fig. 6.9. We begin with point $R(0,y)$

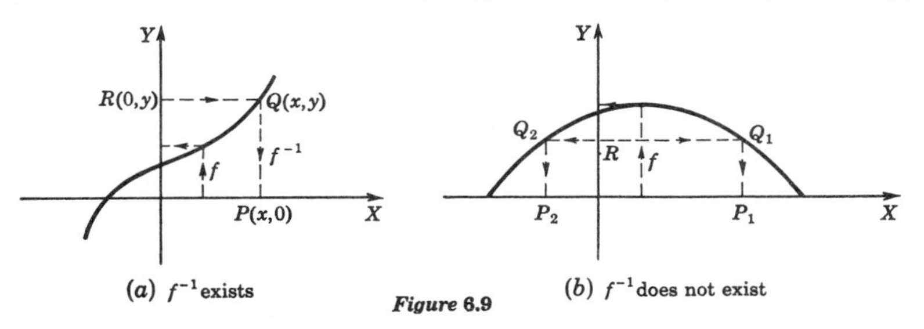

(a) f^{-1} exists (b) f^{-1} does not exist

Figure 6.9

on the Y-axis and draw a horizontal line which meets the graph in one or more points Q. If there is just one such point (as in Fig. 6.9*a*), draw QP perpendicular to the X-axis and determine $P(x,0)$. Thus we have a mapping $Y \to X$, which is the inverse function f^{-1}. If, however, there are several intersections Q (Fig. 6.9*b*), these determine several points P. More than one value of x are then assigned to the given y, and the relationship is not a function. In this case f^{-1} is not defined. Note that, when f^{-1} exists, the graphs of f and f^{-1} are identical, for the X- and Y-axes are not to be interchanged when we graph f^{-1}.

If f is given by a simple formula $y = f(x)$, we can often obtain f^{-1} by solving this for x so that $x = f^{-1}(y)$. There are a number of difficulties with this procedure, which will be clarified by the following illustrations.

Illustration 1. Let f be given by $y = 3x + 1$ and have the domain $\{x \mid 0 \leq x \leq 1\}$. The range of f is then $\{y \mid 1 \leq y \leq 4\}$. Find f^{-1} and its domain and range.

Solution: We solve the given equation for x and find

$$x = \frac{y-1}{3}$$

This is the inverse function $f^{-1}:(y,x)$. Its domain is $\{y \mid 1 \leq y \leq 4\}$; its range is $\{x \mid 0 \leq x \leq 1\}$.

Illustration 2. Let f be defined by $y = \frac{1}{2}\sqrt{4 - x^2}$ with the given domain $\{x \mid -2 \leq x \leq 0\}$ and range $\{y \mid 0 \leq y \leq 1\}$. Find f^{-1}, its domain and range.

Solution: In order to solve for y we square both sides and obtain

$$4y^2 = 4 - x^2$$

This process is risky, for $4y^2 = 4 - x^2$ is also obtained by squaring $y = -\frac{1}{2}\sqrt{4 - x^2}$, which is not the given function. With our fingers crossed, we now solve for x. The result is

$$x = \pm 2\sqrt{1 - y^2}$$

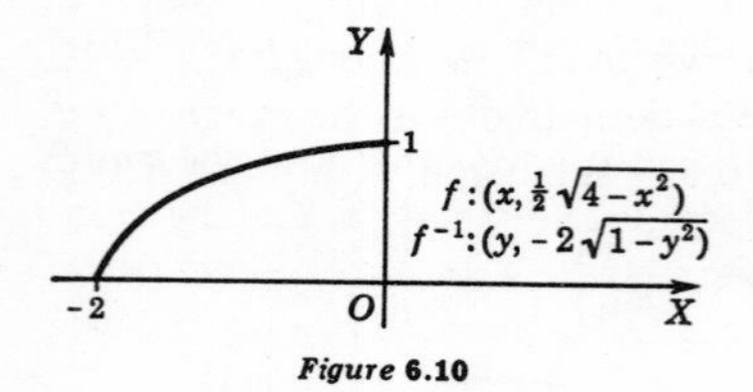

Figure **6.10**

This is not exactly what we want, for we can use only one sign. We recall that the domain of f is $\{x \mid -2 \leq x \leq 0\}$, and this tells us to use the minus sign. The inverse function f^{-1} is therefore given by

$$x = -2\sqrt{1 - y^2}$$

Its domain is $\{y \mid 0 \leq y \leq 1\}$, and its range is $\{x \mid -2 \leq x \leq 0\}$. The common graph of f and f^{-1} is given in Fig. 6.10.

Notice that if we had chosen the maximum domain for f, namely, $\{x \mid -2 \leq x \leq 2\}$, then f^{-1} would not have existed.

This illustration emphasizes an important point which will come up later on: *If f does not have an inverse, we may be able to restrict its domain so that the restricted function does have an inverse* (Sec. 8.11).

Exercise A. The function $y = x^2 (-\infty < x < \infty)$ does not have an inverse. Find a domain for x such that the restricted function does have an inverse. How many domains can you find?

Illustration 3. Let f be defined by $y = 2^x$, where the domain is $\{x \mid -\infty < x < \infty\}$ and the corresponding range is $\{y \mid 0 < y < \infty\}$. Find f^{-1}.

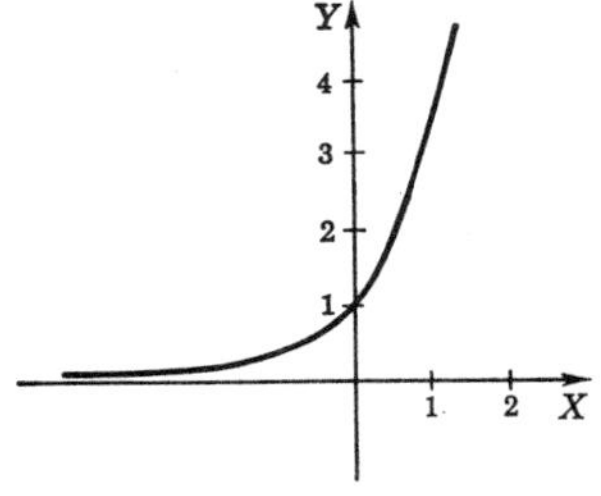

Figure 6.11

Solution: Since the graph of $y = 2^x$ rises steadily as x increases (Fig. 6.11), no two values of x give the same value of y. So the inverse f^{-1} clearly exists. The trouble is that we do not know how to solve for x in terms of y by means of any formula. The function f^{-1} is therefore a new function unnamed at present. If we have frequent occasion to refer to this function, we shall find it convenient to give it a name and to investigate its properties. This is indeed a common method of obtaining new functions from known functions. You may already know the name of f^{-1} when $f(x) = 2^x$. It is called the *logarithm of y to the base* 2, written $\log_2 y$. We shall study this function in detail in Chap. 7.

This illustration emphasizes another point of importance: *If a function f has an inverse which cannot be calculated by elementary means, we shall often give this inverse a name and add it to our list of useful functions.*

Exercise B. Give a definition of $\sqrt[3]{y}\ (y > 0)$ as the inverse of some function.

Exercise C. If you have studied trigonometry, show that $y = \sin x, \left(-\frac{\pi}{2} \le x \le \frac{\pi}{2}\right)$ has an inverse. What is the name of this inverse?

The example of $y = 2^x$ in Illustration 3 suggests the following new term:

Definition: A function f whose domain and range are subsets of the reals is *strictly monotone increasing* if and only if, for every pair x_1 and x_2 such that $x_2 > x_1$, we have $f(x_2) > f(x_1)$. Similarly, f is *strictly monotone decreasing* if and only if, for every pair x_1 and x_2 such that $x_2 > x_1$, we have $f(x_2) < f(x_1)$.

This enables us to state the following theorem.

Theorem 1. A function f whose domain and range are subsets of the reals has an inverse if it is either strictly monotone increasing or strictly monotone decreasing.

Proof: We must show that there are no two pairs (x_1,y) and (x_2,y) with the same second element and different first elements. This, however, is obviously true for strictly monotone functions.

Remark. The converse of this theorem is true if we add the hypothesis that f is defined in an interval in which it is continuous in a sense to be defined later (Chap. 10).

The process of defining new functions as the inverses of known functions introduces a further complication. In Illustration 3, we defined

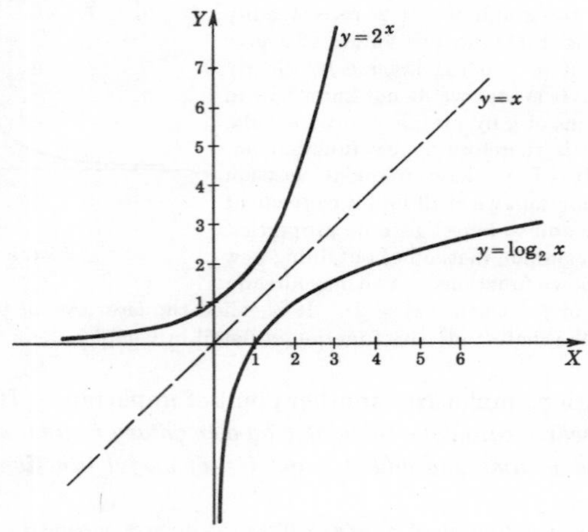

Figure 6.12

$x = \log_2 y$ as the inverse of $y = 2^x$. Now we should like to forget about the origin of this new function and treat it like any other function. Although it is not logically necessary, we generally use x to represent the independent variable and y to represent the dependent variable, but these roles are reversed in the case of $x = \log_2 y$. In order to conform to the usual practice, we therefore interchange the names of our variables and write $y = \log_2 x$. In general, the inverse of $y = f(x)$ is $x = f^{-1}(y)$, which we write as $y = f^{-1}(x)$ after interchanging variables. The graph of $y = \log_2 x$ is obtained from the common graph of $x = \log_2 y$ and $y = 2^x$ by reflection in the line $y = x$ (Fig. 6.12). The same procedure is valid in connection with other inverse functions, $y = f(x)$ and $y = f^{-1}(x)$, and we shall employ it in Chaps. 7 and 8.

In order to introduce a final important property of inverse functions, let us define the identity function whose domain and range are the same set X.

Definition: The function $E:X \to X$ whose elements are the ordered pairs (x,x) is called the identity function. This function maps each x into itself. (See Fig. 6.13.)

Now let us consider a function $f:X \to Y$ and its inverse $f^{-1}:Y \to X$. The composite function $f^{-1}(f)$ maps each x into some y (under f) and then

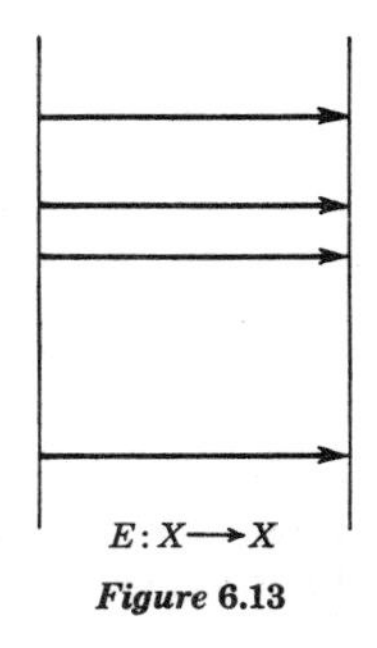

Figure 6.13

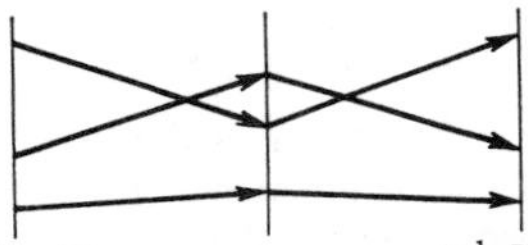

$f: X \longrightarrow Y$ $f^{-1}: Y \longrightarrow X$

Figure 6.14

back into itself (under f^{-1}). We see then that

$$f^{-1}(f) = E \qquad \text{and similarly} \qquad f(f^{-1}) = E$$

Illustration 4. In Illustration 1, $f(x) = 3x + 1$ and $f^{-1}(y) = (y - 1)/3$. Hence

$$f^{-1}(f(x)) = \frac{(3x+1)-1}{3} = x. \quad \text{Also, } f(f^{-1}(x)) = 3\left(\frac{x-1}{3}\right) + 1 = x.$$

Exercise D. In Illustration 2, verify that $f(f^{-1}) = f^{-1}(f) = E$.

This property is reminiscent of the expression in group theory (Chap. 4), which says that

$$a \circ a' = a' \circ a = e$$

It suggests that functions may form a group in which $g \circ f$ is defined to be $g(f)$. This cannot be true, in general, since $g(f)$ is defined only if the range of f is contained in the domain of g and since f^{-1} does not exist for some f's. Special sets of functions, however, may form a group of this sort. One such set is that of the functions which define a 1 to 1 correspondence between the points of the real line X and the real line Y. Such a mapping is said to be a 1 to 1 mapping of the line onto itself.

Exercise E. Verify that the 1 to 1 mappings of the real line onto itself form a group if the group product $g \circ f$ is defined as $g(f)$ and the group inverse f' to be the function f^{-1}.

Problems 6.8

In Probs. 1 to 10 discuss domain and range and plot the graphs of $y = f(x)$ and $x = f^{-1}(y)$, where $y = f(x)$ is defined by the following:

1. $y = 2x - 5$.
2. $y = 3x + 4$.
3. (a) $y = ax$. (b) $y = x$.
4. (a) $y = x + k$. (b) $y = x - k$.
5. $y = f(x) = \sqrt{x^2 - 4}$; $d_f = \{x \mid x \geq 2\}$.
6. $y = f(x) = -\sqrt{x^2 - 4}$; $d_f = \{x \mid x \geq 2\}$.
7. $y = f(x) = \sqrt{4 - x^2}$; $d_f = \{x \mid 0 \leq x \leq 2\}$.
8. $y = f(x) = -\sqrt{4 - x^2}$; $d_f = \{x \mid -2 \leq x \leq 0\}$.
9. $y = f(x) = -\frac{2}{3}\sqrt{9 - x^2}$; $d_f = \{x \mid -3 \leq x \leq 0\}$.
10. $y = f(x) = -\frac{2}{3}\sqrt{x^2 - 9}$; $d_f = \{x \mid x \leq -3\}$.
11. For what function is $y = f(x)$ identical with $y = f^{-1}(x)$?
12. Show that $y = f(x) = |x|$ (d_f is the set of real numbers) has no inverse.
13. Suppose that f has an inverse f^{-1}. Can the graph of $y = f(x)$ be symmetric with respect to the Y-axis?
14*. Consider the set of functions: x, $1/x$, $1 - x$, $1/(1 - x)$, $(x - 1)/x$, $x/(x - 1)$ whose domains are the real numbers, omitting $x = 0$ and $x = 1$. Find the inverse of each function and the two composites of each pair. Hence show that these six functions form a group under the "composite" operation.

6.9 Functions Derived from Equations

We can often convert an equation into an expression which defines a function. This process, however, may involve a number of difficulties, some of which are suggested by the illustrations below.

Illustration 1. The equation $2x - 3y + 1 = 0$ is called a "linear equation" because the pairs (x,y) which satisfy it lie on a straight line (Chap. 9). From the equation we can derive two functions:

$$y = f(x) = \frac{2x + 1}{3} \qquad f:(x,y)$$

$$x = g(y) = \frac{-1 + 3y}{2} \qquad g:(y,x)$$

(We may not know which element is to be considered the first element of the ordered pairs.)

Exercise A. Show that $f^{-1} = g$.

Illustration 2. The equation $s = 16t^2$ gives the distance s in feet through which a body falls from rest under the influence of gravity in t sec. As such it defines a function, namely, $\{(t,s) \mid s = 16t^2\}$. We may ask, however, How long does it take for the body to fall 64 ft? To answer this, we solve for t:

$$t^2 = \frac{s}{16} \qquad \text{or,} \qquad t = \pm\frac{1}{4}\sqrt{s}$$

This gives the two functions $\{(s,t) \mid t = \frac{1}{4}\sqrt{s}\}$ and $\{(s,t) \mid t = -\frac{1}{4}\sqrt{s}\}$. In terms of the physical situation, only the first makes practical sense, but both make mathematical sense. Therefore we choose $t = \frac{1}{4}\sqrt{s}$, put $s = 64$, and find $t = 2$.

Exercise B. Are there any physical situations in which $t = -\frac{1}{4}\sqrt{s}$ makes practical sense?

This illustration makes the point that, although an equation may lead to several functions, not all these necessarily have meaning in a practical situation. You will have to use your head and discard those that are nonsense.

Illustration 3. Consider the equation $x^2 + y^2 = 4$, which represents a circle of radius 2. If we solve for y, we obtain $y = \pm\sqrt{4 - x^2}$. Of the many functions which can be obtained from this, two have outstanding importance:

$$y = f(x) = \sqrt{4 - x^2} \qquad -2 \leq x \leq 2$$
$$y = g(x) = -\sqrt{4 - x^2} \qquad -2 \leq x \leq 2$$

The graph of f is the upper semicircle, and the graph of g is the lower semicircle.

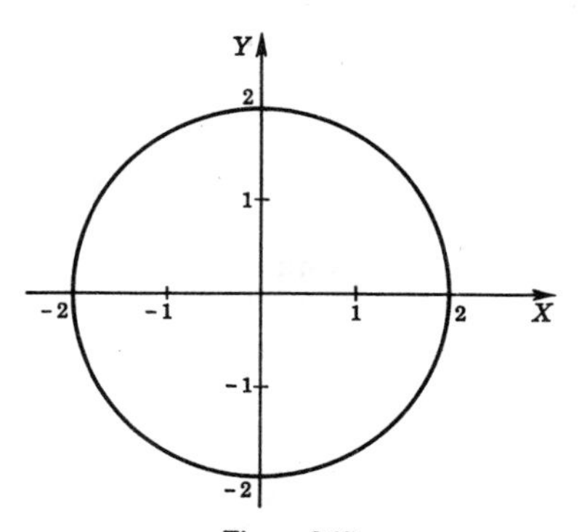

Figure 6.15

Exercise C. Solve for x, and describe the graphs of the two functions so obtained. Call these F and G.

Exercise D (BT). Does F^{-1} equal f or g?

Illustration 4. The equation

$$x^2 + xy + 4 = 0$$

is quadratic in x and thus has the solution

$$x = \frac{-y \pm \sqrt{y^2 - 16}}{2}$$

This yields two functions $f:(y,x)$ and $g:(y,x)$, where

$$x = f(y) = \frac{-y + \sqrt{y^2 - 16}}{2} \qquad |y| \geq 4$$

$$x = g(y) = \frac{-y - \sqrt{y^2 - 16}}{2} \qquad |y| \geq 4$$

When we solve for y, we obtain

$$y = -\frac{x^2 + 4}{x}$$

which gives the function h, where

$$y = h(x) = -\frac{x^2 + 4}{x} \qquad x \neq 0$$

Exercise E. What is the domain of $f \circ h$? Show that in this domain $f \circ h = E$. Answer the same questions for $g \circ h$.

All these functions derived from equations have a common property: their elements ((x,y) or (y,x)) satisfy the equation. This suggests the more general definition:

Definition: If the elements of a function f satisfy an equation, the function f is said to be derived from this equation.

In many textbooks and older works, a function thus derived from an equation is said to be given "implicitly" by the equation. The functions themselves are called *implicit functions.*

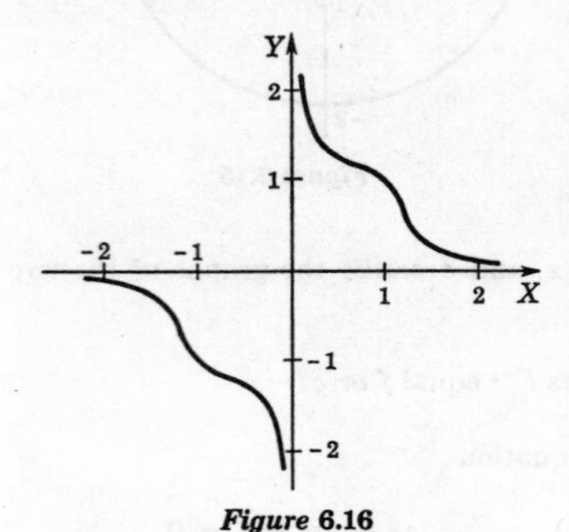

Figure 6.16

In the examples given above, the functions derived from an equation were obtained by solving for one of the variables. It is important to note that derived functions may exist even when we are unable to carry through such a solution. We shall sometimes wish to consider such functions, an example of which is given below.

Illustration 5. The equation $x^5y + xy^5 = 2$ has a graph given by Fig. 6.16. There is no way of solving this equation for x or y, but the graph indicates that functions $\{(x,y) \mid y = f(x)\}$ and $\{(y,x) \mid x = g(y)\}$ exist which are derived from this equation.

Problems 6.9

In Probs. 1 to 10 find functions derived from the given equation, and state their domain and range.

1. $4x + 3y + 3 = 0.$
2. $2x - 5y = 6.$
3. $3x^2 + y^2 = 1.$
4. $x^2 + 2y^2 = 1.$
5. $x^2y - y = 1.$
6. $x^2y - 2y = 2.$
7. $|x| + |y| = 1.$
8. $2|x| + |y| = 1.$
9. $s = -16t^2 + v_0t$, where s is the distance at time t above ground of a particle fired at $t = 0$ vertically upward with initial velocity v_0.
10. $v^2 - 2gs = 0$, where v is the velocity of a body falling from rest, s is the distance fallen, and g is a given positive constant.

In Probs. 11 to 16 find point pairs which satisfy the equation, plot the graph, and describe any derived functions which you can find.

11. $[x] + [y] = 1,\ x \geq 0,\ y \geq 0.$
12. $[x] + [y] = 1,\ -2 \leq x \leq 0.$
13. $[x] - [y] = 0,\ -2 \leq x \leq 2.$
14. $[x] + [y] = 0,\ -2 \leq x \leq 2.$
15. $1^x = 1^y.$
16. $2^x = 2^y.$

6.10 Algebraic Functions

In order to discuss the properties of functions, it is very helpful for us to classify them into convenient categories. Here we shall discuss a broad type of functions called *algebraic functions*, and in Chaps. 7 and 8 we shall meet other important functions of elementary mathematics.

(*a*) *Polynomial Functions.* The simplest algebraic functions are the *polynomial functions*, which are defined by equations of the form $y = P(x)$, where $P(x)$ is a polynomial. When $P(x)$ is defined over the field of real numbers, the domain is X and the range is a subset of Y.

Exercise A. Prove that the sum of two polynomials is a polynomial.

Exercise B. Show that the product of two polynomials is a polynomial.

Exercise C. Show by an example that the quotient of two polynomials may be a polynomial. Find another example in which the quotient is not a polynomial.

Exercise D. Show that the polynomials defined over the reals form a ring and not a field.

Exercise E. Show that the composite $g \circ f$ of two polynomial functions is a polynomial function.

(*b*) *Rational Functions.* The next simplest type of algebraic function is a *rational function,* which is so called because we now permit the use of division along with the other rational operations.

Definition: A function R defined by $y = R(x) = P(x)/Q(x)$, where $P(x)$ and $Q(x)$ are polynomials, is called a rational function.

The remarks made above about the domain and range of a polynomial function apply equally well to a rational function, but here we must be a little careful: the function R is not defined at points where $Q(x) = 0$. This is made clear by the following illustrations.

Illustration 1

(**a**) $y = 1/x$ is not defined at $x = 0$.
(**b**) $y = (x - 1)/(x + 2)$ is not defined at $x = -2$.
(**c**) $y = 3x^3/(x - 1)(x^2 + 1)$ is not defined at either $x = 1$ or at $x = \pm i$.

Illustration 2

(**a**) $y = x/x$ is not defined at $x = 0$. For other values of x, however, $x/x = 1$. The two functions x/x and 1 are consequently not identical. This illustration brings up an important point: the cancellation of a common factor in the numerator and denominator may change the function.
(**b**) As a similar example, consider the two rational functions:

$$y = \frac{x(x-1)}{x-1} \qquad \text{and} \qquad y = x$$

These have the same values for $x \neq 1$, but at $x = 1$ the first is undefined, whereas the second has the value 1. Hence they are different functions.

Illustration 3. Some functions which are not written in the explicit form $y = P(x)/Q(x)$ are nevertheless equivalent to rational functions. Consider

$$\begin{aligned} y &= \left(\frac{1}{x} - \frac{4}{x+1}\right)\left(1 + \frac{1}{x-1}\right) \\ &= \frac{x + 1 - 4x}{x(x+1)} \cdot \frac{x - 1 + 1}{x - 1} \\ &= \frac{(1 - 3x)(x)}{x(x+1)(x-1)} \\ &= \frac{-3x^2 + x}{x^3 - x} \end{aligned}$$

In this bit of algebra we did not cancel out the x in numerator and denominator. For the function

$$y = \frac{-3x + 1}{x^2 - 1}$$

is not equivalent to the given function. Why?

(*c*) *Explicit Algebraic Functions.* Explicit algebraic functions constitute the next important class of functions. They are generated by a finite number of the five algebraic operations, namely, addition, subtraction, multiplication, division, and root extraction. Thus the function whose values are given by

$$\frac{\sqrt{1+x} - \sqrt[3]{x^5}}{\sqrt[6]{(2+x-x^2)^3 - 8}}$$

is an example of an explicit algebraic function. Because of the possible appearance of (even) roots in the equation defining such a function, it may happen that the value y of the function is real only when x is restricted to a very limited subset of the real numbers. For the example above, it is seen first of all that x (real) must be greater than or equal to -1 if $\sqrt{1+x}$ in the numerator is to be real. Similarly, in $\sqrt[6]{(2+x-x^2)^3 - 8}$ it must be true that $(2+x-x^2)^3 > 8$, that is, $2+x-x^2 > 2$ or $x(1-x) > 0$. This says that x must lie between 0 and 1, exclusive. The domain of definition is therefore $0 < x < 1$.

Problems 6.10

In Probs. 1 to 10 state which of the following define polynomial functions.

1. $y = x - 2^x$

2. $y = 1^x$.

3. $y = \sqrt{x}$

4. $y = \sqrt{x^2}$.

5. $y = x^{-1}$.

6. $y = 4/x^2$.

7. $y = 1 - ix + i$.

8. $y = 4i2^x - \frac{4}{i}2^2$.

9. $y = \sqrt{5}x + \pi x^0$.

10. $y = x - \frac{x^3}{3!} + \frac{x^5}{5!}$.

In Probs. 11 to 20 state which of the following define rational functions.

11. $f(x) = 2/x$.

12. $f(x) = \sqrt[3]{x}$.

13. $f(x) = |x - 1|$.

14. $f(x) = x - 2^x$.

15. $f(x) = \frac{2x}{x-1}$.

16. $f(x) = \frac{1 - x + x^2}{x^2 - x^4 + x^5}$.

17. $f(x) = (-2)^x$.

18. $f(x) = \frac{x-i}{x-i}$.

19. $f(x) = \frac{\sqrt{2}\,x + 1}{x}$.

20. $f(x) = \frac{\sqrt{2x} + 1}{x}$.

In Probs. 21 to 30, which are explicit algebraic functions?

21. $(x, |x|)$.

22. $(x, [\![x]\!])$.

23. $(x, (-2)^x)$.

24. $(x, \sqrt{x^2})$.

25. $(x, |x|^2)$.

26. $\left(x, \frac{\sqrt{2x}+1}{x-1}\right)$.

27. $f(x) = \begin{cases} 1, & x \text{ irrational}, \\ 0, & x \text{ rational}. \end{cases}$ **28.** $f(x) = \begin{cases} -1, & x \text{ an integer}, \\ 1, & \text{otherwise}. \end{cases}$

29. $f(x) = x - \dfrac{x^3}{3!} + \dfrac{x^5}{5!} - \cdots + (-1)^{n-1}\dfrac{x^{2n-1}}{(2n-1)!} + \cdots.$

30. $f(x) = 1 - \dfrac{x^2}{2!} + \dfrac{x^4}{4!} - \cdots + (-1)^{n-1}\dfrac{x^{2n-2}}{(2n-2)!} + \cdots.$

In Probs. 31 to 40 state kind of function and domain (we assume x and y both real).

31. $\{(x,y)|y = 4 - \sqrt{5}x + x^8\}$. **32.** $\{(x,y)|y = 4 - \sqrt{5x} + x^8\}$.

33. $f: X \to Y$, where $y = \sqrt{x - x^3}$. **34.** $f: X \to Y$, where $y = \dfrac{1 - \sqrt{5x}}{x^8}$.

35. $\{(x,y) \mid y = \sqrt{x}\}$. **36.** $\{(x,y) \mid y = \sqrt{x^2 - 1}\}$.

37. $f(x) = \dfrac{3x}{x(x-1)(x+2)}$ **38.** $f(x) = 4 - 6x^2 + 7x^5$.

39. $y = \sqrt{x - 2} - \sqrt{x - 2}$. **40.** $y = \dfrac{\sqrt{x^2 - 4}}{\sqrt{4 - x^2}}$. (BT)

References

Apostol, Tom M.: "Calculus", pp. 40–50, 196–200, Blaisdell, New York (1961).
Begle, Edward G.· "Introductory Calculus", chap. 3, Holt, New York (1954).
Courant, Richard: "Differential and Integral Calculus", chap. 1, Interscience, New York (1937).

Exponential and Logarithmic Functions | 7

7.1 Exponential Functions

In your earlier studies you have become acquainted with powers such as 2^3, $(-3)^4$, π^5, and the like. You have also met

$$7^{-2} = \frac{1}{7^2} \qquad 4^0 = 1 \qquad \pi^{-3} = \frac{1}{\pi^3}$$

The general expression for symbols like these is a^n, where a is any real number and n is an integer. You will also recall the use of fractional exponents to represent roots, such as

$$3^{\frac{1}{2}} = \sqrt{3} \qquad 5^{\frac{1}{3}} = \sqrt[3]{5} \qquad 2^{-\frac{1}{7}} = \frac{1}{\sqrt[7]{2}}$$

For our present purposes we are interested in the roots of positive real numbers only, and we know that every positive real number a has a unique real nth root which we write $a^{1/n}$, where n is a positive integer.

Also, we recall that $a^{p/q}$ is defined to be $a^{(1/q)p} = (a^p)^{1/q}$, where p and q are integers and a is positive. Hence we know the meaning of the function defined by

$$y = a^x \qquad a \text{ positive, } x \text{ rational}$$

We should like to extend the domain of definition of this function to the entire set of real numbers and thus give sense to numbers such as 2^π, $\pi^{-\sqrt{3}}$, and $3^{\sqrt{2}}$. The most natural way to obtain $3^{\sqrt{2}}$, for example, is to consider the successive decimal approximations to $\sqrt{2}$, such as 1.4, 1.41, 1.414, 1.4142, etc. (Sec. 2.13). Then $3^{1.4}$, $3^{1.41}$, $3^{1.414}$, $3^{1.4142}$, etc., are successive approximations to $3^{\sqrt{2}}$. Indeed, we can use R16, the Axiom of Completeness (Sec. 2.13), to give us a precise definition.

Definition: Let a be any positive real number, and let $\{b_1, b_2, \ldots\}$ be a bounded set of rational numbers whose least upper bound is the real number b. Then a^b is defined to be the least upper bound of $\{a^{b_1}, a^{b_2}, \ldots\}$.

The Axiom of Completeness assures us that this least upper bound exists, and so a^b is well defined. We can now define the exponential function as follows:

Definition: The function f defined by $y = a^x (a > 0)$ is called the exponential function with base a. Its domain of definition is the set of real numbers. We observe that its range of values is $0 < y < \infty$.

We now wish to develop some of its properties.

Theorem 1. For $a > 0$ and $b > 0$ and x and y real:

(**a**) $a^x \times a^y = a^{x+y}$
(**b**) $(a^x)^y = a^{xy}$
(**c**) $(ab)^x = a^x \times b^x$

These theorems are proved in elementary texts for rational values of x. We do not give the proof for irrational values of x.

Theorem 2

(**a**) $a^x > 1$ for $a > 1$, x real and > 0
(**b**) $a^x = 1$ for $a = 1$, x real and > 0
(**c**) $a^x < 1$ for $0 < a < 1$, x real and > 0

Proof: Part (a) is immediate when x is a positive integer, for the product of two numbers each of which is greater than 1 must itself exceed 1. When $x = 1/n$ (n a positive integer), part (a) also is true. For if $a^{1/n}$ were to be less than 1 in this case, its nth power $(a^{1/n})^n = a$ would be less than 1. This follows from the fact that the product of two numbers each between zero and 1 must itself be less than 1. Finally, part (a) is true for rational x by combining the above cases. We omit the proof for irrational values of x. Part (b) is immediate since all powers and roots of 1 are themselves 1. Part (c) is proved similarly to part (a).

Exercise A. Write out the details of the proof of part (c) for rational x.

Theorem 3. Let x and y be real numbers such that $x < y$. Then

(**a**) $a^x < a^y$ for $a > 1$
(**b**) $a^x = a^y$ for $a = 1$
(**c**) $a^x > a^y$ for $0 < a < 1$

The proof depends on Theorem 2. In all cases we know, from Theorem 1, that $a^y = a^{y-x} \cdot a^x$. By hypothesis, $y - x$ is positive. Thus, if $a > 1$, $a^{y-x} > 1$ and $a^y > a^x$, and similarly for the other cases.

Exercise B. Complete the proof of Theorem 3.

Theorem 3 shows that when $a > 1$, the function a^x is strictly monotone increasing (Sec. 6.8) and that, when $a < 1$, it is strictly monotone decreasing. Typical graphs are given in Figs. 7.1 and 7.2.

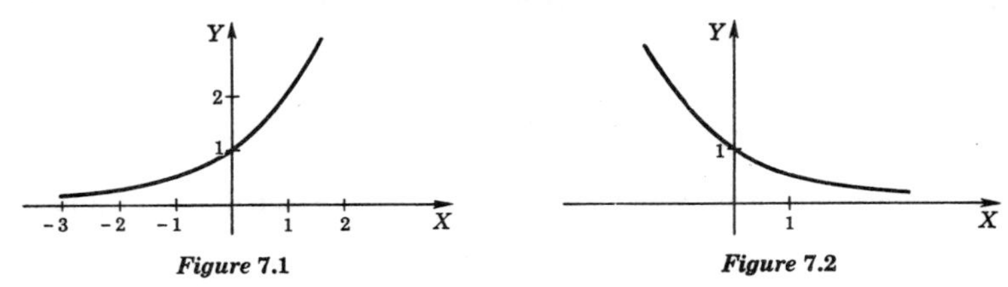

Figure 7.1 *Figure 7.2*

There is an interesting symmetry between the graphs of $y = a^x$ and $y = (1/a)^x$, for $(1/a)^x = a^{-x}$, and the graph of $y = a^{-x}$ is just like the graph of $y = a^x$, with the X-axis reversed in direction.

Exercise C. Draw the graph of $y = 1^x$.

Problems 7.1

In Probs. 1 to 10 simplify.

1. $8^{-3} \cdot 6^2$.

2. $\dfrac{(49)^{\frac{1}{2}}}{3^3 \cdot 7^2}$.

3. $10^{-2} \cdot 4^2$.

4. $\dfrac{1}{10^3 \cdot 5^{-6}}$.

5. $2^6 \cdot 4^{-3}(\sqrt{2})^4$.

6. $9^2 \cdot 9^{-\frac{1}{2}}$.

7. $4^3(16)^{\frac{1}{2}}(64)^2$.

8. $\sqrt{(0.23)10^8}$.

9. $\sqrt[3]{(27.3)^2 \cdot 24}$.

10. $\sqrt[4]{(12.34)10^{12}}$.

In Probs. 11 to 18 simplify, but leave the answer in exponential form.

11. $a^x \cdot a^{-4x} \cdot a^{2x}$.

12. $b^{3y} \cdot b^{\frac{1}{3}y} \cdot b^{y-1}$.

13. $c^{-z} \cdot c^{3z} \cdot c^{\frac{2}{3}} \cdot c^2$.

14. $\dfrac{10^2 \cdot 10^{\frac{1}{2}} \cdot 10^x}{10^3 \cdot 10^{-2x} \cdot 10^{x-2}}$.

15. $8^{\frac{1}{3}} \cdot 8^2 \cdot 8^{-3} \cdot 8^t$.

16. $3^2 \cdot 9^2 \cdot 27^2 \cdot 81^{\frac{1}{4}}$.

17. $\dfrac{4^{\frac{1}{2}}}{4^{\frac{1}{2}} + 4^{\frac{1}{2}}}$.

18. $5^x + 5^{2x}$. (BT)

In Probs. 19 and 20 plot on the same axes and to the same scales.

19. (a) $y = 2^x$. (b) $y = (\frac{1}{2})^x$.
20. (a) $y = 4^x$. (b) $y = (\frac{1}{4})^x$.

21. Plot the graph of $y = 3^x$. Now change the scale on the Y-axis so that the graph you have drawn is that of $y = \frac{1}{2} \cdot 3^x$.

22. Plot the graph of $y = (\frac{1}{2})^x$. Now change the scale on the Y-axis so that the graph you have drawn is that of $y = 3 \cdot (\frac{1}{2})^x$.

7.2 The Number e

We have defined the function $f: (x,y)$ whose values are given by $y = a^x$. Now there is a particular number $a > 1$ of great importance in mathematics; it is an irrational number and approximately equal to 2.71828. It bears the name e; thus

$$e \approx 2.71828$$

It is impossible for us to explain at this time why the number e and the associated exponential function defined by

$$y = e^x$$

are of such importance.

This exponential function is in fact so important that we speak of it as *the* exponential function and neglect to mention its base. Sometimes this function is written

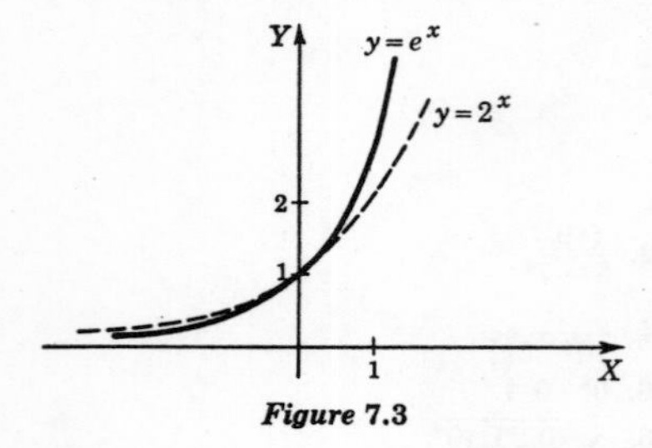

Figure 7.3

$$y = \exp x$$

where no base appears at all; the base is assumed to be e unless otherwise specified. Its values are tabulated in many convenient tables. Its graph is plotted in Fig. 7.3.

A convenient approximation to e^x is given by the polynomial:

$$1 + x + \frac{x^2}{2!} + \frac{x^3}{3!} + \cdots + \frac{x^n}{n!}$$

(By definition $n! = 1 \times 2 \times \cdots \times n$ for n a positive integer, and $0! = 1$.) As n increases, the approximation becomes closer and closer; as a matter of fact, e^x is exactly equal to the infinite sum:

$$e^x = 1 + x + \frac{x^2}{2!} + \cdots + \frac{x^n}{n!} + \cdots$$

Of course, we have not defined such an infinite sum at present, but we shall do so in Chap. 10. Then we can give a proper definition of e and of the function e^x. Table I in the Appendix gives values of e^x and e^{-x}.

The exponential function is also used in the definitions of some other functions which occur in elementary mathematics. These are called the *hyperbolic functions.* Their definitions are:

$$\sinh x = \frac{e^x - e^{-x}}{2}$$
$$\cosh x = \frac{e^x + e^{-x}}{2}$$
$$\tanh x = \frac{e^x - e^{-x}}{e^x + e^{-x}}$$

The function $\sinh x$ is read "the hyperbolic sine of x"; $\cosh x$, "the hyperbolic cosine of x"; and $\tanh x$, "the hyperbolic tangent of x". These are somewhat related to the trigonometric functions with similar names which we shall study in Chap. 8.

Problems 7.2

In Probs. 1 and 2 obtain the numbers from the exponential table in the Appendix.

1. $e^{1.35}, e^{-0.07}, e^{\sqrt{2}}$.

2. $e^{2.15}, e^{-0.71}, e^{\pi}$.

In Probs. 3 to 8 plot the graph.

3. $y = 2e^{-x}$.

4. $y = \frac{1}{2}e^{2x}$.

5. $y = \frac{1}{3}e^{3x}$.

6. $y = -3e^{-2x}$.

7. $y = \dfrac{e^x + e^{-x}}{2}$ $(= \cosh x)$. The curve is called a catenary.

8. $y = \dfrac{e^x - e^{-x}}{2}$ $(= \sinh x)$.

9. From Probs. 7 and 8 above show that $e^x = \cosh x + \sinh x$.

10. From Probs. 7 and 8 above find $\{x \mid \cosh^2 x - \sinh^2 x = 1\}$.

11. Show that

$$\sinh x = x + \frac{x^3}{3!} + \frac{x^5}{5!} + \cdots + \frac{x^{2n-1}}{(2n-1)!} + \cdots$$
$$\cosh x = 1 + \frac{x^2}{2!} + \frac{x^4}{4!} + \cdots + \frac{x^{2n-2}}{(2n-2)!} + \cdots$$

7.3 Logarithmic Functions

Since $y = a^x$ defines a strictly monotone function, for $a \neq 1$, each value of y is obtained from a single x. Therefore the inverse function exists, and this is called the logarithmic function.

Definition: The function inverse to that given by $y = a^x$, $a > 0$, $a \neq 1$, is written $y = \log_a x$ and is called the *logarithm* of x to the base a.

For computational purposes, the base is usually taken to be 10, so that properties of our decimal system may be used to simplify the needed tables. For theoretical purposes, the base is always taken to be e. In advanced books, this base is omitted, and $\log x$ is to be understood to mean $\log_e x$. Frequently, the notation $\ln x$ is used for $\log_e x$. Logarithms to the base 10 are called "common" logarithms; those to the base e are called "natural" logarithms. Tables of both kinds are available in most collections of elementary tables. (See Appendix, Tables II and III.)

Exercise A. Prove that $\log_a a^x = x$ and that $a^{\log_a x} = x$.

The graph of $y = \log_a x$ is obtained from that of $y = a^x$ by reflecting it in the line $y = x$. It is given in Figs. 7.4 and 7.5.

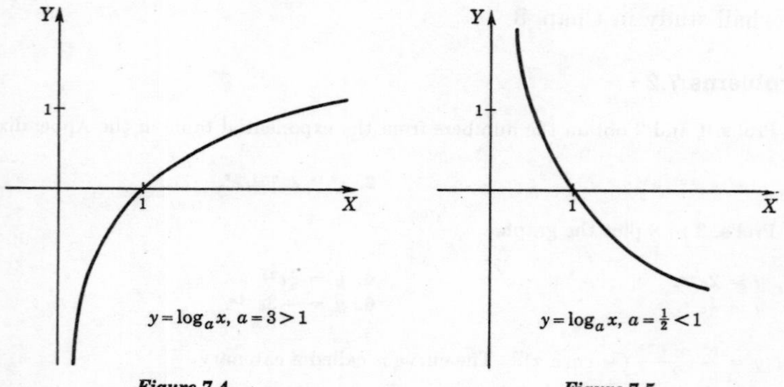

Figure 7.4 *Figure 7.5*

From the graphs we see that the domain and range are as follows.

Domain and *Range*. The domain of definition of $\log_a x$ is the set of positive real numbers. Its range of values is the set of all real numbers.

Note that the logarithms of negative numbers are not defined here. In advanced books, you will learn how to extend the definition of $\log_a x$ so that x can be negative. Its value turns out to be complex in this case. We do not consider this case.

Properties. The logarithmic function to the base a defined by $y = \log_a x$ is strictly monotone increasing for $a > 1$, strictly monotone decreasing for $0 < a < 1$, and not defined for $a = 1$. The following theorems have useful applications.

Theorem 4. $\log_a xy = \log_a x + \log_a y$.

Proof:

Let $z = \log_a xy$; then $a^z = xy$.
$u = \log_a x$; then $a^u = x$.
$v = \log_a y$; then $a^v = y$.

Therefore $$a^z = xy = a^u \cdot a^v = a^{u+v}$$
or $$z = u + v$$

from which the theorem follows.

Theorem 5. $\log_a \dfrac{1}{x} = -\log_a x$.

Proof: Let $z = \log_a \dfrac{1}{x}$; then $a^z = \dfrac{1}{x}$, and $a^{-z} = x$. Therefore $-z = \log_a x$. Hence the theorem follows.

Theorem 6. $\log_a \dfrac{y}{x} = \log_a y - \log_a x$.

Proof: Combine Theorems 3 and 4.

Theorem 7. $\log_a (x^y) = y \log_a x$.

Proof:

Let $z = \log_a (x^y)$; then $a^z = x^y$.
$u = \log_a x$; then $a^u = x$.

Therefore $$(a^u)^y = a^z \qquad \text{or} \qquad uy = z$$

from which the theorem follows.

Exercise B. Prove $\log_b a = \dfrac{1}{\log_a b}$.

Exercise C. Prove $\log_a x = \dfrac{\log_b x}{\log_b a}$. A special case of this is $\log_a x = \dfrac{\log_e x}{\log_e a}$.

Theorems 4 to 6 are useful for numerical computations involving only products and quotients. Logarithms to the base 10 are generally employed.

Illustration 1. Find $\frac{(33.0)(27.2)}{15.8}$.

Solution: We compute

$$\log_{10} \frac{(33.0)(27.2)}{15.8} = \log_{10} 33.0 + \log_{10} 27.2 - \log_{10} 15.8$$

To find these logarithms, we use the table of common logarithms in the Appendix. We find

$$\begin{aligned} \log_{10} 33.0 &= 1.5185 \\ \log_{10} 27.2 &= \underline{1.4346} \\ & 2.9531 \\ -\log_{10} 15.8 &= \underline{-1.1987} \\ \log_{10} \frac{(33.0)(27.2)}{15.8} &= 1.7544 \end{aligned}$$

Working backward from the table, we obtain

$$\frac{(33.0)(27.2)}{15.8} = 56.81$$

The importance of logarithms in problems of this sort is not as great as it was in former years. Calculations such as that above can be performed more rapidly on a slide rule, provided that the numbers involved do not contain more than three essential digits. When the numbers are more complicated, or when greater accuracy is desired, rapid results can be obtained from a desk computing machine. For this reason most students do not need to develop great skill in this use of logarithms.

On the other hand, logarithms must be used to compute exponentials such as $2^{1.42}$ by the use of Theorem 7.

Illustration 2. Compute $2^{1.42}$.

Solution: From Theorem 7,

$$\begin{aligned} \log_{10} 2^{1.42} &= 1.42 \log_{10} 2 \\ &= (1.42)(0.3010) \\ &= 0.4274 \end{aligned}$$

Therefore $$2^{1.42} = 2.675$$

Problems 7.3

In Probs. 1 and 2 pick out those pairs that define mutually inverse functions and state domain and range.

1. 3^{-x}, 4^{6x}, 6^{4x}, $4 \log_6 x$, $\frac{1}{6} \log_4 x$, $\log_3 (-x)$, $\log_3 \frac{1}{x}$.

2. $3 \cdot 2^x$, $4 \cdot 5^{6x}$, $\log_2 \frac{x}{3}$, $\log_3 \frac{x}{2}$, $\frac{1}{6} \log_5 \frac{x}{4}$, ae^{bx}, $\frac{1}{b} \log_e \frac{x}{a}$.

3. Compute, using common logarithms:

(**a**) $3^{2.5}$. (**b**) $10^{1.2}$. (**c**) $e^{0.5}$.

4. Compute, using natural logarithms:

(**a**) $3^{2.5}$. (**b**) $10^{1.2}$. (**c**) $e^{0.5}$.

5. Compute, using natural logarithms:

$$10^{1.2} + e^{0.5}$$

6. Compute, using common logarithms:

$$10^{1.2} + e^{0.5}$$

7. By using seven-place common logarithms, we find that $(1 + \frac{1}{10})^{10} \approx 2.594$, $(1 + \frac{1}{100})^{100} \approx 2.705$, and $\left(1 + \frac{1}{1{,}000}\right)^{1{,}000} \approx 2.717$. Compare with the value of e.

In Probs. 8 to 10 evaluate or simplify.

8. $3^{\log_3 3} + 3^{\log_3 9}$.
9. $5^{\log_{25} 5}$.
10. $e^{\log_e x} + \log_e e^x$.
11. Show that $x^x = a^{x \log_a x}$, $x > 0$.
12. Show that $[f(x)]^{g(x)} = a^{g(x) \log_a f(x)}$, $f(x) > 0$.
13. If $a^x b^{2x} = b^z$, find z.
14. If $a^x b^x = c^z$, find z.
15. If $y = \log_e (x + \sqrt{x^2 + 1})$, find $\sinh y$.
16. If $y = \log_e (x + \sqrt{x^2 - 1})$, find $\cosh y$.
17. If $y = \frac{1}{2} \log_e \frac{1 + x}{1 - x}$, find $\tanh y$.

7.4 Graphs

With a set of the standard mathematical tables at our disposal, we can now make light work of graphing various exponential, logarithmic, and related functions.

Illustration 1. Plot the graph of the function given by $y = xe^{-x}$.

Solution: Again we prepare a table of x's and corresponding values of y and sketch the graph in Fig. 7.6 on page 228.

x	-3	-2	-1	0	1	2	3
e^{-x}	20.09	7.39	2.72	1	0.37	0.14	0.05
$y = xe^{-x}$	-60.27	-14.78	-2.72	0	0.37	0.28	0.15

[By methods of the calculus (Chap. 11) we find that a maximum value of the function occurs at $x = 1$.]

Illustration 2. On the same axes and to the same scale plot the graphs of $y = e^x$, $y = \log_e x$.

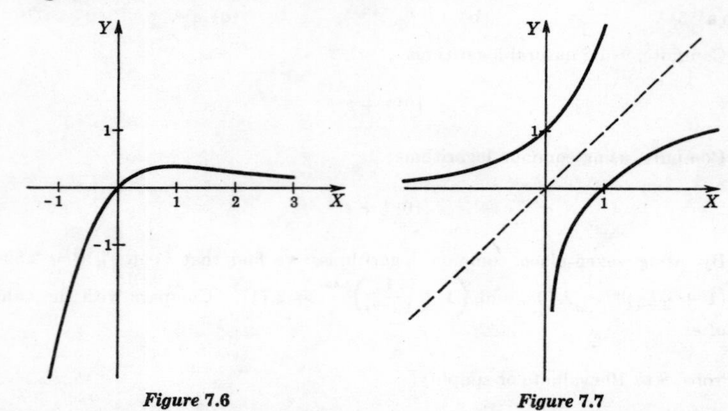

Figure 7.6 *Figure 7.7*

Solution: The functions defined by these equations are inverses of each other. We use a table of natural logarithms to prepare the following entries.

x	-3	-2	-1	0	0.2	0.5	1	2	3
$y = \log_e x$	......	......	......	$-\infty$	-1.61	-0.69	0	0.69	1.10
$y = e^x$	0.05	0.14	0.37	1	1.22	1.65	2.72	7.39	20.09

The graphs are plotted in Fig. 7.7.

7.5 Applications

There are many problems in biology, chemistry, economics, etc., involving growth and decay for which the natural mathematical model is the exponential function. Our basic illustration is from the field of economics.

Illustration 1

(**a**) An amount P dollars (principal) is invested at 100 per cent interest (rate), compounded annually. (The accrued interest is to be added to the principal.) Find the total amount A after 1 year.
(**b**) Same problem compounded monthly.
(**c**) Same problem compounded daily (360 days/year).
(**d**) Same problem compounded continuously.

Solution:

(**a**) $A = P(1 + 1)$.
(**b**) $A = P(1 + \frac{1}{12})^{12}$.

(**c**) $A = P(1 + \frac{1}{360})^{360}$.

(**d**) In order to arrive at something meaningful in this case we should begin with a description of what is meant by compounding "continuously". At this time we can give only an intuitive explanation since a precise explanation involves the theory of limits. We would have an approximate answer if we compounded each second. A year (360 days) has 31,104,000 seconds. The amount, at the end of 1 year, would be

$$A_{31,104,000} = P\left(1 + \frac{1}{31,104,000}\right)^{31,104,000}$$

We should like to know what, if anything, $A = P(1 + 1/n)^n$ would approach with ever-increasing n. The answer (beyond the scope of this text to develop) is Pe. That is, in technical language: "The limit of $(1 + 1/n)^n$, as n grows without bound, is e." Or, in symbols,

$$\lim_{n \to \infty} \left(1 + \frac{1}{n}\right)^n = e$$

If continuous compounding took place over a period of kt years, the amount would be given by

$$A = P \lim_{n \to \infty} \left(1 + \frac{1}{n}\right)^{nkt} = P \lim_{n \to \infty} \left[\left(1 + \frac{1}{n}\right)^n\right]^{kt} = Pe^{kt}$$

The same kind of problem arises in biology, where each of P cells in a given culture splits into two cells in a certain time t.

Illustration 2. The number of bacteria in a culture at time t was given by

$$y = N_0 e^{5t}$$

What was the number present at time $t = 0$? When was the colony double this initial size?

Solution: When $t = 0$, $y = N_0 e^0 = N_0$. The colony will be $2N_0$ in size when t satisfies the equation $2N_0 = N_0 e^{5t}$, that is, when $5t = \log_e 2$ or when $t = \frac{1}{5} \log_e 2$ $= 0.6932/5 \approx 0.1386$ unit of time.

In chemistry, certain disintegration problems are similarly explained.

Illustration 3. Radium decomposes according to the formula $y = k_0 e^{-0.038t}$, where k_0 is the initial amount (corresponding to $t = 0$) and where y is the amount undecomposed at time t (in centuries). Find the time when only one-half of the original amount will remain. This is known as the "half-life" of radium.

Solution: We must solve $\frac{1}{2}k_0 = k_0 e^{-0.038t}$ for t.

$$\log_e \tfrac{1}{2} = -0.038t$$
$$-0.6932 = -0.038t$$
$$t = \frac{693.2}{38} = 18.24 \text{ centuries}$$

Illustration 4. Given that the half-life of a radioactive substance is 10 min, how much out of a given sample of 5 g will remain undecomposed after 20 min?

Solution: The substance decays according to the formula:

$$y = 5e^{-kt}$$

First we must find k. From the given data

$$\frac{5}{2} = 5e^{-10k}$$

or

$$\frac{1}{2} = e^{-10k}$$

Taking natural logarithms of both sides, we have:

$$-\log_e 2 = -10k$$

$$k = \frac{\log_e 2}{10}$$

Substituting back, we find:

$$y = 5e^{-(\log_e 2)(t/10)}$$

$$= 5e^{-0.06932t}$$

When $t = 20$ min,

$$y = 5e^{-1.386}$$

$$= 1.25 \text{ g}$$

We could have seen this at once, for half remains after 10 min and so half of a half, or a quarter, remains after 20 min. The above method, however, will give us the answer for any time t.

Problems 7.5

In Probs. 1 to 10 sketch the graph.

1. $y = |\log_e x|$.
2. $y = \log_e |x|$.
3. $y = \frac{1}{2} \log_e x^2$.
4. $y = \log_e \sqrt{x}$.
5. $y = \log_e (-x)$.
6. $y = x \log_e x$.
7. (a) $y = \log_e x$. (b) $y = \log_{10} x$.
8. $y = e^{-x^2}$. (The graph is called the "probability curve".)
9. $y = \dfrac{e^{-2}2^x}{x!}$, for $x = 0, 1, 2, 3, 4$, etc. (This defines an important function in statistics; domain: the positive integers and zero. It is called the "Poisson distribution function".)
10. $y = \dfrac{e^{-\frac{1}{2}x}(\frac{1}{2})^x}{x!}$, for $x = 0, 1, 2, 3$.
11. The number of bacteria in a certain culture at time t was given by

$$y = N_0 e^{3t}$$

What was the number present at time $t = 0$? When was the colony double this initial size?

12. Let N_0 be the number of π^0 mesons generated at time $t = 0$, and let $y = N_0e^{-at}$ be the number at any subsequent time t. If only $N_0/2$ are present when $t = 3 \times 10^{-16}$ sec, find a.

13. Find the truth set of the open sentence

$$2^x = 10$$

14. Find the real number x if $3^{-x} = 4$.

15. Solve the equation $e^x - e^{-x} + 1 = 0$.

16. Compute $\{x \mid 10^{|x|} = 11.2\}$.

17. Solve simultaneously for x and y.

$$\begin{aligned} 4^x &= 3^y \\ 2(4^x) &= 5^y \end{aligned}$$

18. Solve simultaneously for x and y.

$$\begin{aligned} 3^x &= 5^y \\ 6^x &= 3(5^y) \end{aligned}$$

19. If $x = \log_b a$ and $y = \log_a b$, what is the value of xy?

20. If $x = \log_b a$ and $y = \log_a b$, what is the value of x/y?

References

In addition to the many standard textbooks on algebra, the reader should consult the following articles in the *American Mathematical Monthly*.

Cairns, W. D.: Napier's Logarithms as He Developed Them, vol. 35, p. 64 (1928).

Cajori, Florian: History of the Exponential and Logarithmic Concepts, vol. 20, pp. 5, 20, 35, 75, 107, 148, 173, 205 (1913).

Huntington, E. V.: An Elementary Theory of the Exponential and Logarithmic Functions, vol. 23, p. 241 (1916).

Lenser, W. T.: Note on Semi-logarithmic Graphs, vol. 49, p. 611 (1942).

Sandham, H. F.: An Approximate Construction for e, vol. 54, p. 215 (1947).

8 | Trigonometric Functions

8.1 Introduction

Trigonometry was originally developed in connection with the study of the relationships between the sides and angles of a triangle. You have probably already met some of the trigonometric functions, such as the sine and cosine, and have applied them to simple problems about triangles. This aspect of trigonometry was investigated extensively by the early Greeks, especially by Hipparchus (ca. 180–125 B.C.), who, because of his work in astronomy, actually developed spherical rather than plane trigonometry. The trigonometry of the triangle continues to be of importance in modern technology in such areas as surveying, navigation, and the applications of vectors to mechanics. This chapter is concerned with those portions of this material which deal with the geometry of the plane. You will need to consult other books for material on spherical trigonometry.

It would be a serious error, however, to limit the study of trigonometry to its applications to triangles. Its modern uses are widespread in many theoretical and applied fields of knowledge. The trigonometric functions force themselves on you in a very surprising fashion when you study the calculus of certain algebraic functions. You will also meet them when you study wave motion, vibrations, alternating current, and sound. In none of these subjects, however, do angles appear in any natural fashion. It is therefore essential that we extend the concept of a trigonometric function so that it is a function of a general real variable, and no longer merely a function of an angle. These more general trigonometric functions become, then, members of our arsenal of functions, which have been developed in the previous chapters. Their definitions and properties are given in the following chapter.

The complete set, consisting of the algebraic functions, the exponential function, the logarithmic functions, and the trigonometric functions, is called the set of "elementary functions". Virtually all undergraduate courses in mathematics restrict themselves to these elementary functions. In more advanced work, however, it is necessary to introduce additional functions which carry curious names such as the "gamma function", "Bessel function", "theta function", etc. We shall not need to refer to these hereafter in this book.

8.2 Distance in the Plane

We begin our study of trigonometry by developing certain properties of plane geometry which we shall need. Naturally, we assume that you are already familiar with much of this subject from your study of it in high school. We shall be using all the logical structure of this geometry, including the undefined words, axioms, definitions, and theorems. Of course, we must assume these here, for a review of this material would take us too far afield. As a minimum, you should be familiar with the properties of similar triangles and with the theorem of Pythagoras.

We shall employ the usual system of rectangular axes which was discussed in Secs. 5.2 and 5.3. In our work so far in this book we have permitted you to use quite different units on the two axes according to your immediate needs. Here, however, we must be more particular. In this chapter, coordinates on the X-axis and the Y-axis will represent *distance* in the *same* units of measurement.

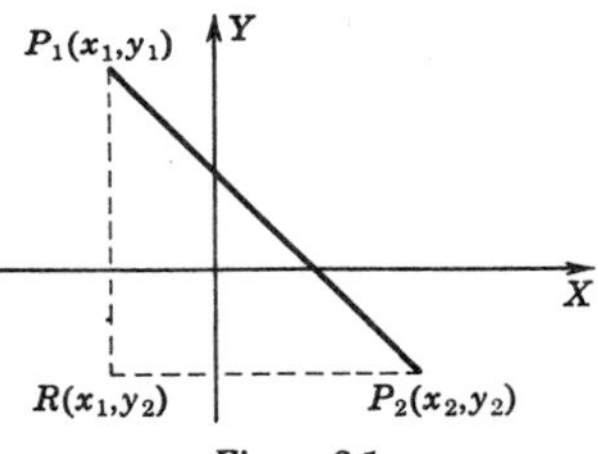

Figure 8.1

Let us now consider two points P_1 and P_2 which do not lie on a line parallel to one of the axes (Fig. 8.1). The length of the segment, or the "distance P_1P_2", can be computed from the Theorem of Pythagoras. Construct the right triangle P_1RP_2 with P_1R parallel to the Y-axis and RP_2 parallel to the X-axis. R has coordinates (x_1,y_2). From the Theorem of Pythagoras,

$$(P_1P_2)^2 = (RP_2)^2 + (P_1R)^2$$

We know (Sec. 5.3) that

$$RP_2 = |x_2 - x_1| \qquad \text{and} \qquad P_1R = |y_1 - y_2|$$

Hence $$(P_1P_2)^2 = (x_2 - x_1)^2 + (y_2 - y_1)^2$$

We observe that this is also true even if the line P_1P_2 is parallel to one of the axes. We have thus proved the general theorem:

Theorem 1. The distance d between any two points in the plane $P_1(x_1,y_1)$ and $P_2(x_2,y_2)$ is given by

$$d = \sqrt{(x_2 - x_1)^2 + (y_2 - y_1)^2}$$

Illustration 1. Find the distance between $A(4, -3)$ and $B(-2,5)$.

Solution: The distance $d = AB$ is given by

$$\begin{aligned} d &= \sqrt{[4-(-2)]^2+[-3-5]^2} \\ &= \sqrt{36+64} \\ &= 10 \end{aligned}$$

Illustration 2. Find the lengths of the diagonals of the quadrilateral $A(1,2)$, $B(-2,1)$, $C(-3,-4)$, $D(5,-7)$.

Solution: The diagonals are AC and BD, and their lengths are given by

$$\begin{aligned} AC &= \sqrt{4^2+6^2} = \sqrt{52} \\ BD &= \sqrt{(-7)^2+8^2} = \sqrt{113} \end{aligned}$$

Problems 8.2

In Probs. 1 to 6 find the distance between the pairs of points.

1. $(5,1)$, $(1,-2)$.
2. $(1,6)$, $(-1,-1)$.
3. $(3,2)$, $(3,0)$.
4. $(4,8)$, $(5,17)$.
5. $(9,-1)$, $(-2,10)$.
6. $(21,1)$, $(-4,101)$.

7. Show that the triangle $A(0,3)$, $B(4,11)$, $C(4,1)$ is a right triangle.
8. Show that the triangle $A(2,0)$, $B(\frac{4}{5},\frac{12}{5})$, $C(4,4)$ is a right triangle.
9. Show that the triangle $A(1,1)$, $B(\frac{7}{2}, 1+\frac{5}{2}\sqrt{3})$, $C(6,1)$ is an equilateral triangle.
10. Show that the triangle $A(0,1)$, $B(1, 1+\sqrt{3})$, $C(2,1)$ is an equilateral triangle.
11. Show that the points $A(2,1)$, $B(1,2)$, $C(-3,0)$, $D(-2,-1)$ are the vertices of a parallelogram.
12. Show that the points $A(1,2)$, $B(-2,-1)$, $C(2,1)$, $D(5,4)$ are the vertices of a parallelogram.

In Probs. 13 and 14 show that the point P is on the perpendicular bisector of the line segment AB.

13. $P(2,\frac{1}{2})$, $A(5,1)$, $B(-1,0)$.
14. $P(4,4)$, $A(3,-1)$, $B(-1,5)$.
15. Show that $A(-2,0)$, $B(0,4)$, $C(1,3)$, $D(-3,1)$ are the vertices of a rectangle.
16. Show that $A(-2,2)$, $B(-1,3)$, $C(-1,1)$, $D(0,2)$ are the vertices of a square.

8.3 General Definitions

We begin the subject of trigonometry by defining the trigonometric functions as functions of a real variable. We believe that you will have less difficulty in grasping the basic notions if we present the subject first without using the concept of angle, reserving the trigonometry of angles as a matter for secondary consideration.

To this end we consider a unit circle with center placed at the origin of a rectangular coordinate system (Fig. 8.2). From Theorem 1 we see that the equation of such a circle is $x^2 + y^2 = 1$, where, of course, the coordinates (x,y) of a point on the circle satisfy the above equation.

Next we lay off on the circle an arc of (positive) length θ beginning at the point (1,0) and running counterclockwise. An arc running clockwise will be called an arc of negative length. The set of all arcs is in 1 to 1 correspondence with the set of all real numbers. Figures 8.2*e* and *f* show arcs that lap over more than one circumference. Before proceeding, let us say that the concept of *arc length,* as yet undefined, is a very profound one and is not to be lightly passed over. Length measured along a

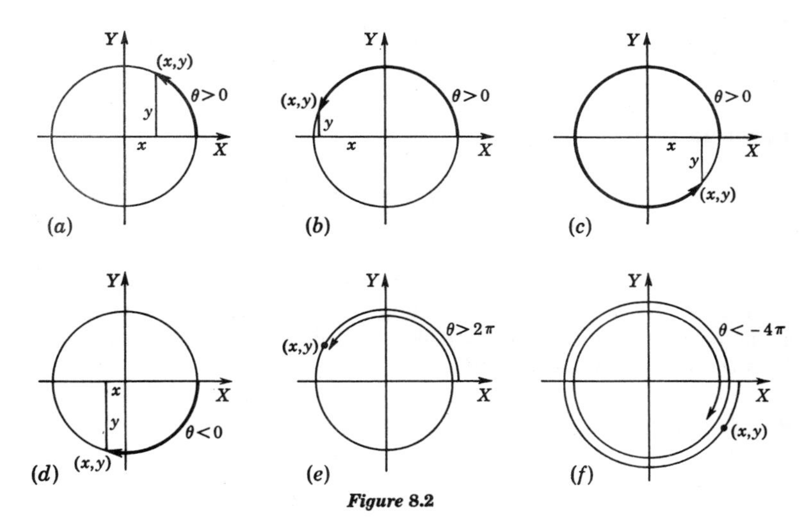

***Figure* 8.2**

straight line is one thing; but what could be the meaning of length measured along a circle or some other curve? If all the known history of mathematics is any indication of the truth, Archimedes was the only one of the ancients who had any clear notion of how to treat arc length, and after his death the subject languished for almost two thousand years! We do not treat this matter and therefore must ask you to rely simply upon your intuition.

The total length of the circumference of the unit circle is 2π units. (For a circle of radius r, the circumference C is given by $C = 2\pi r$.) The number π is irrational, and, approximately, $\pi = 3.14159$.

Returning now to Fig. 8.2, we note that associated with each real number θ there is a unique, ordered pair (x,y) which are the coordinates of the end point of the arc θ whose initial point is (1,0).

Exercise A. Is there a unique arc beginning at (1,0) the coordinates of whose end point are a given pair (x,y)?

Definitions: We define x and y, respectively, to be $\cos\theta$ and $\sin\theta$ and write

$$x = \cos\theta$$
$$y = \sin\theta$$

These are to be read "x is the cosine of the real number θ" and "y is the sine of the real number θ". The sine and cosine are therefore functions of the real number θ and could be expressed in any one of the several notations previously suggested in Sec. 6.2.

As an immediate consequence of these definitions, we have the following theorem:

Theorem 2

$$\forall_\theta: \cos^2\theta + \sin^2\theta = 1 \tag{1}$$

This is read "for every real number θ the square of the cosine of θ plus the square of the sine of θ equals unity". An equation such as (1) that holds for every value of the variable is called an *identity,* but we usually omit the writing of the quantifier $\forall_\theta$.

Proof: This follows since (x,y) is on the circle if and only if

$$x^2 + y^2 = 1$$

Exercise B. Discuss domain of definition for each of sine and cosine. Also discuss the range of these functions.

Directly from Fig. 8.2, we see that $\sin 0 = 0$, $\sin\pi = 0$, $\sin(-\pi) = 0$, $\sin(-2\pi) = 0, \ldots, \sin n\pi = 0$ for any integer n.

Exercise C. For what values of θ is $\cos\theta = 0$?

Exercise D. For what values of θ is

(a) $\sin\theta = 1$? **(b)** $\cos\theta = 1$? **(c)** $\sin\theta = -1$? **(d)** $\cos\theta = -1$?

Relation (1) shows that sine and cosine are not independent: when either is determined for a given θ, the other can be computed to within sign from this identity. But each of these functions is useful as are the following additional functions.

Definition: The trigonometric functions tangent, secant, cosecant, and cotangent are defined as follows:

$$\text{tangent} = \frac{\text{sine}}{\text{cosine}}$$
$$\text{secant} = \frac{1}{\text{cosine}}$$
$$\text{cosecant} = \frac{1}{\text{sine}}$$
$$\text{cotangent} = \frac{\text{cosine}}{\text{sine}}$$

Abbreviated, and when written with the real number θ, these become, respectively,

$$\tan\theta = \frac{\sin\theta}{\cos\theta} \qquad \cos\theta \neq 0$$
$$\sec\theta = \frac{1}{\cos\theta} \qquad \cos\theta \neq 0$$
$$\csc\theta = \frac{1}{\sin\theta} \qquad \sin\theta \neq 0$$
$$\cot\theta = \frac{\cos\theta}{\sin\theta} \qquad \sin\theta \neq 0$$

Exercise E. What is the domain of definition of each of the functions tangent, secant, cosecant, cotangent? (Their ranges will be treated in Sec. 8.6.)

Exercise F. Show that:

(a) $\tan\theta = \frac{y}{x}$, $x \neq 0$.
(b) $\sec\theta = \frac{1}{x}$, $x \neq 0$.
(c) $\csc\theta = \frac{1}{y}$, $y \neq 0$.
(d) $\cot\theta = \frac{x}{y}$, $y \neq 0$.

Exercise G. Prove that for each value of θ for which the functions are defined:

(a) $1 + \tan^2\theta = \sec^2\theta$.
(b) $1 + \cot^2\theta = \csc^2\theta$.

HINT: Use Exercise F and the relation $x^2 + y^2 = 1$.

Problems 8.3

1. For each real number θ below, find the value (if it exists) of $\sin\theta$, $\cos\theta$, and $\tan\theta$.

(a) $\theta = 0$.
(b) $\theta = \pi/2$.
(c) $\theta = \pi$.
(d) $\theta = \frac{3}{2}\pi$.
(e) $\theta = 2\pi$.
(f) $\theta = 3\pi$.
(g) $\theta = -10\pi$.
(h) $\theta = -93\pi$.

2. For each real number θ below, find the value (if it exists) of cot θ, sec θ, and csc θ.

(a) $\theta = 0$. (b) $\theta = \pi/2$. (c) $\theta = \pi$. (d) $\theta = \frac{3}{2}\pi$.
(e) $\theta = 2\pi$. (f) $\theta = -\frac{7}{2}\pi$. (g) $\theta = -14\pi$. (h) $\theta = -61\pi$.

3. Draw a figure indicating approximately every arc θ (where $0 \leq \theta \leq 2\pi$) for which:

(a) $\sin \theta = \frac{3}{4}$. (b) $\sin \theta = \frac{3}{4}$ and $\tan \theta$ is negative.
(c) $\cos \theta = -\frac{1}{3}$. (d) $\cos \theta = -\frac{1}{3}$ and $\tan \theta$ is positive.
(e) $\tan \theta = 2.5$. (f) $\tan \theta = 2.5$ and $\sin \theta$ is negative.

4. Draw a figure indicating approximately every arc θ (where $0 \leq \theta \leq 2\pi$) for which:

(a) $\cot \theta = -3$. (b) $\cot \theta = -3$ and $\sec \theta$ is positive.
(c) $\sec \theta = 2$. (d) $\sec \theta = 2$ and $\csc \theta$ is negative.
(e) $\csc \theta = 3$. (f) $\csc \theta = 3$ and $\cot \theta$ is negative.

5. Prove: $\forall_\theta$: $\sin \theta = -\sin(-\theta)$.
6. Prove: $\forall_\theta$: $\cos \theta = \cos(-\theta)$.
7. Prove: $\exists_\theta \exists_\phi$: $\sin(\theta + \phi) \neq \sin \theta + \sin \phi$.
8. Prove: $\exists_\theta \exists_\phi$: $\cos(\theta - \phi) \neq \cos \theta - \cos \phi$.
9. Prove false: $\forall_\theta$: $\sin 2\theta = 2 \sin \theta$.
10. Prove false: $\forall_\theta$: $\cos 2\theta = 2 \cos \theta$.

In Probs. 11 to 14 draw an approximate figure and prove that, for all θ:

11. $\sin(\theta + 2n\pi) = \sin \theta$.
12. $\sin[\theta + (2n + 1)\pi] = -\sin \theta$.
13. $\cos(\theta + 2n\pi) = \cos \theta$.
14. $\cos[\theta + (2n + 1)\pi] = -\cos \theta$.
15. If $\tan \theta = a$, find $\cot(\theta + \pi/2)$.
16. Find all numbers θ for which $\sin \theta = \cos \theta$.
17. Refer to Exercise G above and show that

$$\sec^2 \theta - \csc^2 \theta = \tan^2 \theta - \cot^2 \theta$$

18. For $0 < \theta < \pi/2$, show that $\sin \theta < \tan \theta$.

8.4 Special Real Numbers

We wish to discuss the trigonometric functions sine, cosine, and tangent for the special real numbers $\theta = \pi/4$, $\theta = \pi/6$, and $\theta = \pi/3$.

Case I: $\theta = \pi/4$. From Fig. 8.3 we wish to compute the value of $x = \cos(\pi/4)$ and $y = \sin(\pi/4)$. Arc BA is given to be equal to $\pi/4$. Since arc BC is one-fourth of the complete circumference, it is clear that arc $BC = \pi/2$. It is easy to prove that triangle OAE is isosceles.

Hence $x = y$. Since $x^2 + y^2 = 1$, we have

$$x^2 + x^2 = 1$$
$$2x^2 = 1$$
$$x^2 = \tfrac{1}{2}$$
$$x = \sqrt{\frac{1}{2}} = \frac{\sqrt{2}}{2}$$

Also

$$y = \sqrt{\frac{1}{2}} = \frac{\sqrt{2}}{2}$$

(We do not use the $-\frac{1}{2}\sqrt{2}$ since, in the figure, x is positive.)

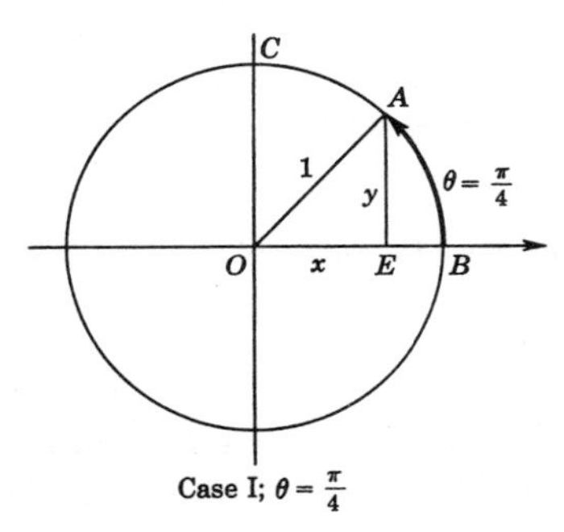

Case I; $\theta = \frac{\pi}{4}$

Figure 8.3

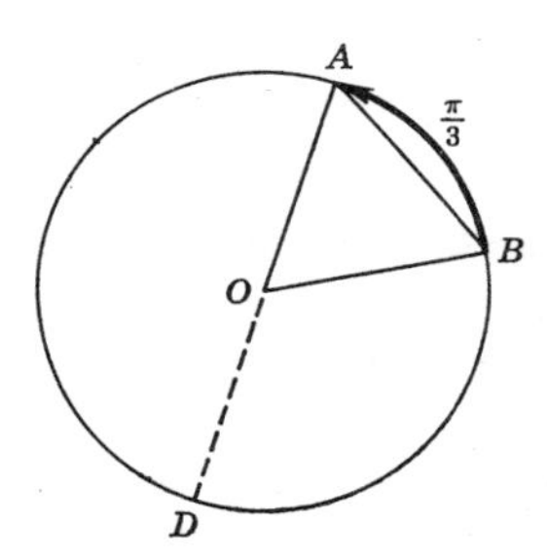

Figure 8.4

Therefore $\sin(\pi/4) = \frac{1}{2}\sqrt{2} = 0.7071$, $\cos(\pi/4) = \frac{1}{2}\sqrt{2} = 0.7071$, and $\tan(\pi/4) = 1.0000$.

Exercise A. Prove that triangle OAE in Fig. 8.3 is isosceles.

Exercise B. Draw illustrative figures and write down the values of $\sin\theta$, $\cos\theta$, and $\tan\theta$, where:

(a) $\theta = \frac{3}{4}\pi$. (b) $\theta = \frac{5}{4}\pi$. (c) $\theta = \frac{7}{4}\pi$.

In treating case II ($\theta = \pi/6$) and case III ($\theta = \pi/3$), we make use of the following lemma, referring to Fig. 8.4:

Lemma. In a circle the triangle BOA formed by the two radii BO and AO and the chord AB subtending an arc of $\pi/3$ is an equilateral triangle.

Exercise C. Prove the above lemma.

Case II: $\theta = \pi/6$. Place the triangle BOA in the unit circle as in Fig. 8.5*a*. Draw the perpendicular bisector OE of AB. Thus we see that $2y = 1$,

$y = \frac{1}{2}$. From $x^2 + y^2 = 1$, it follows that $x = \frac{1}{2}\sqrt{3}$. Therefore

$$\sin\frac{\pi}{6} = \frac{1}{2} = 0.5000 \quad \cos\frac{\pi}{6} = \frac{1}{2}\sqrt{3} = 0.8660 \quad \tan\frac{\pi}{6} = \frac{1}{\sqrt{3}} = 0.5774$$

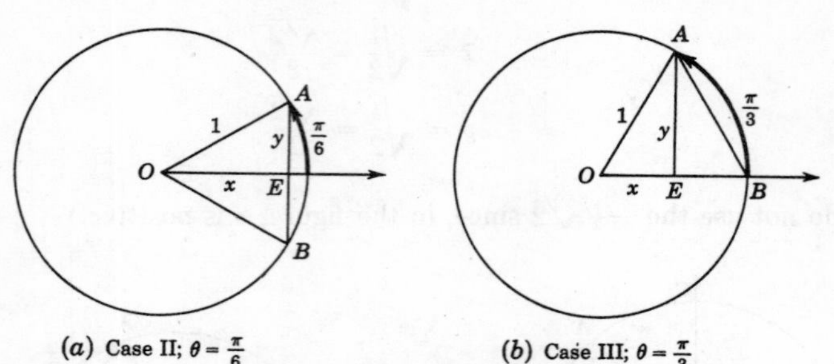

(*a*) Case II; $\theta = \frac{\pi}{6}$ (*b*) Case III; $\theta = \frac{\pi}{3}$

***Figure* 8.5**

Exercise D. Draw illustrative figures and write down the values of $\sin\theta$, $\cos\theta$, and $\tan\theta$, where:

(a) $\theta = \frac{5}{6}\pi$. (b) $\theta = \frac{7}{6}\pi$. (c) $\theta = \frac{11}{6}\pi$.

Case III: $\theta = \pi/3$. Place the triangle BOA in the unit circle as in Fig. 8.5*b*, and drop the perpendicular bisector AE of OB. Thus $x = \frac{1}{2}$ and $y = \frac{1}{2}\sqrt{3}$. Hence

$$\sin\frac{\pi}{3} = \frac{1}{2}\sqrt{3} = 0.8660 \qquad \cos\frac{\pi}{3} = \frac{1}{2} = 0.5000 \qquad \tan\frac{\pi}{3} = \sqrt{3} = 1.7321$$

Exercise E. Draw illustrative figures, and write down the values of $\sin\theta$, $\cos\theta$, and $\tan\theta$, where:

(a) $\theta = \frac{2}{3}\pi$. (b) $\theta = \frac{4}{3}\pi$. (c) $\theta = \frac{5}{3}\pi$.

Problems 8.4

In Probs. 1 to 6, for each real number θ indicated, find the value (if it exists) of $\sin\theta$, $\cos\theta$, and $\tan\theta$. Draw a figure for each θ.

1. $\theta = -\pi/4, -\frac{3}{4}\pi, -\frac{5}{4}\pi, -\frac{7}{4}\pi$.
2. $\theta = \frac{19}{4}\pi, \frac{151}{4}\pi, -\frac{160}{4}\pi, \frac{161}{4}\pi$.
3. $\theta = \frac{1}{4}\pi + 2n\pi, \frac{3}{4}\pi + 2n\pi$, ($n$ an integer).
4. $\theta = \frac{5}{4}\pi + 2n\pi, \frac{7}{4}\pi + 2n\pi$, ($n$ an integer).
5. $\theta = \frac{1}{6}\pi + 2n\pi, \frac{5}{6}\pi + 2n\pi, \frac{7}{6}\pi + 2n\pi, \frac{11}{6}\pi + 2n\pi$, ($n$ an integer).
6. $\theta = \frac{1}{3}\pi + 2n\pi, \frac{2}{3}\pi + 2n\pi, \frac{4}{3}\pi + 2n\pi, \frac{5}{3}\pi + 2n\pi$, ($n$ an integer).
7. Find the values of $1/\cos\theta$, $1/\sin\theta$, and $1/\tan\theta$ (if they exist) for $\theta = 0, \pi/2, \pi, \frac{3}{2}\pi, 2\pi$. Draw figures.

8. Draw a figure indicating every arc θ (where $0 \leq \theta \leq 2\pi$) for which:

(a) $\tan \theta$ fails to exist.   (b) $1/\cot \theta$ fails to exist.

9. Is there a number θ such that $\cos \theta = 2$? Why?
10. Is there a number θ such that $\csc \theta = 0.5$? Why?
11. For $\theta = \frac{2}{7}\pi$, find the value of $\sin^2 \theta + \cos^2 \theta$.
12. If $\sec \theta = 2.5$, find the value of $1 + \tan^2 \theta$.

8.5 General Real Numbers

In the preceding section we dealt with the special numbers $\theta = \pi/6$, $\theta = \pi/4$, $\theta = \pi/3$ (and also the numbers $0, \pi/2, \pi, \frac{3}{2}\pi$) because, by very elementary methods, we can compute the values of the trigonometric functions of them and of their counterparts $\left(\frac{\pi}{6} + n\frac{\pi}{2}, \text{etc.}\right)$. There are methods involving higher mathematics for obtaining approximate values of the functions for any real number θ (Sec. 10.7). Many such computations have been made and put in tabular form. Table IV in the Appendix is adequate for our purposes. From it we read:

$$\begin{aligned} \sin 0.24 &= 0.23770 \\ \cos 0.82 &= 0.68222 \\ \tan 1.47 &= 9.8874 \\ \cot 1.95 &= -0.39849 \end{aligned}$$

Exercise A. Why is $\cot 1.95$ negative?

Exercise B. Use a few values from Table IV to plot $y = \sin x$, $0 \leq x \leq 2$.

Exercise C. In advanced mathematics it is shown that, for small values of θ, $\sin \theta$ can be obtained approximately from the formula:

$$\sin \theta = \theta - \frac{\theta^3}{6} + \frac{\theta^5}{120}$$

Put $\theta = 0.1$, and find $\sin \theta$ from this formula. The result is 0.09983. Compare this with the value for $\sin 0.1$ in the table.

Exercise D. Compute $\sin 0.2$ by the formula of Exercise C, and compare with the tabular value.

Exercise E. For small θ,

$$\cos \theta = 1 - \frac{\theta^2}{2} + \frac{\theta^4}{24}$$

approximately. From this compute $\cos 0.1$, and compare with the value in the table.

As a matter of fact, we can define sin θ and cos θ by the following infinite sums (to be explained in Chap. 10):

$$\sin\theta = \theta - \frac{\theta^3}{3!} + \frac{\theta^5}{5!} - \cdots + (-1)^{n-1}\frac{\theta^{2n-1}}{(2n-1)!} + \cdots$$

$$\cos\theta = 1 - \frac{\theta^2}{2!} + \frac{\theta^4}{4!} - \cdots + (-1)^{n-1}\frac{\theta^{2n-2}}{(2n-2)!} + \cdots$$

Compare these expressions with those in Prob. 11, Sec. 7.2, for sinh θ and cosh θ.

Problems 8.5

In Probs. 1 to 6 find sin θ from an appropriate table.

1. $\theta = 0.25$.
2. $\theta = 0.50$.
3. $\theta = 1.00$.
4. $\theta = 1.39$.
5. $\theta = \pi + 0.20$.
6. $\theta = \pi - 0.20$.

In Probs. 7 to 12 find cos θ from an appropriate table.

7. $\theta = 0.30$.
8. $\theta = 1.30$.
9. $\theta = 1.57$.
10. $\theta = 1.58$.
11. $\theta = \pi + 0.20$.
12. $\theta = \pi - 0.20$.

In Probs. 13 to 18 find tan θ from an appropriate table.

13. $\theta = 1.57$.
14. $\theta = 1.58$.
15. $\theta = \pi + 0.90$.
16. $\theta = 2\pi - 0.20$.
17. $\theta = \pi - 1.20$.
18. $\theta = 3\pi - 0.60$.

19. From the approximations $\sin\theta = \theta - (\theta^3/6) + (\theta^5/120)$ and $\cos\theta = 1 - (\theta^2/2) + (\theta^4/24)$, compute $\sin^2\theta + \cos^2\theta$.

20. From the approximations $\sinh\theta = \theta + (\theta^3/6) + (\theta^5/120)$ and $\cosh\theta = 1 + (\theta^2/2) + (\theta^4/24)$, compute $\cosh^2\theta - \sinh^2\theta$. Check your result by using the exponential definitions of sinh θ and cosh θ given in Sec. 7.2.

8.6 Range and Graphs of the Functions

As the arc θ increases from 0 to 2π, the trigonometric functions vary. It is relatively simple to see how each varies; we discuss only the cases of sin, cos, and tan. Draw the unit circle, and consider several arcs such that $0 < \theta_1 < \cdots < \theta_3 < \pi/2$ (Fig. 8.6). Remember that, for a given θ, which is a real number, the abscissa x *is* the cos θ and the ordinate y *is* sin θ. Now erect a line tangent to the circle at $P(1,0)$; also draw the several lines from the origin to the end points of $\theta_1, \theta_2, \ldots, \theta_3$, extending these to intersect the tangent line at $A_1, A_2, \ldots, A_3$, respectively. It is seen that the length of the segment of the tangent PA_1 is actually equal to

$\tan \theta_1$, by proportional parts of the triangles OPA_1 and OCB. Indeed, this is the source of the name "tangent" of θ. By observing the variation of the length of PA as θ varies, we can obtain the behavior of $\tan \theta$ as θ varies. From Fig. 8.2 we can read off the variations of $\sin \theta$ and $\cos \theta$. The results are tabulated below.

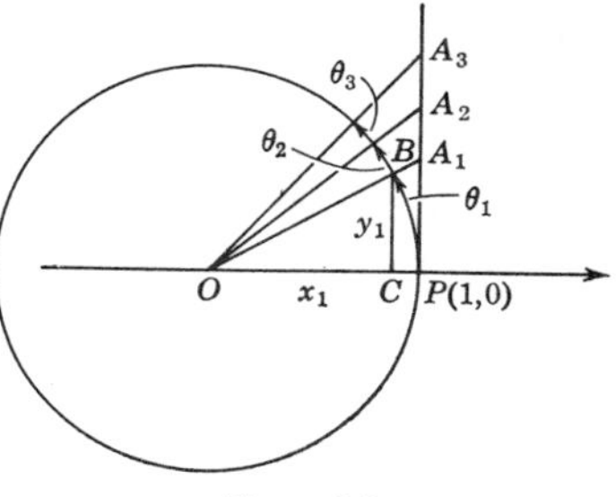

Figure **8.6**

The entries $\pm\infty$ under $\tan \theta$ need further explanation. It is quite evident that, as θ $(0 < \theta < \pi/2)$ gets nearer and nearer to $\pi/2$, $\tan \theta$ gets larger and larger. When the arc is exactly $\pi/2$, the value of the tan ceases to exist. We indicate this here by writing $\tan(\pi/2) = \infty$. However, in the second quadrant, the tan is negative; hence the entry of $-\infty$ on the second line. Similarly for III and IV quadrant entries.

Quadrant	As θ varies from:	$\sin \theta$ varies from:	$\cos \theta$ varies from:	$\tan \theta$ varies from:
I	0 to $\pi/2$	0 to 1	1 to 0	0 to ∞
II	$\pi/2$ to π	1 to 0	0 to -1	$-\infty$ to 0
III	π to $\frac{3}{2}\pi$	0 to -1	-1 to 0	0 to ∞
IV	$\frac{3}{2}\pi$ to 2π	-1 to 0	0 to 1	$-\infty$ to 0

Exercise A. Draw a figure similar to Fig. 8.6 but for arcs θ where

$$\pi/2 < \theta < \pi$$

Moreover, the variations are "essentially" the same for a given function, quadrant by quadrant: a function may increase or decrease, be positive or turn negative, but the range is the same if sign is disregarded. This will become clearer as we begin to graph the functions. We are already in a position to make up the following detailed table for the first quadrant. The entries below were found in the problems at the end of Sec. 8.4.

θ	$\sin \theta$	$\cos \theta$	$\tan \theta$
0	0	1	0
$\pi/6$	$\frac{1}{2} = 0.500$	0.866	$\frac{1}{3}\sqrt{3} = 0.577$
$\pi/4$	$\frac{1}{2}\sqrt{2} = 0.707$	0.707	1
$\pi/3$	$\frac{1}{2}\sqrt{3} = 0.866$	0.5	$\sqrt{3} = 1.732$
$\pi/2$	1	0	∞

For your own understanding you should extend this table through the other three quadrants. Figure 8.7 is helpful. With the aid of this information we sketch the graphs as in Figs. 8.8 to 8.10, which also include the graphs of the sec, csc, and cot. To plot a more accurate graph, we could

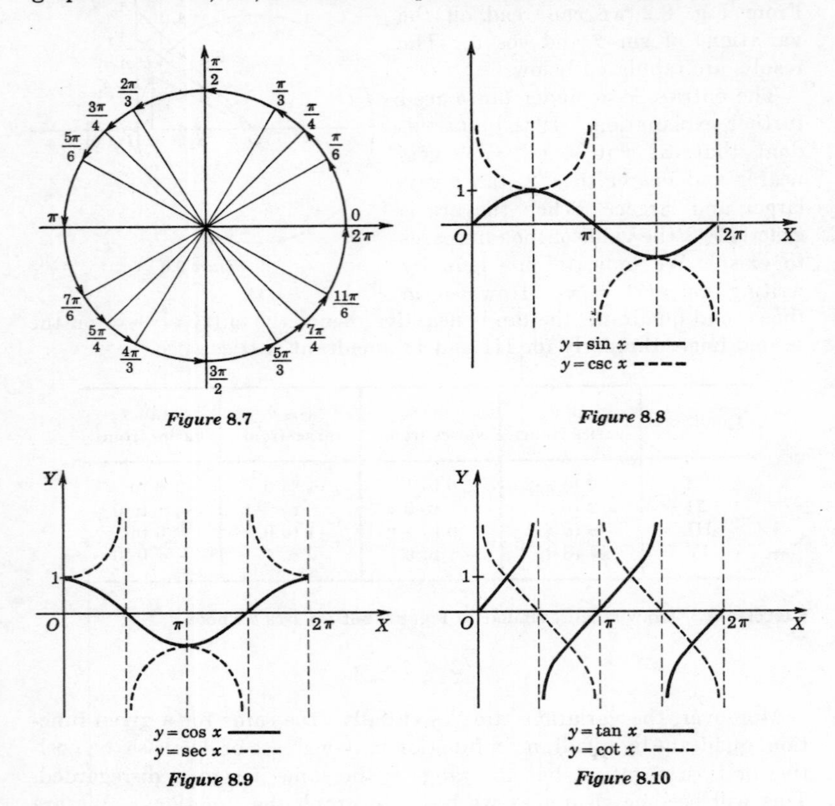

Figure 8.7

Figure 8.8

Figure 8.9

Figure 8.10

obtain other elements from a table of these functions. Note that sin, cos, sec, csc repeat after 2π, but that tan and cot repeat after π; that is, sin, cos, sec, csc are periodic functions of period 2π, and tan and cot are periodic functions of period π, according to the following definitions.

Definitions: A function f such that $f(x + a) = f(x)$ for some positive a and all x is said to be a periodic function. The least positive number a for which this is true is called the period of the function.

Exercise B. Prove that a constant function is a periodic function.

Exercise C. Prove that no periodic function can be a rational function. HINT: Use the theorem that if a polynomial of degree n has more than n zeros, then the polynomial is identically zero (see Sec. 5.13, Prob. 42).

Exercise D. Prove that $f(x + 2a) = f(x)$ if f is a periodic function of period a.

Illustration 1. Sketch on the same axes and to the same scale the graphs of $y = \sin x$ and $y = 3 \sin (\frac{1}{2}x)$, $0 \leq x \leq 4\pi$.

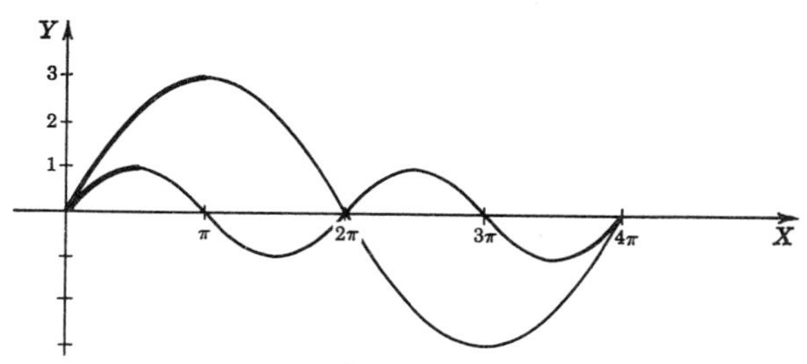

Figure 8.11

Solution: We compute the following entries, treating $\sin x$ and $3 \sin \frac{1}{2}x$ separately and making use only of the special values we know about. (For a more accurate graph, we should make use of a table.)

x	$\frac{1}{2}x$	$y = \sin x$	$\sin \frac{1}{2}x$	$y = 3 \sin \frac{1}{2}x$
0	0	0.000	0.000	0.000
$\pi/6$		0.500		
$\pi/4$		0.707		
$\pi/3$	$\pi/6$	0.866	0.500	1.500
$\pi/2$	$\pi/4$	1.000	0.707	2.121
$2\pi/3$	$\pi/3$	0.866	0.866	2.598
$3\pi/4$		0.707		
$5\pi/6$		0.500		
π	$\pi/2$	0.000	1.000	3.000

Now the graph of $y = \sin x$ is "essentially" the same in the second, third, and fourth quadrant as it is in the first quadrant. By this we mean that it is the same shape but placed differently. Also, $\sin x$ is periodic of period 2π while $\sin \frac{1}{2}x$ is periodic of period 4π. With this information we sketch the graphs in Fig. 8.11. The whole of a given curve is sketched by means of a template with which the heavy portion was drawn.

Illustration 2. Sketch the graph of $y = \sin x + \frac{1}{3} \sin 3x$.

Solution: We should now know enough to sketch

$$Y_1 = \sin x \qquad \text{and} \qquad Y_2 = \tfrac{1}{3} \sin 3x$$

They are dotted and labeled in Fig. 8.12. Then we sketch $y = Y_1 + Y_2$ by adding the ordinates of the two curves.

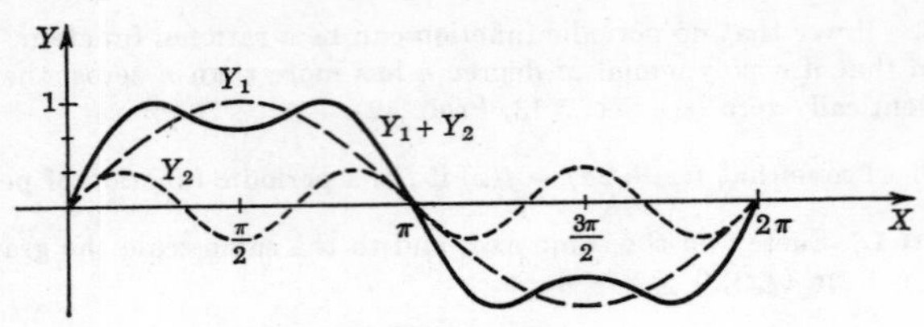

Figure 8.12

Problems 8.6

1. Figures 8.2 and 8.6 give the so-called line values of the three functions sin, cos, tan. Prepare a similar figure for sec, csc, and cot.

In Probs. 2 to 13 sketch on the same axes and, to the same scale, the following pairs of graphs (complete period of each).

2. $y = \sin x$, $y = \sin 2x$.
3. $y = \sin x$, $y = \frac{1}{2} \sin 2x$.
4. $y = \sin x$, $y = 2 \sin x$.
5. $y = \sin x$, $y = 2 \sin \frac{1}{2}x$.
6. $y = \cos x$, $y = \cos 2x$.
7. $y = \cos x$, $y = \frac{1}{2} \cos 2x$.
8. $y = \cos x$, $y = 2 \cos x$.
9. $y = \cos x$, $y = 2 \cos \frac{1}{2}x$.
10. $y = \sin x$, $y = \sin 3x$.
11. $y = \cos x$, $y = \cos 3x$.
12. $y = \sin x$, $y = \sin (x + \pi/2)$.
13. $y = \cos x$, $y = \frac{1}{2} \cos (x - \pi/4)$.

In Probs. 14 to 25 sketch a complete period.

14. $y = \sin x + \cos x$. (Consider this as a sum $f + g$, plot the graphs of f and g separately, and then add coordinates.)
15. $y = \sin x - \cos x$.
16. $y = \sin x - \frac{1}{3} \sin 3x$.
17. $y = \cos x + \frac{1}{3} \cos 3x$.
18. $y = 2 \sin x - \cos x$.
19. $y = 2 \cos x - \sin x$.
20. $y = \sin x + \frac{1}{2} \sin 2x$.
21. $y = \cos x + \frac{1}{2} \cos 2x$.
22. $y = \sin^2 x + \cos^2 x$. (BT)

23. What is the range of sec, quadrant by quadrant?
24. What is the range of csc, quadrant by quadrant?
25. What is the range of cot, quadrant by quadrant?

8.7 Amplitude, Period, Phase

One of the most important trigonometric concepts is that of a "sinusoidal wave". It occurs in innumerable ways and places in astronomy, mathematics, and all the sciences including the social sciences. It is, simply, the graph of $y = A \sin (Bx + C)$, where A, B, C are positive constants.

We begin with a comparison of the graphs of

$$y = \sin x$$
$$y = A \sin x$$

These are exhibited superimposed on the same axes and drawn to the same scales, in Fig. 8.13.

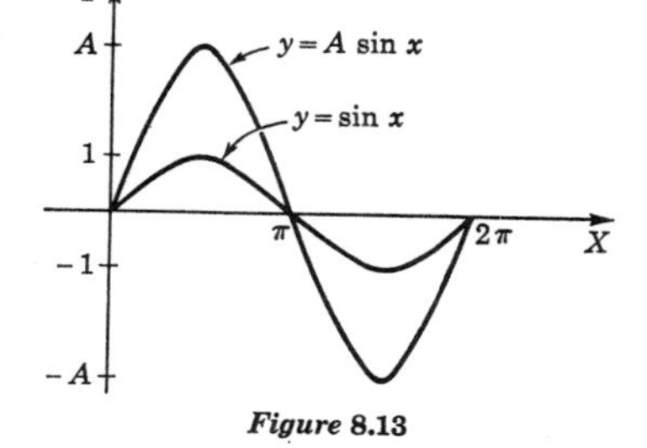

Figure 8.13

Since the maximum value of $\sin x$ is 1 and occurs when

$$x = \frac{\pi}{2} + 2k\pi$$

it is evident that A is the maximum value of $A \sin x$. The constant A is called the *amplitude* of the sine wave (sinusoidal wave). The *period* p of $y = \sin x$ (and of $y = A \sin x$) is $p = 2\pi$.

Next we compare

$$y = \sin x$$
$$y = \sin Bx$$

Now when $Bx = 0$, $x = 0$, and when $Bx = 2\pi$, $x = 2\pi/B$. Therefore the period p of $y = \sin Bx$ is $p = 2\pi/B$ (Fig. 8.14). Combining these two ideas, we have $y = A \sin Bx$ (Fig. 8.15).

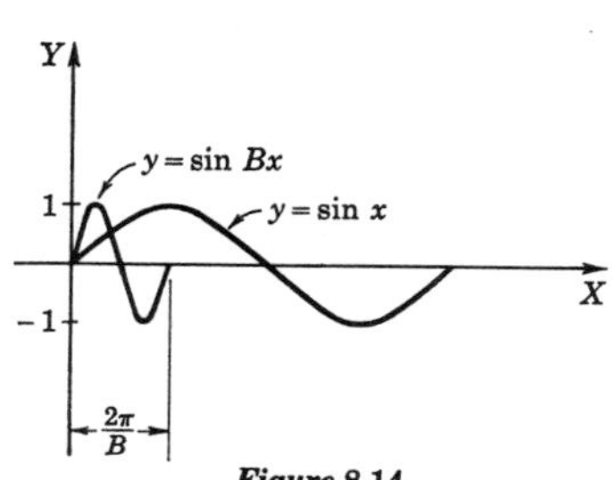

Figure 8.14

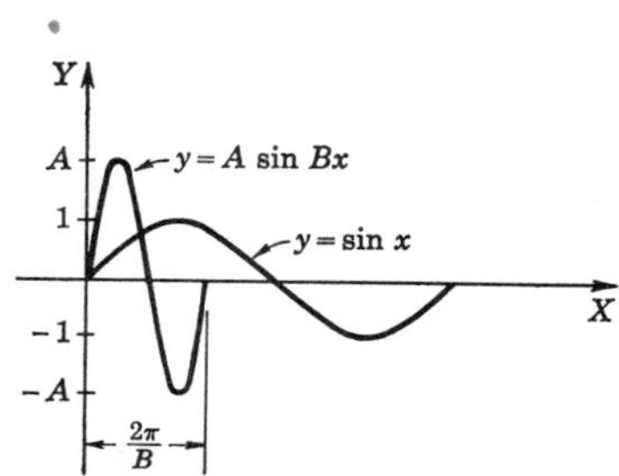

Figure 8.15

The frequency of an oscillation is the number of periods which take place in some standard interval of time, usually one second. Frequency is measured in periods per second, or, in more usual language, in cycles per second. A cycle is the same thing as a period. In radio broadcasting the convenient unit is kilocycles (one thousand cycles) per second. These are the numbers on your radio dial. The usual house current is "60-cycle" current, and this means a frequency of 60 cycles/sec.

If an oscillation has a period of p sec, it has a frequency of

$$\omega = \frac{1}{p} \text{ cycles/sec}$$

The frequency of $A \sin Bx$ is $\omega = B/2\pi$, and the frequency of $A \sin (2\pi\omega t)$ is ω cycles/sec. In radio waves the period is usually expressed in terms of the distance which the wave travels in p sec; this distance is called the *wavelength* λ. Since the velocity of a radio wave is 3×10^{10} cm/sec, the wavelength corresponding to one period of p sec is $3p \times 10^{10}$ cm. Thus, in terms of frequency,

$$\lambda = \frac{3 \times 10^{10}}{\omega} \quad \text{cm}$$

Using this formula, one can convert from frequencies to wavelengths and back again. For instance, a broadcast signal with a frequency of 600 kc/sec has a wavelength

$$\lambda = \frac{3 \times 10^{10}}{600 \times 10^3} \text{ cm} = 500 \text{ meters}$$

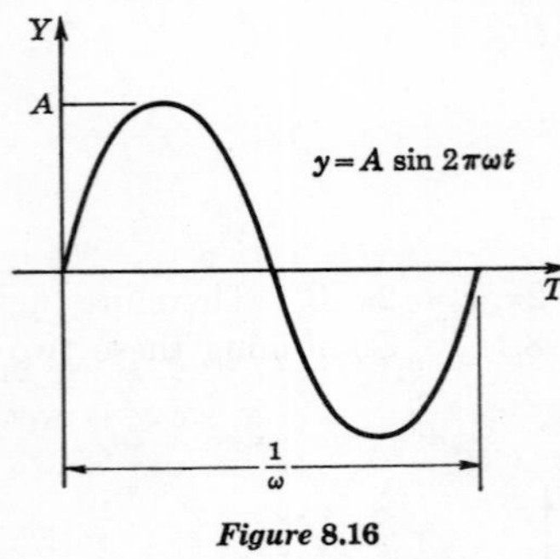

Figure 8.16

As you well know, radio waves are used to transmit the sounds of human voices and musical instruments. To keep things simple, let us restrict ourselves to a pure musical tone. Such pure tones are represented by a sine curve in which the amplitude corresponds to the loudness and the frequency ω to the pitch (Fig. 8.16). For example, the frequency of middle C is 512 cycles/sec. As we have seen, the equation of such a sine curve is

$$(1) \qquad y = A \sin (2\pi\omega t)$$

The radio station transmits a *carrier wave* which is also a sine curve whose frequency is that assigned to the particular station. These frequencies are much higher than those of sound waves, for example, 800,000 cycles/sec for standard broadcasts and 80,000,000 cycles/sec for the sound portion of a television signal. Let us write the equation of this carrier wave as

$$(2) \qquad y = A_0 \sin (2\pi\omega_0 t)$$

where A_0 is its amplitude, and ω_0 its frequency.

The carrier wave can now be used to transmit the pure tone of (1) above by imposing on it a periodic change in its amplitude A_0. Instead of holding A_0 constant, the station *modulates* A_0 so that it has the value

$$A_0(t) = A_0 + mA_0 \sin (2\pi\omega t)$$

where ω is the frequency of the pure tone in Eq. (2), and m is a factor of proportionality called the *degree of modulation.* The transmitted wave then has the equation

$$y = A_0[1 + m \sin (2\pi\omega t)] \sin (2\pi\omega_0 t)$$

whose graph is sketched in Fig. 8.17. The receiver then converts this signal back into the original pure tone, which is sent out through the loud speaker. This process is called *amplitude modulation*, or AM.

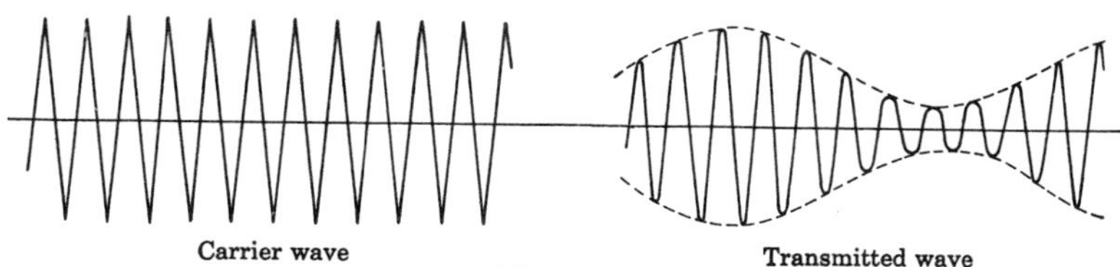

Figure 8.17

Alternatively, the frequency of the carrier wave can be modulated in conformity with the tone to be transmitted. In order to do this, a band of frequencies centered on the carrier frequency is selected, say, those in the interval $[\omega - a, \omega + a]$. The frequency ω_0 of the carrier wave is now forced to vary according to the equation

$$\omega_0(t) = \omega_0 + \frac{a}{2\pi\omega t} \sin (2\pi\omega t)$$

The transmitted wave then has the equation

$$\begin{aligned} y &= A_0 \sin \left[2\pi t \left(\omega_0 + \frac{a}{2\pi\omega t} \sin 2\pi\omega t \right) \right] \\ &= A_0 \sin \left[2\pi\omega_0 t + \frac{a}{\omega} \sin 2\pi\omega t \right] \end{aligned}$$

Its graph is plotted in Fig. 8.18. This process is called *frequency modulation*, or FM.

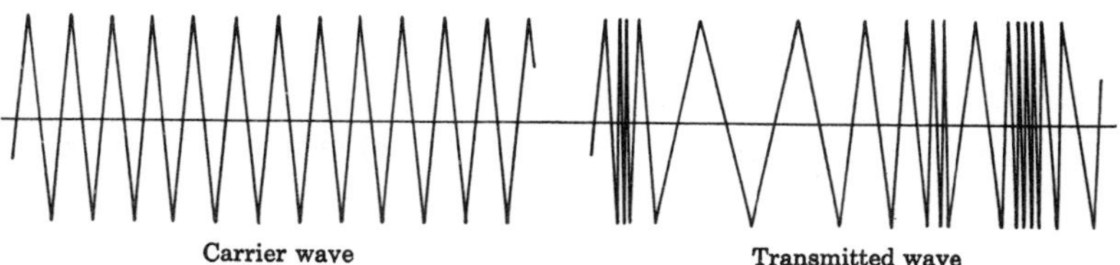

Figure 8.18

Consider, now, the graph of $y = \sin(x + C)$. When $x + C = 0$, $x = -C$, and when $x + C = 2\pi$, $x = 2\pi - C$. The graph, indicated in Fig. 8.19, is therefore a sine wave shifted to the *left* by an amount C.

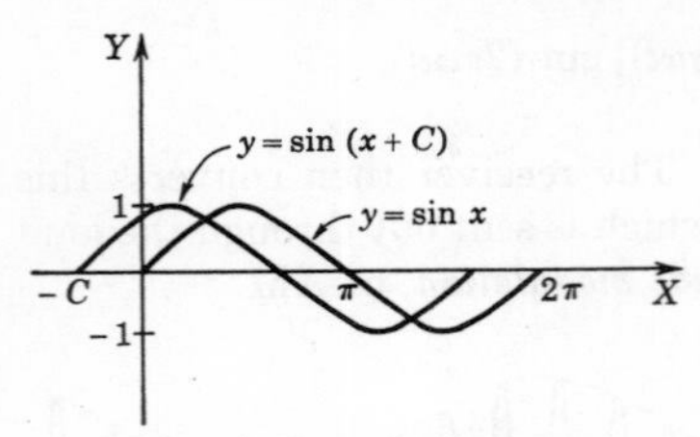

Figure 8.19

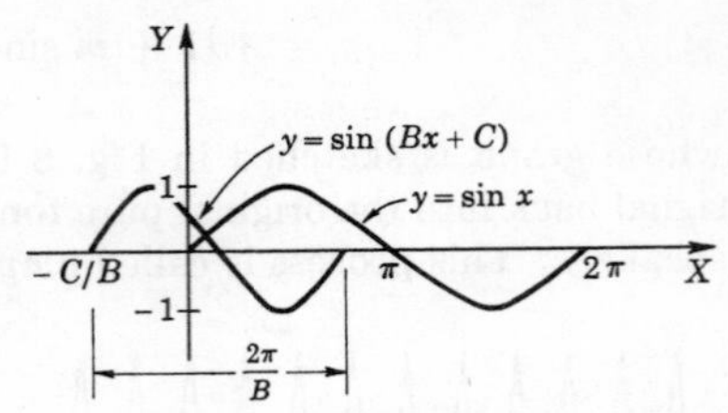

Figure 8.20

The constant C is called the *phase shift*, or *phase angle*.

For the wave $y = \sin(Bx + C)$, we note that, when $Bx + C = 0$, $x = -C/B$, and when $Bx + C = 2\pi$, $x = (2\pi - C)/B$ (Fig. 8.20). Here the *phase shift* is represented by the number $-C/B$.

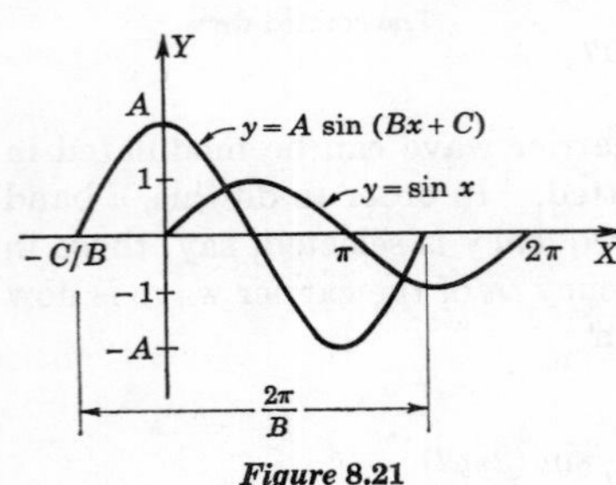

Figure 8.21

Finally, we combine all these ideas in the representation of the most general sine wave:

$$y = A \sin(Bx + C)$$

The *amplitude* is A, the *period* is $2\pi/B$, and the *phase shift* is $-C/B$ (Fig. 8.21).

Similar remarks apply to the graphs of the other functions.

8.8 Addition Theorems

Let us consider a function f and its values $f(x_1)$ and $f(x_2)$, where x_1 and x_2 are to be thought of as any two x's in the domain of definition such that $x_1 + x_2$ is also in the domain. The following general question arises, What can we say about $f(x_1 + x_2)$ in terms of $f(x_1)$ and $f(x_2)$ separately? Such a theorem is referred to as an *addition theorem for the function f*. Both classical and modern mathematics place emphasis on the discovery and use of such theorems. They are of special importance in trigonometry where we should like to know the answers to the following questions:

(a) Can we express $\sin(\theta \pm \phi)$ and $\cos(\theta \pm \phi)$ in terms of $\sin\theta$, $\sin\phi$, $\cos\theta$, and $\cos\phi$?

(b) If so, what are the formulas?

There are many derivations of these formulas, and they all have some degree of artificiality. This can be said about most of the theorems of plane geometry where certain construction lines—once they have been thought of and drawn in—aid in the analysis of the problem. We want to find a formula for one of the four quantities $\sin(\theta \pm \phi)$, $\cos(\theta \pm \phi)$, and therefore we begin by drawing two unit circles as in Fig. 8.22. In

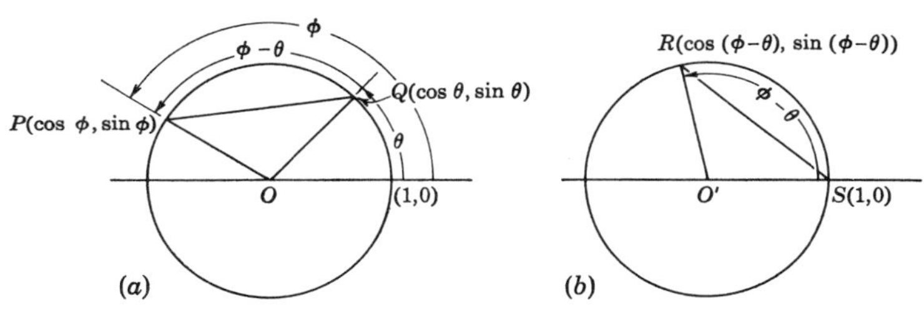

Figure **8.22**

Fig. 8.22*a*, we have laid off the arcs θ and ϕ and indicated the resulting arc $\phi - \theta$ as the arc QP. In Fig. 8.22*b* we have laid off the arc $\phi - \theta$ as the arc SR. The coordinates of the points P, Q, R, and S are indicated on the figure; their values follow from the definitions of sine and cosine.

The key to our result is the remark that the segments PQ and RS are equal, for they are corresponding sides of the congruent triangles OPQ and $O'RS$. If we express this equality by the use of the distance formula (Sec. 8.2), we find

$$
\begin{aligned}
(PQ)^2 &= (\cos\phi - \cos\theta)^2 + (\sin\phi - \sin\theta)^2 \\
&= 2 - 2(\cos\phi\cos\theta + \sin\phi\sin\theta) \qquad (3)\\
(RS)^2 &= [\cos(\phi - \theta) - 1]^2 + [\sin(\phi - \theta) - 0]^2 \\
&= 2 - 2\cos(\phi - \theta) \qquad (4)
\end{aligned}
$$

Equating these, we obtain

$$2 - 2\cos(\phi - \theta) = 2 - 2(\cos\phi\cos\theta + \sin\phi\sin\theta) \qquad (5)$$

Therefore we have the formula

$$\cos(\phi - \theta) = \cos\phi\cos\theta + \sin\phi\sin\theta \qquad (6)$$

In Fig. 8.22 we have implicitly assumed that $\phi - \theta$ is positive. If ϕ and θ are such that $\phi - \theta$ is negative, the figure will be different but the argument remains the same. Moreover, if $\phi - \theta = 2n\pi$, where n is an integer, points P and Q (and hence R and S) will coincide, so that the figure is misleading. Again, however, the algebra is the same and the result is true.

This is exactly the kind of formula we are seeking. It is an identity which expresses the cosine of the difference of two real numbers ϕ and θ in terms of the sines and cosines of ϕ and θ separately. You must memorize formula (6); it is one of the "addition theorems" for the trigonometric functions. We now derive three others, namely, (9), (13), and (14) below.

Directly from the definitions of $\sin\theta$ and $\cos\theta$ it follows that

$$\sin(-\theta) = -\sin\theta \tag{7}$$

and

$$\cos(-\theta) = \cos\theta \tag{8}$$

In (6) put $\theta = -\alpha$. [We may do this since (6) is an identity.] We get $\cos(\phi + \alpha) = \cos\phi\cos(-\alpha) + \sin\phi\sin(-\alpha)$. Using (7) and (8), this simplifies to the second important "addition theorem":

$$\cos(\phi + \alpha) = \cos\phi\cos\alpha - \sin\phi\sin\alpha \tag{9}$$

Next, in (6), let us put $\theta = \pi/2$. We get

$$\cos\left(\phi - \frac{\pi}{2}\right) = \cos\phi\cos\frac{\pi}{2} + \sin\phi\sin\frac{\pi}{2}$$

or

$$\cos\left(\phi - \frac{\pi}{2}\right) = \sin\phi \tag{10}$$

If we put $\alpha = \phi - \pi/2$, or $\phi = \alpha + \pi/2$ in (10), we can write (10) in the form:

$$\cos\alpha = \sin\left(\alpha + \frac{\pi}{2}\right) \tag{11}$$

Similarly, by putting $\theta = -\pi/2$ in (6), we obtain

$$\cos\left(\phi + \frac{\pi}{2}\right) = -\sin\phi \tag{12}$$

We are now ready to derive the addition theorem for $\sin(\phi - \theta)$. We use (10) to write

$$\sin(\phi - \theta) = \cos\left[(\phi - \theta) - \frac{\pi}{2}\right] = \cos\left[\phi - \left(\theta + \frac{\pi}{2}\right)\right]$$

Now apply (6) to the right-hand expression, and obtain

$$\sin(\phi - \theta) = \cos\phi \cos\left(\theta + \frac{\pi}{2}\right) + \sin\phi \sin\left(\theta + \frac{\pi}{2}\right)$$

Using (11) and (12), we can simplify this to

$$\sin(\phi - \theta) = -\cos\phi \sin\theta + \sin\phi\cos\theta$$

or

$$\sin(\phi - \theta) = \sin\phi\cos\theta - \cos\phi\sin\theta \tag{13}$$

This is the desired result.

Finally, putting $\theta = -\alpha$ in (13), we obtain

$$\sin(\phi + \alpha) = \sin\phi\cos\alpha + \cos\phi\sin\alpha \tag{14}$$

We now collect (6), (9), (13), and (14) and write them in the form:

$$\sin(\phi \pm \theta) = \sin\phi\cos\theta \pm \cos\phi\sin\theta \tag{I}$$

$$\cos(\phi \pm \theta) = \cos\phi\cos\theta \pm \sin\phi\sin\theta \tag{II}$$

You should study this material until you understand it thoroughly. Be sure to note that (I) and (II) are identities which hold for all values of θ and ϕ. They should be memorized.

A large part of analytic trigonometry is imbedded in Probs. 8.8, and you should work through these in detail.

Problems 8.8

In Probs. 1 to 8 make use of (I) and (II) above to show that:

1. $\tan(\phi \pm \theta) = \dfrac{\tan\phi \pm \tan\theta}{1 \mp \tan\phi\tan\theta}$.
2. $\tan 2\theta = \dfrac{2\tan\theta}{1 - \tan^2\theta}$.
3. $\sin 2\theta = 2\sin\theta\cos\theta$.
4. $\sin\theta = 2\sin(\theta/2)\cos(\theta/2)$.
5. $\cos 2\theta = \cos^2\theta - \sin^2\theta$.
6. $\cos 2\theta = 2\cos^2\theta - 1$.
7. $\cos 2\theta = 1 - 2\sin^2\theta$.
8. $\cos\theta = 1 - 2\sin^2(\theta/2)$.

We define a first, second, third, and fourth quadrantal arc θ as one such that, when θ is laid off from (1,0), its end point lies in the first, second, third, and fourth quadrant, respectively.

9. Use Prob. 7 to show that

$$\sin\frac{\theta}{2} = \begin{cases} \sqrt{\dfrac{1-\cos\theta}{2}} & \dfrac{\theta}{2} \text{ a 1st or 2d quadrantal arc} \\ -\sqrt{\dfrac{1-\cos\theta}{2}} & \dfrac{\theta}{2} \text{ a 3d or 4th quadrantal arc} \end{cases}$$

10. Use Prob. 6 to show that

$$\cos\frac{\theta}{2} = \begin{cases} \sqrt{\dfrac{1+\cos\theta}{2}} & \dfrac{\theta}{2} \text{ a 1st or 4th quadrantal arc} \\ -\sqrt{\dfrac{1+\cos\theta}{2}} & \dfrac{\theta}{2} \text{ a 2d or 3d quadrantal arc} \end{cases}$$

11. Use Probs. 9 and 10 to show that

$$\tan\frac{\theta}{2} = \begin{cases} \sqrt{\dfrac{1-\cos\theta}{1+\cos\theta}} & \dfrac{\theta}{2} \text{ a 1st or 3d quadrantal arc} \\ -\sqrt{\dfrac{1-\cos\theta}{1+\cos\theta}} & \dfrac{\theta}{2} \text{ a 2d or 4th quadrantal arc} \end{cases}$$

12. Use Prob. 11 to show that

$$\tan\frac{\theta}{2} = \frac{1-\cos\theta}{\sin\theta} \qquad \text{for all } \theta \text{ for which } \tan\frac{\theta}{2} \text{ exists}$$

13. Use Prob. 11 to show that

$$\tan\frac{\theta}{2} = \frac{\sin\theta}{1+\cos\theta} \qquad \text{for all } \theta \text{ for which } \tan\frac{\theta}{2} \text{ exists}$$

14. Use Probs. 4 and 9 to show that

$$\tan\frac{\theta}{2} = \frac{1-\cos\theta}{\sin\theta} \qquad \text{for all } \theta \text{ for which } \tan\frac{\theta}{2} \text{ exists}$$

15. Use (I) to show that

$$\sin x \cos y = \tfrac{1}{2}\sin(x-y) + \tfrac{1}{2}\sin(x+y)$$

16. Use (II) to show that

$$\sin x \sin y = \tfrac{1}{2}\cos(x-y) - \tfrac{1}{2}\cos(x+y)$$

17. Use (II) to show that

$$\cos x \cos y = \tfrac{1}{2}\cos(x-y) + \tfrac{1}{2}\cos(x+y)$$

In Probs. 18 to 21 derive the identities. HINT: Use Probs. 15 to 17, and set $x + y = A$, $x - y = B$.

18. $\sin A + \sin B = 2\sin\frac{1}{2}(A+B)\cos\frac{1}{2}(A-B)$.
19. $\sin A - \sin B = 2\cos\frac{1}{2}(A+B)\sin\frac{1}{2}(A-B)$.

20. $\cos A + \cos B = 2 \cos \frac{1}{2}(A + B) \cos \frac{1}{2}(A - B)$.
21. $\cos A - \cos B = -2 \sin \frac{1}{2}(A + B) \sin \frac{1}{2}(A - B)$.
22. Find a formula for $\sin 3\theta$ in terms of $\sin \theta$ and $\cos \theta$.
23. Find a formula for $\cos 3\theta$ in terms of $\sin \theta$ and $\cos \theta$.

In problems 24 to 29 prove the identity.

24. $\sin\left(\frac{\pi}{4} - x\right) = \frac{\cos x - \sin x}{\sqrt{2}}$.
25. $\sin\left(\frac{\pi}{3} + x\right) - \sin x = \sin\left(\frac{\pi}{3} - x\right)$.
26. $\frac{\sin x + \sin y}{\cos x + \cos y} = \tan \frac{1}{2}(x + y)$. HINT: Use Probs. 18 and 20.
27. $\cot x + \cot y = \frac{\sin (x + y)}{\sin x \sin y}$.
28. $\left(\sin \frac{x}{2} - \cos \frac{x}{2}\right)^2 = 1 - \sin x$.
29. $1 + \tan x \tan \frac{1}{2}x = \sec x$.
30. Find the value of $\sin^2 (x/2) + \cos^2 (x/2)$. (BT)
31. Show that it is false that $\forall_x\ [2 \cos (x/2) = \cos x]$.
32. Show that it is false that $\exists_x\ [\sin 2x + \cos 2x = 4]$.

In Probs. 33 to 48 compute the value of sine, cosine, and tangent of each real number, leaving your answer in radical form.

33. $\frac{1}{12}\pi$. **34.** $\frac{5}{12}\pi$. **35.** $\frac{7}{12}\pi$.
36. $\frac{11}{12}\pi$. **37.** $\frac{13}{12}\pi$. **38.** $\frac{17}{12}\pi$.
39. $\frac{19}{12}\pi$. **40.** $\frac{23}{12}\pi$. **41.** $\frac{1}{8}\pi$
42. $\frac{3}{8}\pi$. **43.** $\frac{5}{8}\pi$. **44.** $\frac{7}{8}\pi$.
45. $\frac{9}{8}\pi$. **46.** $\frac{11}{8}\pi$. **47.** $\frac{13}{8}\pi$.
48. $\frac{15}{8}\pi$.

In Probs. 49 to 52 compute the value of each of the six trigonometric functions for each real number, leaving your answer in radical form.

49. $\frac{1}{24}\pi$. **50.** $\frac{1}{16}\pi$. **51.** $\frac{5}{24}\pi$. **52.** $-\frac{1}{6}\pi$.

53. Show that $\sin \frac{32}{180}\pi + \sin \frac{28}{180}\pi = \cos \frac{2}{180}\pi$. (See Prob. 18.)
54. Show that $\sin \frac{40}{180}\pi - \cos \frac{70}{180}\pi = 3 \sin \frac{10}{180}\pi$.

8.9 Identities

Many of Probs. 8.8 are identities, that is, equations that are true for every value of the variable or variables present (for which both sides are defined). Those of Probs. 1 to 21 (Sec. 8.8) are generally considered basic in a thorough and complete course in trigonometry, and often the student is asked to memorize them. These and many similar identities arise quite naturally in pure and applied mathematics and engineering. It is therefore worthwhile to work through a few more in order to fix more firmly in mind the fundamental relations existing among the trigonometric functions. Before discussing the logic of the proof of identities, let us illustrate by an example.

Illustration 1. Prove that

$$\frac{\sin^2 x}{1 + \cos x} = 1 - \cos x$$

is an identity.

Solution: We recall that we are to show that the two sides are equal for *all* values of x for which both are defined. Since $1 + \cos x = 0$ when $x = (2n + 1)\pi$, we are to show that:

For all $x \neq (2n + 1)\pi$:

$$\frac{\sin^2 x}{1 + \cos x} = 1 - \cos x$$

We give three possible procedures:

(1) **Direct Proof after Reversing Steps.** At first we proceed informally and multiply both sides by $1 + \cos x$. The result is:

$$\sin^2 x = (1 - \cos x)(1 + \cos x)$$
$$\sin^2 x = 1 - \cos^2 x$$
$$\sin^2 x = \sin^2 x$$

So far this proves nothing about the given identity, but it motivates the following steps, which are obtained by reserving the order of those above. The proof of the identity is then contained in the following argument.

For all x:

$\sin^2 x = \sin^2 x$

$\sin^2 x = 1 - \cos^2 x$ [Substitution from elementary identity]

$\sin^2 x = (1 - \cos x)(1 + \cos x)$ [Factorization]

For all $x \neq (2n + 1)\pi$:

$\dfrac{\sin^2 x}{1 + \cos x} = 1 - \cos x$ [Division by nonzero expression]

which is the required identity.

(2) **Conversion of One Side to the Other**

For all $x \neq (2n + 1)\pi$:

$\dfrac{\sin^2 x}{1 + \cos x} = \dfrac{1 - \cos^2 x}{1 + \cos x}$ [Substitution from elementary identity]

$= \dfrac{(1 + \cos x)(1 - \cos x)}{1 + \cos x}$ [Factorization]

$= 1 - \cos x$ [Division by nonzero expression]

(3) Indirect Proof. Assume that there is an $x \neq (2n + 1)\pi$ for which

$$\frac{\sin^2 x}{1 + \cos x} \neq 1 - \cos x$$

Then, for this x:

$$\sin^2 x \neq (1 - \cos x)(1 + \cos x)$$

for we have multiplied an inequality by a nonzero factor and have thus obtained an inequality. Finally, we conclude the absurd statement:

For this x:

$$\sin^2 x \neq \sin^2 x$$

From this contradiction we derive the truth of our identity.

Exercise A. Use the algebra of method (2) to construct another indirect proof.

The general methods of procedure are then as follows:

(1) Direct proof after reversing steps

(a) Through substitution from known identities and algebraic simplification involving one or both sides, transform the given identity into one whose two sides are the same.

(b) Reverse the steps in (a), and thus derive the given identity from known identities. You must verify, however, that the steps in (a) are indeed reversible. As in Illustration 1, it may be necessary to exclude certain values of the variable in order to reverse these steps, but if these are already excluded in the statement of the given identity, there is no harm. As in Illustration 2 below, however, it may be necessary to exclude values of the variable for which the identity is asserted to be true. In this case the proof fails.

Illustration 2. Prove the identity

$$\sqrt{1 - \cos^2 x} = \sin x$$

Squaring both sides, we obtain

$$\begin{aligned} 1 - \cos^2 x &= \sin^2 x \\ \sin^2 x &= \sin^2 x \end{aligned}$$

Now we must reverse the steps:
For all x:

$$\begin{aligned} \sin^2 x &= \sin^2 x \\ 1 - \cos^2 x &= \sin^2 x \\ \sqrt{1 - \cos^2 x} &= |\sin x| \end{aligned}$$

For all x in first or second quadrant:

$$\sqrt{1 - \cos^2 x} = \sin x$$

Since we must exclude x in the third or fourth quadrant, the proof fails. Actually, the identity is false for such x.

Remark. You may be tempted to try to prove an identity by using part (a) of the procedure only. This, however, is not a proof unless you make certain that the steps can be reversed. Frequently we short-cut the process by writing out (a) only and doing (b) mentally. This is satisfactory if you actually do (b) and, as in Illustration 3, make a note on your paper that you have done so.

Illustration 3. Prove the identity

$$\frac{\csc x + 1}{\cot x} = \frac{\cot x}{\csc x - 1}$$

Multiplying both sides by $\cot x\ (\csc x - 1)$, we obtain

$$\begin{aligned}(\csc x + 1)(\csc x - 1) &= \cot^2 x\\ \csc^2 x - 1 &= \cot^2 x\\ \cot^2 x &= \cot^2 x\end{aligned}$$

The steps are reversible unless $\cot x = 0$ or $\csc x = 1$. In other words, we must exclude $x = (\pi/2) + n\pi$ and $x = (\pi/2) + 2n\pi$. Since these are excluded in the given identity, the proof is complete.

(2) Conversion of one side to the other. By substitution of known identities and algebraic simplification on one side only, this side is shown to equal the other side for all required values of the variable. All work must be done on just one side of the identity, and the other side must be left untouched.

(3) Indirect proof. Assume that there is an x for which the identity is false. Reasoning from this inequality through the use of known identities and careful algebra, arrive at an absurd conclusion. Then the identity is established.

Problems 8.9

Prove the following identities.

1. $\sin x \cot x \sec x = 1.$

2. $\dfrac{\cos x \csc x}{\cot x} = 1.$

3. $\dfrac{1}{\tan x + \cot x} = \sin x \cos x.$

4. $\sec x - \tan x \sin x = \cos x.$

5. $\tan x \sin x + \cos x = \sec x.$

6. $\dfrac{\sin x}{1 + \cos x} + \dfrac{1 + \cos x}{\sin x} = 2 \csc x.$

7. $\dfrac{\sin x \sec x}{\tan x + \cot x} = 1 - \cos^2 x.$

8. $\dfrac{1 + \tan x}{1 - \tan x} + \dfrac{1 + \cot x}{1 - \cot x} = 0.$

9. $\dfrac{\cos x}{1 - \sin x} - \dfrac{1 - \sin x}{\cos x} = 2 \tan x.$

10. $\dfrac{\sin x + \cos x}{\sec x + \csc x} = \dfrac{\sin x}{\sec x}.$

11. $\dfrac{\sin x + \sin y}{\sin x - \sin y} = \dfrac{\csc y + \csc x}{\csc y - \csc x}.$

12. $4 \sin^2 x \cos^2 x = 1 - \cos^2 2x.$

13. $\sec x - \tan x = \dfrac{\cos x}{1 + \sin x}.$

14. $\sec x + \tan x = \tan\left(\dfrac{x}{2} + \dfrac{\pi}{4}\right).$

15. $\csc x = \cot x + \tan (x/2).$

16. $\sin 3x \sin 2x = \frac{1}{2}(\cos x - \cos 5x).$

17. $2 \sin 4x \cos 3x = \sin 7x + \sin x.$

18. $\sin 4x \cos 3x + \cos 4x \sin 3x = \sin 7x.$

19. $3 \cos x \cos 3x = \frac{3}{2} \cos 4x + \frac{3}{2} \cos 2x.$

20. $\cos \frac{3}{5}x \cos \frac{2}{5}x - \sin \frac{2}{5}x \sin \frac{3}{5}x = \cos x.$

8.10 Equations

In the illustrations of Sec. 8.9 we encountered equations such as: $1 + \cos x = 0$, $\cot x = 0$, and $\csc x - 1 = 0$. These are quite evidently not identities, but are conditional equations which we wish to solve. We saw that the solutions of $1 + \cos x = 0$ are $x = (2n + 1)\pi$, where n is any integer. Similarly the solutions of $\cot x = 0$ are $x = (\pi/2) + n\pi$; and the solutions of $\csc x - 1 = 0$ are $x = (\pi/2) + 2n\pi$. A given equation might have no solution; $\sin x = 3$ is an example. In case an equation is complicated, we may not be able to tell offhand whether it is a conditional equation or an identity.

There are practically no general rules which, if followed, will lead to the roots of a trigonometric equation. You might try to factor or to solve by quadratic formula where appropriate. Or again, you might reduce each and every trigonometric function present to one and the same function of one and the same independent variable. In this section we exhibit some of the obvious ways of solving such an equation.

Illustration 1. Solve the equation

$$2 \sin^2 x + \sin x - 1 = 0$$

for all roots.

Solution: The left-hand member is quadratic in the quantity $\sin x$; that is, it is a polynomial of the second degree in $\sin x$. It is factorable:

$$(2 \sin x - 1)(\sin x + 1) = 0$$

Thus from the first factor we get

$$\begin{aligned} 2 \sin x - 1 &= 0 \\ \sin x &= \tfrac{1}{2} \\ x &= \frac{\pi}{6} + 2n\pi \\ x &= \tfrac{5}{6}\pi + 2n\pi \end{aligned}$$

If you have studied congruences (Sec. 2.8), you will note that these answers can be written

$$x \equiv \frac{\pi}{6}, \text{ mod } 2\pi$$
$$x \equiv \tfrac{5}{6}\pi, \text{ mod } 2\pi$$

The second factor yields

$$\sin x + 1 = 0$$
$$\sin x = -1$$
$$x = \tfrac{3}{2}\pi + 2n\pi$$

There are no other roots.

Illustration 2. Find all values of x in the interval 0 to 2π satisfying the equation

$$\cos^2 2x + 3 \sin 2x - 3 = 0$$

Solution: This appears to offer some difficulty at first thought because of the presence of both sine and cosine. We use the identity $\sin^2 2x + \cos^2 2x = 1$ and rewrite the equation in the form

$$1 - \sin^2 2x + 3 \sin 2x - 3 = 0$$

which factors into

$$(1 - \sin 2x)(2 - \sin 2x) = 0$$

The first factor yields

$$1 - \sin 2x = 0$$
$$\sin 2x = 1$$
$$2x = \frac{\pi}{2} + 2n\pi$$

Whence

$$x = \frac{\pi}{4} + n\pi$$

The second factor leads to the equation

$$\sin 2x = 2$$

which has no solution.

Illustration 3. Solve the equation $\tan x + 2 \sec x = 1$.

Solution: You should be able to follow each of the steps:

$$\begin{aligned}
\frac{\sin x}{\cos x} + \frac{2}{\cos x} &= 1 \\
\frac{\sin x + 2}{\cos x} &= 1 \\
\sin x + 2 &= \cos x \qquad \text{provided } \cos x \neq 0 \\
&= \pm 1 \sqrt{1 - \sin^2 x} \\
(\sin x + 2)^2 &= 1 - \sin^2 x \\
\sin^2 x + 4 \sin x + 4 &= 1 - \sin^2 x \\
2 \sin^2 x + 4 \sin x + 3 &= 0 \\
\sin x &= \frac{-4 \pm \sqrt{16 - 24}}{4}
\end{aligned}$$

Since we are dealing with the real numbers, we conclude that the original equation is satisfied by no real number.

Illustration 4. Solve the equation $\sin 2x + \sin x = 0$.

Solution: We first write $\sin 2x = 2 \sin x \cos x$ (see Prob. 3, Sec. 8.8).

$$\begin{aligned}
2 \sin x \cos x + \sin x &= 0 \\
\sin x\,(2 \cos x + 1) &= 0
\end{aligned}$$

$$\begin{array}{ll}
\sin x = 0 & \cos x = -\frac{1}{2} \\
x = n\pi & x = \frac{2}{3}\pi + 2n\pi \\
 & x = \frac{4}{3}\pi + 2n\pi
\end{array}$$

Illustration 5. Solve the equation $\tan^2 x - 5 \tan x - 4 = 0$.

Solution: This is a quadratic equation in the quantity $\tan x$. Solving this by formula, we get

$$\begin{aligned}
\tan x &= \frac{5 \pm \sqrt{25 + 16}}{2} \\
&= \tfrac{5}{2} \pm \tfrac{1}{2}\sqrt{41} \\
&= 2.50000 \pm 3.20656 \\
&= 5.70656 \text{ and } -0.70656
\end{aligned}$$

Since these values do not correspond to any of the special numbers, we make use of Table IV of the Appendix. We find that $\tan 1.39 = 5.4707$ and $\tan 1.40 = 5.7979$; hence we must interpolate.

We write

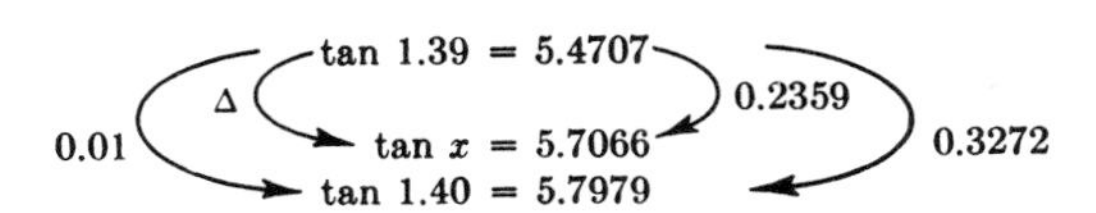

and set up the proportionality

$$\frac{\Delta}{0.01} = \frac{0.2359}{0.3272}$$

whence

$$\Delta = \frac{2359}{3272}(0.01)$$
$$\approx 0.0072$$

Therefore

$$\tan(1.39 + 0.0072) = 5.7066$$

and

$$x = 1.3972 + 2n\pi \qquad \text{[1st quadrantal arc]}$$
$$x = (1.3972 + \pi) + 2n\pi$$
$$= 4.5388 + 2n\pi \qquad \text{[3d quadrantal arc]}$$

Exercise A. Draw figures for these arcs.

Now we must use $\tan x = -0.70656$; but negative values do not occur in the table. This is no handicap, however, since the sign merely tells us the arcs are second and fourth quadrantal arcs. Temporarily we write

$$\tan x' = 0.70656$$

From the table we find

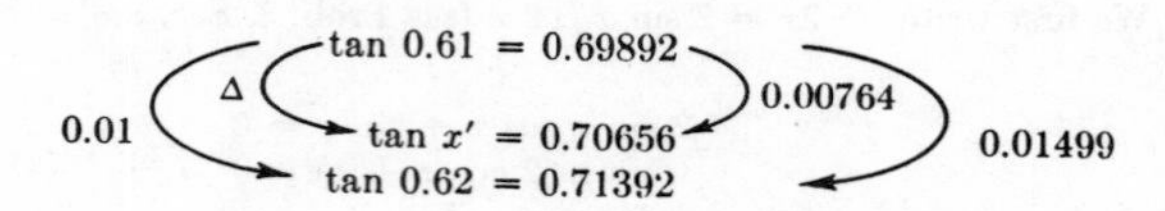

The proportionality is

$$\frac{\Delta}{0.01} = \frac{0.00764}{0.01499}$$

from which

$$\Delta = \tfrac{764}{1499}(0.01)$$
$$\approx 0.0051$$

Therefore

$$\tan(0.61 + 0.0051) = 0.70656$$

and

$$x' = 0.6151$$

But now we must use the minus sign since at present what we have is

$$\tan 0.6151 = 0.70656$$

whereas we seek x such that $\tan x = -0.70656$. This means that either

$$x = \pi - x'$$

or

$$x = 2\pi - x'$$

Finally, therefore, we have

$$x = 2.5265 + 2n\pi \qquad \text{[2d-quadrantal arc]}$$
$$x = 5.6681 + 2n\pi \qquad \text{[4th-quadrantal arc]}$$

Exercise B. Draw figures for these arcs.

Problems 8.10

Solve the following equations for all roots.

1. $2 \sin^2 x - \sin x = 0.$
2. $\cos^2 x - \cos x = 0.$
3. $2 \sin^2 x - 3 \sin x + 1 = 0.$
4. $\cos^2 x + 2 \cos x - 3 = 0.$
5. $\sin x + \cos x = 0.$
6. $\sin 2x - \cos x = 0.$
7. $3 \tan x = 2 \cos x.$
8. $\sqrt{3}\,(\tan x + \cot x) = 4.$
9. $\sin^3 x - \cos^3 x = 0.$
10. $4 \tan^2 x = 3 \sec^2 x.$
11. $\tan^2 x - 4 \tan x + 1 = 0.$
12. $\tan^4 x - 9 = 0.$
13. $\sin 3x = \frac{1}{2}.$
14. $\cos 3x = \frac{1}{2}.$
15. $\csc^2 x = \frac{4}{3}.$
16. $\tan x - 2 \sin x = 0.$
17. $\tan 2x = \tan x.$
18. $2 \sin^2 2x - \sin 2x - 1 = 0.$
19. $\sin^2 x - \sin x = \frac{1}{4}.$
20. $\cos^2 2x + \cos 2x = \frac{1}{2}.$

8.11 Inverse Functions

If we are given the function $y = \sin x$, we may well ask, For what real numbers x is $\sin x$ equal to a given real number y? We recognize this as the problem of finding the function which is the inverse of $\sin x$ (Sec. 6.8). In order to approach this question, let us reexamine the graph of $y = \sin x$ (Fig. 8.23).

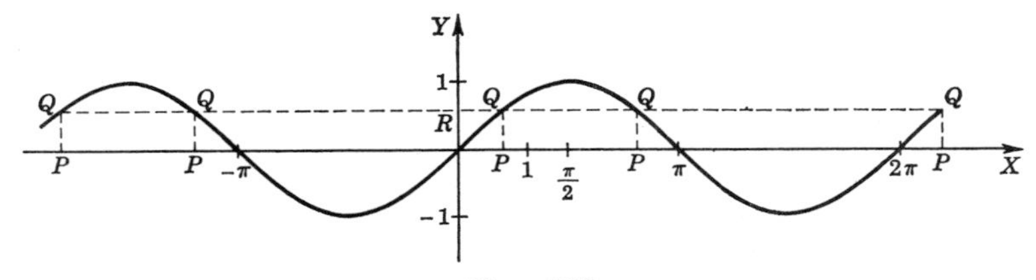

Figure 8.23

We wish to pick a point R on the Y-axis and hence determine a unique point P on the X-axis as in Fig. 6.9. We know, of course, that we must pick R in the interval $-1 \leq y \leq 1$ if a horizontal line through R is to meet the graph. Choosing such a point as in Fig. 8.23, we find ourselves in immediate trouble, for the line through R meets the graph in infinitely many points Q, which yield infinitely many points P. Hence there is no inverse in the sense we have defined.

We have seen, however, in Sec. 6.8, that when a function does not have an inverse we can sometimes restrict its domain so that the restricted function does have an inverse. Theorem 1 of Sec. 6.8 suggests that we should restrict the domain of $\sin x$ to an interval in which it is strictly monotone increasing or monotone decreasing. There are many choices for such an interval, but it is customary to choose the one in which $|x|$ is as small as possible and which still maintains the full range for y; this is the interval $-(\pi/2) \leq x \leq (\pi/2)$ (Fig. 8.24). In this interval $\sin x$ is strictly monotone increasing, and so the restricted function does have an inverse. So that we may be specific, let us call $y = \text{Sin } x$ (read "Cap-Sin x") the restriction of $y = \sin x$ to the interval $-(\pi/2) \leq x \leq (\pi/2)$. Its range is clearly $-1 \leq y \leq 1$.

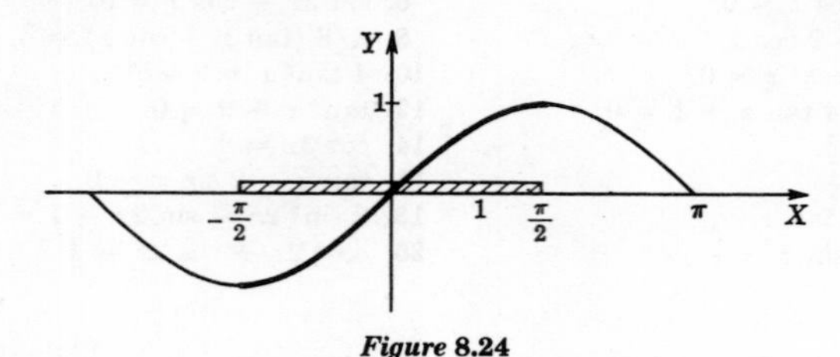

Figure **8.24**

According to our procedure, we should now solve $y = \text{Sin } x$ for x, but we have no direct way of doing so. Hence the inverse of $y = \text{Sin } x$ is a new function to which we must assign a name. Two names are used for this function, namely:

$$x = \text{Sin}^{-1}\, y \qquad \text{and } x = \text{arc Sin } y$$

These may be read: "x is the inverse Cap-Sine of y", or "x is the real number whose Cap-Sine is y". The domain of this function is $-1 \leq y \leq 1$, and the range is $-(\pi/2) \leq x \leq (\pi/2)$. As in the case of $x = \log y$ in Sec. 7.3, we wish to write x as the independent variable and y as the dependent variable, so finally we have:

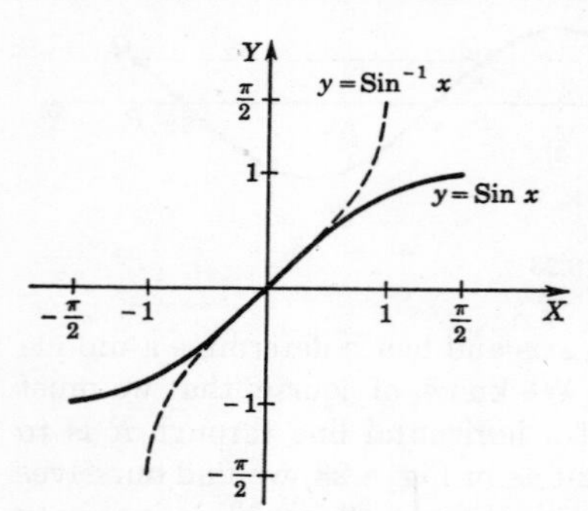

Figure **8.25**

$$y = \text{Sin}^{-1}\, x$$

whose domain is $-1 \leq x \leq 1$ and whose range is $-(\pi/2) \leq y \leq (\pi/2)$. Its graph is given in Fig. 8.25.

Similarly, we define other restricted trigonometric functions:

$y = \text{Cos } x$, $0 \leq x \leq \pi$, $-1 \leq y \leq 1$, as $\cos x$ restricted to this domain.

$y = \text{arc Cos } x$, $-1 \leq x \leq 1$, $0 \leq y \leq \pi$, as the inverse of $y = \text{Cos } x$.

$y = \text{Tan } x$, $-\pi/2 < x < \pi/2$, $-\infty < y < \infty$, as $\tan x$ restricted to this domain.

$y = \text{arc Tan } x$, $-\infty < x < \infty$, $-\pi/2 < y < \pi/2$, as the inverse of $y = \text{Tan } x$.

Problems 8.11

In Probs. 1 to 18 find the value of each expression.

1. $\text{Cos}^{-1} \frac{1}{2}\sqrt{2}$.
2. $\text{Sin}^{-1} \frac{1}{2}\sqrt{3}$.
3. $\text{Tan}^{-1} 1$.
4. $\text{Sin}^{-1}(-1)$.
5. $\text{Cos}^{-1}(0)$.
6. $\text{Tan}^{-1}\sqrt{3}$.
7. $\text{arc Sin}(-\frac{1}{2}\sqrt{3})$.
8. $\text{arc Tan}(\cos 270°)$.
9. $\text{arc Cos}[\tan(-\pi/4)]$.
10. $\text{arc Sin}(\sin \frac{3}{2}\pi)$.
11. $\text{Sin}(\text{Tan}^{-1} 2)$.
12. $\text{Sin}^{-1}(\tan 2)$. BT
13. $\cos(\text{Sin}^{-1} \frac{3}{5})$.
14. $\sin(2\,\text{Sin}^{-1} \frac{4}{5})$.
15. $\cos(\text{arc Cos} \frac{2}{3} + \pi/2)$.
16. $\sin(2\,\text{arc Tan}\, 2 + \pi)$.
17. $\tan(2\,\text{Tan}^{-1} \frac{3}{4} + \text{Tan}^{-1} \frac{5}{12})$.
18. $\text{Tan}^{-1}[\tan(\text{Cos}^{-1} \frac{1}{2})]$.

Probs. 19 to 21 were used to compute (and check) π to 100,000 decimals. [See, for calculation of π to 100,000 decimals, Daniel Shanks and John W. Wrench, Jr., Mathematics of Computation, vol. 16, no. 77 (January, 1962).] Verify each.

19. $\pi = 16\,\text{Tan}^{-1} \frac{1}{5} - 4\,\text{Tan}^{-1} \frac{1}{239}$.
20*. $\pi = 24\,\text{Tan}^{-1} \frac{1}{8} + 8\,\text{Tan}^{-1} \frac{1}{57} + 4\,\text{Tan}^{-1} \frac{1}{239}$.
21*. $\pi = 48\,\text{Tan}^{-1} \frac{1}{18} + 32\,\text{Tan}^{-1} \frac{1}{57} - 20\,\text{Tan}^{-1} \frac{1}{239}$.

In Probs. 22 to 27 sketch the graph.

22. $y = \text{Cos}\, x$.
23. $y = \text{Cos}^{-1} x$.
24. $y = \text{Tan}\, x$.
25. $y = \text{Tan}^{-1} x$.
26. $y = \frac{1}{2}\,\text{arc Sin}\, 2x$.
27. $y = 2\,\text{arc Sin}(x/2)$.

In Probs. 28 to 30 restrict the domain of definition and define, stating domain and range:

28. Cap-Sec x, $\text{Sec}^{-1} x$.
29. Cap-Csc x, $\text{Csc}^{-1} x$.
30. Cap-Cot x, $\text{Cot}^{-1} x$.

8.12 Trigonometric Functions of Angles

In your high school geometry an angle AOB was defined as the union of the two half-lines OA and OB. Provided that OA and OB do not lie in the same straight line, the *interior* of angle AOB was defined to be the portion of the plane included between these two half-lines, i.e., the shaded regions in Fig. 8.26. In order to assign a measure to angle AOB, we draw

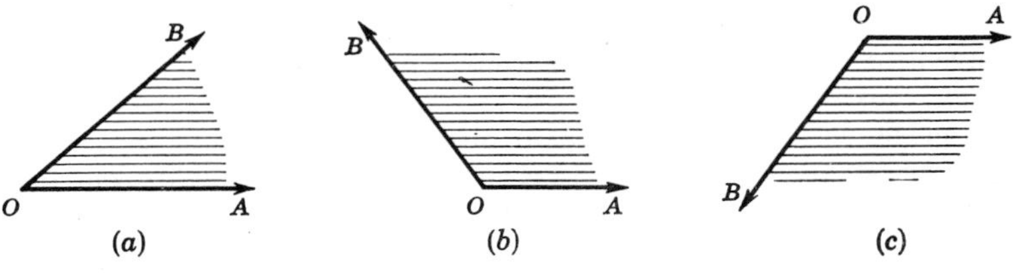

Figure **8.26**

a circle of radius 1 with center at the origin O, place OA along the positive X-axis, and let OB fall into whatever position it takes (Fig. 8.27). Angle AOB is then said to be in *standard position.* OA intersects this circle in the point $Q(1,0)$ and OB in a point which we call $P(x,y)$. Now consider

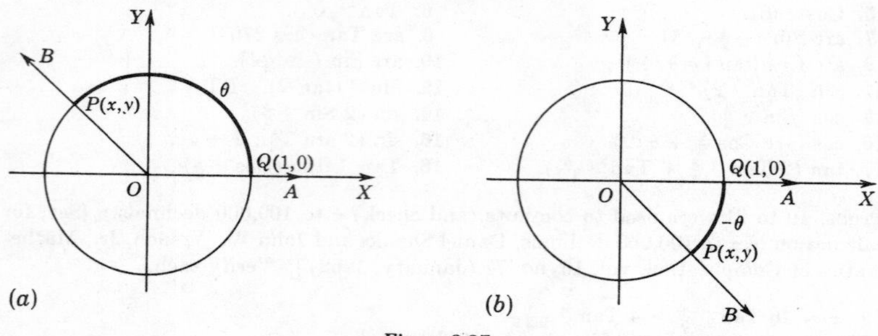

Figure 8.27

that circular arc PQ which lies in the interior of the angle AOB, and suppose that its length is θ. Then we say that the measure of angle AOB is:

(**a**) θ radians if B lies in the upper half plane.
(**b**) $-\theta$ radians if B lies in the lower half plane.

To complete the picture when B is on the X-axis, we define the measure of AOB in these cases to be:

(**c**) 0 radians if B lies on the positive half of the X-axis.
(**d**) π radians if B lies on the negative half of the X-axis.

Another unit of measure of an angle is a *degree,* which is defined by the equation:

$$180° = \pi \text{ radians}$$

or

$$1° = \frac{\pi}{180} \text{ radians} \qquad 1 \text{ radian} = \frac{180°}{\pi}$$

Convenient tables for making this conversion are available.

This construction establishes a 1 to 1 correspondence between angles AOB in standard position and real numbers θ in the interval $-\pi < \theta \leq \pi$, and hence we may speak either of the angle AOB or of the angle θ (radians). Similarly, there is a 1 to 1 correspondence between angles in standard position and their measures in degrees in the interval $-180° < \phi° \leq 180°$, and hence we may refer to the angle $\phi°$ (e.g., 56° or $-152°$, etc.).

When two undirected lines meet in a point O, four angles are determined, and so it is not proper to refer to *the angle between two undirected lines.* On the other hand, two *directed* lines meeting at O do determine a unique angle, namely, the angle defined by the two half-lines beginning at O and pointing in the two given directions. Since the two given lines play equivalent roles in this situation, this is an unsigned angle, and we take its measure to be positive. We may, however, be interested in the angle *from the first directed line to the second directed line.* In this case we attach the appropriate sign.

In our definitions of angles we have restricted ourselves to angles whose measures lie in the intervals $-\pi < \theta \leq \pi$, or $-180° < \phi° \leq 180°$. It is frequently desirable, however, to enlarge this concept of an angle to include angles whose measures lie outside of these ranges. In view of the above discussion, it is natural to use the following definition of a generalized angle:

Definition: A generalized angle is the plane figure consisting of: (1) two half-lines with common end point O; (2) a directed arc of a unit circle (with center at O) whose ends lie on the two given half-lines. This arc may be of any length and hence may wrap around the circle any number of times in either direction.

The length θ of this arc (together with its sign) is defined to be the measure of the generalized angle in radians. The measure in degrees is obtained by using the above conversion formula.

Hereafter, we shall drop the term *generalized angle* and use *angle* to refer to both the angles of plane geometry and to the generalized angles just defined.

It is now a simple matter to define the trigonometric functions of an angle:

Definition: The trigonometric functions of an angle whose measure is θ radians are equal to the trigonometric functions of the real number θ.

Exercise A. Show that the trigonometric functions of an angle whose measure is $\phi°$ are equal to the trigonometric functions of the real number $\pi\phi/180$.

Exercise B. Show that $\sin \phi° = \sin (\phi + n360)°$, where n is any integer. Similar results hold for the other functions.

Exercise C. Show that, in Fig. 8.27, $\sin \theta = y$ and $\cos \theta = x$.

In Fig. 8.27, we might have drawn a circle of arbitrary radius r instead of a unit circle (Fig. 8.28). Then OB intersects the circle of radius r at point S with coordinates (a,b). Since the triangles PRO and STO

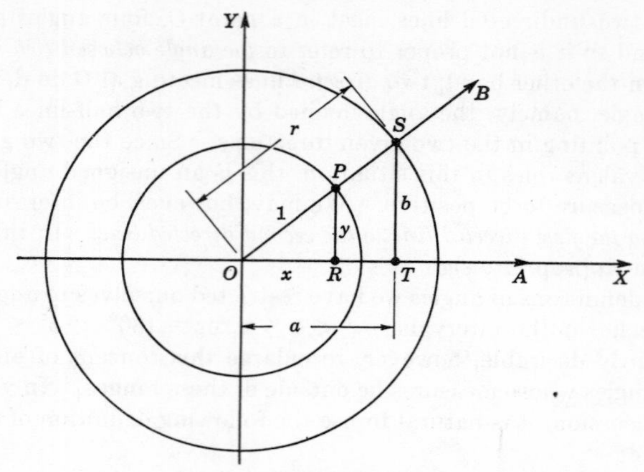

Figure 8.28

are similar, it follows that $PR/OP = ST/OS$, $OR/OP = OT/OS$, and $PR/OR = ST/OT$. From Exercise C above we see that

$$\sin AOB = \frac{y}{1} = \frac{PR}{OP} = \frac{ST}{OS} = \frac{b}{r}$$

$$\cos AOB = \frac{x}{1} = \frac{OR}{OP} = \frac{OT}{OS} = \frac{a}{r}$$

$$\tan AOB = \frac{y}{x} = \frac{PR}{OR} = \frac{ST}{OT} = \frac{b}{a}$$

This brings our definitions of these functions into agreement with those you undoubtedly learned earlier, namely:

Definition: If α is an acute angle of a right triangle

$$\sin \alpha = \frac{\text{opposite side}}{\text{hypotenuse}}$$

$$\cos \alpha = \frac{\text{adjacent side}}{\text{hypotenuse}}$$

$$\tan \alpha = \frac{\text{opposite side}}{\text{adjacent side}}$$

The formulas of this definition are true only when α is an acute angle of a right triangle. For the more general situation in which α is any angle of an arbitrary triangle, we shall rely upon the Law of Cosines instead.

To derive this law, place the triangle on the axes as indicated in Fig. 8.29. The angle α at A may be acute or obtuse. From the discussion above it

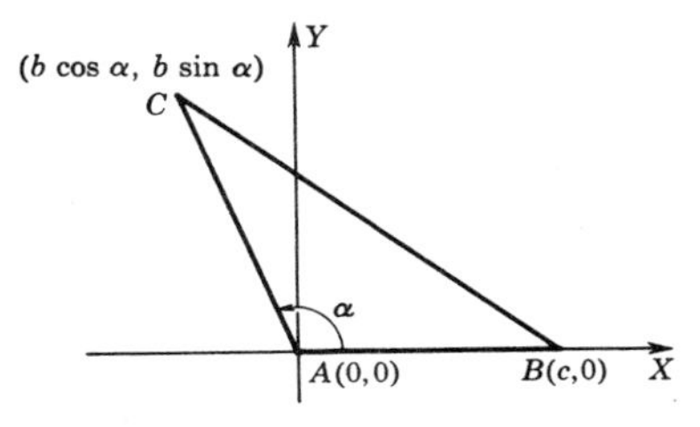

Figure **8.29**

follows that C has coordinates $(b \cos \alpha, b \sin \alpha)$. The distance BC is equal to a and is given by the distance formula:

$$\begin{aligned} a^2 &= (b \cos \alpha - c)^2 + (b \sin \alpha - 0)^2 \\ &= b^2 \cos^2 \alpha - 2bc \cos \alpha + c^2 + b^2 \sin^2 \alpha \\ &= b^2 + c^2 - 2bc \cos \alpha \end{aligned}$$

Using other letters, this formula may be written

$$\begin{aligned} b^2 &= a^2 + c^2 - 2ac \cos \beta \\ c^2 &= a^2 + b^2 - 2ab \cos \gamma \end{aligned}$$

We may also write these in the form:

$$\cos \alpha = \frac{b^2 + c^2 - a^2}{2bc}$$

$$\cos \beta = \frac{a^2 + c^2 - b^2}{2ac}$$

$$\cos \gamma = \frac{a^2 + b^2 - c^2}{2ab}$$

Any of these formulas is called the Law of Cosines.

From the first of these and the identity $\sin \alpha = \sqrt{1 - \cos^2 \alpha}$, we find that

$$\begin{aligned} \sin \alpha &= a \left[\frac{\sqrt{(a^2 + b^2 + c^2)^2 - 2(a^4 + b^4 + c^4)}}{2abc} \right] \\ &= a \left[\frac{2\sqrt{s(s-a)(s-b)(s-c)}}{abc} \right] \end{aligned}$$

where $2s = a + b + c$.

From this we see that

$$\frac{\sin \alpha}{a} = \frac{\sin \beta}{b} = \frac{\sin \gamma}{c}$$

a relationship called the Law of Sines.

Exercise D. Why do we choose only the + sign above when we write $\sin \alpha = +\sqrt{1 - \cos^2 \alpha}$?

Exercise E. Carry out the details of our derivation of the above formula for $\sin \alpha$.

Exercise F. Complete the proof of the Law of Sines.

Exercise G. From the above formula for $\sin \alpha$, show that

$$\sqrt{s(s-a)(s-b)(s-c)}$$

is equal to the area of the triangle.

Problems 8.12

In Probs. 1 to 9 find the sine of the angle. Use Table V of the Appendix.

1. 6°6′. **2.** 45°20′. **3.** 67°12′.
4. 130°40′. **5.** 250°30′. **6.** 335°10′.
7. −47.3°. **8.** −156.2°. **9.** −37.7°.

In Probs. 10 to 18 find the cosine of the angle. Use Table V.

10. 6°6′. **11.** 45°20′. **12.** 67°12′.
13. 130°40′. **14.** 250°30′. **15.** 335°10′.
16. −47.3°. **17.** −156.2°. **18.** −37.7°.

In Probs. 19 to 27 find the tangent of the angle. Use Table V.

19. 6°6′. **20.** 45°20′. **21.** 67°12′.
22. 130°40′. **23.** 250°30′. **24.** 335°10′.
25. −47.3°. **26.** −156.2°. **27.** −37.7°.

8.13 Complex Numbers

We have already met (Sec. 2.7) a complex number represented in rectangular form: $a + ib$ (or $a + bi$). There is a 1 to 1 correspondence between such numbers and points in the plane. Now, since $a = r \cos \theta$

and $b = r \sin \theta$ (Fig. 8.30),

(1) $$a + ib = r(\cos \theta + i \sin \theta)$$
where $$r = |a + ib| = \sqrt{a^2 + b^2}$$
and $$\tan \theta = \frac{b}{a}$$

The real, nonnegative number $r\ (= \sqrt{a^2 + b^2})$ is called the absolute value (or modulus) of the complex number and is written $|a + ib|$. The angle θ associated with the number $a + ib$ is called the argument (or

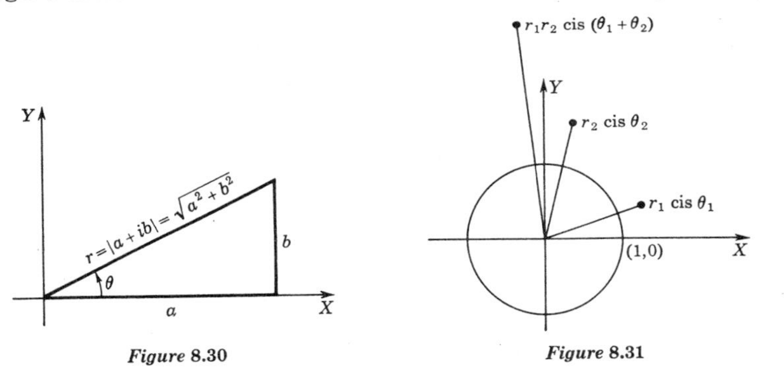

Figure 8.30 *Figure* 8.31

amplitude) of $a + ib$. The left-hand side of (1) is the *rectangular form*, and the right-hand side is the *polar form*, of a complex number. A complex number is therefore a vector having both magnitude (absolute value) and direction (argument).

Addition (and subtraction) of complex numbers is best accomplished in rectangular $(a + ib)$ form. Thus

$$(a + ib) \pm (c + id) = (a \pm c) + i(b \pm d)$$

But multiplication and division are conveniently treated in polar $[r(\cos \theta + i \sin \theta)]$ form. Often, to simplify the notation, we write $r(\cos \theta + i \sin \theta)$ in the form $r \text{ cis } \theta$.

Multiplication. Consider $r_1 \text{ cis } \theta_1$ and $r_2 \text{ cis } \theta_2$, two complex numbers. Their product (Fig. 8.31) is given by

$$\begin{aligned}
&r_1 \text{ cis } \theta_1 \cdot r_2 \text{ cis } \theta_2 \\
&= r_1 r_2 \text{ cis } \theta_1 \text{ cis } \theta_2 \\
&= r_1 r_2 (\cos \theta_1 + i \sin \theta_1)(\cos \theta_2 + i \sin \theta_2) \\
&= r_1 r_2 [(\cos \theta_1 \cos \theta_2 - \sin \theta_1 \sin \theta_2) + i(\sin \theta_1 \cos \theta_2 + \cos \theta_1 \sin \theta_2)] \\
&= r_1 r_2 [\cos (\theta_1 + \theta_2) + i \sin (\theta_1 + \theta_2)] \\
&= r_1 r_2 \text{ cis } (\theta_1 + \theta_2)
\end{aligned}$$

Therefore the absolute value of the product is the product of the absolute values, and the argument of the product is the sum of the arguments (plus or minus a multiple of 2π).

By similar reasoning

$$r_1 \text{ cis } \theta_1 \cdot r_2 \text{ cis } \theta_2 \cdot r_3 \text{ cis } \theta_3 = r_1 r_2 r_3 \text{ cis } (\theta_1 + \theta_2 + \theta_3)$$

If the three numbers θ_1, θ_2, θ_3 are all equal to θ, and if r_1, r_2, r_3 are all equal to r, we have

$$[r \text{ cis } \theta]^3 = r^3 \text{ cis } 3\theta$$

And similarly,

$$(2) \qquad [r \text{ cis } \theta]^n = r^n \text{ cis } n\theta \qquad n \text{ a positive integer}$$

With proper interpretations, (2) is true for any real number n, but we shall not give the proof. This is known as De Moivre's Theorem.

Theorem 1. De Moivre's Theorem. $[r \text{ cis } \theta]^n = r^n \text{ cis } n\theta$, n real.

Exercise A. Prove De Moivre's Theorem for the case where n is a positive integer. HINT: Use induction.

Division. To find the quotient of two complex numbers, write

$$\begin{aligned}\frac{r_1 \text{ cis } \theta_1}{r_2 \text{ cis } \theta_2} &= \frac{r_1 \text{ cis } \theta_1}{r_2 \text{ cis } \theta_2} \times \frac{r_2 \text{ cis } (-\theta_2)}{r_2 \text{ cis } (-\theta_2)} \\ &= \frac{r_1 r_2 \text{ cis } (\theta_1 - \theta_2)}{r_2^2 \text{ cis } 0} \\ &= \frac{r_1}{r_2} \text{ cis } (\theta_1 - \theta_2)\end{aligned}$$

Thus we see that the absolute value of the quotient of two complex numbers is the quotient of their absolute values, and the argument of the quotient is the argument of the numerator minus the argument of the denominator (Fig. 8.32).

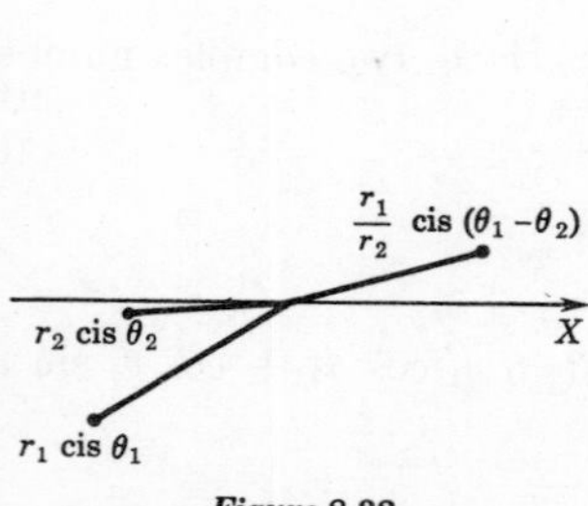

Figure 8.32

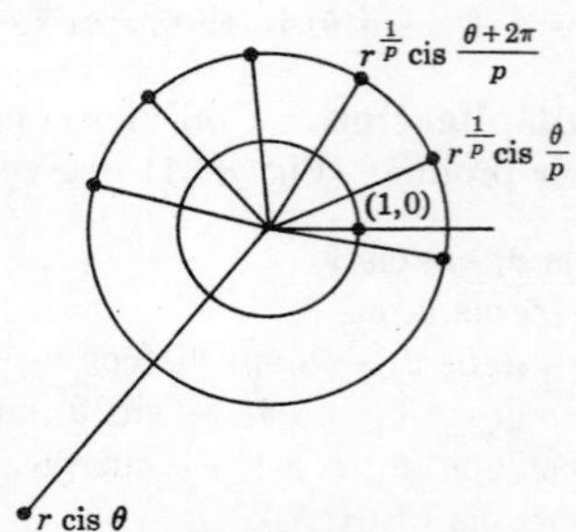

Figure 8.33

Roots of complex numbers. First, we note that the argument of a complex number is not uniquely defined. If

$$a + ib = r \text{ cis } \theta$$

it is also equal to $r[\text{cis } (\theta + 2\pi n)]$ for any integer n. Up to now this was not important, but we must use it here.

Given the complex number r cis θ, we seek to find all complex numbers whose pth powers are equal to r cis θ. These are called its pth roots. From De Moivre's Theorem we see at once that, for every n,

$$\left[r^{1/p} \text{ cis } \left(\frac{\theta + 2\pi n}{p}\right)\right]^p = r \text{ cis } (\theta + 2\pi n) = r \text{ cis } \theta$$

Therefore the numbers (Fig. 8.33)

$$r^{1/p} \text{ cis } \left(\frac{\theta + 2\pi n}{p}\right) \tag{3}$$

are pth roots of r cis θ. It can be shown that these comprise all the pth roots.

Exercise B. If $[R \text{ cis } \phi]^p = r \text{ cis } \theta$, show that R cis ϕ must have the form (3) for some value of n.

Let us examine (3) for various values of n: Letting $n = 0$, we have

$$(r \text{ cis } \theta)^{1/p} = r^{1/p} \text{ cis } \frac{\theta}{p}$$

Letting $n = 1$, we have

$$(r \text{ cis } \theta)^{1/p} = r^{1/p} \text{ cis } \frac{\theta + 2\pi}{p}$$

Each of these two (distinct) numbers is a pth root of r cis θ. By letting $n = 2, 3, \ldots, p - 1$, we obtain $p - 2$ other distinct pth roots. Letting $n = p$, we have

$$\begin{aligned}(r \text{ cis } \theta)^{1/p} &= r^{1/p} \text{ cis } \frac{\theta + 2\pi p}{p} \\ &= r^{1/p} \text{ cis } \frac{\theta}{p}\end{aligned}$$

which yields the same result as did $n = 0$. And $n = p + 1$ yields the same result as $n = 1$, etc. Therefore there are p (distinct), and only p, pth roots of a complex number, $a + ib = r \text{ cis } \theta$. These are given by

$$(4) \qquad (r \text{ cis } \theta)^{1/p} = r^{1/p} \text{ cis} \left(\frac{\theta + 2\pi n}{p} \right) \qquad n = 0, 1, 2, \ldots, p - 1$$

You should memorize (4).

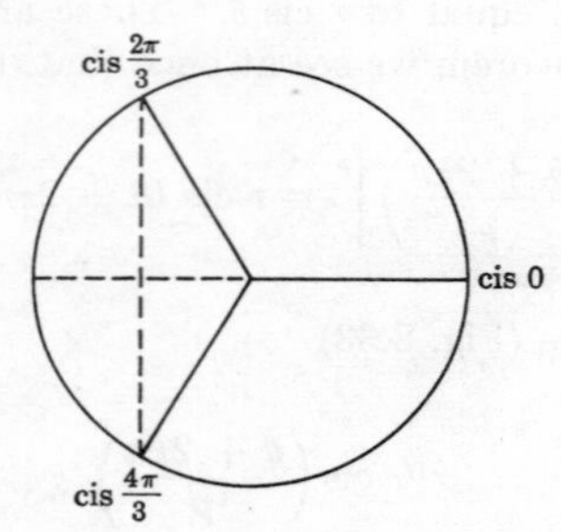

Figure **8.34**

Illustration 1. Find the three cube roots of 1 (Fig. 8.34). Since

$$1 = 1 + i0$$

$r = 1$ and $\theta = 0$. Thus $1 = \text{cis } (0 + 2\pi n)$. The cube roots are

$$1^{\frac{1}{3}} \cos \frac{0 + 2\pi n}{3} \qquad n = 0, 1, 2$$

or

$$1 \cos 0 = 1$$

$$1 \text{ cis } \frac{2\pi}{3} = \cos \frac{2\pi}{3} + i \sin \frac{2\pi}{3} = -\frac{1}{2} + i \frac{\sqrt{3}}{2}$$

$$1 \text{ cis } \frac{4\pi}{3} = \cos \frac{4\pi}{3} + i \sin \frac{4\pi}{3} = -\frac{1}{2} - i \frac{\sqrt{3}}{2}$$

To check the result, multiply out $\left(-\frac{1}{2} + i \frac{\sqrt{3}}{2} \right)^3$ and $\left(-\frac{1}{2} - i \frac{\sqrt{3}}{2} \right)^3$. The results should be 1.

This example is equivalent to solving the equation

$$x^3 - 1 = 0$$

or

$$(x - 1)(x^2 + x + 1) = 0$$

The roots are

$$x = 1$$

$$x = \frac{-1 \pm \sqrt{1 - 4}}{2} = -\frac{1}{2} \pm i \frac{\sqrt{3}}{2}$$

Problems 8.13

In Probs. 1 to 6 change to polar form.

1. $\sqrt{2} + i\sqrt{2}$.
2. $\sqrt{2} - i\sqrt{2}$
3. $2 + i2\sqrt{3}$.
4. $2 - i2\sqrt{3}$.
5. $-i$.
6. -1.

In Probs. 7 to 14 change to rectangular form.

7. cis 0°.
8. 3 cis 150°.
9. 2 cis 180°.
10. cis 120°.
11. cis $\frac{7}{6}\pi$.
12. cis $\frac{4}{3}\pi$.
13. cis $(-\pi/2)$.
14. cis $(-\frac{3}{2}\pi)$.

In Probs. 15 to 20 write the product in cis form.

15. (3 cis 45°)(2 cis 60°).
16. (4 cis 30°)(4 cis 330°).
17. (2 cis 210°)(2 cis 210°).
18. (5 cis $\frac{7}{6}\pi$)(cis $\frac{4}{3}\pi$).
19. cis $(-\pi/2)$ cis $(-\frac{3}{2}\pi)$.
20. cis $\frac{5}{6}\pi$ cis $\frac{5}{6}\pi$ cis $\frac{5}{6}\pi$.

In Probs. 21 to 26 write the quotient in cis form.

21. 3 cis 45°/2 cis 60°.
22. 2 cis 30°/4 cis 210°.
23. 5 cis 180°/5 cis 360°.
24. 4 cis 20°/3 cis 10°.
25. 3 cis $\frac{4}{3}\pi$/2 cis $(-\frac{3}{2}\pi)$.
26. 7 cis $\frac{2}{3}\pi$/49 cis $\frac{5}{6}\pi$.

27. Find the two square roots of 3 cis 150°. Plot these and also the number 3 cis 150°.
28. Find the two square roots of 4 cis $\frac{3}{2}\pi$. Plot these and also the number 4 cis $\frac{3}{2}\pi$.
29. Find the three cube roots of 8 cis π. Plot these and also the number 8 cis π.
30. Find the three cube roots of 27 cis $(\pi/3)$. Plot these and also the number 27 cis $(\pi/3)$.
31. Find the four fourth roots of 1. Plot these and also the number 1.
32. Find the four fourth roots of -1. Plot these and also the number -1.
33. Find the four fourth roots of i. Plot these and also the number i.
34. Find the four fourth roots of $-i$. Plot these and also the number $-i$.
35. Find all roots of the equation $x^5 - 1 = 0$.
36. Find all roots of the equation $x^5 + 1 = 0$.
37. Find all roots of the equation $x^6 - 1 = 0$.
38. Show that the four fourth roots of 1 form a group under ordinary multiplication.

References

Dubisch, Roy: "Trigonometry", Ronald, New York (1955).
Terman, F. E.: "Electronic and Radio Engineering", McGraw-Hill, New York (1955).

In addition to the many standard textbooks on trigonometry, the reader should consult the following articles in the *American Mathematical Monthly:*

Burton, L. J., and E. A. Hedberg: Proofs of the Addition Formulae for Sines and Cosines, vol. 56, p. 471 (1949).
Burton, L. J.: The Laws of Sines and Cosines, vol. 56, p. 550 (1949).
Carver, W. B.: Trigonometric Functions—of What?, vol. 26, p. 243 (1919).
Householder, A. S.: The Addition Formulas in Trigonometry, vol. 49, p. 326 (1942).
McShane, E. J.: The Addition Formulas for the Sine and Cosine, vol. 48, p. 688 (1941).
Vance, E. P.: Teaching Trigonometry, vol. 54, p. 36 (1947).

9 | Analytic Geometry

9.1 Introduction

René Descartes (1596–1650) introduced the subject of analytic geometry with the publishing of his "La Géométrie" in 1637. Accordingly, it is often referred to as *cartesian* geometry; it is, essentially, merely a method of studying geometry by means of a coordinate system and an associated algebra. The application of this basic idea enabled the mathematicians of the seventeenth century to make the first noteworthy advances in the field of geometry since the days of Euclid. The next great advance came with the invention of the calculus (Chaps 10 and 11).

There are two central problems in plane analytic geometry:

(a) Given an equation in x and y, to plot its graph, or to represent it geometrically as a set of points in the plane.
(b) Given a set of points in the plane, defined by certain geometric conditions, to find an equation whose graph will consist wholly of this set of points.

The second problem is frequently called a *locus problem.* A locus is the geometric counterpart of a relation, and we define it as follows:

Definition: A *locus* is a subset of the set of points in the plane.

A locus is defined by some geometric conditions, usually expressed in words. Let P represent an arbitrary point in the plane; then the following are examples of loci:

(1) $\{P \mid P$ is at a fixed distance r from a point $C\}$; this locus is then a circle with radius r and center C.
(2) $\{P \mid PA = PB$, where A and B are fixed points$\}$; this locus is the perpendicular bisector of the segment AB.
(3) $\{P \mid P$ is a fixed point on the rim of a wheel which rolls along a line$\}$; this locus is called a cycloid.

Many loci are defined in terms of a physical notion like example (3). For this reason you may run across statements like: "The locus of a point which moves so that . . . ." Since there is no motion in geometry, we prefer to avoid this language except in applications to mechanics.

When we are given such a locus, the problem before us is to find the corresponding relation. That is, we seek an equation whose graph is the given locus. We call this an *equation of the locus.* Having found such an equation, we study its properties by algebraic means and thus derive properties of the locus.

We have studied the notion of distance between two points (length of a line segment) when the points have given coordinates with respect to rectangular axes. We now wish to consider some related problems.

9.2 Mid-point of a Line Segment

Consider a line segment $P_1(x_1,y_1)$, $P_2(x_2,y_2)$. We seek the coordinates $(\bar{x},\bar{y})$ of the mid-point P in terms of x_1, y_1, x_2, and y_2. From Fig. 9.1 it is

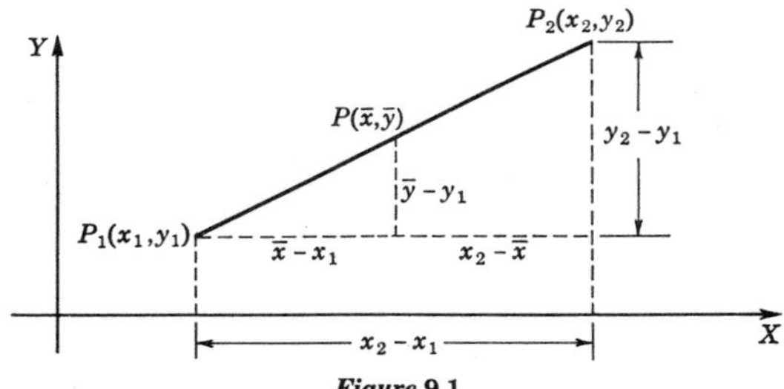

Figure 9.1

evident that

$$\frac{x_2 - x_1}{P_1P_2} = \frac{\bar{x} - x_1}{P_1P} \tag{1}$$

But $P_1P_2 = 2P_1P$. Therefore (1) becomes

$$\frac{x_2 - x_1}{2} = \frac{\bar{x} - x_1}{1}$$

from which we get

$$2\bar{x} - 2x_1 = x_2 - x_1$$

$$2\bar{x} = x_1 + x_2$$

or

$$\bar{x} = \frac{x_1 + x_2}{2}$$

Similarly,

$$\bar{y} = \frac{y_1 + y_2}{2}$$

Exercise A. From Fig. 9.1 derive the expression for $\bar{y}$.

Thus the X-coordinate of the mid-point is the average of the X-coordinates of the end points; the Y-coordinate is the average of the Y-coordinates of the end points. For example, the mid-point of the segment whose end points are $(-1,5)$, $(4,-7)$ has coordinates $(\frac{3}{2},-1)$.

Exercise B. Find the coordinates of the mid-points of the sides of the triangle whose vertices are $A(4,7)$, $B(-3,-3)$, $C(2,-5)$.

Exercise C. The above derivation of this formula assumes that $x_1 \neq x_2$ and that $y_1 \neq y_2$. Prove that the formula is correct when $x_1 = x_2$ or when $y_1 = y_2$.

9.3 Directed Line Segment

Often it is desirable to associate with a line segment (or line) the notion of *direction* or *sense* (Sec. 5.2). When sense becomes important, it will be specified by the order in which the end points are given, by an arrow appropriately placed in the figure, or in some other unambiguous way. Thus in Fig. 9.2*a* the line segment P_1P_2 is to be considered without sense,

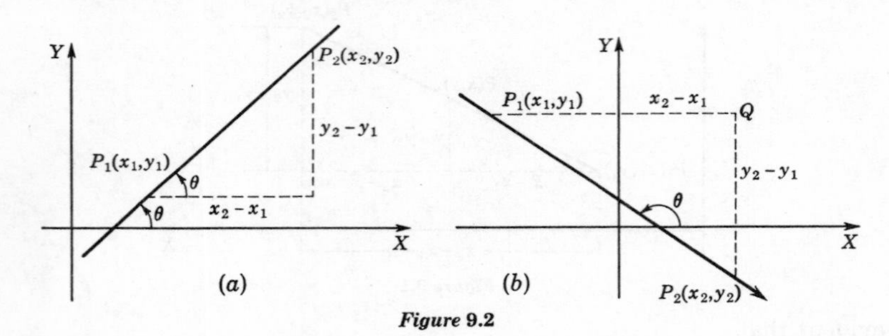

Figure 9.2

whereas the segment P_1P_2, with the arrow attached as in Fig. 9.2*b*, has the positive direction $\overrightarrow{P_1P_2}$ and the negative direction $\overrightarrow{P_2P_1}$. We write $\overrightarrow{P_1P_2} = -\overrightarrow{P_2P_1}$. If sense is unimportant, no mention will be made of it.

Definition: The *senses* of the X-axis and the Y-axis are in the positive directions.

9.4 Rise, Run, Slope, Inclination

A line which is not parallel to a coordinate axis may *rise from lower left to upper right*, as in Fig. 9.2*a*, or it may *fall from upper left to lower right*, as in Fig. 9.2*b*. This language implies that the observer is oriented from left to right even though the line or line segment P_1P_2 may not be! Intuitively, we are looking uphill in the first instance and downhill in the second. In order to clarify these ideas, we lay down the following definitions with respect to the line determined by $P_1(x_1,y_1)$ and $P_2(x_2,y_2)$, where $x_2 > x_1$ and $y_2 > y_1$, which is the situation in Fig. 9.2*a*.

Definitions: The positive number $y_2 - y_1$ is called the *rise*, and the positive number $x_2 - x_1$ is called the *run*.

Remark. When P_1 and P_2 are in other positions, the words *rise* and *run* are, strictly speaking, no longer appropriate, although the quantities $x_2 - x_1$ and $y_2 - y_1$ are well defined in all cases. That which is important here is given in the following definition:

Definition: When $x_2 - x_1 \neq 0$, the number

$$m = \frac{y_2 - y_1}{x_2 - x_1}$$

is called the *slope* of the line.

Remarks. The slope of a line parallel to the Y-axis $(x_2 - x_1 = 0)$ is not defined. The slope of a line parallel to the X-axis $(y_2 - y_1 = 0)$ is zero. Where rise and run apply, slope = rise/run. Since

$$\frac{y_2 - y_1}{x_2 - x_1} = \frac{y_1 - y_2}{x_1 - x_2}$$

it makes no difference how we label the points when computing slope.

If the same units and scales are used on the X- and Y-axes, still another notion is of use, according to the following definition:

Definition: If x and y are measured in the same units, we call θ, where $m = \tan \theta$, the *inclination;* θ is measured counterclockwise from the positive X-axis and $0° \leq \theta < 180°$.

The inclination of a line parallel to the X-axis is zero, and the inclination of a line parallel to the Y-axis is 90° from other considerations. The notion of inclination is of no value if x and y are in different units such as, for example, if x represents "calendar year" and y represents "dollars per ton-mile". On the other hand, slope defined by

$$\text{Slope} = m = \frac{y_2 - y_1}{x_2 - x_1} = \frac{\text{rise}}{\text{run}} \qquad x_2 \neq x_1$$

is useful regardless of the units employed. If units and scales are the same on the two axes, it is meaningful to say that

$$\tan \theta = \frac{y_2 - y_1}{x_2 - x_1} \qquad x_2 \neq x_1$$

In analytic geometry we always assume equal scales on the two axes.

Directly from Fig. 9.1 or Fig. 9.2 and the Pythagorean theorem, it follows that the positive distance

$$d = P_1P_2 = \sqrt{(x_2 - x_1)^2 + (y_2 - y_1)^2}$$

This was also developed in Sec. 8.2.

Problems 9.4

In Probs. 1 to 8 find the coordinates of the mid-point of the line segment joining the given points.

1. (3,2), (1,4).
2. (5,9), (7,5).
3. (2,−3), (−2,7).
4. (−8,4), (7,−6).
5. $(0,a)$, $(a,0)$.
6. $(2k,4)$, $(2 - 2k, 4b)$.
7. $(\frac{1}{2} + \frac{1}{2}\sqrt{3}, 0)$, $(\frac{1}{2} - \frac{1}{2}\sqrt{3}, a)$.
8. (a,b), (c,d).

In Probs. 9 to 16 find (*a*) rise, (*b*) run, where meaningful, and (*c*) slope of the line joining the given points.

9. (2,5), (4,1).
10. (4,6), (2,8).
11. (6,11), (−11,6).
12. (30,−15), (−10,5).
13. (4,8), (−16,8).
14. (9,3), (−2,−7).
15. (10,2), (10,−2). (BT)
16. (−3,−4), (−3,4). (BT)

In Probs. 17 to 24 find (*a*) slope and (*b*) inclination of the line joining the given points.

17. $(0,0)$, (a,b).
18. $(0,0)$, $(a,0)$.
19. (4,20), (100,−10).
20. $(10^3,-10^2)$, $(-10^2,10^3)$.
21. $(\sqrt{3},-2)$, $(\sqrt{3},2)$.
22. $(1 + \sqrt{2}, a)$, $(1 - \sqrt{2}, b)$.
23. (a,b), (c,d).
24. (a,b), (a,c).

In Probs. 25 to 30 find (*a*) the rise and (*b*) the run, where meaningful, and (*c*) the distance for the line segment P_1P_2.

25. (4,−3), (2,8).
26. (1,2), (3,4).
27. (1,−2), (4,−6).
28. (−1,−5), (7,1).
29. (13,7), (−1,2).
30. (3,−9), (−2,4).

In Probs. 31 to 34 the point P is the mid-point of P_1P_2. Find the coordinates of:

31. P_1, given $P(3,2)$, $P_2(-4,7)$.
32. P_1, given $P(2,-1)$, $P_2(-2,-3)$.
33. P_2, given $P(-1,10)$, $P_1(2,-4)$.
34. P_2, given $P(a,1)$, $P_1(1,b)$.

In Probs. 35 to 40 prove that:

35. The triangle $A(-4,-3)$, $B(4,3)$, $C(-3,4)$ is a right triangle.
36. The triangle $A(2,-1)$, $B(10,5)$, $C(9,-2)$ is a right triangle.
37. The triangle $A(-1,-1)$, $B(0,3)$, $C(\frac{7}{2},0)$ is isosceles but not equilateral.
38. The triangle $A(1,-2)$, $B(2\sqrt{3} - \frac{1}{2}, \frac{3}{2}\sqrt{3})$, $C(-2,2)$ is equilateral.
39. The quadrilateral $A(1,4)$, $B(7,\frac{7}{2})$, $C(8,0)$, $D(2,\frac{1}{2})$ is a parallelogram.
40. The mid-points of the sides of the (nonconvex) quadrilateral $ABCD$ with vertices $A(0,0)$, $B(4,-1)$, $C(3,4)$, $D(2,1)$ are themselves the vertices of a parallelogram.
41. Show that $A(5,-1)$, $B(3,-4)$, $C(7,2)$ are on the same straight line.
42. Are $A(100,-50)$, $B(-300,20)$, $C(-700,80)$ on the same straight line?
43. Find the slopes of the medians of the triangle $A(1,5)$, $B(-2,3)$, $C(0,-4)$.
44. Write an equation which states that $P(x,y)$ is twice as far from $(2,1)$ as it is from $(0,-3)$.
45. The points $A(4,-1)$, $B(6,0)$, $C(7,-2)$ are vertices of a square. Find the coordinates of the fourth vertex.
46. For the directed line segment $\overrightarrow{P_1P_2}$, $P_1(x_1,y_1)$, $P_2(x_2,y_2)$, find the coordinates of $P(x,y)$ such that $P_1P/PP_2 = r_1/r_2$.

9.5 Direction Cosines

Since distance, slope, and inclination are related to a right triangle (P_1P_2Q in Fig. 9.3), it is desirable to make further use of trigonometry as in the following definitions.

Definitions: The angles α and β, between the positive direction $\overrightarrow{P_1P_2}$ and the positive directions of the axes, are called the *direction angles* of the directed line. The two numbers given by $\lambda = \cos\alpha$ and $\mu = \cos\beta$ are called the *direction cosines* of the line. Any two numbers proportional, respectively, to the direction cosines are called *direction numbers* of the line. Thus $a = k\lambda = k\cos\alpha$ and $b = k\mu = k\cos\beta$, where $k \neq 0$, are direction numbers. A line without direction has two sets of direction angles: α, β and $180° - \alpha$, $180° - \beta$, corresponding to the two possible directions.

For a sensed line, since α and β are unique, so are λ and μ. Hence a sensed line has unique direction cosines. But a line without sense has two sets of direction cosines, namely, $\lambda = \cos\alpha$, $\mu = \cos\beta$ and $-\lambda = \cos(180° - \alpha)$, $-\mu = \cos(180° - \beta)$. Note that in any case $m = \tan\theta = \mu/\lambda$, provided that the line is not perpendicular to the X-axis, which would make $\lambda = 0$. The slope of a line perpendicular to the Y-axis is zero; the direction cosines of such a line are $\lambda = \pm 1$, $\mu = 0$. The slope of a line perpendicular to the X-axis does not exist; the direction cosines are $\lambda = 0$, $\mu = \pm 1$. These concepts are of very great importance in higher mathematics.

Now

$$d = \sqrt{(x_2 - x_1)^2 + (y_2 - y_1)^2}$$

Directly from Fig. 9.3, which is typical, we see that

(2) $$m = \tan\theta = \frac{y_2 - y_1}{x_2 - x_1}$$

(3) $$\cos\alpha = \frac{x_2 - x_1}{d} \qquad \cos\beta = \frac{y_2 - y_1}{d}$$

Hence

(4) $$x_2 - x_1 = d\cos\alpha \qquad y_2 - y_1 = d\cos\beta$$

Therefore, for any two points on a line the differences in the respective coordinates, namely, $x_2 - x_1$ and $y_2 - y_1$, are direction numbers of the

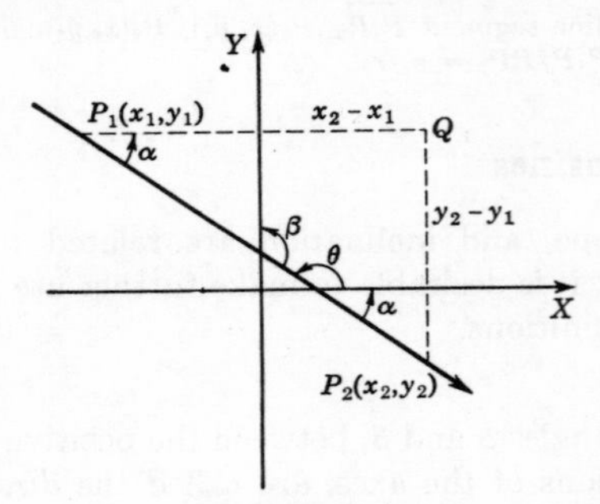

Figure **9.3**

line. The constant of proportionality is, in this case, the distance d between the two points.

If we square and add in (3), we get

$$\cos^2\alpha + \cos^2\beta = \frac{(x_2 - x_1)^2}{d^2} + \frac{(y_2 - y_1)^2}{d^2} = \frac{(x_2 - x_1)^2 + (y_2 - y_1)^2}{d^2}$$

which becomes, since $d^2 = (x_2 - x_1)^2 + (y_2 - y_1)^2$,

(5) $$\cos^2\alpha + \cos^2\beta = 1$$

We can also write this as

(6) $$\lambda^2 + \mu^2 = 1$$

and this relation holds for every pair of direction angles α, β.

Exercise A. Show that if a and b are direction numbers of a line L, then

$$\lambda^2 = \frac{a^2}{a^2 + b^2} \qquad \text{and} \qquad \mu^2 = \frac{b^2}{a^2 + b^2}$$

In Exercise A care must be taken in order to obtain the direction cosines λ and μ themselves. The trouble is apparent when you write

$$\lambda = \pm \frac{a}{\sqrt{a^2 + b^2}} \qquad \text{and} \qquad \mu = \pm \frac{b}{\sqrt{a^2 + b^2}}$$

There are essentially just two cases:

(I) The line goes from lower left to upper right. If the sense is in the upward direction, the direction cosines are both positive; for the opposite sense, λ and μ are both negative. If the line is not directed, then either $\lambda = +$, $\mu = +$ or $\lambda = -$, $\mu = -$ may be used.

(II) The line goes from lower right to upper left. If the sense is in the upward direction, then λ is negative, μ positive; for the opposite sense, λ is positive, μ negative. If the line is not directed, then either $\lambda = -$, $\mu = +$ or $\lambda = +$, $\mu = -$ may be used.

Illustration 1. Find the slope and direction cosines of the sensed line cutting the X-axis at 32°.

Solution: Here $\alpha = 32°$, $\beta = 90° - 32° = 58°$. Also $\theta = 32°$. It follows from a table of the natural trigonometric functions that

$$\text{Slope} = m = \tan 32° = 0.62487$$

and $\lambda = \cos 32° = 0.84805$, $\mu = \cos 58° = 0.52992$.

Illustration 2. Find the inclination, slope, and direction cosines of the line joining the two points $(2,-3)$, $(-5,1)$.

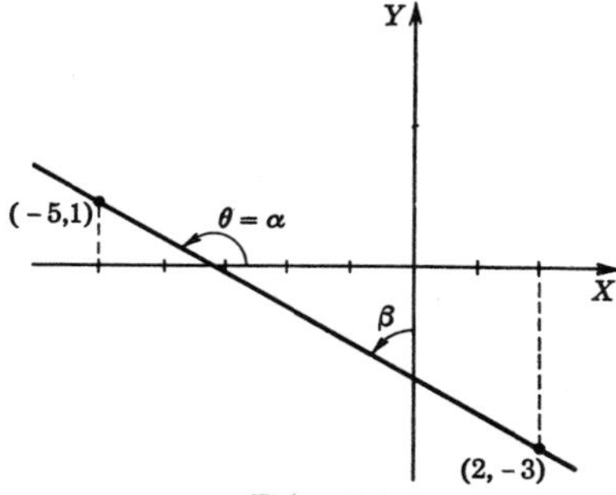

Figure **9.4**

Solution: This line is not sensed. We may take α and β as shown in Fig. 9.4; they correspond to the upward sense of the line.

$$\text{Slope} = m = \tan\theta = \frac{1-(-3)}{-5-2} = -\frac{4}{7}$$

and θ is a second quadrantal angle. Looking up θ in a table of natural tangents, we find, first of all, that the angle θ', in the first quadrant, whose tangent is $+\frac{4}{7} = 0.57143$, is

$$\theta' = \text{arc Tan}\,(0.57143) = 29°44.7'$$

Therefore

$$\text{Inclination} = \theta = 180° - \theta' = 150°15.3'$$

The direction cosines may be taken as

$$\lambda = \cos\alpha = -\cos\theta' = -\cos 29°44.7' = -0.86823$$

and

$$\mu = \cos\beta = \cos(90° - \theta') = \cos 60°15.3' = 0.49614$$

Illustration 3. Show that the line AB joining (2,0) and (0,6) and the line PQ joining (1, −3) and (−2,6) have the same slope.

Solution:

$$m_{AB} = \frac{6-0}{0-2} = -3$$

$$m_{PQ} = \frac{-3-6}{1-(-2)} = -3$$

Theorem 1. If two lines have the same slope, then they are parallel.

The proof is immediate since, if two lines have the same slope, they have the same inclination, and hence they are parallel.

Exercise B. State and prove the converse theorem.

Sometimes directed lines which have the same inclination but opposite sense are called *antiparallel*.

9.6 Angle between Two Directed Lines

In Sec 8.12 we defined the angle between two directed lines. We now prove the following theorem:

Theorem 2. The angle θ between the positive directions of two directed lines L_1 and L_2 with direction cosines λ_1, μ_1 and λ_2, μ_2, respectively, is given by

$$\text{(7)} \qquad \cos\theta = \lambda_1\lambda_2 + \mu_1\mu_2 \qquad 0 \le \theta \le 180°$$

Proof: There is no loss in generality if we suppose that L_1 and L_2 meet at the origin (Fig. 9.5). Choose P_1 on L_1 at a distance 1 from 0. Then P_1 has coordinates (λ_1,μ_1). Choose P_2 on L_2 so that $OP_2 = 1$. Similarly, P_2 has coordinates (λ_2,μ_2).

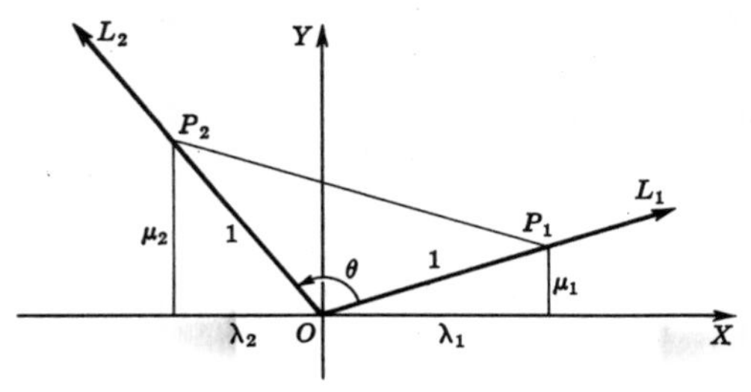

Figure 9.5

We apply the Law of Cosines (Sec. 8.12) to the triangle P_1OP_2; this gives

$$(P_1P_2)^2 = 1 + 1 - 2\cos\theta \tag{8}$$

By the distance formula we compute $(P_1P_2)^2$. We get

$$(\lambda_1 - \lambda_2)^2 + (\mu_1 - \mu_2)^2 = 2 - 2\cos\theta$$

$$\lambda_1^2 - 2\lambda_1\lambda_2 + \lambda_2^2 + \mu_1^2 - 2\mu_1\mu_2 + \mu_2^2 = 2 - 2\cos\theta \tag{9}$$

Since $\lambda_1^2 + \mu_1^2 = \lambda_2^2 + \mu_2^2 = 1$, (9) reduces to

$$-2\lambda_1\lambda_2 - 2\mu_1\mu_2 = -2\cos\theta$$

or finally,

$$\cos\theta = \lambda_1\lambda_2 + \mu_1\mu_2$$

which was to be proved.

Corollary 1. Two lines are perpendicular if and only if their direction cosines satisfy

$$\lambda_1\lambda_2 + \mu_1\mu_2 = 0 \tag{10}$$

This follows since $\cos 90° = 0$.

Corollary 2. Two lines are perpendicular if and only if their direction numbers satisfy

$$a_1a_2 + b_1b_2 = 0$$

for $a_1 = k_1\lambda_1$, $b_1 = k_1\mu_1$, $a_2 = k_2\lambda_2$, and $b_2 = k_2\mu_2$. Hence the result follows at once from Corollary 1.

Corollary 3. If neither L_1 nor L_2 is parallel to an axis and if L_1 is perpendicular to L_2, then the slope of one is the negative reciprocal of the slope of the other.

Proof: Since none of $\lambda_1, \mu_1, \lambda_2, \mu_2$ is zero, we are allowed to write (10) in either of the forms:

$$\frac{\mu_1}{\lambda_1} = -\frac{1}{\mu_2/\lambda_2} \quad \text{or} \quad \frac{\mu_2}{\lambda_2} = -\frac{1}{\mu_1/\lambda_1}$$

We have seen above that $m_1 = \mu_1/\lambda_1$ and $m_2 = \mu_2/\lambda_2$. Hence

$$m_1 = -\frac{1}{m_2} \quad \text{or} \quad m_2 = -\frac{1}{m_1}$$

Sometimes we write this in the form: $m_1 m_2 = -1$.

Illustration 1. Find the slope of a line which is (*a*) parallel to and (*b*) perpendicular to the line joining $A(4,-3), B(6,1)$.

Solution:

(**a**) The slope of a parallel line is the same as the slope of the line AB.

$$m_{AB} = \frac{1-(-3)}{6-4} = \frac{4}{2} = 2$$

(**b**) The slope of a perpendicular line by Corollary 3 is $-\frac{1}{2}$.

Illustration 2. Find the cosine of the angle B of the triangle $A(0,0)$, $B(2,-1)$, $C(9,2)$.

Solution: In order to obtain an angle (interior) of a triangle, we *think* of the sides as being directed *away* from that particular vertex. To obtain angle B, therefore, we impose the directions $\overrightarrow{BA}$ and $\overrightarrow{BC}$. We compute

$$d_{BA} = \sqrt{5} \qquad d_{BC} = \sqrt{58}$$

$$\lambda_{\overrightarrow{BA}} = -\frac{2}{\sqrt{5}} \qquad \lambda_{\overrightarrow{BC}} = \frac{7}{\sqrt{58}}$$

$$\mu_{\overrightarrow{BA}} = \frac{1}{\sqrt{5}} \qquad \mu_{\overrightarrow{BC}} = \frac{3}{\sqrt{58}}$$

Therefore

$$\cos\theta = \cos B = \lambda_{\overrightarrow{BA}}\lambda_{\overrightarrow{BC}} + \mu_{\overrightarrow{BA}}\mu_{\overrightarrow{BC}}$$

$$= \frac{-14}{\sqrt{5}\sqrt{58}} + \frac{3}{\sqrt{5}\sqrt{58}}$$

$$= \frac{-11}{\sqrt{5}\sqrt{58}}$$

The angle B is obtuse.

Exercise A. Prove that the acute angle between two undirected lines is given by

$$\cos\theta = |\lambda_1\lambda_2 + \mu_1\mu_2|$$

Problems 9.6

In Probs. 1 to 8 find direction cosines of the line joining the given points.

1. $(1,-2)$, $(0,5)$.
2. $(-2,0)$, $(3,-2)$.
3. $(1,-4)$, $(4,-1)$.
4. $(4,5)$, $(3,-4)$.
5. $(6,-7)$, $(8,-6)$.
6. $(1,4)$, $(-3,2)$.
7. $(5,2)$, $(-2,4)$.
8. $(1,-3)$, $(3,-1)$.

In Probs. 9 to 12, (*a*) by using direction cosines, (*b*) by using slopes, show that triangle ABC is a right triangle.

9. $A(-3,0)$, $B(4,14)$, $C(3,-3)$.
10. $A(-3,-3)$, $B(2,7)$, $C(1,-5)$.
11. $A(-5,2)$, $B(-1,-4)$, $C(5,0)$.
12. $A(3,-6)$, $B(1,2)$, $C(5,3)$.

In Probs. 13 to 16 show that $ABCD$ is a parallelogram.

13. $A(0,-3)$, $B(0,-1)$, $C(4,0)$, $D(4,-2)$.
14. $A(2,0)$, $B(6,4)$, $C(7,2)$, $D(3,-2)$.
15. $A(6,8)$, $B(10,7)$, $C(0,-5)$, $D(-4,-4)$.
16. $A(-4,-3)$, $B(4,1)$, $C(5,-4)$, $D(-3,-8)$.

In Probs. 17 to 20 find the cosine of the smaller angle made by the two lines AB and CD.

17. $A(1,1)$, $B(5,0)$; $C(2,3)$, $D(-1,3)$.
18. $A(2,2)$, $B(-2,5)$; $C(6,3)$, $D(-3,-1)$.
19. $A(3,4)$, $B(-2,2)$; $C(-3,1)$, $D(3,2)$.
20. $A(0,6)$, $B(-3,8)$; $C(0,-2)$, $D(-2,-3)$.

In Probs. 21 to 26 find the cosine of the angle at B of the triangle.

21. $A(0,3)$, $B(4,2)$, $C(-2,-1)$.
22. $A(4,5)$, $B(5,0)$, $C(-3,-4)$.
23. $A(1,2)$, $B(0,4)$, $C(-3,1)$.
24. $A(2,-11)$, $B(0,-3)$, $C(4,-2)$.
25. $A(7,1)$, $B(7,3)$, $C(3,2)$.
26. $A(-2,0)$, $B(4,8)$, $C(1,-1)$.

In Probs. 27 to 32 find the slope of a line which is (*a*) parallel to and (*b*) perpendicular to the line joining the mid-points of the segments AB and CD.

27. $A(4,1)$, $B(2,-3)$; $C(5,-2)$, $D(3,0)$.
28. $A(-1,2)$, $B(3,6)$; $C(-3,-5)$, $D(-1,3)$.
29. $A(6,8)$, $B(4,6)$; $C(2,4)$, $D(0,2)$.
30. $A(5,-3)$, $B(3,-1)$; $C(6,2)$, $D(2,6)$. (BT)
31. $A(8,9)$, $B(6,5)$; $C(10,2)$, $D(4,12)$. (BT)
32. $A(96,-40)$, $B(35,40)$; $C(4,100)$, $D(27,-63)$.

In Probs. 33 to 36 find the cosine of the acute angle θ made by the two lines with given direction cosines.

33. $\lambda_1 = \frac{1}{2}$, $\mu_1 = \frac{1}{2}\sqrt{3}$, and $\lambda_2 = \frac{3}{5}$, $\mu_2 = -\frac{4}{5}$.
34. $\lambda_1 = \frac{2}{3}$, $\mu_1 = \sqrt{5}/3$, and $\lambda_2 = \frac{1}{2}$, $\mu_2 = -\frac{1}{2}\sqrt{3}$.
35. $\lambda_1 = \frac{1}{2}\sqrt{2}$, $\mu_1 = -\frac{1}{2}\sqrt{2}$, and $\lambda_2 = \sqrt{5}/5$, $\mu_2 = 2\sqrt{5}/5$.
36. $\lambda_1 = 2\sqrt{2}/3$, $\mu_1 = \frac{1}{3}$, and $\lambda_2 = \frac{1}{4}$, $\mu_2 = -\sqrt{15}/4$.

In Probs. 37 to 40 find the cosine of the acute angle θ made by the two lines with given direction numbers.

37. $a_1 = 1, b_1 = 1$, and $a_2 = 1, b_2 = -1$.
38. $a_1 = 2, b_1 = -1$, and $a_2 = 1, b_2 = -2$.
39. $a_1 = 2, b_1 = 3$, and $a_2 = 5, b_2 = 4$.
40. $a_1 = 10, b_1 = 15$, and $a_2 = -5, b_2 = 20$.
41. Apply Eq. (7) to find the angle between two antiparallel lines.
42. Given $P_1(1,-2)$, $P_2(3,a)$, $Q_1(-3,1)$, $Q_2(a,4)$, determine a if P_1P_2 is perpendicular to Q_1Q_2.
43. Given $P_1(1,-2)$, $P_2(4,a)$, $Q_1(7,2)$, $Q_2(3,a)$, determine a if P_1P_2 and Q_1Q_2 are parallel.
44. Given $P_1(1,2)$, $P_2(-3,1)$, $Q_1(0,5)$, $Q_2(a,b)$, find numbers a and b such that the angle made by $\overrightarrow{P_1P_2}$ and $\overrightarrow{Q_1Q_2}$ is 90°.
45. Show that the angle θ_{12} (read, "theta sub 1, 2") *from* line L_1 *to* line L_2 is given by

$$\tan \theta_{12} = \frac{m_2 - m_1}{1 + m_2m_1}$$

HINT: Use the formula for $\tan(\theta_2 - \theta_1)$.
46. Use the result of Prob. 45 above to find the tangent of the smallest angle in the triangle $A(4,7)$, $B(-3,1)$, $C(2,0)$.
47. In Fig. 9.5 show that $\theta = |\alpha_2 - \alpha_1|$. Hence prove Theorem 2 by using the addition formula for $\cos(\alpha_2 - \alpha_1)$.

9.7 Applications to Plane Geometry

The properties of a given geometric configuration usually found in Euclidean plane geometry do not in any way depend upon a related coordinate system. It often happens, however, that the introduction of a coordinate system will help to simplify the work of proving a theorem and especially if axes are chosen properly. But the axes must be chosen so that there will be no loss in generality. For example, if the problem is to prove some proposition relating to a triangle, then a coordinate axis can be chosen coincident with a side, and one vertex can then be taken as the origin. Consider the following illustration.

Illustration 1. Prove: The line segment joining the mid-points of two sides of a triangle is parallel to the third side and equal to one-half its length.

Solution: We choose axes as in Fig. 9.6. The mid-points D and E have the coordinates $D\left(\frac{b}{2}, \frac{c}{2}\right)$ and $E\left(\frac{a+b}{2}, \frac{c}{2}\right)$. The slope of DE is

$$m_{DE} = \frac{(c/2) - (c/2)}{(a+b)/2 - (b/2)} = \frac{0}{a/2} = 0$$

Since AB also has slope zero, it follows that DE is parallel to AB. The length AB is a. The length of DE is

$$\begin{aligned} DE &= \sqrt{\left(\frac{b}{2} - \frac{a+b}{2}\right)^2 + \left(\frac{c}{2} - \frac{c}{2}\right)^2} \\ &= \sqrt{\frac{a^2}{4}} \\ &= \frac{a}{2} \end{aligned}$$

Thus the theorem is proved.

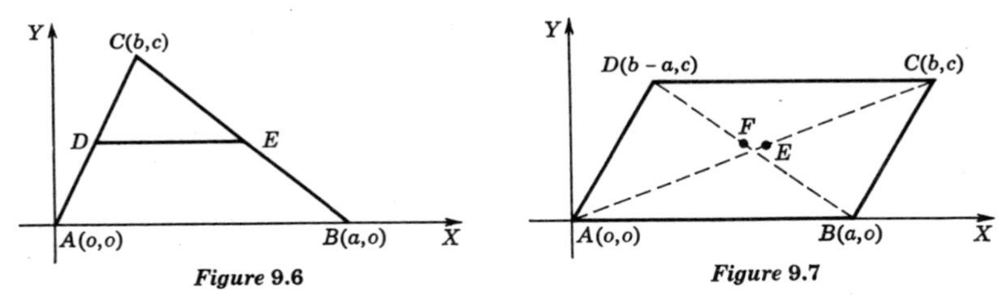

Figure 9.6

Figure 9.7

Illustration 2. Prove: The diagonals of a parallelogram bisect each other.

Solution: Choose axes as in Fig. 9.7, letting the coordinates of three vertices be $A(0,0)$, $B(a,0)$, and $C(b,c)$. Then, since the figure $ABCD$ is a parallelogram, the coordinates of D are determined. It is easy to see that they must be $D(b - a, c)$. The mid-point of AC has coordinates $E\left(\frac{b}{2}, \frac{c}{2}\right)$. Let F be the mid-point of BD. The coordinates of F are $F\left(\frac{b-a-a}{2}, \frac{c}{2}\right)$, that is, $F\left(\frac{b}{2}, \frac{c}{2}\right)$. Since E and F have the same coordinates, they must coincide. Hence the proposition is proved.

Problems 9.7

Draw a figure and prove by analytic geometry:

1. The diagonals of a square are perpendicular.
2. The diagonals of a rectangle are equal.
3. The diagonals of a rhombus are perpendicular. (A rhombus is an equilateral parallelogram.)
4. The diagonals of a parallelogram are equal only if the parallelogram is a rectangle.
5. The mid-point of the hypotenuse of a right triangle is equidistant from the vertices.
6. The line joining the mid-point of the hypotenuse of a right triangle and the mid-point of one side is parallel to the other side and one-half its length.
7. The lines joining the mid-points of the sides of a triangle divide it into four congruent triangles. For the purposes of this problem, two triangles are congruent if and only if their corresponding sides are equal.
8. The distance between the mid-points of the nonparallel sides of a trapezoid is one-half the sum of the parallel sides.

9. The diagonals of a trapezoid are equal if the trapezoid is isosceles. (The non-parallel sides of an isosceles trapezoid are equal.)
10. The diagonals of a trapezoid are equal only if the trapezoid is isosceles.
11. The line segments joining the mid-points of adjacent sides of a quadrilateral form a parallelogram.
12. The sum of the squares of the sides of a parallelogram equals the sum of the squares of the diagonals.
13. The medians of a triangle meet in a point. HINT: Show that the point $[\frac{1}{3}(x_1 + x_2 + x_3), \frac{1}{3}(y_1 + y_2 + y_3)]$ lies on each median.
14. With respect to skewed axes where the X-axis and Y-axis make an angle $\theta°$ $(0 < \theta° < 180°)$, the distance formula is $d^2 = (x_2 - x_1)^2 + 2(x_2 - x_1)(y_2 - y_1) \cos\theta + (y_2 - y_1)^2$.

9.8 The Straight Line

We now wish to study certain curves defined by special equations. About the simplest algebraic relation is that given by the equation

$$Ax + By + C = 0$$

where A, B, and C are real numbers. We exclude the case where $A = B = 0$, $C \neq 0$, and also the case where $A = B = C = 0$, as they are not sensible ones from our present point of view. The equation $Ax + By + C = 0$ is called a linear equation because its graph is a straight line, as is proved in Theorem 3.

Theorem 3. The graph of a linear equation is a straight line.

Proof: Choose a point $P_0(x_0,y_0)$ whose coordinates satisfy the given equation, i.e., such that

$$Ax_0 + By_0 + C = 0$$

Hence $$C = -Ax_0 - By_0$$

and we can write the given equation in the form:

$$A(x - x_0) + B(y - y_0) = 0 \tag{11}$$

Construct the line L through P_0 with direction numbers A, B. Let P be the point $P(x,y)$, where (x,y) satisfies (11). Then Eq. (11) tells us that the segment PP_0 is perpendicular to the line L (Sec. 9.6, Corollary 2). We know from geometry that there is a unique line M passing through P_0 and perpendicular to L. Hence P must lie on M. Our argument also shows that the coordinates of any point on M satisfy (11). Therefore M is the graph of the given equation.

Theorem 4. There exists a linear equation whose graph is a given straight line.

Proof: Let the given line be M (Fig. 9.8), and choose a fixed point $P_0(x_0,y_0)$ on it. Let $P(x,y)$ be any other point on M. Construct L perpendicular

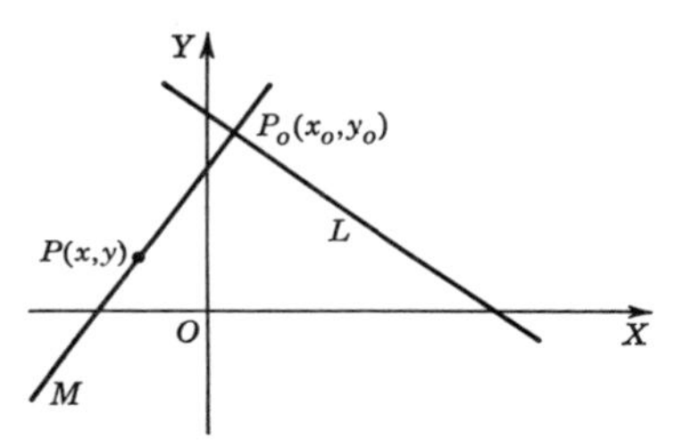

Figure **9.8**

to M at P_0. Let L have direction numbers (A,B). Since the segment P_0P is perpendicular to L, we can write:

$$A(x - x_0) + B(y - y_0) = 0$$

But this is a linear equation which is satisfied by the coordinates of all points on M. Since, moreover, all solutions (x,y) of this equation correspond to points on M, it is the desired equation.

Corollary. A and B are direction numbers of any line perpendicular to the line whose equation is $Ax + By + C = 0$.

When the given line is defined by a pair of points on it $P_1(x_1,y_1)$ and $P_2(x_2,y_2)$, we can find its equation by the following theorem:

Theorem 5. Let $P_1(x_1,y_1)$ and $P_2(x_2,y_2)$ be two points on a given line. Then one equation of this line is:

$$(y_1 - y_2)x + (x_2 - x_1)y + (x_1y_2 - x_2y_1) = 0 \qquad (12)$$

Proof: First, if the line is perpendicular to the X-axis, $x_1 = x_2$ and the above equation reduces to

$$(y_1 - y_2)x + x_1(y_2 - y_1) = 0$$

or to

$$x = x_1$$

which is a suitable equation for this line.

Second, on any other line we can find two distinct points $P_1(x_1,y_1)$ and $P_2(x_2y_2)$, where $x_1 \neq x_2$. Directly from Fig. 9.9 we have

$$\text{(13)} \qquad \frac{y - y_1}{x - x_1} = \frac{y_2 - y_1}{x_2 - x_1}$$

where $P(x,y)$ is a point different from P_1 on the line joining P_1 and P_2.

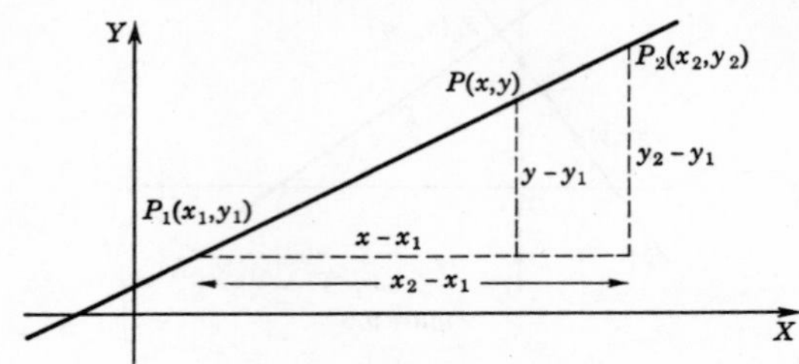

Figure **9.9**

From (13) we derive (12) at once. Equation (12) is called the *two-point* form of the equation of a straight line.

Exercise A. Derive (12) from (13).

Exercise B. Show that (x_1,y_1) and (x_2,y_2) satisfy (12).

Exercise C. Can you combine Theorems 3 and 4, using "necessary and sufficient" language?

It is desirable to find the direction cosines of a line when we are given its equation. Let the equation be

$$Ax + By + C = 0$$

Let us suppose $B \neq 0$. Otherwise we must have $A \neq 0$, and a similar discussion follows. Two points on this line are

$$\left(x_1, -\frac{C + Ax_1}{B}\right) \qquad \left(x_2, -\frac{C + Ax_2}{B}\right)$$

Therefore $\qquad a = x_2 - x_1 \qquad b = -\frac{A}{B}(x_2 - x_1)$

are direction numbers of this line. We can get another set of direction numbers by multiplying these by $B/(x_2 - x_1)$. These are

$$a = B \qquad b = -A$$

We have therefore proved the following theorem:

Theorem 6. The direction cosines of the line whose equation is $Ax + By + C = 0$ are

$$\lambda = \frac{B}{\sqrt{A^2 + B^2}} \qquad \mu = \frac{-A}{\sqrt{A^2 + B^2}} \tag{14}$$

or

$$\lambda = \frac{-B}{\sqrt{A^2 + B^2}} \qquad \mu = \frac{A}{\sqrt{A^2 + B^2}}$$

The slope $m = \mu/\lambda$ is therefore equal to $-A/B$, $B \neq 0$.

Illustration 1. Find the direction cosines and slope of the line

$$2x - 3y + 5 = 0$$

Solution:

$$\lambda = \frac{-3}{\sqrt{4 + 9}} = \frac{-3}{\sqrt{13}} \qquad \mu = \frac{-2}{\sqrt{13}} \qquad m = \frac{2}{3}$$

or

$$\lambda = \frac{3}{\sqrt{13}} \qquad \mu = \frac{2}{\sqrt{13}}$$

Theorem 7. Let λ and μ be the direction cosines of a line segment OP of positive length p issuing from the origin. Then $\lambda x + \mu y - p = 0$ is an equation of the line L perpendicular to OP and passing through P.

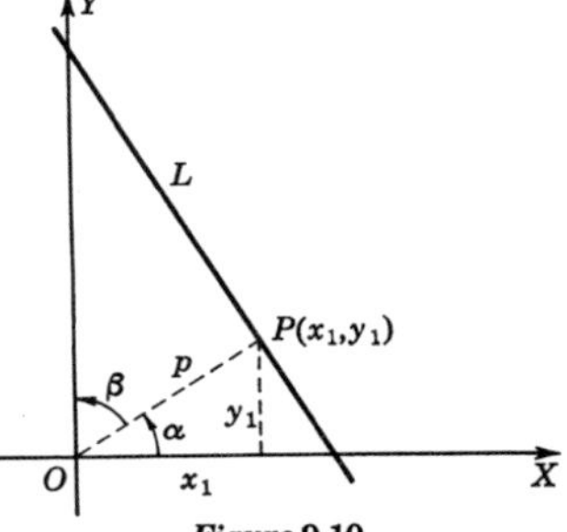

Figure 9.10

Proof: The line perpendicular to OP passing through P will have an equation of the form

$$\lambda x + \mu y + k = 0 \tag{15}$$

by the Corollary to Theorem 4. Keep in mind that λ and μ here are the direction cosines of any line perpendicular to the line whose equation is (15). They are *not* the direction cosines of the line (15) itself. The coordinates of P, namely, $x_1 = \lambda p$, $y_1 = \mu p$, must satisfy (15). Hence

$$\lambda^2 p + \mu^2 p + k = 0$$
$$p(\lambda^2 + \mu^2) + k = 0$$

or, since $\lambda^2 + \mu^2 = 1$,

$$p = -k$$

Therefore the equation of L is

$$\lambda x + \mu y - p = 0 \tag{16}$$

and the theorem is proved.

You can easily deduce the following rule for reducing

$$Ax + By + C = 0$$

to (16), which is called the *normal form* of the equation of the straight line.

Rule: Divide $Ax + By + C = 0$ by $\pm\sqrt{A^2 + B^2}$, using the sign opposite to that of $C(\neq 0)$. If $C = 0$, the sign does not matter. Thus

$$\frac{Ax}{\pm\sqrt{A^2 + B^2}} + \frac{By}{\pm\sqrt{A^2 + B^2}} + \frac{C}{\pm\sqrt{A^2 + B^2}} = 0 \tag{17}$$

is in normal form if the sign of $\sqrt{A^2 + B^2}$ is chosen so that

$$\frac{C}{\pm\sqrt{A^2 + B^2}}$$

is negative.

Illustration 2. Reduce $3x - 2y + 7 = 0$ to normal form.

Solution:

$$\frac{3}{-\sqrt{13}}x + \frac{-2}{-\sqrt{13}}y + \frac{7}{-\sqrt{13}} = 0$$

or

$$\frac{-3}{\sqrt{13}}x + \frac{2}{\sqrt{13}}y - \frac{7}{\sqrt{13}} = 0$$

Here $\lambda = -3/\sqrt{13}$, $\mu = 2/\sqrt{13}$, and $p = 7/\sqrt{13}$. Again note that λ and μ are direction cosines of any line perpendicular to $3x - 2y + 7 = 0$.

Exercise D. Find the direction cosines of the line $3x - 2y + 7 = 0$.

Illustration 3. Find the perpendicular distance from the origin to the line

$$x + y - 6 = 0$$

Solution: The normal form is

$$\frac{x}{\sqrt{2}} + \frac{y}{\sqrt{2}} - \frac{6}{\sqrt{2}} = 0$$

The distance is $p = 6/\sqrt{2}$ units.

Problems 9.8

In Probs. 1 to 12 sketch the straight line.

1. $x - 5y + 5 = 0$.
2. $2x - 3y + 1 = 0$.
3. $x + y - 4 = 0$.
4. $x - 2y - 3 = 0$.
5. $x - 2 = 0$.
6. $y + 3 = 0$.
7. $(x/2) + (y/3) = 1$.
8. $(x/3) - (y/4) = 1$.
9. $3y = 2x$.
10. $y = 3x + 2$.
11. $y - 1 = 4(x - 2)$.
12. $y + 2 = -3(x - 1)$.

In Probs. 13 to 22 write down an equation for each straight line.

13. Through $(1,7)$ and $(-2,3)$.
14. Through $(4,-3)$ and $(-5,0)$.
15. Through $(0,0)$ and rising 3 units for every forward horizontal unit.
16. Through $(-1,2)$ and rising 2 units for every forward horizontal unit.
17. Through $(4,-3)$ with rise 5 and run 2.
18. Through $(1,0)$ with rise 2 and run 3.
19. Through $(2,-5)$ and falling 2 units for every forward horizontal unit.
20. Through $(-1,1)$ and falling 1 unit for every forward horizontal unit.
21. Parallel to $3x + 4y = 0$ and such that each ordinate exceeds the corresponding ordinate of $3x + 4y = 0$ by 2 units.
22. Perpendicular to $2x - y + 1 = 0$ and passing through $(0,1)$.

23. Discuss the problem of graphing $(Ax + By + C)^n = 0$, n a positive integer.
24. Graph $(x^2 + y^2 + 4)(2x + y + 1) = 0$.
25. Graph $x^2 - y^2 = 0$.
26. Show that Eq. (13) can be written in the form $y - y_1 = m(x - x_1)$, where m is the slope of the line. This is called the *point-slope* form and is useful where the slope m and the point (x_1,y_1) are given.
27. Find an equation of the line passing through $(4,-3)$ and making 30° with the X-axis.
28. Show that the line $y = mx + b$ has slope m. What lines have no such equation?
29. Show that Eq. (12) reduces to $(x/a) + (y/b) = 1$ if the given points are $(a,0)$ and $(0,b)$.
30. Plot the lines $4x - y - 9 = 0$ and $3x + y - 5 = 0$ and find, by solving the equations simultaneously, the coordinates of the point of intersection.
31. Given the two straight lines $A_1x + B_1y + C_1 = 0$ and $A_2x + B_2y + C_2 = 0$, with A_1, B_1, C_1, A_2, B_2, C_2 in the field of real numbers. Assuming that the two lines intersect in one and only one point, are the coordinates of this point in the field of real numbers? Explain.
32. Given the two straight lines $A_1x + B_1y + C_1 = 0$ and $A_2x + B_2y + C_2 = 0$ with A_1, B_1, C_1, A_2, B_2, C_2 in the field of rational numbers. Assuming that the two lines intersect in one and only one point, are the coordinates of this point in the field of rational numbers? Explain.

33. Given the two straight lines $A_1x + B_1y + C_1 = 0$ and $A_2x + B_2y + C_2 = 0$ with A_1, B_1, C_1, A_2, B_2, C_2 in the integral domain of the integers. Assuming that the two lines intersect in one and only one point, are the coordinates of this point in the integral domain of the integers? Explain.

34. Show that, for each value of k, the graph of the equation

$$(A_1x + B_1y + C_1) + k(A_2x + B_2y + C_2) = 0$$

is a straight line through the point of intersection of $A_1x + B_1y + C_1 = 0$ and $A_2x + B_2y + C_2 = 0$ if there is such a point. What is the situation in case there is no such point of intersection?

35. Show that $2x - y + 8 = 0$, $3x - y + 11 = 0$, $5x - y + 17 = 0$ meet in a common point.

36. Show that $3x + y - 1 = 0$, $x - y - 3 = 0$, $5x + 3y + 1 = 0$ meet in a common point.

In Probs. 37 to 42 reduce to normal form.

37. $x - y - 3 = 0$.

38. $2x + 3y - 4 = 0$.

39. $4x - 3y + 5 = 0$.

40. $3x + 7y + 4 = 0$.

41. $6x + 8y + 20 = 0$.

42. $8x - 6y - 15 = 0$.

43. Show that the distance from the line $\lambda x + \mu y - p = 0$ to the point (x_1,y_1) is $|\lambda x_1 + \mu y_1 - p|$.

In Probs. 44 to 47 find the distance indicated.

44. From $x + y - 1 = 0$ to $(3,4)$.

45. From $3x - 4y + 10 = 0$ to $(2,-5)$.

46. From $x - 4y - 2\sqrt{17} = 0$ to $(-5,-3)$.

47. From $3x - 3y + 6\sqrt{2} = 0$ to $(-4,2)$.

48. Find the equations of the bisectors of the angles between the lines $\lambda_1x + \mu_1y - p_1 = 0$ and $\lambda_2x + \mu_2y - p_2 = 0$. (See Prob. 43.)

49. Find the equations of the bisectors of the angles between the lines $x + y - 5 = 0$ and $2x - y + 4 = 0$. (See Prob. 48.)

50. Find the equation of the locus of points P which are at a distance of 2 units from the line $4x - 3y + 10 = 0$.

9.9 Conic Sections

One way of generalizing $Ax + By + C = 0$, which represents a straight line, is to add all possible quadratic terms (terms of the second degree in x and y). Where an obvious shift has been made in renaming the coefficients, such an equation can be written in the form

$$Ax^2 + Bxy + Cy^2 + Dx + Ey + F = 0 \tag{18}$$

It is the general equation of the second degree in each variable (provided it is not true that $A = B = C = 0$).

We shall consider some special cases of (18). The treatment of the general case is complicated; but the total set of points corresponding to the ordered pairs (x,y) satisfying the relation defined by (18) is called a *conic section.* This is because, geometrically, the curve can be obtained by cutting a cone with a plane. This fact was known to the Greek mathematicians of 300 B.C.; we shall give the appropriate geometric illustration as we treat each case.

9.10 Case I. The Circle

Definition: A *circle* is the locus of points P which are at a fixed distance from a fixed point.

Thus consider a fixed point $C(h,k)$. Now the point $P(x,y)$ will be r units from C if and only if the distance PC equals r, that is, if and only if (Fig. 9.11)

$$\sqrt{(x-h)^2+(y-k)^2}=r$$

This becomes, upon squaring,

$$(19)\quad (x-h)^2+(y-k)^2=r^2$$

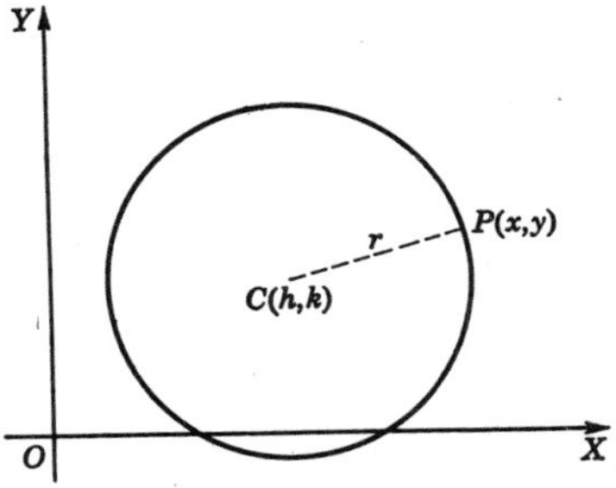

Figure 9.11

which is the equation whose graph is the circle with center at $C(h,k)$ and with radius r since (19) expresses the condition that the point P, with coordinates x and y, shall always be exactly r units from C.

Equation (19) reduces, after a little rearranging, to

$$(20)\qquad x^2+y^2-2hx-2ky+h^2+k^2-r^2=0$$

This is a special case of (18) where $A = C$ and $B = 0$ [which indeed constitutes a necessary condition that (18) represent a circle].

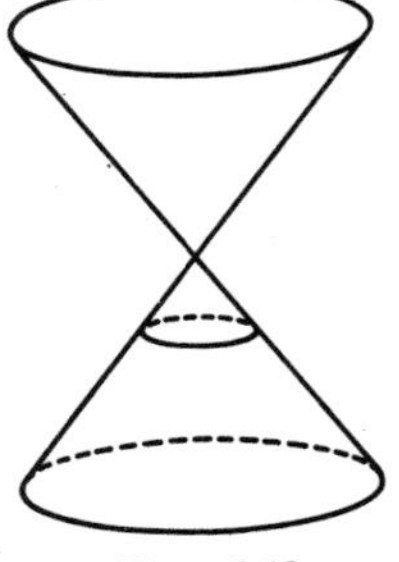

Figure 9.12

Exercise A. In (20) what coefficients correspond to A, B, C, D, E, and F in (18)?

Exercise B. Is the necessary condition that (18) represent a circle, namely, $A = C$ and $B = 0$, also sufficient?

The circle is a conic section. Geometrically it is obtained by cutting a right circular cone with a plane parallel to the base (Fig. 9.12).

Illustration 1. Write down the equation of the circle with center at $(-2,1)$ and with radius 3.

Solution: It is

$$(x + 2)^2 + (y - 1)^2 = 9$$

Illustration 2. Plot the curve given by

$$x^2 + y^2 - 3x + 6y - 5 = 0$$

Solution: We complete the square separately on the x terms and on the y terms as follows:

$$x^2 - 3x + [\tfrac{9}{4}] + y^2 + 6y + [9] = 5 + [\tfrac{9}{4}] + [9]$$

The "5" was transposed, and the brackets merely indicate the terms added to complete the square (Sec. 5.10). This can be rewritten as

$$(x^2 - 3x + \tfrac{9}{4}) + (y^2 + 6y + 9) = \tfrac{65}{4}$$

or again, as

$$(x - \tfrac{3}{2})^2 + (y + 3)^2 = \tfrac{65}{4}$$

This is precisely in the form (19), so that the graph is a circle with center at $(\frac{3}{2}, -3)$ and $r = \sqrt{\frac{65}{4}} = \frac{1}{2}\sqrt{65}$.

The equation

$$x^2 + y^2 = r^2 \tag{21}$$

is that of a circle of radius r with center at the origin.

Exercise C. In (21) what coefficients correspond to A, B, C, D, E, and F in (18)?

Problems 9.10

In Probs. 1 to 12 sketch and find the equation of the circle:

1. Center at $(3,5)$, radius 6.
2. Center at $(-1,2)$, radius 3.
3. Center at $(2,-2)$, radius 5.
4. Center at $(-3,-1)$, radius 4.
5. Ends of diameter at $(-2,-1)$, $(3,4)$.
6. Ends of diameter at $(4,1)$, $(-1,0)$.
7. Tangent to the X-axis, center at $(2,1)$.
8. Tangent to the Y-axis, center at $(2,-3)$.
9. Has for diameter the portion of $x - 2y + 1 = 0$ lying in second quadrant.
10. Has radius 4 and is concentric with $x^2 + y^2 + 2x - 6y = 0$.
11. Has radius 4 and is tangent to both axes.
12. Tangent to both axes, and center lies on $y = -x$.

In Probs. 13 to 20 find center and radius. Sketch.

13. $x^2 + y^2 - 4x + 6y = 0.$
14. $x^2 + y^2 + 2x + 8y - 1 = 0.$
15. $16 + 6x - x^2 - y^2 = 0.$
16. $24 - 8x - 6y - x^2 - y^2 = 0.$
17. $x^2 + y^2 - 2x + 4y + 21 = 0.$ (BT)
18. $x^2 + y^2 + x - 3y - 7 = 0.$
19. $2x^2 + 2y^2 + 5x - 7y - 10 = 0.$
20. $3x^2 + 3y^2 + ax = 0.$

21. Find the locus of points P such that the sum of squares of the distances from P to $(1,-4)$ and to $(2,3)$ is 57.
22. Find the locus of points P such that the sum of squares of the distances from P to $(-2,-3)$ and to $(1,3)$ is 23.
23. Write the equation of every circle of radius 1.
24. Write the equation of every circle of radius 1 with center on $y = -x$.
25. Write the equation of every circle of radius 1 tangent to the circle $x^2 + y^2 = 1$.
26. The locus of points P such that the sum of the squares of the distances from P to n fixed points (x_1,y_1), (x_2,y_2), . . . , (x_n,y_n) is (a sufficiently large) constant k is a circle. Prove it.

9.11 Case II. The Parabola

We are already somewhat familiar with the parabola (Sec. 5.12).

Definitions: A *parabola* is the locus of points P such that the distance of P from a fixed point is always equal to its distance from a fixed line. The fixed point is called the *focus;* the fixed line is called the *directrix.* The line perpendicular to the directrix and passing through the focus is called the *axis of the parabola.*

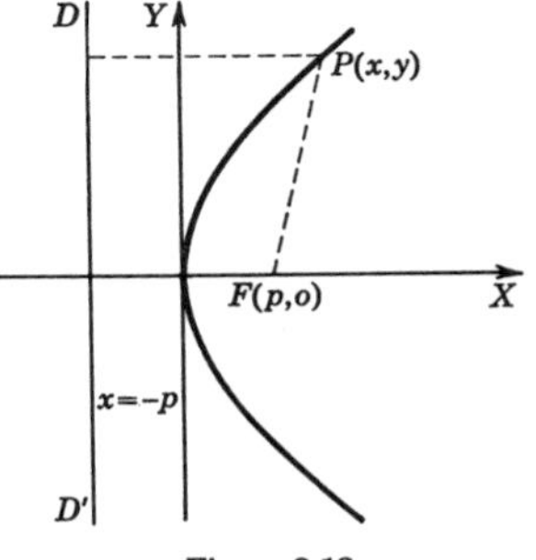

Figure 9.13

In order to arrive at an equation for this locus, we choose the coordinate axes so that the focus F has the coordinates $F(p,0)$ and the directrix line DD' has the equation $x = -p$ (Fig. 9.13). (This choice of axes leads to the simplest equation, although this is not immediately apparent.) By definition, the distance PF must equal the (perpendicular) distance from P to DD'. The distance from P to DD' is $|x + p|$. We have

$$|x + p| = \sqrt{(x - p)^2 + (y - 0)^2}$$

which yields, upon squaring,

$$x^2 + 2px + p^2 = x^2 - 2px + p^2 + y^2$$

This reduces to

$$y^2 = 4px \tag{22}$$

This is the equation sought; it defines a relation. It is a special case of (18).

Exercise A. In (22) what coefficients correspond to A, B, C, D, E, and F in (18)?

The parabola is a conic section (Fig. 9.14). Geometrically, the parabola can be obtained by cutting a right circular cone with a plane parallel to a generator.

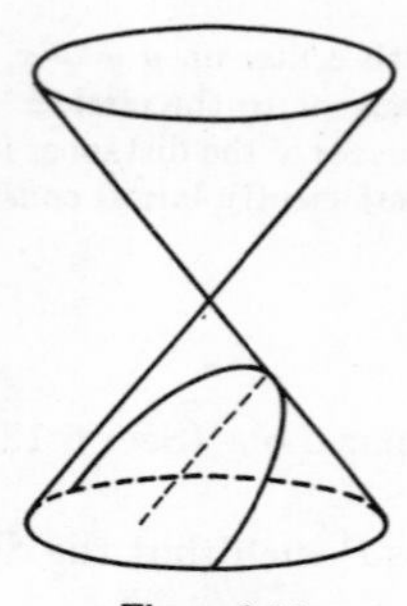

Figure **9.14**

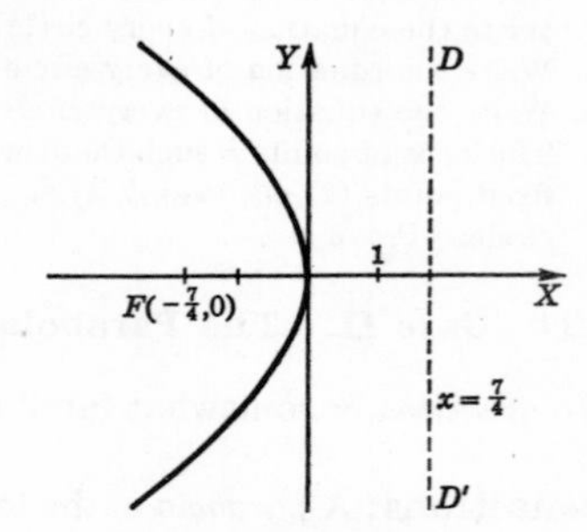

Figure **9.15**

Illustration 1. Write down the equation of the parabola with $F(3,0)$ and directrix $x = -3$.

Solution: In this case $p = 3$, and the equation is consequently $y^2 = 12x$.

Illustration 2. Sketch the parabola whose equation is $y^2 = -7x$. Find the coordinates of the focus and the equation of the directrix.

Solution: Here $4p = -7$. Hence the focus has coordinates $F(-\frac{7}{4},0)$, and the equation of the directrix DD' is $x = \frac{7}{4}$ (Fig. 9.15).

Problems 9.11

In Probs. 1 to 10 find the coordinates of the focus and the equation of the directrix. Sketch.

1. $y^2 = 4x$.
2. $y^2 = 8x$.
3. $y^2 = -x$.
4. $y^2 = -3x$.
5. $x^2 = y$.
6. $x^2 = 5y$
7. $x^2 = -12y$.
8. $x^2 = -7y$.
9. $(y - 2)^2 = 8(x + 1)$. HINT: Plot the lines $y - 2 = 0$ and $x + 1 = 0$, and think of these as new axes. That is, make the transformations $y' = y - 2$, $x' = x + 1$.
10. $(x + 3)^2 = 4(y - 5)$. (See hint, Prob. 9.)

In Probs. 11 to 22 find the equation of the parabola and sketch.

11. Focus at (5,0), directrix $x = -5$.
12. Focus at $(\frac{1}{2},0)$, directrix $x = -\frac{1}{2}$.
13. Focus at $(-3,0)$, directrix $x = 3$.
14. Focus at $(-2,0)$, directrix $x = 2$.
15. Focus at (0,2), directrix $y = -2$.
16. Focus at (0,4), directrix $y = -4$.
17. Focus at $(0,-3)$, directrix $y = 3$.
18. Focus at $(0,-\frac{3}{2})$, directrix $y = \frac{3}{2}$.
19. Focus at (0,2), directrix $y = 0$.
20. Focus at (4,0), directrix $x = -2$.
21. Focus at (3,2), directrix $y = 0$.
22*. Focus at $(a,2b)$, vertex at (a,b).

23. Find the points of intersection of $y^2 = 2x$, $x^2 = 2y$.
24. Find the points of intersection of $x^2 = 8y$, $x^2 = -(y - 4)$.
25. A point has the property that the sum of its distances from $F(4,1)$, $F'(-4,1)$ is 10. Find the equation of the locus of such points.
26. A point has the property that the sum of its distances from $F(2,2)$, $F'(-2,2)$ is $2\sqrt{5}$. Find the equation of the locus of such points.
27. Each circle of a set of circles passes through (3,0) and is tangent to the line $x = -1$. Find the equation of the locus of the centers of the circles.

9.12 Case III. The Ellipse

Definitions: An *ellipse* is the locus of points P such that the sum of the distances from P to two fixed points is a constant. The two fixed points are called *foci*.

A very simple equation results from choosing the axes and scales so that the foci F and F' have the coordinates $F(c,0)$, $F'(-c,0)$. We let the sum of the distances be the constant $2a$. Note that $2a > 2c$; hence $a > c$ (Fig. 9.16). The definition requires that

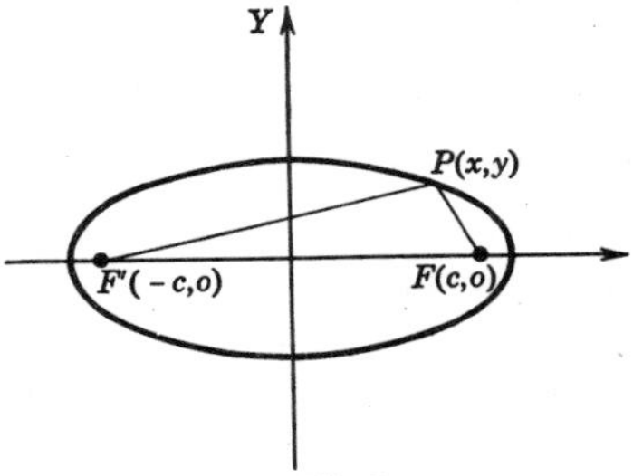

$$\sqrt{(x+c)^2 + y^2} + \sqrt{(x-c)^2 + y^2} = 2a \tag{23}$$

Y
P(x,y)
F'(−c,o)
F(c,o)
$|PF + PF'| = 2a$

Figure **9.16**

We transpose the second radical and square, getting

$$x^2 + 2cx + c^2 + y^2 = 4a^2 - 4a\sqrt{(x-c)^2 + y^2} + x^2 - 2cx + c^2 + y^2$$

which simplifies to

$$4cx - 4a^2 = -4a\sqrt{(x-c)^2 + y^2}$$

We can now cast out the 4, and the reason for choosing $2a$ as the sum of the distances instead of a is now apparent. Square again. Thus

$$c^2x^2 - 2a^2cx + a^4 = a^2(x^2 - 2cx + c^2 + y^2)$$

which reduces to

$$(24) \qquad \begin{aligned}(a^2 - c^2)x^2 + a^2y^2 &= a^4 - a^2c^2 \\ &= a^2(a^2 - c^2)\end{aligned}$$

Since $a > c$, it follows that $a^2 > c^2$ and $a^2 - c^2 > 0$. Let us call $a^2 - c^2 = b^2$ (a positive number). We can then write (24) in the form

$$b^2x^2 + a^2y^2 = a^2b^2$$

or, finally,

$$(25) \qquad \frac{x^2}{a^2} + \frac{y^2}{b^2} = 1$$

This is the equation of the ellipse.

Exercise A. In (25) what coefficients correspond to A, B, C, D, E, and F in (18)?

Exercise B. Show that the points $V(a,0)$ and $V'(-a,0)$ are on the ellipse.

Exercise C. Show that the points $(0,b)$ and $(0,-b)$ are on the ellipse.

Definitions: The points V and V' are called the *vertices* of the ellipse. The segment joining V and V' is called the *major axis;* its length is $2a$. The segment joining $(0,b)$ and $(0,-b)$ is called the *minor axis;* its length is $2b$. The *center* of the ellipse is the mid-point of the major axis.

The ellipse is a conic section (Fig. 9.17). Geometrically, the ellipse can be obtained by cutting a right circular cone with a plane inclined (but not parallel to a generator) so that it cuts only one nappe of the cone. This permits an ellipse to reduce to a circle if the cutting plane is perpendicular to the axis of the cone. Algebraically, this is the case where $a = b$ and where, therefore, Eq. (25) reduces to

$$\frac{x^2}{a^2} + \frac{y^2}{a^2} = 1$$

which represents a circle of radius a.

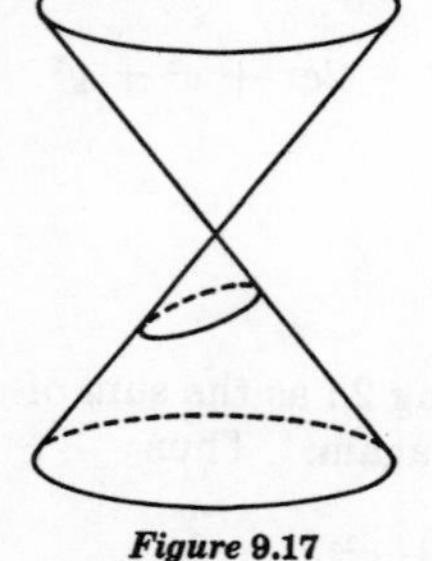

Figure 9.17

Illustration 1. Plot the graph of

$$(26) \qquad \frac{x^2}{9} + \frac{y^2}{4} = 1$$

Solution: The total graph (Fig. 9.18) will be made up of the graphs of the two algebraic functions f and g derived from (26) and defined by the equations

$$f(x) = \tfrac{2}{3}\sqrt{9 - x^2} \begin{cases} \text{domain, } -3 \le x \le 3 \\ \text{range, } 0 \le y \le 2 \end{cases}$$
$$g(x) = -\tfrac{2}{3}\sqrt{9 - x^2} \begin{cases} \text{domain, } -3 \le x \le 3 \\ \text{range, } -2 \le y \le 0 \end{cases}$$

The zeros of both f and g are $x = \pm 3$; that is, the vertices of the ellipse are $V(3,0)$ and $V'(-3,0)$. The point $(0,2)$ is on the graph of f, $(0,-2)$ is on the graph of g; each graph is a semiellipse. The coordinates of the foci are $F(\sqrt{5},0)$ and $F'(-\sqrt{5},0)$ since $c^2 = a^2 - b^2 = 5$. You should compute a few elements of f and g.

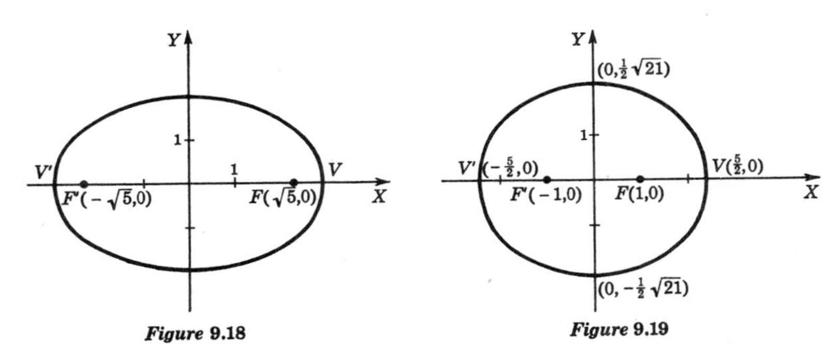

Figure 9.18

Figure 9.19

Illustration 2. Write down the equation of the ellipse with $F(1,0)$, $F'(-1,0)$, and major axis 5.

Solution: Now $c = 1$ and $2a = 5$. Therefore $a = \frac{5}{2}$; and since $b^2 = a^2 - c^2$, we have $b^2 = \frac{25}{4} - 1 = \frac{21}{4}$. Therefore the equation is

$$\frac{x^2}{\frac{25}{4}} + \frac{y^2}{\frac{21}{4}} = 1$$

The graph is drawn in Fig. 9.19.

Problems 9.12

In Probs. 1 to 10 find the coordinates of the foci and of the vertices and sketch.

1. $(x^2/9) + (y^2/4) = 1$.
2. $(x^2/25) + (y^2/9) = 1$.
3. $(x^2/100) + (y^2/25) = 1$.
4. $(x^2/144) + (y^2/16) = 1$.
5. $(x^2/5) + (y^2/2) = 1$.
6. $x^2 + 4y^2 = 4$.
7. $(x^2/16) + (y^2/36) = 1$. (BT)
8. $(x^2/9) + (y^2/25) = 1$. (BT)
9. $\dfrac{(x-2)^2}{4} + \dfrac{(y+1)^2}{2} = 1$. HINT: Plot the lines $x - 2 = 0$ and $y + 1 = 0$, and think of these as new axes. That is, make the transformation $x' = x - 2$, $y' = y + 1$.
10. $\dfrac{(x+3)^2}{49} + \dfrac{(y-2)^2}{16} = 1$. (See hint in Prob. 9.)

In Probs. 11 to 22 find the equation of the ellipse and sketch.

11. $F(1,0)$, $F'(-1,0)$, and major axis 6.
12. $F(4,0)$, $F'(-4,0)$, and major axis 10.
13. $F(1,0)$, $F'(-1,0)$, and minor axis 6.
14. $F(4,0)$, $F'(-4,0)$, and minor axis 10.

15. Major axis (along X-axis) 10, minor axis 6, and center at (0,0).
16. Major axis (coincident with the X-axis) 10, minor axis 6, and center at (3,0).
17. $V(5,0)$, $V'(-5,0)$, $F(2,0)$, $F'(-2,0)$.
18. $V(6,0)$, $V'(-6,0)$, $F(3,0)$, $F'(-3,0)$.
19. $V(5,2)$, $V'(-5,2)$, $F(2,2)$, $F'(-2,2)$.
20. $V(6,-1)$, $V'(-6,-1)$, $F(3,-1)$, $F'(-3,-1)$.
21. $V(0,10)$, $V'(0,-10)$, $F(0,4)$, $F'(0,-4)$.
22. $V(7,0)$, $V'(-13,0)$, $F(3,0)$, $F'(-9,0)$.
23. A point has the property that the numerical difference of its distances from $F(4,2)$, $F'(-4,2)$ is 4. Find the equation of the locus of such points.
24. A point has the property that the numerical difference of its distances from $F(1,3)$, $F'(1,-3)$ is 6. Find the equation of the locus of such points. (BT)
25. The hypotenuse of each of a set of right triangles is the segment joining (0,0) and (1,0). Find the equation of the locus of the third vertices.

9.13 Case IV. The Hyperbola

Definitions: A *hyperbola* is the locus of points P such that the numerical difference of the distances from P to two fixed points is a constant. The two fixed points are called *foci*.

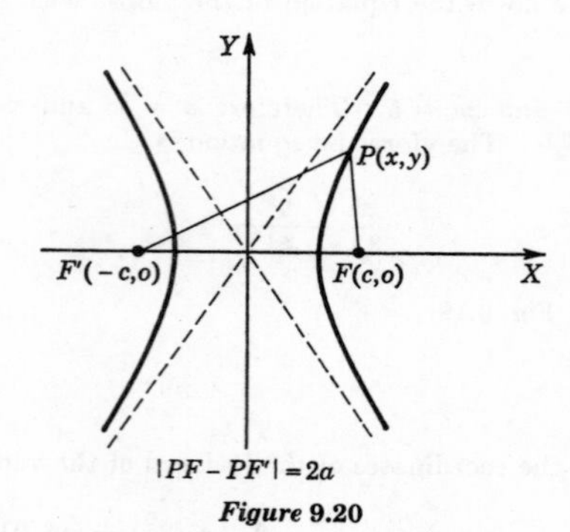

Figure 9.20

We choose axes as we did for the ellipse, writing $F(c,0)$ and $F'(-c,0)$ (Fig. 9.20). We let the numerical difference of the distances be the constant $2a(a > 0)$. The definition requires

$$\left|\sqrt{(x+c)^2+y^2}-\sqrt{(x-c)^2+y^2}\right| = 2a$$

This is equivalent to

$$\sqrt{(x+c)^2+y^2} - \sqrt{(x-c)^2+y^2} = +2a \tag{27}$$

if

$$\sqrt{(x+c)^2+y^2} > \sqrt{(x-c)^2+y^2}$$

and to

$$\sqrt{(x+c)^2+y^2} - \sqrt{(x-c)^2+y^2} = -2a \tag{28}$$

if

$$\sqrt{(x+c)^2+y^2} < \sqrt{(x-c)^2+y^2}$$

In either case, (27) or (28), if we square (twice) and simplify as we did in the case of the ellipse, we shall arrive again at Eq. (24). For the hyperbola, however, $2a < 2c$, as can be seen directly from the figure. This means that $a^2 - c^2 < 0$; we set $a^2 - c^2 = -b^2$ (a negative number). Continuing the simplification of (24), we get, on this basis,

$$b^2x^2 - a^2y^2 = a^2b^2$$

or, finally,

$$\frac{x^2}{a^2} - \frac{y^2}{b^2} = 1 \tag{29}$$

This is the equation of the hyperbola.

Exercise A. In (29) what coefficients correspond to A, B, C, D, E, and F in (18)?

Exercise B. Show that the points $V(a,0)$, $V'(-a,0)$ are on the hyperbola.

Definitions: The points V and V' are called the *vertices* of the hyperbola. The segment VV' is called the *transverse axis;* its length is $2a$. The segment joining $(0,b)$ and $(0,-b)$ is called the *conjugate axis;* its length is $2b$. The *center* of the hyperbola is the mid-point of the transverse axis.

If we divide (29) by x^2, we get, after a little simplification,

$$\frac{y^2}{x^2b^2} = \frac{1}{a^2} - \frac{1}{x^2}$$

$$\frac{y^2}{x^2} = \frac{b^2}{a^2} - \frac{b^2}{x^2}$$

$$\frac{y}{x} = \pm\sqrt{\frac{b^2}{a^2} - \frac{b^2}{x^2}}$$

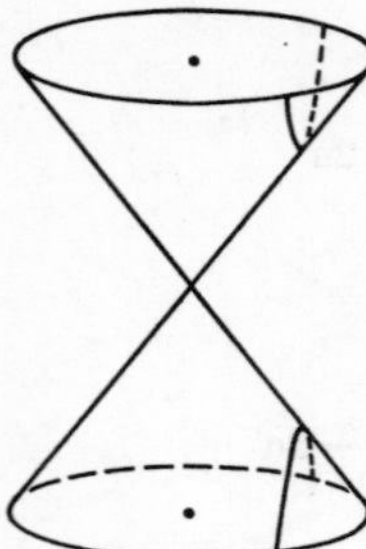

Figure 9.21

This says that, for large x (since b^2/x^2 is then small), the numerical ratio $|y/x|$ is just a very little less than b/a. (See Chap. 10, where the notion of *limit* is treated.) With this information at hand we sketch in the two lines (called the *asymptotes* of the *hyperbola*) whose equations are $y = \pm(b/a)x$ to act as guides in sketching the graph of the hyperbola itself. The asymptotes are *not* part of the locus; they merely serve as aids in plotting.

The hyperbola is a conic section (Fig. 9.21). Geometrically, the hyperbola can be obtained by cutting a right circular cone with a plane that is inclined so as to cut both nappes of the cone but not placed so as to pass through the vertex of the cone.

Illustration 1. Sketch the graph of the equation

$$\frac{x^2}{9} - \frac{y^2}{4} = 1 \tag{30}$$

Solution: The total graph will be made up of the graphs of the two algebraic functions f and g derived from (30) and defined by the equations

$$f(x) = \tfrac{2}{3}\sqrt{x^2 - 9} \begin{cases} \text{domain, } -\infty < x \le -3,\ 3 \le x < \infty \\ \text{range. } 0 \le y < \infty \end{cases}$$

$$g(x) = -\tfrac{2}{3}\sqrt{x^2 - 9} \begin{cases} \text{domain, } -\infty < x \le -3,\ 3 \le x < \infty \\ \text{range, } -\infty < y \le 0 \end{cases}$$

The zeros are $x = \pm 3$ for each function; that is, the vertices of the hyperbola are $V(3,0)$ and $V'(-3,0)$. The coordinates of the foci are $F(\sqrt{13},0)$ and $F'(-\sqrt{13},0)$, since $c^2 = b^2 + a^2 = 13$. You should compute a few elements of f and g.

The length of the transverse axis is 6; the length of the conjugate axis is 4. The equations of the asymptotes are $y = \pm\frac{2}{3}x$ (Fig. 9.22).

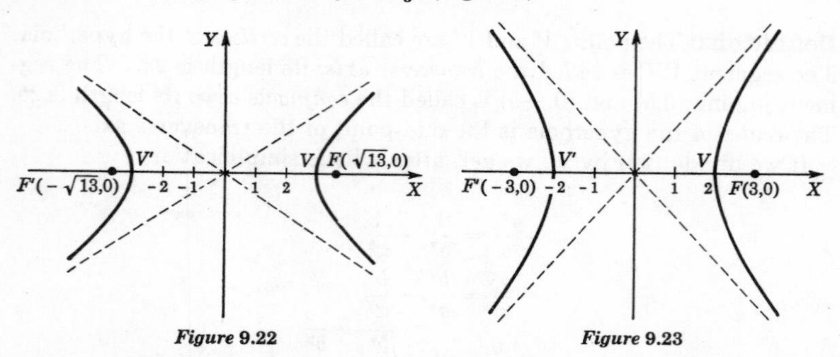

Figure 9.22 Figure 9.23

Illustration 2. Write down the equation of the hyperbola with vertices $V(2,0)$, $V'(-2,0)$ and with foci $F(3,0)$, $F'(-3,0)$.

Solution: We are given $a = 2$ and $c = 3$. Therefore, since $c^2 = b^2 + a^2$, we find $b^2 = c^2 - a^2 = 9 - 4 = 5$. The equation is

$$\frac{x^2}{4} - \frac{y^2}{5} = 1$$

The equations of the asymptotes are

$$y = \pm \frac{\sqrt{5}}{2} x \quad \text{[Fig. 9.23]}$$

Problems 9.13

In Probs. 1 to 12 sketch. In case the figure is a hyperbola, find the equations of the asymptotes, the coordinates of the vertices, and the coordinates of the foci. In case the figure is a circle, find the center and radius. In case the figure is a parabola, find the equation of the directrix and the coordinates of the vertex and of the focus. In case the figure is an ellipse, find the coordinates of the center, of the vertices, and of the foci.

1. $x^2 + y^2 - x + 2y - 20 = 0$.
2. $x^2 + 2y^2 + 2x + 4y - 16 = 0$.
3. $x^2 + x + y - 1 = 0$.
4. $y^2 + 2x + 2y + 1 = 0$.
5. $x^2 - y^2 + 5x - 10 = 0$.
6. $2x^2 - y^2 + 8x - 2y = 0$.
7. $(x + y + 1)(x + 2y - 2) = 0$.
8. $(x - y)(x + y) + y + x = 0$.
9. $x(x - 1) + y + 1 = 0$.
10. $x + (y - 1)(y + 2) = 0$.
11. $(a^2/x^2) + (b^2/y^2) = 1$ [the cruciform].
12. $(a^2/x^2) - (b^2/y^2) = 1$ [the arc light].

13. Find the equation of the locus of points P such that the distance from (h,k) is always 5 units.
14. Find the equation of the locus of points P such that the numerical distance from the line $x = 2$ is always 1 unit.
15. Find the equation of the locus of points P such that the distance from P to the line $x = 3$ is always equal to its distance from the point (1,0). Rationalize and simplify your answer.
16. Find the equation of the locus of points P such that the sum of the distances from P to the two points $(-6,0)$ and $(6,0)$ is always 20 units. Rationalize and simplify your answer.
17. Find the equation of the locus of points P such that the numerical difference of the distances from P to the two points $(-8,0)$ and $(8,0)$ is always 12 units. Rationalize and simplify your answer.
18. What geometric configurations, other than ellipse, parabola, and hyperbola, can be obtained by cutting a cone with a plane? Illustrate with figures.
19. $a^2x^2 + 2axy + y^2 + ax + y = 0$ is an equation of type (18), Sec. 9.9. Does it represent a *conic section*, i.e., a curve obtained by cutting a cone with a plane? (BT)
20. What are the coordinates of the center, of the vertices, and of the foci of the ellipse $x^2/9 + y^2/9 = 1$?
21. Sketch and discuss: $(y - k)^2 = 4p(x - h)$.
22. Sketch and discuss: $\dfrac{(x - h)^2}{a^2} + \dfrac{(y - k)^2}{b^2} = 1$.
23. Sketch and discuss: $\dfrac{(x - h)^2}{a^2} - \dfrac{(y - k)^2}{b^2} = 1$.

24. Sketch and discuss: $x^2 = 4py$.

25. Sketch on same axes: $y = x^2$, $y = 2mx - 1$. Find the x-coordinates of the points of intersection of this parabola and straight line. Find the condition that there is only one point of intersection. Discuss the geometry for the lines $y = 2x - 1$ and $y = -2x - 1$.

26. Sketch: $2x^2 - xy - y^2 + 3x + 3y - 2 = 0$. HINT: Factor.

27. Sketch: $3x^2 + 4xy + y^2 - 3x + y - 6 = 0$. HINT: Factor.

28. Show that an ellipse is the locus of points P such that the ratio of the distances of P from a fixed point and from a fixed line is a constant e less than unity. HINT: Take the fixed point $F(ae,0)$ and the fixed line $x = a/e$. Show that the equation of the locus is then

$$\frac{x^2}{a^2} + \frac{y^2}{a^2(1 - e^2)} = 1$$

The constant e is called the eccentricity of the ellipse. What is the eccentricity of a circle?

29. Show that a hyperbola is the locus of points P such that the ratio of the distances of P from a fixed point and from a fixed line is a constant e greater than unity. HINT: Take the fixed point $F(ae,0)$ and the fixed line $x = a/e$. Show that the equation of the locus is then

$$\frac{x^2}{a^2} - \frac{y^2}{a^2(e^2 - 1)} = 1$$

The constant e is called the eccentricity of the hyperbola. (See the definition of the parabola where e, defined similarly, would be equal to unity.)

9.14 Applications

In order to treat in detail many of the scientific applications of the theory of conic sections, we need especially the methods of the calculus (Chaps. 10 and 11). Therefore, at this time, we shall just mention some of them briefly.

Parabola

(**a**) Path of a projectile (neglecting air resistance).
(**b**) Cable of a suspension bridge (uniformly loaded along the bridge).
(**c**) Parabolic reflector [surface generated by revolving a parabola about its axis has the property that each light ray coming in parallel to the axis will be reflected to (through) the focus].
(**d**) Graphs of many equations in physics.
(**e**) The antenna of a radio telescope involves parabolas.

Ellipse

(**a**) Orbit of a planet (sun at one focus).
(**b**) Orbits of planetary moons, satellites, some comets.

(c) Elliptic gears for certain machine tools.
(d) Focal property: a ray emanating at one focus is reflected to the other.
(e) Many scientific formulas are equations which plot into ellipses.

Hyperbola

(a) Used in the construction of certain telescopic lenses.
(b) Some comets trace hyperbolas.
(c) Formulas taken from the field of the physical sciences are often of hyperbolic type.

9.15 Polar Coordinates

So far we have used rectangular coordinates exclusively, although there are other coordinate systems that are most useful with certain types of problems. One of these is called polar coordinates and is defined below.

Definitions: Consider a (horizontal) line called the *polar axis* and a point O on it called the *pole*, or *origin*. From the pole to an arbitrary point P, draw the line segment r, called the *radius vector;* the radius vector makes a directed angle θ with the positive direction of the polar axis, which is taken to the right. We assign to P the ordered pair (r,θ); call them the *polar coordinates of* P, and write $P(r,\theta)$.

We permit θ to be positive (counterclockwise) or negative (clockwise). If no restrictions are imposed upon us, we may use any convenient angular unit, radian measure and degree measure being the most common. Likewise we permit r to be positive or negative, as we shall explain.

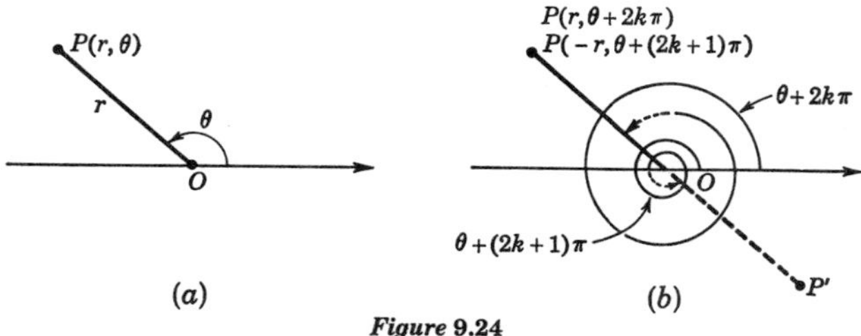

Figure **9.24**

First, note that a fixed point P has several sets of polar coordinates. Indeed, if P has coordinates (r,θ), then it also has coordinates $(r, \theta + 2k\pi)$ for every integer k. Regardless of what integer k is used in the second element of the number pair $(r, \theta + 2k\pi)$, r itself is positive. It is desirable to permit r to be negative. We agree to call the number pair $(-r, \theta + \pi)$ coordinates of P; likewise for the pair $(-r, \theta + (2k + 1)\pi)$, k an integer.

The geometric interpretation is evident from Fig. 9.24b: we extend PO in the direction OP' and measure θ to the extension. For this case r is negative.

In summary, the polar coordinates of the point P are

$$(r, \theta + 2k\pi) \qquad k \text{ an integer} \tag{31}$$

or

$$(-r, \theta + (2k + 1)\pi) \qquad k \text{ an integer} \tag{32}$$

The pole itself is a very special point since, when $r = 0$, there is no unique angle θ.

Comment. There is no 1 to 1 correspondence between points in the plane and polar number pairs. To a given pair (r,θ) there corresponds a unique point, but to a given point there corresponds no unique pair (r,θ). This is in contrast to the situation in rectangular coordinates.

There is evidently a connection between the rectangular coordinates and the polar coordinates of a point P. By superposition of the two systems (Fig. 9.25) we find that

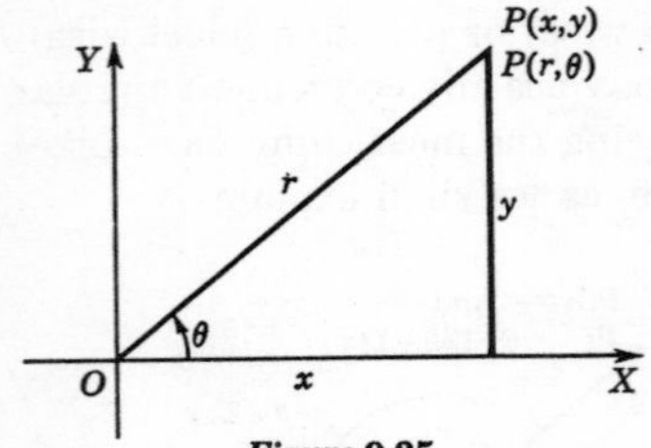

***Figure* 9.25**

$$\text{(I)} \qquad x = r\cos\theta \qquad y = r\sin\theta$$

Exercise A. Figure 9.25 assumes that r is positive. Draw a figure with r negative, and use it to establish formulas (I).

From these equations we can readily find the rectangular coordinates when we know the polar coordinates. To arrive at polar coordinates when we are first given rectangular coordinates is harder because of the ambiguity in the former. We generally settle upon the least angle θ, where $0 \leq \theta < 360°$, to be associated with P. In this case r is positive. But there are times when we need to consider expressions (31) and (32), from both of which we shall then pick out appropriate coordinates (perhaps more than one set) to suit our purpose. Thus, usually,

$$\text{(II)} \qquad r = \sqrt{x^2 + y^2} \qquad \tan\theta = \frac{y}{x}$$

where θ is determined as the least angle $\theta \geq 0$, which satisfies the above equation and the equations:

$$\text{(III)} \qquad \sin\theta = \frac{y}{\sqrt{x^2 + y^2}} \qquad \cos\theta = \frac{x}{\sqrt{x^2 + y^2}}$$

By means of (I), (II), and (III) we can transform equations from one system to another. Sometimes one system is more suitable to a given problem than another.

Illustration 1. Transform the polar equation

$$Ar\cos\theta + Br\sin\theta + C = 0$$

to rectangular coordinates.

Solution: We make use of (III) and write the transformed equation

$$A\sqrt{x^2 + y^2} \cdot \frac{x}{\sqrt{x^2 + y^2}} + B\sqrt{x^2 + y^2} \cdot \frac{y}{\sqrt{x^2 + y^2}} + C = 0$$

or

$$Ax + By + C = 0$$

In either system the graph is a straight line.

Exercise A. Show that $r\cos(\theta - a) = b$ is the equation of a straight line, and compare with $r(A\cos\theta + B\sin\theta) + C = 0$.

Exercise B. Sketch the graph of $r = \sec\theta$.

Exercise C. Write the polar equation of an arbitrary line passing through the pole.

Illustration 2. Transform the rectangular equation $x^2 + y^2 = a^2$ (of a circle of radius a with center at the origin) to polar coordinates.

Solution: Using (I), we write $x^2 + y^2 = a^2$ in the form

$$\begin{aligned}(r\cos\theta)^2 + (r\sin\theta)^2 &= a^2\\ r^2(\cos^2\theta + \sin^2\theta) &= a^2\\ r^2 &= a^2\end{aligned}$$

The graph of $r = a$ is a circle of radius a with center at the pole; $r = -a$ plots the same circle. Hence there are two different equations in polar coordinates for this circle. This is not an isolated example; certain curves may have several distinct polar equations. This is due to the fact that the polar coordinates of a point are not unique. It is important to note that the *coordinates* (a,θ), satisfying $r = a$, do *not* satisfy $r = -a$.

Exercise D. Write down the polar equation of the circle with unit radius and center at the point ($r = \frac{1}{2}$ $\theta = \pi/2$).

Problems 9.15

In Probs. 1 to 14 transform to polar coordinates.

1. $x^2 + y^2 - 2x = 0$.
2. $x^2 + y^2 + 3y = 0$.
3. $x - 2y = 0$.
4. $2x + 3y = 0$.
5. $y^2 = 8x$.
6. $x^2 = 5y$.
7. $(x^2/16) + (y^2/4) = 1$.
8. $(x^2/4) + (y^2/16) = 1$.
9. $(x^2/25) - (y^2/9) = 1$.
10. $(x^2/9) - (y^2/25) = 1$.
11. $-(x^2/4) + (y^2/1) = 1$.
12. $-(x^2/1) + (y^2/4) = 1$.
13. $(x^2 + y^2 - x)^2 = x^2 + y^2$.
14. $(x^2 + y^2 + y)^2 = x^2 + y^2$.

In Probs. 15 to 30 transform to rectangular coordinates.

15. $r = 2 \sin \theta$.
16. $r = -2 \sin \theta$.
17. $r = 3 \cos \theta$.
18. $r = -3 \cos \theta$.
19. $r = 1 + \cos \theta$.
20. $r = -1 + \cos \theta$.
21. $r = 1 + 2 \sin \theta$.
22. $r = 1 - 2 \sin \theta$.
23. $r = \dfrac{1}{1 - \cos \theta}$.
24. $r = \dfrac{-1}{1 + \cos \theta}$.
25. $r = 4$.
26. $\theta = 45°$.
27. $\theta = \dfrac{\pi}{3}$.
28. $r\theta = a$.
29. $r = \cos 2\theta$.
30. $r = \sin 2\theta$.

9.16 Polar Coordinates (Continued)

In Probs. 28 and 29, Sec. 9.13, we gave new definitions for the ellipse and hyperbola. For a simple treatment of the conic sections in polar coordinates, we need the following definitions of a conic:

Definitions: The locus of points P such that the ratio of the distances from P to a fixed point F and to a fixed line DD' is a constant e is called a *conic section*. The point F is called the *focus*, DD' is called the *directrix*, and e is called the *eccentricity*.

If $e = 1$, the locus is a *parabola*
$e < 1$, the locus is an *ellipse*
$e > 1$, the locus is a *hyperbola*

Of course, these definitions must be consistent with our previous definitions. They are, but we shall not prove it. As you know, it turns out that the ellipse and hyperbola have two foci. They also have two directrices.

To derive the equation of a conic in polar coordinates is quite simple if we choose the focus F for the pole and the line through F and perpendicular to DD' for the polar axis. Consult Fig. 9.26; we let p be the distance from

F to DD'. By definition, for every point $P(r,\theta)$, it must be true that

$$\frac{\text{Dist. } PF}{\text{Dist. } P \text{ to } DD'} = e$$

that is,

$$\frac{r}{p + r \cos \theta} = e$$

This reduces to

$$r = \frac{ep}{1 - e \cos \theta} \tag{33}$$

which is the equation of the conic.

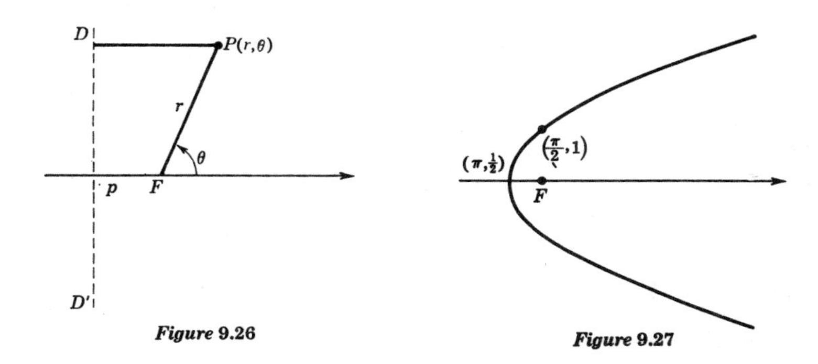

Figure 9.26

Figure 9.27

Illustration 1. Sketch the graph of the parabola $r = 1/(1 - \cos \theta)$.

Solution: We make out a table of values.

θ	0	$\pi/6$	$\pi/4$	$\pi/3$	$\pi/2$	π
$1 - \cos \theta$	0	0.134	0.293	0.500	1	2
r	...	7.47	3.41	2	1	$\frac{1}{2}$

You can easily make out another table for third and fourth quadrants (for the lower half of the curve). It is important to be careful in plotting points for values of θ near 0 since r is not defined for $\theta = 0$ (Fig. 9.27).

Exercise A

(**a**) Plot the graph of $r = -1/(1 + \cos \theta)$.
(**b**) Find all pairs (r,θ) which satisfy $r = -1/(1 + \cos \theta)$ and $r = 1/(1 - \cos \theta)$ simultaneously.

Illustration 2. Sketch the graph of $r = \sin 2\theta$.

Solution: Again we prepare a table of values. This time, for practice, we shall use degree measure.

θ	0°	15°	22.5°	30°	45°	60°	67.5°	75°	90°
2θ	0°	30°	45°	60°	90°	120°	135°	150°	180°
r	0	0.500	0.707	0.866	1	0.866	0.707	0.500	0

The graph of this much is given in Fig. 9.28*a*.

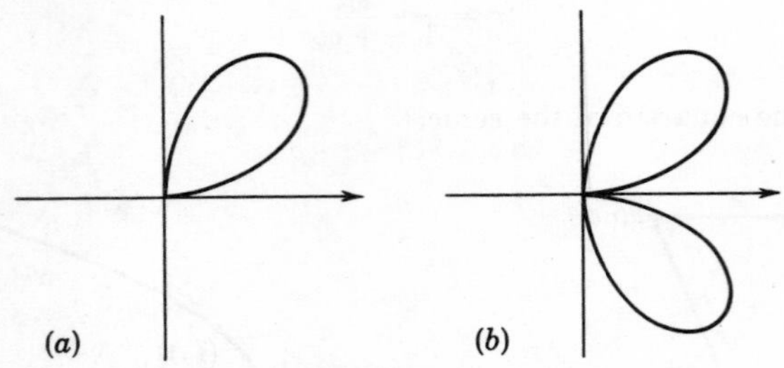

Figure 9.28

We continue the table:

θ	105°	112.5°	120°	135°	150°	157.5°	165°	180°
2θ	210°	225°	240°	270°	300°	315°	330°	360°
r	−0.500	−0.707	−0.866	−1	−0.866	−0.707	−0.500	0

The graph of the preceding two tables is given in Figure 9.28*b*. So far, no portion of the graph has repeated. Indeed, θ must run the full course of 360° before repetition. The total graph is the "four-leaved rose" exhibited in Fig. 9.29.

Exercise B. Make out the remaining portion of the above table for the complete graph (drawn in Fig. 9.29).

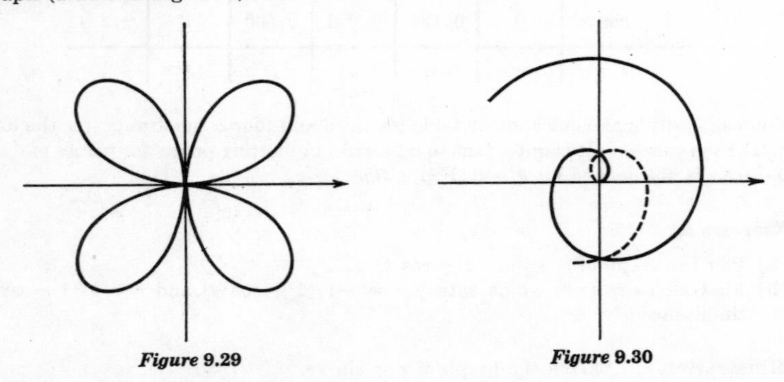

Figure 9.29 *Figure* 9.30

You should study this illustration to see that really all we need is the set of the first five entries in the first portion of the table. With a knowledge of the behavior of the trigonometric function sine, we can draw the total graph. The essential variations of $\sin 2\theta$ take place in the first half of the first quadrant. This is typical of much of the graph work in polar coordinates involving the trigonometric functions where periodicity plays an important role.

Illustration 3. Plot the graph of $r = \theta$.

Solution: Here we must use radian measure, but there is no need of making out a table. Take note of the portion corresponding to negative values of θ in Fig. 9.30. The curve is known as the Spiral of Archimedes.

Illustration 4. Sketch the graph of $r = \cos \theta$.

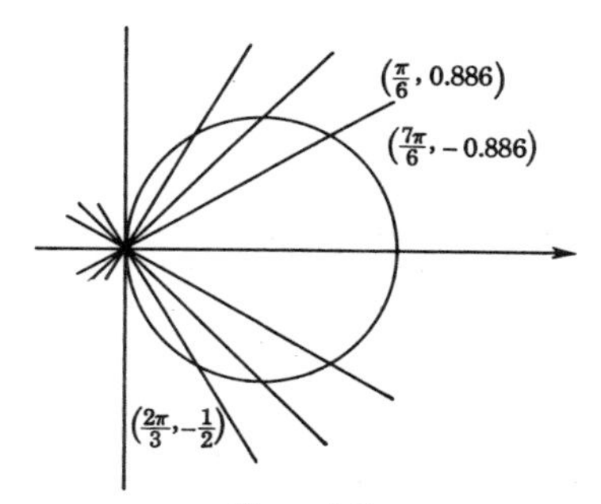

Figure 9.31

Solution: We prepare the following table:

θ	0	$\pi/6$	$\pi/4$	$\pi/3$	$\pi/2$	$2\pi/3$	$3\pi/4$	$5\pi/6$	π	$7\pi/6$
r	1	0.866	0.707	0.500	0	−0.500	−0.707	−0.866	−1	−0.866

The geometric point whose coordinates are $(\pi/6, 0.866)$ is the same as that with coordinates $(7\pi/6, -0.866)$. These are the second and last entries in the preceding table. Extension of the table through third and fourth quadrantal angles shows that the curve is being traced a second time. Therefore the description of the curve is complete after θ runs through the first two quadrants (Fig. 9.31). The curve is actually a circle, as you can immediately verify by transforming $r = \cos \theta$ to the rectangular form

$$(x - \tfrac{1}{2})^2 + y^2 = \tfrac{1}{4}$$

Notice that $(x - \frac{1}{2})^2 + y^2 = \frac{1}{4}$ defines a *relation*, whereas the corresponding polar equation $r = \cos \theta$ defines a *function*.

Problems 9.16

In Probs. 1 to 29 sketch. Also transform to rectangular coordinates.

1. $r = \dfrac{7}{2 - \cos\theta}$.

2. $r = \dfrac{4}{1 - 2\cos\theta}$.

3. $r = \dfrac{5}{5 - \sin\theta}$.

4. $r = \dfrac{3}{1 - 3\sin\theta}$.

5. $r^2 = 2r\cos\left(\theta - \dfrac{\pi}{3}\right)$.

6. $r^2 = 4r\sin\left(\theta - \dfrac{\pi}{3}\right)$.

7. $r = \sin\theta$.

8. $r = \sin 3\theta$ [three-leaved rose].

9. $r = \sin 4\theta$ [eight-leaved rose].

10. $r = \cos\theta$.

11. $r = \cos 2\theta$ [four-leaved rose].

12. $r = \cos 3\theta$ [three-leaved rose].

13. $r = \cos 4\theta$ [eight-leaved rose].

14. $r = 1 - \sin\theta$ [cardioid].

15. $r = 1 - \cos\theta$.

16. $r = 1 - 2\sin\theta$ [limaçon of Pascal].

17. $r = 1 - 2\cos\theta$.

18. $r^2 = \sin\theta$.

19. $r^2 = \cos\theta$.

20. $r^2 = \sin 2\theta$ [the lemniscate].

21. $r^2 = \cos 2\theta$.

22. $r = 2 - \sin\theta$.

23. $r = 2 - \cos\theta$.

24. $r\theta = \pi$ [hyperbolic spiral].

25. $r\theta = -\pi$.

26. $r^2\theta = \pi$ [the lituus].

27. $r^2\theta = -\pi$.

28. $r(\cos\theta + \sin\theta) = 1$.

29. $r = \sec\theta$.

30. Sketch and find the points of intersection: $r = \sin\theta$ and $r = \cos\theta$.

31. Sketch and find the points of intersection: $r = \sin\theta$ and $r = 1 - \sin\theta$.

32. Find the equation of the locus of points P such that P is a fixed distance a from $P(r_1,\theta_1)$.

33. Find the equation of the locus of the mid-points of chords of a circle of radius a drawn from a fixed point Q on the circle.

34. Find the equation of the locus of points P such that the radius vector of P is proportional to the square of its vectorial angle.

9.17 Parametric Equations

It is often desirable to express each element of a pair, such as (x,y), in terms of a third variable, say t. When this is done, we find that we need a pair of equations of the form

$$x = f(t) \qquad y = g(t) \tag{34}$$

to represent a given curve analytically. We refer to (34) as *parametric equations;* by eliminating the *parameter* t we obtain the *cartesian equation* of the curve. Many loci problems are best treated in terms of parametric equations. Since the parameter can be chosen in many ways, we expect to find a great variety of parametric equations representing a given locus. In some cases a set of parametric equations will represent only a portion of the locus, and several such sets will be needed to represent it completely.

Illustration 1. Write the equation of a straight line in parametric form.

Solution: In Eqs. (4) we saw that

$$x_2 - x_1 = d \cos \alpha \qquad y_2 - y_1 = d \cos \beta$$

where (x_1,y_1) and (x_2,y_2) are points on the line, and d is the distance between them. If we write (x,y) for (x_2,y_2) and t for the distance between (x,y) and (x_1,y_1), these can be written

$$x = x_1 + t \cos \alpha \qquad y = y_1 + t \cos \beta$$

These are parametric equations of the line. They may also be written in the form

$$x = x_1 + u(x_2 - x_1) \qquad y = y_1 + u(y_2 - y_1)$$

where $\mu = t/d$. The graph of them is the whole line.

Exercise A

(**a**) Plot the line whose parametric equations are $x = 1 + 3t$, $y = 2 - 2t$.
(**b**) Eliminate t, and find the cartesian equation.

Illustration 2. Find parametric equations for the ellipse

$$b^2x^2 + a^2y^2 = a^2b^2$$

Solution: We choose the parameter t as the angle shown in Fig. 9.32. In terms of the angle t we can write down the equations immediately since $x/a = \cos t$ and $y/b = \sin t$. They are therefore

$$x = a \cos t \qquad y = b \sin t \tag{35}$$

and the graph is the complete ellipse.

Exercise B. Eliminate t from Eqs. (35).

Exercise C. Write down parametric equations for the circle $x^2 + y^2 = a^2$.

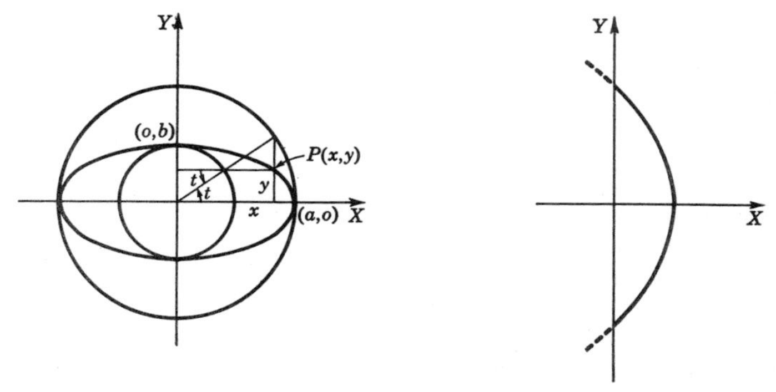

Figure 9.32 *Figure* 9.33

Illustration 3. Show that $x = \sin^2 t$, $y = 2 \cos t$ represents only a portion of the parabola whose cartesian equation is $y^2 = 4(1 - x)$(Fig. 9.33).

Solution: The given parametric equations permit x to vary from 0 to 1 only and y to vary from -2 to $+2$. We eliminate t as follows:

$$x = \sin^2 t$$
$$\tfrac{1}{4}y^2 = \cos^2 t$$

Adding, we get

$$x + \tfrac{1}{4}y^2 = 1$$

or

$$y^2 = 4(1 - x)$$

Illustration 4. A circle of radius a rolls along a line. Find the locus described by a point on the circumference.

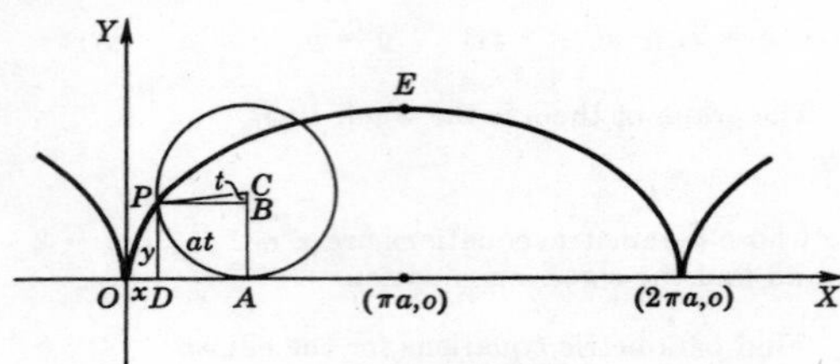

***Figure* 9.34**

Solution: Take the X-axis as coincident with the line and the initial position of the tracing point as the origin (Fig. 9.34). The positive angle

$$PCA = t \text{ radians}$$

will be chosen as the parameter. The arc $PA = at$ (Sec. 8.3).

Clearly,
$$OA = PA = at$$

Hence
$$\begin{aligned} x &= OA - DA \\ &= OA - PB \\ &= at - a \sin t \end{aligned}$$

Further,
$$\begin{aligned} y &= PD = CA - CB \\ &= a - a \cos t \end{aligned}$$

Parametric equations of the locus, called a *cycloid*, are therefore

$$x = a(t - \sin t)$$
$$y = a(1 - \cos t)$$

This curve is very important in physics, where it is called the *brachistochrone*, or the *curve of quickest descent*. This means that, if we think of the curve as turned upside down, then, out of all possible curves connecting O and E, the brachistochrone is the one down which a frictionless particle will slide in least time. As a matter of fact, this time is independent of the point on the curve from which the particle is released; the particle will slide from O to E in the same time that it will slide from any other point (such as P) to E. The cartesian equation obtained by eliminating t is troublesome; therefore we do not consider it.

Problems 9.17

In Probs. 1 to 26, eliminate the parameter t and identify the curve if possible.

1. $x = 1 - 2t,\ y = 3 - t.$

2. $x = 4 - 3t,\ y = 1 + 2t.$

3. $x = 2t^2,\ y = t.$

4. $x = 3t^2,\ y = 1 - t.$

5. $x = t,\ y = t^2 - 2t.$

6. $x = 2t - t^2,\ y = t.$

7. $x = 2t^2 - 3t,\ y = 2t^2 + 3t.$

8. $x = t^2 + t,\ y = t^2 - t.$

9. $x = 1 - \cos t,\ y = 1 + \sin t.$

10. $x = \cos t,\ y = \sin^2 t.$

11. $x = \cos t + \sin t,\ y = \cos t - \sin t.$

12. $x = a \sin t + b \cos t,\ y = a \cos t - b \sin t.$

13. $x = 1/t,\ y = t.$

14. $x = 1/2t,\ y = 1 - t.$

15. $x = \dfrac{1}{1 - t},\ y = \dfrac{1}{1 - t}.$

16. $x = \dfrac{1}{\sqrt{1 + t^2}},\ y = \dfrac{2t}{\sqrt{1 + t^2}}.$

17. $x = 1 - t,\ y = t^2 + 2t.$

18. $x = t^2 - t,\ y = 1 + t.$

19. $x = 2 \sin t,\ y = \cos t.$

20. $x = 1 - 3 \cos t,\ y = 2 + 2 \sin t.$

21. $x = 4 - t,\ y = \sqrt{-t}.$

22. $x = \dfrac{t}{1 + t^3},\ y = \dfrac{t^2}{1 + t^3}$ [folium of Descartes].

23. $x = \tan t,\ y = \sec t.$

24. $x = \sin 2t,\ y = \cos t.$

25. $x = \cos 2t,\ y = \sin t.$

26. $x = -2at/(1 + t^2),\ y = a(1 - t^2)/(1 + t^2).$

27. A circle of radius $a/4$ rolls inside a circle of radius a. Show that parametric equations of the locus described by a point on the circumference of the rolling circle are $x = a \cos^3 t,\ y = a \sin^3 t$. [The parameter t is the angle through which the line of centers turns, the center of the stationary circle being placed at the origin. The initial position of the line of centers coincides with the X-axis, and $(a,0)$ is the initial position of the tracing point.] The curve is called the hypocycloid.

28. Find the parametric equations of the parabola whose equation is $y^2 = 4px$, in terms of the parameter t, which is the slope of the line $y = tx$.

References

In addition to the many standard textbooks on analytic geometry, the reader should consult the following articles in the *American Mathematical Monthly:*

Boyer, C. B.: The Equation of an Ellipse, vol. 54, p. 410 (1947).
Boyer, C. B.: Newton as an Originator of Polar Coordinates, vol. 56, p. 73 (1949.)
Hammer, D. C.: Plotting Curves in Polar Coordinates, vol. 48, p. 397 (1941).
Hawthorne, Frank: Derivation of the Equations of Conics, vol. 54, p. 219 (1947).
Johns, A. E.: The Reduced Equation of the General Conic, vol. 54, p. 100 (1947).
Wagner, R. W.: Equations and Loci in Polar Coordinates, vol. 55, p. 360 (1948).

10 | Limits

10.1 Introduction

Most of the subjects treated so far in this book have dealt with processes which can be carried out in a finite number of steps. These processes, however, are not adequate for the applications of mathematics to practical science, and indeed they are not adequate for the purposes of mathematics itself. As we shall see, it is essential that we know how to manage *infinite* processes, and it is to these that we now turn our attention. The most important new idea which must be introduced is the notion of *limit*. We have already hinted at this idea, and we shall meet it over and over again in this and the next chapter. Therefore we must take great pains to make the idea clear and precise. You should be warned that this concept is not easy and that you will have to pay careful attention to what is said if you are going to understand it fully.

10.2 Historical Notes

The modern theory of limits is the result of a long series of developments beginning with the crude notions first stated by Newton and Leibniz at the end of the seventeenth century. The basic idea, however, is much older and was discussed by Euclid (365?–? B.C.), Archimedes (287–212 B.C.), and other early mathematicians. Although these ancient mathematicians never obtained a definition of a limit which was suitable for widespread application, they must have had an intuitive grasp of the essential idea. This idea grew out of their need to solve a host of practical problems. So that you may see the importance of limits in such connections before we go into their detailed treatment, we shall give Archimedes' solution of the following problem.

What is the area of the region of the plane bounded by the parabola $y = x^2$, the portion of the X-axis between 0 and 1, and the ordinate at $x = 1$ as in Figs. 10.1 to 10.3?

Before solving this problem, we should note that the "area" of such a region has not been defined so far in this book, and indeed Archimedes did not have a careful definition. We shall define "area" in detail later, but for the present your intuitive idea will suffice. Archimedes reasoned as follows. The area sought, call it A, is larger than the combined areas of

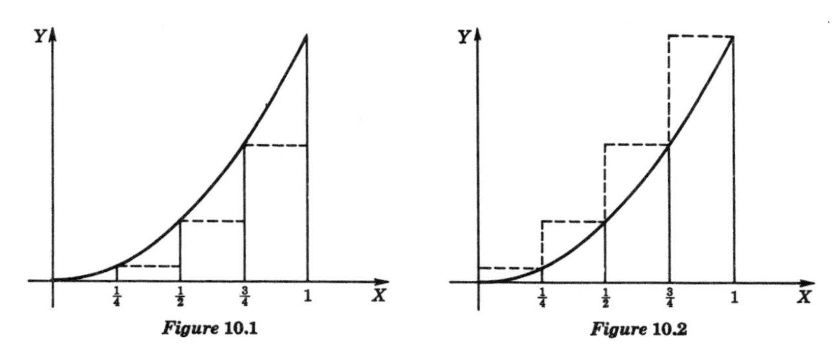

Figure 10.1

Figure 10.2

the rectangles formed as in Fig. 10.1, where ordinates have been erected at the quarter marks. That is,

$$\tfrac{1}{4}(\tfrac{1}{4})^2 + \tfrac{1}{4}(\tfrac{2}{4})^2 + \tfrac{1}{4}(\tfrac{3}{4})^2 < A$$

or

$$\frac{1}{4^3}[1^2 + 2^2 + 3^2] < A$$

This reduces, numerically, to

$$\tfrac{14}{64} < A$$

Similarly, from Fig. 10.2, A is smaller than the sum of the rectangles which have been drawn. That is,

$$\tfrac{1}{4}(\tfrac{1}{4})^2 + \tfrac{1}{4}(\tfrac{2}{4})^2 + \tfrac{1}{4}(\tfrac{3}{4})^2 + \tfrac{1}{4}(\tfrac{4}{4})^2 > A$$

or

$$\frac{1}{4^3}[1^2 + 2^2 + 3^2 + 4^2] > A$$

that is,

$$\tfrac{30}{64} > A$$

Hence

$$\tfrac{14}{64} < A < \tfrac{30}{64}$$

If ordinates had been erected at the eighth marks, the corresponding inequalities would have been (you should draw a figure for each case)

$$\tfrac{1}{8}(\tfrac{1}{8})^2 + \tfrac{1}{8}(\tfrac{2}{8})^2 + \cdots + \tfrac{1}{8}(\tfrac{7}{8})^2 < A$$

$$\frac{1}{8^3}[1^2 + 2^2 + \cdots + 7^2] < A$$

$$\tfrac{140}{512} < A$$

and $$\tfrac{1}{8}(\tfrac{1}{8})^2 + \tfrac{1}{8}(\tfrac{2}{8})^2 + \cdots + \tfrac{1}{8}(\tfrac{8}{8})^2 > A$$

$$\frac{1}{8^3}[1^2 + 2^2 + \cdots + 8^2] > A$$

$$\tfrac{204}{512} > A$$

or $$\tfrac{140}{512} < A < \tfrac{204}{512}$$

Observe that the bounding interval in the second case lies wholly within that of the first case; i.e.,

$$\tfrac{14}{64} < \tfrac{140}{512} < A < \tfrac{204}{512} < \tfrac{30}{64}$$

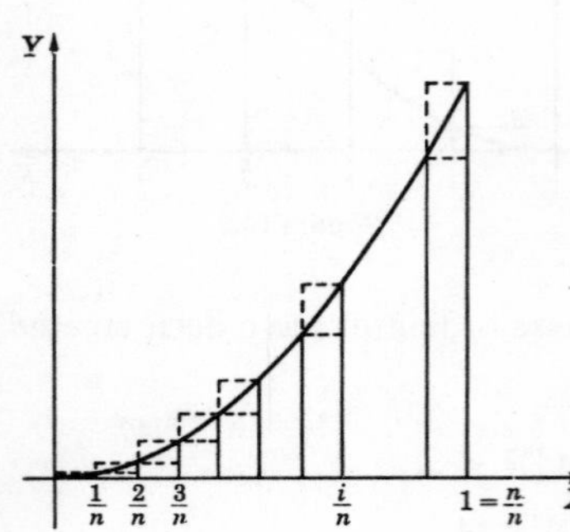

Figure **10.3**

Now let us *imagine* that ordinates are erected at abscissa marks which are multiples of $1/n$ (Fig. 10.3). Then we would have A "boxed in" by

$$(1) \quad \frac{1}{n^3}[1^2 + \cdots + (n-1)^2] < A < \frac{1}{n^3}[1^2 + \cdots + (n-1)^2 + n^2]$$

Let us write L_n for the left term of (1) and R_n for its right term. Hence (1) is rewritten

$$L_n < A < R_n$$

From Prob. 7, Sec. 3.2, we learn by mathematical induction that

$$1^2 + 2^2 + \cdots + n^2 = \tfrac{1}{6}n(n+1)(2n+1)$$

Hence $$L_n = \frac{(n-1)n(2n-1)}{6n^3} = \frac{(n-1)(2n-1)}{6n^2}$$

$$R_n = \frac{(n+1)(2n+1)}{6n^2}$$

Also $$L_{n+1} = \frac{n(2n+1)}{6(n+1)^2}$$

$$R_{n+1} = \frac{(n+2)(2n+3)}{6(n+1)^2}$$

From these it follows that

$$\begin{aligned} L_{n+1} - L_n &= \frac{n(2n+1)}{6(n+1)^2} - \frac{(n-1)(2n-1)}{6n^2} \\ &= \frac{n^3(2n+1) - (n+1)^2(n-1)(2n-1)}{6(n+1)^2n^2} \\ &= \frac{3n^2 + n - 1}{6(n+1)^2n^2} > 0 \end{aligned}$$

or

$$L_n < L_{n+1}$$

and similarly, $R_{n+1} < R_n$. Hence we have

$$L_n < L_{n+1} < A < R_{n+1} < R_n$$

Further, $R_n - L_n = 1/n$, which tends to zero as n tends to ∞. Hence it appears that we can approach the true value of A by taking larger and larger values of n, and A should equal the limiting value of L_n or R_n as n gets larger and larger. [Archimedes did not quite say it this way, but he must have had some such notion in his mind; it was Cavalieri (1598–1647) who first carried this out in this way in the year 1630.] Pictorially also, this seems reasonable, although the student should be cautioned against relying too heavily on his geometric intuition (see Prob. 2 below). The question now is to find this limiting value.

We have seen that

$$\begin{aligned} R_n &= \frac{(n+1)(2n+1)}{6n^2} \\ &= \frac{1}{6}\left(1 + \frac{1}{n}\right)\left(2 + \frac{1}{n}\right) \end{aligned}$$

It is reasonable to say that $1/n$ tends to zero as n tends to ∞. Actually, this is a fuzzy statement which will have to be clarified later, but for the moment let it stand. From this it is reasonable to conclude that $(1 + 1/n)$ tends to 1 as n tends to ∞ and that $(2 + 1/n)$ tends to 2 as n tends to ∞. Consequently, R_n tends to $\frac{1}{6}(1)(2) = \frac{1}{3}$.

Similarly, we have seen that

$$\begin{aligned} L_n &= \frac{(n-1)(2n-1)}{6n^2} \\ &= \frac{1}{6}\left(1 - \frac{1}{n}\right)\left(2 - \frac{1}{n}\right) \end{aligned}$$

By the same argument we conclude that L_n tends to $\frac{1}{3}$ as n tends to ∞. Since A is between L_n and R_n, we conclude that $A = \frac{1}{3}$.

Problems 10.2

1. Plot the curve $y = x^3$ for the values of x between 0 and 1. Make use of Archimedes' method described above (for $y = x^2$) and the formula of Prob. 8, Sec. 3.2, and find the area bounded by this curve, the X-axis, and the ordinate at $x = 1$. (It may surprise you to learn that you have just done a problem in the integral calculus.)
2. Consider a unit square, its diagonal AB, and a zigzag path from A to B made up of segments parallel to the sides, as in Fig. 10.4. Now the sum of the lengths of the zigzag is surely just two units. Making the zigs and the zags smaller and smaller indefinitely, we approach the diagonal closer and closer; thus we see that the length of the diagonal of the square is two units. Point out the flaw in the reasoning.

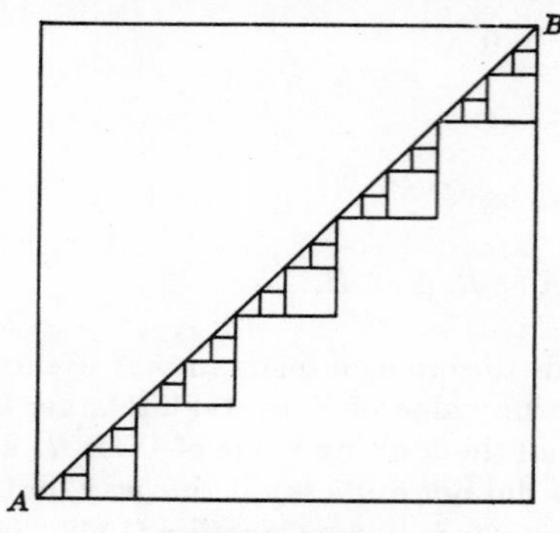

Figure 10.4

3. The following "argument" is ascribed to Zeno* (495–435 B.C.). See if you can detect a flaw in the reasoning, and write it down in a few words. Achilles cannot catch the tortoise in a race if the tortoise has a headstart; for, before he catches the tortoise, he must get up to the place from which the tortoise started. In the meantime, however, the tortoise has gone ahead and so has another headstart. Repeating this argument indefinitely, we see that the tortoise will never be caught.
4. In the "Works of Archimedes", by T. L. Heath, read pages 91–98 on Measurement of a Circle. Archimedes proves that the area of a circle is equal to the area of a right triangle with one leg equal to the radius and the other equal to the circumference. He also computes the ratio of circumference to diameter ($= \pi$) to within the inequality $3\frac{10}{71} < \pi < 3\frac{10}{70}$. The approximation $\pi = 3\frac{1}{7}$ is in current use.

10.3 Sequences

We turn now to the formulation of some general principles in the theory of limits. To this end we may sometimes make use of the language of algebra and sometimes that of geometry, but this should cause no confusion, since we have already seen that there is a 1 to 1 correspondence between the real numbers and points on a line.

We begin with the notion of an infinite sequence, which is defined to be a set of numbers in 1 to 1 correspondence with the positive integers. That is, it is an ordered array of the form $S_1, S_2, S_3, \ldots, S_n, \ldots$. Each S_i ($i = 1, 2, 3, \ldots, n, \ldots$) is called a term of the sequence, S_n the general term; but of course any general subscript index would serve as well, and we might speak of the general term S_k or S_{N+1}, etc. An alter-

* Look up a fine account of the history of Zeno's Paradoxes by Florian Cajori in the *American Mathematical Monthly*, vol. 22, pp. 1, 39, 77, 109, 143, 179, 215, 253, 292 (1915).

nate, but equivalent, definition is: A sequence is a (numerical valued) function of n defined over the set of positive integers. A sequence is given when and only when it is possible to write out the terms in order as far as we like. Tacit assumption of what the elements are is not enough; we must know the function of n or, equivalently, a rule whereby the terms can be constructed. For example, $1, \frac{1}{2}, \frac{1}{3}, \frac{1}{4}, \ldots$ is not a sequence. We do not know what the next term is; it might be -6 or $\sqrt{3}$ or something else. On the other hand,

$$1, \frac{1}{2}, \frac{1}{3}, \frac{1}{4}, \ldots, \frac{1}{n}, \ldots \tag{2}$$

is a sequence, for we are given the *general term*, or function, such that the term occupying the nth position is given by $1/n$. Again,

$$1, \tfrac{1}{2}, \tfrac{1}{3}, \tfrac{1}{4}, \sqrt{3}, \sqrt{4}, \sqrt{5}, \ldots, \sqrt{n-2}, \ldots \tag{3}$$

with $n > 4$ is another sequence. Note that (2) and (3) have the same first four terms. The "general term" of (3) is $\sqrt{n-2}$, but the first four terms are not derived from it. It is permissible for any *finite* number of terms to be given without reference to the general term. All terms following the last of this finite number of terms, however, must be derived from the general term.

The general term must be given and cannot be deduced from any finite set of terms. If we start a sequence with the terms $1, \frac{1}{2}, \frac{1}{3}, \frac{1}{4}$, we may be tempted to say that the general term is $1/n$. To see how foolish it would be to yield to this temptation, consider the following sequence with these first four terms:

$$1, \tfrac{1}{2}, \tfrac{1}{3}, \tfrac{1}{4}, 0, -\tfrac{2}{3}, \ldots, \left(\tfrac{25}{12} - \tfrac{35}{24}n + \tfrac{5}{12}n^2 - \tfrac{1}{24}n^3\right), \ldots \tag{4}$$

Exercise A. In sequence (4), verify that the first five terms are derived from the general term. (For ease in computation, multiply the general term by 24 and then use synthetic division.)

Problems 10.3

In Probs. 1 to 14 write down the first five terms of the sequences whose general terms are given below.

1. $n/(n+1)$.
2. $(n-1)/n$.
3. $1/(n^2+1)$.
4. $2/(n^2+n)$.
5. $n^2/(n+1)$.
6. 1^n.
7. 3^n.
8. $(-1)^n$.
9. $(-1)^n/n$.
10. $n^2/2^n$.
11. $(-1)^{n+1}\dfrac{x^{2n-1}}{(2n-1)!}$.
12. $(-1)^{n+1}\dfrac{x^{2n-2}}{(2n-2)!}$.
13. $\sin nx$.
14. $\log nx$.

10.4 Limits of Sequences

Previously we said that $1/n$ tends to zero as n tends to ∞. Now we must clarify the meaning of this and similar statements. If we examine the sequence,

$$1, \frac{1}{2}, \frac{1}{3}, \frac{1}{4}, \frac{1}{5}, \ldots, \frac{1}{n}, \ldots \tag{5}$$

we see that the terms are getting smaller and are coming closer and closer to zero. We should like to say that "the limit of the sequence (5) is zero". But what does this mean? Let us reword our idea as follows: "$1/n$ can be made as small as we please by choosing n sufficiently large." In this language we can actually prove that sequence (5) does have this property.

Let us think of this as a game between players 1 and 2. Player 1 names a small number, say, 1/1,000, and challenges player 2 to find a real number n_0, such that for all n larger than n_0 it is true that $1/n < 1/1{,}000$. In this case player 2 can choose $n_0 = 1{,}000$, and the result follows. If player 2 can find a suitable n_0 for each choice of a small number by player 1, we see that $1/n$ can be made smaller than any given positive quantity by choosing n sufficiently large. Thus it is reasonable to say that $1/n$ tends to 0.

This argument is usually expressed in more technical language as follows. Suppose that player 1 chooses a small positive number which hereafter we shall call ϵ. Then player 2 must find an n_0 such that for $n > n_0$, $1/n < \epsilon$. If we can prove this to be possible for every choice of ϵ, our result follows. The proof of this is contained in Theorem 1.

Theorem 1. For every ϵ, there exists an n_0 such that $1/n < \epsilon$ for all $n > n_0$.

Proof: We do this by working backward. For a chosen ϵ, we want $1/n < \epsilon$, or $n > 1/\epsilon$. So if we choose $n_0 = 1/\epsilon$, it follows that, for $n > n_0$, we have $n > 1/\epsilon$; hence $1/n < \epsilon$. We observe that $1/\epsilon$ is not necessarily an integer. Some writers require n_0 to be an integer, but this is not essential. However, if we make this requirement, we can obtain our previous result with an integral n_0 by taking n_0 to be the first integer just larger than $1/\epsilon$; that is, $n_0 = [\![1/\epsilon]\!] + 1$. For example, if we take $\epsilon = \frac{2}{99}$, then $n_0 = 50 > \frac{99}{2}$.

As a second illustration we prove that the quantity $\left(1 + \frac{1}{n}\right)$ can be made arbitrarily near 1 by choosing n sufficiently large.

Choose $\epsilon > 0$. We wish to make $\left|\left(1 + \frac{1}{n}\right) - 1\right| < \epsilon$ for n greater than a suitable n_0. This amounts to making $1/n < \epsilon$ for $n > n_0$. Hence

$n_0 = 1/\epsilon$ is a satisfactory choice of n_0. Therefore $1 + 1/n$ tends to 1 as n tends to ∞.

In Theorem 1 we had a sequence $S_n = 1/n$ and saw that there is associated with it a number $S(= 0$ in this case) such that, by choosing n sufficiently large, we could make S_n arbitrarily close to S. In the second example, $S_n = 1 + 1/n$ and $S = 1$. We formalize this idea in the following definition.

Definition: If for every positive number ϵ there exists a real number n_0 such that

$$|S_n - S| < \epsilon \qquad \text{for } n > n_0$$

then the sequence S_n is said to *converge* to the *limit* S. The number n_0 will, in general, depend upon ϵ, and we sometimes stress this relationship by writing $n_0(\epsilon)$. If a sequence does not converge, it is said to diverge.

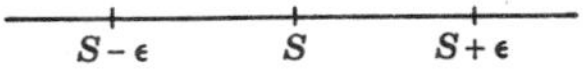

$S-\epsilon$ S $S+\epsilon$

Figure 10.5

Geometrically, when we say that a sequence S_n converges to S, we are saying that if S is the mid-point of an interval of length 2ϵ (Fig. 10.5), then all the terms of the sequence are to be found inside the interval, with the exception of at most a finite number (whose subscripts are less than n_0).

We also use the notations $\lim_{n \to \infty} S_n = S$ and, even more simple, $S_n \to S$ ("as $n \to \infty$" is often omitted when there could be no confusion) to express the fact that S_n converges to S. The former is read "the limit of S_n, as n approaches infinity, is S"; the latter is read "S_n approaches S, as n approaches infinity". The phrase "as n approaches infinity" may be modified to read "as n becomes infinite", "as n increases without bound", etc. But regardless of what other language is used, what is really meant is given in the above definition. The *limit of a sequence* is the basic and underlying concept of that branch of mathematics called *analysis* (calculus).

Problems 10.4

In Probs. 1 to 16 (*a*) write down the first five terms; (*b*) guess a number S which you think is the limit of the sequences; (*c*) find $n_0(\epsilon)$ as in the illustrations above, which will prove that $\lim_{n \to \infty} S_n = S$ according to the definition.

1. $S_n = 1/n^2$.
2. $S_n = 1/n^3$.
3. $S_n = n/(2n - 1)$.
4. $S_n = n/(2n + 1)$.
5. $S_n = 3/(n + 1)$.
6. $S_n = n/(n - 1),\ n > 1$.
7. $S_n = n/(n - 2),\ n > 2$.
8. $S_n = 3 + (2/n)$.
9. $S_n = 3 - (2/n)$.
10. $S_n = 1 - (1/n^2)$.
11. $S_n = 3/2^n$.
12. $S_n = (-1)^n/2^n$.
13. $S_n = 1/(1 + n + n^2)$.
14. $S_n = n/(1 + n^2)$.
15. $S_n = n/(n^2 - 1),\ n > 1$.
16. $S_n = \log (1/n)$.

10.5 Further Examples of Sequences

To clarify the notion of a sequence and its limit, let us consider the following additional examples.

(6) $$S_n = n\text{th prime}$$

(7) $$S_n = \begin{cases} 1 & \text{if } n \text{ is odd} \\ 0 & \text{if } n \text{ is even} \end{cases}$$

(8) $$S_1 = 1 \qquad S_2 = 1 \qquad S_n = S_{n-1} + S_{n-2} \qquad (n \geq 3)$$

(9) $$S_n = \frac{1}{n^4}(1^3 + 2^3 + \cdots + n^3)$$

(10) $$S_n = 1^3 + 2^3 + \cdots + n^3$$

Exercise A. Write out the first six terms in each of the above sequences. Sequences (9) and (10) are tricky.

Note that in (6) there is given a rule whereby the terms may be written down which is not expressed as a formula. (After more than two thousand years it is still an outstanding unsolved problem to find a formula which gives the nth prime in terms of n.) This sequence obviously diverges, for, as we say, the terms grow indefinitely large with n, or, as we write, $\lim_{n\to\infty} S_n = \infty$. But ∞, "infinity", is not to be regarded as a number.

Sequence (7) we write out: $\overline{1, 0}, \ldots$, using the bar over 1, 0 to indicate repetition, as we did in discussing decimal representation of rational numbers. There is no S for this sequence, since $\lim_{n\to\infty} S_n$ does not exist. Sequences (6) and (7) illustrate two different types of divergent sequences. Sequence (6) is said to diverge by becoming infinite; sequence (7) is said to diverge by oscillating (it is a bounded sequence but no limit is approached). There are, of course, many other kinds of divergent sequences.

If sequence (8) is written out correctly, the first six terms will be 1, 1, 2, 3, 5, 8. It is a famous sequence named after its Italian user, Fibonacci. It is a problem of considerable difficulty to show that, expressed as a function of n,

(11) $$S_n = \frac{(1 + \sqrt{5})^n - (1 - \sqrt{5})^n}{2^n \sqrt{5}}$$

As (8) is given, S_n is in terms of the two preceding terms. This is called a recurrence relation. By using this recurrence relation $S_n = S_{n-1} + S_{n-2}$ and the two given values $S_1 = 1$, $S_2 = 1$, we are able to arrive at (11) by higher mathematical methods.

Exercise B. Compute the first four terms of sequence (8) directly from (11).

Sequence (9) is actually the one that was associated with the problem of finding the area under the curve $y = x^2$ (see Sec. 10.2, Prob. 1). It is of a slightly different character since the general term

$$S_n = \frac{1}{n^4}(1^3 + 2^3 + \cdots + n^3)$$

itself includes a sum of n cubes and the number of cubes increases with n. This situation can be handled easily, however, since we know that

$$S_n = \frac{1}{n^4}(1^3 + 2^3 + \cdots + n^3) = \frac{n^2(n+1)^2}{4n^4} = \frac{1}{4}\left(1 + \frac{2}{n} + \frac{1}{n^2}\right)$$

In order to find $\lim S_n$, we note that

$$\lim S_n = \lim \frac{1}{4}\left(1 + \frac{2}{n} + \frac{1}{n^2}\right)$$

Now $\lim (2/n) = 0$ and $\lim (1/n^2) = 0$. Thus it is reasonable to conclude that $\lim S_n = \frac{1}{4}$. To justify this statement, we need to use theorems appearing in the next section.

Sequence (10) is of a roughly similar nature, for S_n is itself expressed as a sum of terms. The difference between (9) and (10) is that (9) involves the factor $1/n^2$, whereas (10) is a pure sum. Sequences like (10) are treated under the heading "series" in a later section.

10.6 Theorems on Limits of Sequences

Here we collect a number of important properties of sequences and their limits. We have tacitly referred to *the* limit of a sequence and have assumed that when there is a limit it is unique. This is proved in Theorem 2.

Theorem 2. The limit S for a given convergent sequence S_n is unique.

Proof: First, let us rephrase the statement of the theorem in the following manner: If $\lim\limits_{n\to\infty} S_n = S$ and if $T \neq S$, then $\lim\limits_{n\to\infty} S_n \neq T$. Proceeding by the indirect method, we suppose $\lim\limits_{n\to\infty} S_n = T$. Then for every positive number ϵ there exists an n_1 such that

$$|S_n - S| < \frac{\epsilon}{2} \qquad \text{when } n > n_1$$

and similarly, for the same ϵ there exists an index n_2 such that

$$|S_n - T| < \frac{\epsilon}{2} \qquad \text{when } n > n_2$$

Let n_0 be the larger of n_1 and n_2. Keeping in mind that

$$|S_n - S| = |S - S_n|$$

and also that the absolute value of a sum is equal to or less than the sum of the absolute values, we may write

$$|S - T| = |(S - S_n) + (S_n - T)| \leq |S - S_n| + |S_n - T| < \epsilon$$

for all $n > n_0$.

This says that the numerical difference between S and T is less than every arbitrarily small quantity ϵ. In other words, this means that $S = T$, which contradicts our assumption. Hence a convergent sequence has one and only one limit.

Remark. In this proof we asked that for a given ϵ,

$$|S_n - S| < \frac{\epsilon}{2} \qquad \text{for } n > n_1$$

Why did we select $\epsilon/2$ here instead of ϵ? To answer this, we must have an overall look at the proof and then work backward from the result that

$$|S - T| < \epsilon$$

The ϵ method of proving that $S_n \to S$ is essential and fundamental. It is unfortunately too complicated to be used in most examples. Its general use can be avoided by applying the following theorems, which we state without proof. Let C be a constant, and let $S_n \to S$ and $T_n \to T$. Then we have the following theorems:

Theorem 3. If $S_n = C$ for every n, then $\lim S_n = C$.

Theorem 4. $\lim (S_n \pm T_n) = S \pm T$.

Theorem 5. $\lim S_nT_n = ST$.

Corollary. $\lim CS_n = CS$.

Theorem 6. $\lim S_n/T_n = S/T$, provided $T \neq 0$, and $T_n \neq 0$ for all $n >$ some n_0.

Illustration 1. In sequence (9) of Sec. 10.5,

$$S_n = \frac{1}{n^4}(1^3 + 2^3 + \cdots + n^3) = \frac{1}{4}\left(1 + \frac{2}{n} + \frac{1}{n^2}\right)$$

We have shown that $\lim \frac{1}{n} = 0$. From Theorem 5, $\lim \frac{2}{n} = 0$ and $\lim \frac{1}{n^2} = 0$. From Theorem 4, $\lim \left(1 + \frac{2}{n} + \frac{1}{n^2}\right) = 1$. Finally, from Theorem 5, $\lim \frac{1}{4}\left(1 + \frac{2}{n} + \frac{1}{n^2}\right) = \frac{1}{4}$. Hence $\lim S_n = \frac{1}{4}$.

Exercise A. Apply these theorems to the problems at the end of Sec. 10.4.

Another useful method for obtaining limits of sequences is the "domination principle" stated in Theorem 7.

Theorem 7. If $A_n \leq S_n \leq B_n$ for all $n >$ some n^* and

$$\lim_{n \to \infty} A_n = \lim_{n \to \infty} B_n = L$$

then $\lim\limits_{n \to \infty} S_n$ exists and is equal to L.

Proof: We need to show that for sufficiently large n_0 all S_n $(n > n_0)$ are close to L. More explicitly, choose $\epsilon > 0$. Since $\lim A_n = L$, there exists an n_1 such that, for $n > n_1$ we have $|A_n - L| < \epsilon$. Similarly, there is an n_2 such that, for $n > n_2$, we have $|B_n - L| < \epsilon$. Choose n_3 to be the larger of n_1 and n_2. Then, for $n > n_3$, all A_n and all B_n lie in the interval $L - \epsilon \leq x \leq L + \epsilon$. Let n_0 be the larger of n_3 and n^*. Since $A_n \leq S_n \leq B_n$ for $n > n^*$, all S_n for $n > n_0$ lie in this same interval $L - \epsilon \leq x \leq L + \epsilon$, or $|S_n - L| < \epsilon$. This completes the proof.

Illustration 2. We use this theorem to show that $\frac{1}{n^2} \to 0$. For all n, $0 < \frac{1}{n^2} \leq \frac{1}{n}$. (What value could be used as n^*?) We put $A_n = 0$, $B_n = \frac{1}{n}$, and hence, for all n, $A_n \leq \frac{1}{n^2} \leq B_n$. We know that $\lim A_n = 0$, and $\lim B_n = \lim 1/n = 0$. Applying the theorem, we see that $\lim \frac{1}{n^2}$ exists and equals zero.

Similarly, $\frac{\cos n}{n} \to 0$, since $-\frac{1}{n} < \frac{\cos n}{n} < \frac{1}{n}$ for all n.

Problems 10.6

In Probs. 1 to 12 find $\lim_{n\to\infty} S_n$.

1. $S_n = \dfrac{n}{n+2}$. $\left(\text{First show that } S_n = \dfrac{1}{1+(2/n)}.\right)$

2. $S_n = \dfrac{3n+2}{n+1}$. $\left(\text{First show that } S_n = \dfrac{3+(2/n)}{1+(1/n)}.\right)$

3. $S_n = \dfrac{\sin n}{n}$.

4. $S_n = \dfrac{(-1)^{n+1}}{2n}$.

5. $S_n = \left(\dfrac{1}{n+1}\right)\left(2 - \dfrac{1}{n}\right)$.

6. $S_n = \left(\dfrac{1}{n+1}\right)\Big/\left(2 - \dfrac{1}{n}\right)$.

7. $S_n = \left(1 + \dfrac{2}{n}\right)\left(2 + \dfrac{1}{n}\right)$.

8. $S_n = \left(1 + \dfrac{2}{n}\right)\Big/\left(2 + \dfrac{1}{n}\right)$.

9. $S_n = 1/\sqrt{n}$. (Use the method of Sec. 10.4.)

10. $S_n = \dfrac{1}{\sqrt{n+1} + \sqrt{n}}$. $\left(\text{Show that } 0 < \dfrac{1}{\sqrt{n+1}+\sqrt{n}} < \dfrac{1}{2\sqrt{n}}, \text{ and use Theorem 7.}\right)$

11. $S_n = \sqrt{n+1} - \sqrt{n}$. $\left(\text{First show that } S_n = \dfrac{1}{\sqrt{n+1}+\sqrt{n}}; \text{ then use Prob. 10.}\right)$

12. $S_n = \sqrt{n-2} - \sqrt{n}$.

In Probs. 13 to 16 explain why the sequence does not have a limit.

13. $S_n = n^{n-1}$.

14. $S_n = (n-1)^n$.

15. $S_n = \cos n$.

16. $S_n = \cot n$.

17. $\log (n+1)$.

18. $\log \tan n$.

19*. Write down the positive rational numbers in the form of a sequence. Does this sequence have a limit? HINT: Write the rationals a/b in the order determined by $a + b = n$, $n = 1, 2, 3, \ldots$.

20*. Cauchy's test of convergence for a sequence is the following: If, for every $\epsilon > 0$, an n_0 exists such that $|S_n - S_m| < \epsilon$, for all n, $m > n_0$, then the sequence S_n converges to a limit S. State the converse and prove it. (Cauchy's test itself is more difficult to prove, but you might like to try it.)

21. How is π to be considered as a limit of a sequence?

22. Prove Theorem 3.

23. Prove Theorem 4 (for + only).

24. Prove the corollary to Theorem 5.

25. A (point) penny is to be pushed across the width of a table 2 ft wide. Some time during the first second it is pushed to the 1-ft mark and left there for 1 sec. Then, during the third second, it is pushed to the $1\frac{1}{2}$-ft mark and left for a second. If this process is repeated indefinitely, always leaving the penny at rest at the one-half of the remaining distance mark after a push, how far does the penny travel in reaching the far side of the table? (Do not assume anything that is not explicit in the above conditions!) (BT)†

† See E. J. Moulton, Two Teasers for Your Friends, *American Mathematical Monthly*, vol. 55, p. 342 (1948).

26. Again, the penny is to be pushed across the width of a table 2 ft wide. The penny is pushed and rests at the end of 1 sec on the 1-ft mark. It is again pushed and rests at the end of an additional $\frac{1}{2}$ sec at the $1\frac{1}{2}$-ft mark. Another push, and it rests at the end of an additional $\frac{1}{4}$ sec on the $1\frac{3}{4}$-ft mark. The process is repeated indefinitely, always halving the time and the remaining distance. How long did it take for the penny to reach the far edge of the table? (Again, do not assume anything not explicitly stated.)

10.7 Series

Definition: An expression of the form

$$u_1 + u_2 + \cdots + u_n + \cdots$$

is called an infinite series. The general term is u_n and

$$S_n = u_1 + u_2 + \cdots + u_n$$

is called the partial sum to n terms.

There is a sequence of partial sums $S_1, S_2, \ldots, S_n, \ldots$ whose general term is S_n. If this sequence converges to S, then the series is said to *converge* to S and S is called the *sum* of the infinite series; it is not the usual kind of sum obtained by adding together a finite number of terms. It is a limit, where

$$S = \lim_{n \to \infty} S_n$$

If this limit fails to exist (in any of many possible ways), the series is said to *diverge*. An infinite series is specified if and only if its general term is specified.

You have already met the (infinite) geometric series

$$1 + r + r^2 + \cdots + r^n + \cdots$$

or, more generally,

$$a + ar + ar^2 + \cdots + ar^n + \cdots \qquad a \neq 0$$

First, we wish to develop a simple formula for S_n, which may be familiar from high school algebra. (It was also assigned as a problem in mathematical induction, Prob. 15, Sec. 3.2.) We write

$$S_n = a + ar + ar^2 + \cdots + ar^{n-1}$$

and

$$rS_n = ar + ar^2 + \cdots + ar^{n-1} + ar^n$$

The last line came from the one immediately above by multiplying it through by r. By subtraction we obtain

$$S_n - rS_n = a - ar^n$$

or

$$S_n = \frac{a(1 - r^n)}{1 - r} \tag{12}$$

QUESTION: Does S_n have a limit? The answer is "yes", provided $|r| < 1$, and "no" otherwise. To prove this, we first show that $r^n \to 0$ as $n \to \infty$, provided $|r| < 1$.

When $|r| < 1$, $|r|$ can always be written in the form

$$|r| = \frac{1}{1 + p}$$

where p is positive. Hence

$$|r^n - 0| = |r^n| = \frac{1}{(1 + p)^n}$$

(Prove that $|r|^n = |r^n|$.) In the binomial expansion

$$(1 + p)^n = 1 + np + \frac{n(n - 1)}{2!} p^2 + \cdots + p^n$$

every term is positive and obviously $(1 + p)^n > np$, just one of its terms.

Therefore $0 < |r^n| < 1/np$. Since $1/np \to 0$ as $n \to \infty$, we find, from Theorem 7, that $|r^n| \to 0$, and hence $r^n \to 0$. Now apply Theorems 3 to 6 as needed to

$$\lim S_n = \lim \frac{a(1 - r^n)}{1 - r}$$

The result is $\lim S_n = a/(1 - r)$. This is the sum of the geometric series.

When $r = 1$, $a + ar + ar^2 + \cdots + ar^n + \cdots$ reduces to $a + a + \cdots + a + \cdots$, which is obviously divergent. Finally, if $|r| > 1$, $|r|^n \to \infty$, and hence, from the form of (12), we see that $\lim S_n$ does not exist; thus again the series diverges.

Exercise A. Prove $|r^n| \to \infty$ if $|r| > 1$. What happens to r^n?

One of the most important uses of series is in connection with the values of certain functions. For example, the series

$$1 + x + \frac{x^2}{2!} + \frac{x^3}{3!} + \cdots + \frac{x^n}{n!} + \cdots$$

is not a polynomial since it contains infinitely many terms. However, it can be proved that this series converges for all real values of x, and thus it can be used to give the values of a function whose domain is the X-axis. This function can be shown to be identical with e^x. Hence we write

$$e^x = 1 + x + \frac{x^2}{2!} + \frac{x^3}{3!} + \cdots + \frac{x^n}{n!} + \cdots$$

Putting $x = 1$, we obtain

$$e = 1 + 1 + \frac{1}{2!} + \frac{1}{3!} + \cdots + \frac{1}{n!} + \cdots$$

Exercise B. Use the first five terms of the series for e to obtain an approximation to its decimal expansion.

In a similar fashion the following series converge for all real x and give the values of the familiar functions:

$$\sin x = x - \frac{x^3}{3!} + \frac{x^5}{5!} + \cdots + (-1)^{n+1}\frac{x^{2n-1}}{(2n-1)!} + \cdots$$

$$\cos x = 1 - \frac{x^2}{2!} + \frac{x^4}{4!} + \cdots + (-1)^{n+1}\frac{x^{2n-2}}{(2n-2)!} + \cdots$$

We also have that

$$\sinh x = x + \frac{x^3}{3!} + \frac{x^5}{5!} + \cdots + \frac{x^{2n-1}}{(2n-1)!} + \cdots$$

$$\cosh x = 1 + \frac{x^2}{2!} + \frac{x^4}{4!} + \cdots + \frac{x^{2n-2}}{(2n-2)!} + \cdots$$

converge for all real x. (See Sec. 7.2.)

Problems 10.7

In Probs. 1 to 10 find the sum, when the sum exists.

1. $\frac{1}{2} + \frac{1}{4} + \frac{1}{8} + \frac{1}{16} + \cdots + \frac{1}{2^n} + \cdots.$

2. $2 + 1 + \frac{1}{2} + \frac{1}{4} + \frac{1}{8} + \cdots + \frac{1}{2^{n-2}} + \cdots.$

3. $\frac{1}{2} - \frac{1}{4} + \frac{1}{8} - \frac{1}{16} + \cdots + (-1)^{n+1}\frac{1}{2^n} + \cdots.$

4. $1 + \frac{1}{3} + \frac{1}{9} + \frac{1}{27} + \cdots + \frac{1}{3^{n-1}} + \cdots.$

5. $-1 + \frac{1}{3} - \frac{1}{9} + \frac{1}{27} + \cdots + (-1)^n\frac{1}{3^{n-1}} + \cdots.$

6. $5 - 3 + 1 + \frac{1}{4} + \frac{1}{4^2} + \cdots + \frac{1}{4^n} + \cdots$.
7. $4 + 4\left(\frac{1}{5}\right) + 4\left(\frac{1}{5^2}\right) + \cdots + 4\left(\frac{1}{5^n}\right) + \cdots$.
8. $2^7 + 2^7\left(\frac{1}{2}\right) + 2^7\left(\frac{1}{2^2}\right) + \cdots + 2^7\left(\frac{1}{2^n}\right) + \cdots$.
9. $3^6 + 3^6(\frac{2}{3}) + 3^6(\frac{2}{3})^2 + \cdots + 3^6(\frac{2}{3})^n + \cdots$.
10. $5 - 5 + 5 - 5 + \cdots + (-1)^{n+1}5 + \cdots$.
11. Show that the repeating decimal $0.\overline{3}$ can be represented by the infinite series

$$\frac{3}{10} + \frac{3}{10^2} + \cdots + \frac{3}{10^n} + \cdots$$

and find its sum as a rational number by using the methods of this chapter.

In Probs. 12 to 15 find p and q as in Prob. 11. (See Sec. 2.13 for notation.)

12. $p/q = 0.15\overline{0}$.
13. $p/q = 0.14\overline{9}$.
14. $p/q = 1.2\overline{1}$.
15. $p/q = 2.\overline{134}$.

16. Given $S_n = 1 + \frac{1}{1!} + \frac{1}{2!} + \cdots + \frac{1}{n!}$, assuming that $\lim S_n$ exists, show that it is less than 3. HINT: $S_n < 1 + 1 + \frac{1}{2} + \frac{1}{2^2} + \cdots + \frac{1}{2^{n-1}}$. The $\lim_{n\to\infty} S_n$ is called e, and it is irrational; it is the natural base of logarithms. Also

$$e = \lim_{n\to\infty}\left(1 + \frac{1}{n}\right)^n \approx 2.71828$$

17. $\cos x \approx 1 - (x^2/2) + (x^4/24)$. From this compute cos 0.1, and compare with the value as given by the table.
18. $\sin x \approx x - (x^3/6) + (x^5/120) - (x^7/5{,}040)$. From this compute sin 1, and compare with the value given by the table.
19. A theorem states that if $\sin x$ and $\cos x$ are approximated by using only a few terms, say,

$$\sin x \approx x - \frac{x^3}{3!} + \frac{x^5}{5!} + \cdots + (-1)^{n+1}\frac{x^{2n-1}}{(2n-1)!}$$

$$\cos x \approx 1 - \frac{x^2}{2!} + \frac{x^4}{4!} + \cdots + (-1)^{n+1}\frac{x^{2n-2}}{(2n-2)!}$$

then the error committed does not exceed half the absolute value of the first term omitted. Show that a complete four-place table of sines and cosines can be prepared by using only four terms in each case. HINT: Examine $\pi/(8!\cdot 8)$.

10.8 Limits of Functions

In the preceding paragraphs we were occupied with the notion of the limit of the values of a function of n where n assumed only integral values; we were concerned with sequences and limits of sequences. The concept of limit, however, is of utmost importance in the theory of functions defined for all values of x in an interval of the real line. Such functions will be called "functions of a continuous variable".

We shall consider only real-valued functions f of the real variable x and give a definition of the limit of its values, $f(x)$, as $x \to a$.

At first thought we might describe the operation $x \to a$ by choosing a sequence $x_1, x_2, \ldots, x_n, \ldots$ such that $\lim_{n \to \infty} x_n = a$. Then there is an associated sequence of values $f(x_1), f(x_2), \ldots, f(x_n), \ldots$ whose limit (if any) is defined as in Sec. 10.4. Suppose this exists and is equal to L. This can be called the limit of the values $f(x)$, however, only if we get the same L when we choose any other basic sequence $\bar{x}_1, \ldots, \bar{x}_n, \ldots$ such that $\lim_{n \to \infty} \bar{x}_n = a$.

To illustrate this difficulty, let

$$f(x) = \begin{cases} 1 & x \text{ rational} \\ 0 & x \text{ irrational} \end{cases}$$

what is $\lim_{x \to 0} f(x)$? If we choose $x_1, \ldots, x_n, \ldots$ to be rational, the limit is 1; if we choose them irrational, the limit is zero. In such a case we naturally say that *the* limit does not exist.

Another difficulty is illustrated by the example:

Find $$\lim_{x \to 0} f(x) \text{ when } f(x) = \begin{cases} 1 & x \neq 0 \\ 2 & x = 0 \end{cases}$$

If we choose any set of x_i none of which is zero, the limit is 1. But if we choose all $x_i = 0$ (a perfectly good sequence), the limit is 2. And if the set of x_i contains 0 infinitely many times, but also has infinitely many values not equal to zero, the limit of the sequence does not exist at all.

One solution to the problem is to say that $\lim_{x \to 0} f(x) = L$ if

$$\lim_{n \to \infty} f(x_n) = L$$

for all sequences $x_1, \ldots, x_n, \ldots$ which have a as their limit. From this set of all possible sequences we shall wish to exclude, however, the sequences which have infinitely many values (or all values) equal to a; for our intuitive idea of $x \to a$ is "*x approaches a*" and does not include "$x = a$". It is entirely possible to base the definition of $\lim_{x \to a} f(x)$ upon this idea, but its application to specific problems involves unnecessary difficulties, which we prefer to avoid.

Before introducing you to a better approach, let us define some terms and notation which we shall need hereafter.

Definitions: The *closed interval* $[a, b]$ is the set of real numbers x such that $a \leq x \leq b$. In this notation we assume that $a < b$. The *open interval* $]a, b[$ is the set of real numbers x such that $a < x < b$. Again the notation implies that $a < b$.

In view of these definitions, the meanings of symbols such as $]a, b]$, $[a, b[$, $]-\infty, a]$, $]-\infty, \infty[$, etc., should be self-evident.

Now let us assume that a is a point of an open interval $]c, d[$ and that $f(x)$ is defined at every point of this interval except possibly at $x = a$. The intuitive meaning we give to the phrase

$$\lim_{x \to a} f(x) = L$$

is that $f(x)$ gets close to L as x gets close to a. Let us give a precise formulation of this idea.

Definition: The value $f(x)$ of the function f is said to have the constant L as a *limit* when x approaches a if and only if:

For every positive ϵ, there exists a positive number δ such that $|f(x) - L| < \epsilon$ for all x having the property that $0 < |x - a| < \delta$.

When this limit exists, we write

$$\lim_{x \to a} f(x) = L$$

which is read, "The limiting value of $f(x)$, as x approaches a, equals L."

Remarks

(1) The definition above is *not* a statement about the value of the function *at* a. It is a statement about the values of $f(x)$ for x close to a.

(2) In order to apply the definition, the constant L must be known. If we do not know L, we try to make a reasonable guess and see how this works. In some cases this is relatively easy; in most cases, difficult or impossible.

When $x \to \infty$, where we assume that $f(x)$ is defined at every point of an interval $]a, \infty[$ the definition needs the following modification:

Definition: The value $f(x)$ of the function f is said to have the constant L as a limit when x approaches ∞ if and only if:

For every positive ϵ there exists a positive number A such that $|f(x) - L| < \epsilon$ for all x such that $x > A$.

We write this

$$\lim_{x \to \infty} f(x) = L$$

Let us look at the situation geometrically. In Fig. 10.6*a* the function is defined by the curve except at $x = a$; $f(a)$ is indicated on the graph as the ordinate up to the black dot. The open circle on the curve indicates that that particular point is missing. Clearly, if x ranges between $a - \delta_1$ and $a + \delta_2$ (but does not take the value a itself), then the function ranges between $L - \epsilon$ and $L + \epsilon$. We say the points on the X-axis for which $|x - a| < \delta$, $x \neq a$, are mapped into the points on the Y-axis for which $|f(x) - L| < \epsilon$. In this case $\lim_{x \to a} f(x) = L$, but this has nothing to do with the value $f(a)$.

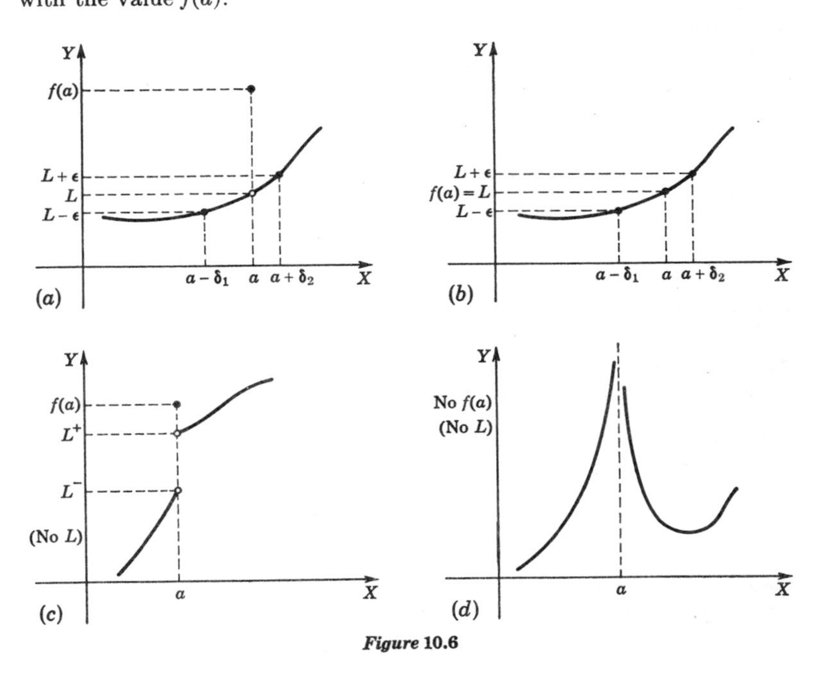

Figure 10.6

In Fig. 10.6*b* we have a similar case except that the $\lim_{x \to a} f(x)$ exists ($= L$), and this is, this time, the value $f(a)$.

In Fig. 10.6*c* the curve representing the function is broken. By L^- we indicate that if x approaches a from the left there will be a limit L^-; similarly, for L^+, where x approaches a from the right.

Exercise A. Formulate a precise definition of L^- and L^+.

For a limit L to exist, L^- must be equal to L^+. Hence, in Fig. 10.6*c*, there is no limit L. In Fig. 10.6*d* we give a case where the function increases without bound as $x \to a$ and there is neither limit L nor value $f(a)$. These figures do not exhaust the possibilities.

Exercise B. Give a geometric illustration of an essentially different example.

We now apply this idea to some important examples.

Illustration 1. Let $f(x) = x$. $-\infty < x < \infty$. Find $\lim_{x \to 2} f(x)$.

Solution: Fron an inspection of $f(x)$ for x near 2, we guess that $f(x) \to 2$ as $x \to 2$. Note that we must not set $x = 2$, for this is not the idea of a limit. Now we prove that our guess is correct.

We choose an ϵ (>0) and try to find a δ such that

$$|f(x) - 2| < \epsilon$$

when $|x - 2| < \delta$. Since $f(x) = x$, this is an easy matter: just choose $\delta = \epsilon$. Similarly,

$$\lim_{x \to a} f(x) = a$$

Illustration 2. Let $f(x) = x^2$, $-\infty < x < \infty$. Find $\lim_{x \to 2} f(x)$.

Solution: The procedure is first to guess the value of this limit by considering a sequence of numerical values, and then to prove that this limit is correct by showing that it satisfies the definition.

To guess the value of the limit, we compute the entries in the following table.

x	1	1.5	1.9	1.99
$f(x)$	1	2.25	3.61	3.9601

We are not allowed to use $x = 2$ in finding the limit (see the definition). From the table it appears that as $x \to 2$, $f(x) \to 4$. Therefore we guess that $\lim_{x \to 2} x^2 = 4$. First, we note that $|x^2 - 4| = |x - 2| \times |x + 2|$. Consider values of x such that $|x - 2| < 1$; that is, $1 < x < 3$. For such x, $|x + 2| < 5$. We wish to have $|x^2 - 4| < \epsilon$, and this will certainly be true if simultaneously

$$|x - 2| < 1 \quad \text{and} \quad |x - 2| < \frac{\epsilon}{5}$$

for then the product

$$|x - 2| \times |x + 2| < \frac{\epsilon}{5} \cdot 5 = \epsilon$$

Therefore we choose δ to be the smaller of the numbers 1 and $\epsilon/5$, for $|x - 2| < \delta$ then implies that $|x^2 - 4| < \epsilon$. Note that our choice of δ depends upon our choice of ϵ, but that for every choice of ϵ, we know how to find a suitable δ. Thus we have shown that

$$\lim_{x \to 2} x^2 = 4$$

Similarly, $\lim_{x \to a} x^2 = a^2$.

It is doubtless evident to you that this proof involves considerable ingenuity and that other tricks may be needed to evaluate other limits. Fortunately, there are general and relatively simple methods available for finding the limits of most of the common types of functions. These depend upon theorems to be stated in the next section. Occasionally, however, no simple method will work, and you must go back to the basic definition and use your ingenuity.

Problems 10.8

Find:

1. $\lim_{x \to 2} 4x$.

2. $\lim_{x \to 1} (-3x)$.

3. $\lim_{x \to -1} (3x + 2)$.

4. $\lim_{x \to 2} (3x - 2)$.

5. $\lim_{x \to 3} (\frac{1}{3}x^2)$.

6. $\lim_{x \to -3} (3x^2)$.

7. $\lim_{x \to 2} (x^2 + x)$.

8. $\lim_{x \to 2} (x^2 + x - 6)$.

9. $\lim_{x \to 3} x^3$.

10. $\lim_{x \to 2} x^4$.

11. $\lim_{x \to 2} (1/x)$.

12. $\lim_{x \to -2} (1/x)$.

13. $\lim_{x \to 5} \sqrt{x - 4}$.

14. $\lim_{x \to -2} \sqrt{2 - x}$.

15. $\lim_{x \to 0} f(x)$, where $f(x) = \begin{cases} 0, & x \leq 0, \\ x, & x > 0. \end{cases}$

16. $\lim_{x \to 0} f(x)$, where $f(x) = \begin{cases} 0, & x \leq 0, \\ x^2, & x > 0. \end{cases}$

17. $\lim_{x \to 0} |x|$.

18. $\lim_{x \to 1} |x - 1|$.

19. $\lim_{x \to -1} |x^3|$.

20. $\lim_{x \to 1.6} [\![x]\!]$.

21. $\lim_{x \to 2.2} [\![x + 1]\!]$.

22. $\lim_{x \to 3} [\![x]\!]$.

10.9 Theorems on Limits of Functions

The process of obtaining limits from the basic definition is so complex that we must have simpler methods if we are to deal with complicated functions. Some of the more useful methods are treated in this section.

Theorems 3 to 6, Sec. 10.6, applying to the limits of sums, products, and quotients of sequences, also hold in the case of functions of a continuous variable x. Let $\lim_{x \to a} f(x) = F$ and $\lim_{x \to a} g(x) = G$. We state the following theorems, but omit their proofs.

Theorem 8. For the constant function f, where $f(x) = C$, $\lim_{x \to a} f(x) = C$, for all a.

Theorem 9. $\lim_{x \to a} (f(x) \pm g(x)) = \lim_{x \to a} f(x) \pm \lim_{x \to a} g(x)$
$= F \pm G.$

Theorem 10. $\lim_{x \to a} f(x) \cdot g(x) = (\lim_{x \to a} f(x)) \cdot (\lim_{x \to a} g(x))$
$= F \cdot G.$

Corollary. $\lim_{x \to a} k \cdot f(x) = k \cdot \lim_{x \to a} f(x) = kF$, k constant.

Theorem 11. $\lim_{x \to a} \dfrac{f(x)}{g(x)} = \dfrac{\lim_{x \to a} f(x)}{\lim_{x \to a} g(x)} = \dfrac{F}{G}$, if $G \neq 0$.

Illustration 1. Find $\lim_{x \to a} x^2$. From Sec. 10.8, Illustration 1, we know that $\lim_{x \to a} x = a$. Now $x^2 = x \cdot x$. Therefore Theorem 10 tells us that

$$\lim_{x \to a} x^2 = a^2$$

Exercise A. Find $\lim_{x \to a} x^n$.

Exercise B. Find $\lim_{x \to a} Cx^n$, where C is a constant.

Illustration 2. Find $\lim_{x \to a} (3x^2 - 2x^2 + 4)$. From Theorem 9,

$$\lim_{x \to 2} (3x^3 - 2x^2 + 4) = \lim_{x \to 2} 3x^3 - \lim_{x \to 2} (2x^2) + \lim_{x \to 2} 4$$

Finally, from Exercise B,

$$\lim_{x \to 2} (3x^3 - 2x^2 + 4) = 24 - 8 + 4 = 20$$

Illustration 3. Find $\lim_{x \to 1} \dfrac{x^2 + 3}{x + 2}$. From Theorem 11,

$$\lim_{x \to 1} \frac{x^2 + 3}{x + 2} = \frac{\lim_{x \to 1} (x^2 + 3)}{\lim_{x \to 1} (x + 2)} = \frac{4}{3}$$

Illustration 4. Find $\lim_{x \to 1} \dfrac{2 - \sqrt{5 - x}}{x - 1}$. We could appeal to the definition, but there is an easier way of determining this limit. We write

$$\begin{aligned}\frac{2-\sqrt{5-x}}{x-1} &= \frac{2-\sqrt{5-x}}{x-1}\cdot\frac{2+\sqrt{5-x}}{2+\sqrt{5-x}}\\ &= \frac{4-(5-x)}{(x-1)(2+\sqrt{5-x})}\\ &= \frac{x-1}{(x-1)(2+\sqrt{5-x})}\\ &= \frac{1}{2+\sqrt{5-x}} \qquad x \neq 1\end{aligned}$$

In the last line we have tossed out a factor of $x - 1$ from numerator and denominator, so that now x must be different from unity. But, with $x \neq 1$, the function $\frac{2-\sqrt{5-x}}{x-1}$ is identical with the function $\frac{1}{2+\sqrt{5-x}}$; the two expressions are just different looking representations of the same thing.

We now assume that we have shown from the basic definition of a limit that $\lim_{x\to 1} \sqrt{5-x} = 2$. Then, from Theorems 9 and 11, we see that

$$\lim_{x\to 1} \frac{1}{2+\sqrt{5-x}} = \frac{1}{4}$$

Hence we conclude that

$$\lim_{x\to 1} \frac{2-\sqrt{5-x}}{x-1} = \frac{1}{4}$$

Exercise C. From the basic definition of a limit, prove

$$\lim_{x\to 1} \sqrt{5-x} = 2$$

Now just a word as to why we multiplied numerator and denominator of $\frac{2-\sqrt{5-x}}{x-1}$ by $2+\sqrt{5-x}$ (the quadratic surd conjugate to $2-\sqrt{5-x}$). It is never easy to explain the logic of discovery, but in this case we could plead that we had seen this sort of thing done before in problems of rationalizing—generally in the denominator—and we thought of this past experience and tried it, and it worked. Obviously, you keep a mental record of the methods you have learned. When you devise one of your own, in a terrific burst of imagination, you will have done some original mathematical thinking.

An important aid in some proofs is the "domination principle", which, in this case, has the following statement.

Theorem 12. If $F(x) \leq f(x) \leq G(x)$ for all x in an interval containing $x = a$, except possibly at $x = a$, and if $\lim_{x\to a} F(x) = L$ and $\lim_{x\to a} G(x) = L$, then $\lim_{x\to a} f(x) = L$.

This theorem is analogous to Theorem 7 and has a similar proof. We use this principle in the next three illustrations.

Exercise D. Following the ideas of the proof of Theorem 7, write a proof of Theorem 12.

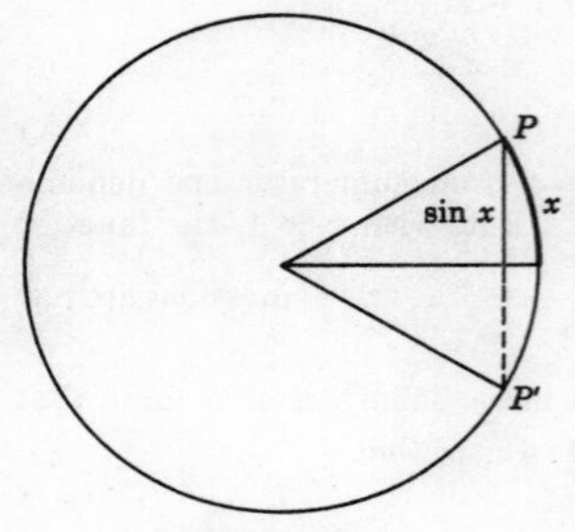

Figure 10.7

Illustration 5. Show that $\lim_{x \to 0} \sin x = 0$.

Solution: We recall the definition of $\sin x$ and note that for small x, $|\sin x| \leq |x|$. This follows from the fact that, in Fig. 10.7,

$$2|\sin x| = PP'$$
$$2|x| = \text{arc } PP'$$

But a chord has a length less than the corresponding arc; thus $PP' < \text{arc } PP'$ or $|\sin x| \leq |x|$. The equality occurs at $x = 0$. Therefore

$$-|x| \leq \sin x \leq |x|$$

in a small interval about $x = 0$. We apply the domination principle with $F(x) = -|x|$ and $G(x) = |x|$. Since $\lim_{x \to 0} |x| = 0$, we conclude that

$$\lim_{x \to 0} \sin x = 0$$

Illustration 6. Show that $\lim_{x \to 0} \cos x = 1$.

Solution: Write:

$$1 - \cos^2 x = \sin^2 x$$

Then $\lim_{x \to 0} (1 - \cos^2 x) = \lim_{x \to 0} \sin^2 x = 0$.

Moreover

$$1 - \cos x = \frac{1 - \cos^2 x}{1 + \cos x} < 1 - \cos^2 x$$

for x near zero. Hence, for x near zero,

$$0 \leq 1 - \cos x \leq 1 - \cos^2 x$$

Therefore $\lim_{x \to 0} (1 - \cos x) = 0$

and

$$\lim_{x \to 0} \cos x = 1$$

Illustration 7. Consider $f(x) = (\sin x)/x$. This function is defined when x is any real number except 0. Does $L = \lim_{x \to 0} f(x)$ exist? If so, what is this limit L?

Solution: We begin by guessing what L might be, but we make some systematic guesses by first computing the entries in the table below by using a table of sines of real numbers (Appendix, Table IV).

x	0.2	0.1	0.02	0.01
$\sin x$	0.19867	0.09983	0.02000 –	0.01000 –
$f(x) = \dfrac{\sin x}{x}$	0.9933	0.9983	1.000 –	1.000 –

We have not included negative values of x, since $\sin x$ would then be negative and the ratio again positive. It looks as if L might be equal to unity. We now prove this, namely,

$$\lim_{x \to 0} \frac{\sin x}{x} = 1$$

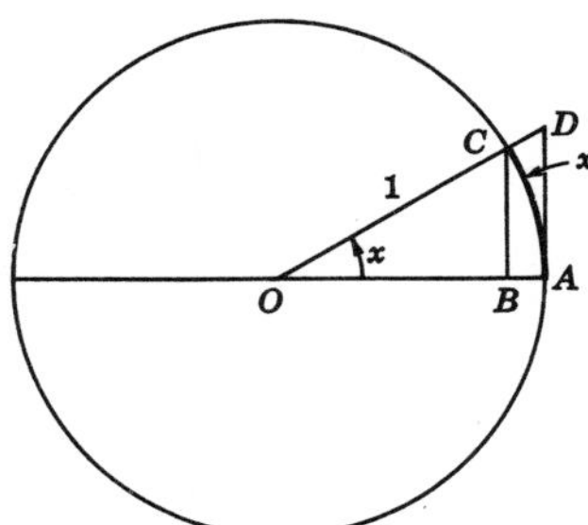

Figure **10.8**

Look at Fig. 10.8, which is drawn with x positive. We might have drawn a similar figure with x negative. It is evident that when $|x|$ is not too large and is not equal to zero,

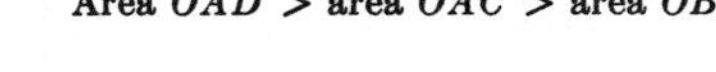

$$\text{Area } OAD > \text{area } OAC > \text{area } OBC$$

From the figure,

$$\text{Area } OAD = \tfrac{1}{2}OA \cdot OD = \tfrac{1}{2}|\tan x|$$

We must use $|\tan x|$, and not $\tan x$, since x may be negative. Also, since angle x is in radian measure,

$$\text{Area } OAC = \tfrac{1}{2}r|x| = \tfrac{1}{2}|x|$$

Finally,

$$\text{Area } OBC = \tfrac{1}{2}OB \cdot BC = \tfrac{1}{2}(\cos x) \cdot |\sin x|$$

Since $\cos x$ is positive, we need not write its absolute value. Therefore

$$|\tan x| \geq |x| > (\cos x) \cdot |\sin x|$$

Dividing this inequality through by $|\sin x|$, we obtain

$$\frac{1}{\cos x} > \left| \frac{x}{\sin x} \right| > \cos x$$

We already know that $\lim_{x \to 0} \cos x = 1$. Therefore, from the domination principle,

$$\lim_{x \to 0} \left| \frac{x}{\sin x} \right| = 1$$

or, from Theorem 11,

$$\lim_{x\to 0}\left|\frac{\sin x}{x}\right| = 1$$

Now x and $\sin x$ have the same signs for small x, and hence $\frac{\sin x}{x}$ is positive. Therefore, for small x ($x \neq 0$),

$$\frac{\sin x}{x} = \left|\frac{\sin x}{x}\right|$$

and thus

$$\lim_{x\to 0}\frac{\sin x}{x} = 1$$

Keep in mind that $\frac{\sin 0}{0} = \frac{0}{0}$ is indeterminate.

The $\lim_{x\to a} f(x)$ may fail to exist, just as in the case of a function of a discrete variable n, in any one of several ways such as by oscillating or by becoming infinite. For example, $\lim_{x\to 0}\cos\frac{1}{x}$ does not exist; as $x \to 0$, $\cos(1/x)$ oscillates between -1 and $+1$; $\lim_{x\to\pi/2}\tan\frac{1}{x-(\pi/2)}$ does not exist; as $x \to \pi/2$, $\tan\frac{1}{x-(\pi/2)}$ oscillates between $-\infty$ and $+\infty$; $\lim_{x\to 0}\frac{1}{x}$ does not exist; as $x \to 0$, $\left|\frac{1}{x}\right|$ grows larger without bound.

We end this section with another discussion of the meaning, if any, to be attached to the expressions $\frac{0}{c}, \frac{c}{0}, \frac{0}{0}$, where $c \neq 0$ (see Sec. 2.3).

First, $\frac{0}{c} = 0$.

Second, $\frac{c}{0}$ is not a possible operation; we have ruled out division by zero. Again we might think of this as $\lim_{x\to a}\frac{c}{g(x)} = \frac{c}{0}$, if $\lim_{x\to a} g(x) = 0$, but this does not help. It is true, however, that $\left|\frac{c}{g(x)}\right|$ can be made arbitrarily large by choosing x sufficiently near a. Sometimes we write $\left|\frac{c}{g(x)}\right| \to \infty$ as $x \to a$, but we still mean that $\lim_{x\to a}\frac{c}{g(x)} = \frac{c}{0}$ does not exist.

Third, $\frac{0}{0}$ is an indeterminate symbol as it stands. But again we could consider this as $\lim_{x\to a}\frac{f(x)}{g(x)} = \frac{0}{0}\cdot$ If $\lim_{x\to a} f(x) = 0$ and if $\lim_{x\to a} g(x) = 0$, some meaning may possibly be attached to $\lim_{x\to a} \frac{f(x)}{g(x)}$. It may be a perfectly good real number.

We have already met this situation in the case of

$$\lim_{x\to 0} \frac{\sin x}{x} = \left(\frac{0}{0}\right) = 1$$

(Illustration 7) and also in Illustration 4 where

$$\lim_{x\to 1} \frac{2 - \sqrt{5 - x}}{x - 1} = \left(\frac{0}{0}\right) = \frac{1}{4}$$

In the next chapter we shall return to this same problem, which plays a major role in the differential calculus.

Problems 10.9

In Probs. 1 to 18 make use of the theorems on limits and results in Sec. 10.9 to find the limit indicated.

1. $\lim_{x\to 1} (1 + 2x^2 - 3x^3 + x^4)$.

2. $\lim_{x\to -1} (\sqrt{2} + 3x^2 - x^4 + x^6)$.

3. $\lim_{x\to 2} x(2 + 3x)(3 - 2x)$.

4. $\lim_{x\to 3} x(4 - x^2)(x + 5)$.

5. $\lim_{x\to 1} \frac{7x^2(1 - x^2)}{4 + x}$.

6. $\lim_{x\to 1} \frac{x^3(1 - x^2)}{1 - x}$.

7. $\lim_{x\to 10} |x + 1|$.

8. $\lim_{x\to 20} \frac{|x + 1|}{|x|}$.

9. $\lim_{x\to 1} \left(x - \frac{x^3}{3!} + \frac{x^5}{5!} - \frac{x^7}{7!}\right)$.

10. $\lim_{x\to k} (a_0 + a_1x + \cdots + a_nx^n)$.

11. $\lim_{x\to (\pi/6)} (x + \sin x)$.

12. $\lim_{x\to (\pi/4)} (1 + \tan x)(1 + \cot x)$.

13. $\lim_{x\to (\pi/3)} \frac{1 + \cos 2x}{x}$.

14. $\lim_{x\to (\pi/2)} \tan 2x$.

15. $\lim_{x\to 0} \frac{x}{\sin x}$.

16. $\lim_{x\to 0} \frac{x \cos 2x}{\sin x}$.

17. $\lim_{x\to 0} \frac{4 \sin 3x}{x}$.

18. $\lim_{x\to 0} \sec x$.

19. $\lim_{x\to 5} \frac{1 - \sqrt{x - 4}}{x - 5}$. HINT: Rationalize numerator.

20. $\lim_{x\to -2} \frac{2 - \sqrt{2 - x}}{x + 2}$. HINT: Rationalize numerator.

In Probs. 21 to 25 sketch the graph.

21. $y = \sin \frac{1}{x}$, $x \neq 0$.

22. $y = \begin{cases} \sin \frac{1}{x}, & x \neq 0, \\ 1, & x = 0. \end{cases}$

23. $y = x \sin \frac{1}{x}, x \neq 0.$

24. $y = \begin{cases} x \sin \frac{1}{x}, & x \neq 0, \\ 0, & x = 0. \end{cases}$

25. $y = \begin{cases} x^2 \sin \frac{1}{x}, & x \neq 0, \\ 0, & x = 0. \end{cases}$

In Probs. 26 to 30 find the limit if it exists.

26. $\lim_{x \to \infty} \sin x.$

27. $\lim_{x \to 0} \sin \frac{1}{x}.$

28. $\lim_{x \to 0} x \sin \frac{1}{x}.$

29. $\lim_{x \to 0} x^2 \sin \frac{1}{x}.$

30. $\lim_{x \to \infty} \frac{1}{x} \sin x.$

31. Find the limit of the postage function (Sec. 6.1):

(a) As x approaches 2 from above.
(b) As x approaches 2 from below.

32. Consider a circular pie of radius R. A piece, with central angle θ, is cut out and is to be placed on a circular plate of least possible diameter D. How does D vary with θ? Show that

$$D(\theta) = \begin{cases} 0 & \theta = 0 \\ R \sec \frac{\theta}{2} & 0 < \theta \leq 90^\circ \\ 2R \sin \frac{\theta}{2} & 90^\circ < \theta \leq 180^\circ \\ 2R & 180^\circ < \theta \leq 360^\circ \end{cases}$$

Can you show that $\lim_{\theta \to 0} R \sec (\theta/2) = R$? Note that $D(0) = 0$, which is not the same as $\lim_{\theta \to 0} D(\theta)$.*

10.10 Continuity

In the previous sections we have found limits such as $\lim_{x \to 2} x$, $\lim_{x \to 2} x^2$, $\lim_{x \to 3} \frac{2x + 3}{x - 1}$, but you may have thought this process very strange. Why should we bother to show that $\lim_{x \to 2} x = 2$? Why not substitute $x = 2$ and get the answer at once? Similarly, why not substitute in $\sin x$ and get $\sin 0 = 0$? The answer to this is straightforward: finding the limiting value of $f(x)$ as $x \to a$ and finding $f(a)$ are two completely distinct operations. The results need not be equal even though they are equal in the illustrations! Even worse—consider Sec. 10.9, Illustration 7. Here

* See J. P. Ballantine, A Peculiar Function, *American Mathematical Monthly*, vol. 37, p. 350 (1930).

$\lim_{x \to 0} \frac{\sin x}{x} = 1$, but $\frac{\sin 0}{0}$ is complete nonsense. We can find the limit as $x \to 0$, but the function is not defined at $x = 0$. If we wished, we could have defined a new function F, where

$$F(x) = \begin{cases} \frac{\sin x}{x} & x \neq 0 \\ 6\pi & x = 0 \end{cases}$$

Then F is defined everywhere, and as before, $\lim_{x \to 0} F(x) = 1$. But, by our choice, $F(0) = 6\pi$.

Do not think this is just an artificiality. We may define a function any way we please, interval by interval or point by point, as we like (see Sec. 10.9, Prob. 32). The function f, where

$$f(x) = \begin{cases} \sin x & -\infty < x \leq -\pi \\ 0 & -\pi < x \leq 0 \\ x & 0 < x < \infty \end{cases}$$

is a perfectly good function.

Exercise A. Sketch the above function for $-2\pi < x < 2\pi$.

For an important class of functions, however, the two processes of finding $\lim_{x \to a} f(x)$ and of finding $f(a)$ give the same result. This class is the class of continuous functions defined below.

Definition: Let a be a point lying in an interval $]c, d[$, and let $f(x)$ be defined in this interval, except possibly at $x = a$. Then $f(x)$ is said to be *continuous* at a if $\lim_{x \to a} f(x)$ exists, $f(a)$ exists, and $\lim_{x \to a} f(x) = f(a)$. If $f(x)$ is not continuous at $x = a$, it is said to be *discontinuous* there.

If $f(x)$ is continuous at every point of an open or closed interval, it is said to be continuous in that interval.

Continuity of a function in an interval implies that the graph of the function in this interval is an uninterrupted curve. Hence knowledge about the continuity of functions is of great assistance to us in plotting their graphs.

In Illustration 7, where $f(x) = (\sin x)/x$, $x \neq 0$, f is not defined at the origin; f is therefore not continuous at $x = 0$ and hence must be discontinuous there. The function F, where

$$F(x) = \begin{cases} \frac{\sin x}{x} & x \neq 0 \\ 6\pi & x = 0 \end{cases}$$

is discontinuous at $x = 0$ since $\lim_{x\to 0} F(x) = 1 \neq F(0) = 6\pi$. The function G, where

$$G(x) = \begin{cases} \dfrac{\sin x}{x} & x \neq 0 \\ 1 & x = 0 \end{cases}$$

is continuous at $x = 0$, because $\lim_{x\to 0} G(x) = 1 = G(0) = 1$. For the function G, which is continuous at $x = 0$, we can compute $\lim_{x\to 0} G(x)$ by substituting $x = 0$. Reread the definition of continuity. This explains why, in the case where $f(x) = x^2$, we got the same answer whether we computed $\lim_{x\to 2} f(x)$ or $f(2)$. The function x^2 is continuous at $x = 2$. But continuity at $x = 2$ must be proved first before the operation of finding $f(2)$ can be used to evaluate $\lim_{x\to 2} f(x)$. Be sure you understand that, in proving a function continuous at $x = a$, you must do two things: first, compute $\lim_{x\to a} f(x)$; second, compute $f(a)$. If and only if these exist and are equal is the function continuous at $x = a$. After and only after we have learned or proved that a particular function has the continuity property are we entitled to assume that $\lim_{x\to a} f(x) = f(a)$. Of these two expressions, $f(a)$ is usually easier to evaluate.

From Theorems 8 to 11 we can conclude the truth of Theorem 13.

Theorem 13. The sum, difference, and product of two functions $f(x)$ and $g(x)$ which are continuous at $x = a$ are continuous at $x = a$. The quotient $f(x)/g(x)$ is continuous at $x = a$ unless $g(a) = 0$.

Theorem 14. A polynomial function

$$P(x) = a_0x^n + a_1x^{n-1} + \cdots + a_n$$

defined over the real numbers is continuous for all values of x. A rational function

$$R(x) = \frac{P(x)}{Q(x)}$$

defined over the real numbers is continuous for all real values of x except for the zeros of $Q(x)$.

Proof: To show that a polynomial is continuous, we begin by recalling that $f(x) = x$ is continuous for all x.

Exercise B. Prove that the constant function $f(x) = C$ is continuous for all x.

Then we appeal to Theorem 13 and prove the continuity of x^2, x^3, . . . , x^n, where n is a positive integer. Similarly, the functions a_0x^n, a_1x^{n-1}, etc., are continuous, and so is the function which is the sum of these. Therefore $P(x)$ is continuous for all x. By Theorem 13, $R(x)$ is continuous except at points where $Q(x) = 0$.

Although we shall not give the proofs, let us state the continuity properties of other functions of a real variable which we have met in earlier chapters.

Function	**Domain of continuity**
$\sqrt{x}$	All $x > 0$
$\lvert x\rvert$	All x
$[x]$	All x except x equal to an integer
e^x	All x
$\log x$	All $x > 0$
$\sin x$	All x
$\cos x$	All x
$\tan x$	All x except $x = (2n + 1)(\pi/2)$
$\cot x$	All x except $x = n\pi$
$\sec x$	All x except $x = (2n + 1)(\pi/2)$
$\csc x$	All x except $x = n\pi$
arc Sin x	$-1 < x < 1$
arc Cos x	$-1 < x < 1$
arc Tan x	All x
$\sinh x$	All x
$\cosh x$	All x

In order to deal with more complicated functions, we need the following theorem, which we state without proof.

Theorem 15. If f and g are two continuous functions of x, then the composite function h defined by $h = g[f(x)]$ is continuous and we write

$$\lim_{x\to a} g[f(x)] = g[\lim_{x\to a} f(x)] = g[f(a)]$$

For example:

(**a**) $\displaystyle \lim_{x\to 0} \sin\left(x + \frac{\pi}{2}\right) = \sin\left[\lim_{x\to 0}\left(x + \frac{\pi}{2}\right)\right] = \sin\frac{\pi}{2} = 1.$

(**b**) $\displaystyle \lim_{x\to 1} \log (x^2 + 2) = \log [\lim_{x\to 1} (x^2 + 2)] = \log 3.$

Finally, we state without proof two more theorems which are intuitively reasonable and which are useful in later mathematical work.

Theorem 16. A continuous function on a closed interval $[a, b]$ attains a maximum value and a minimum value (and every value between the maximum and the minimum) somewhere in the interval.

Theorem 17. If f is continuous in $[a, b]$ and if $f(a) < 0$ and $f(b) > 0$, then for some x_1 in $[a, b]$, $f(x_1) = 0$.

Geometrically, the last says that a continuous curve which starts below the X-axis and ends up above the X-axis must cross it somewhere in between.

We illustrate with the following functions and their graphs.

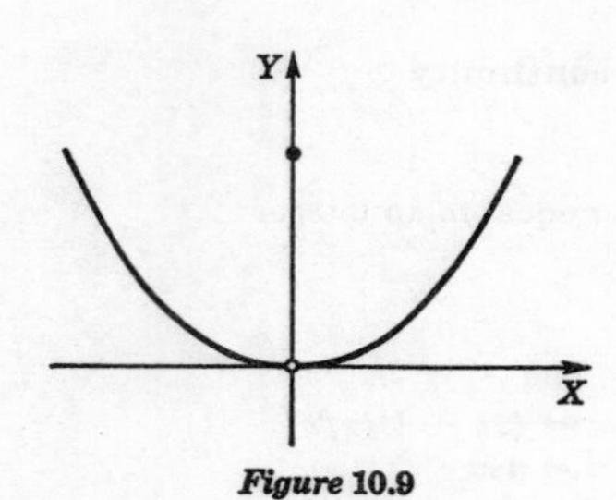

Figure 10.9

$$f(x) = \begin{cases} x^2 & x \neq 0 \\ 1 & x = 0 \end{cases}$$

Domain of definition: $-\infty < x < \infty$.
Point of discontinuity: $x = 0$.
f has neither maximum nor minimum.

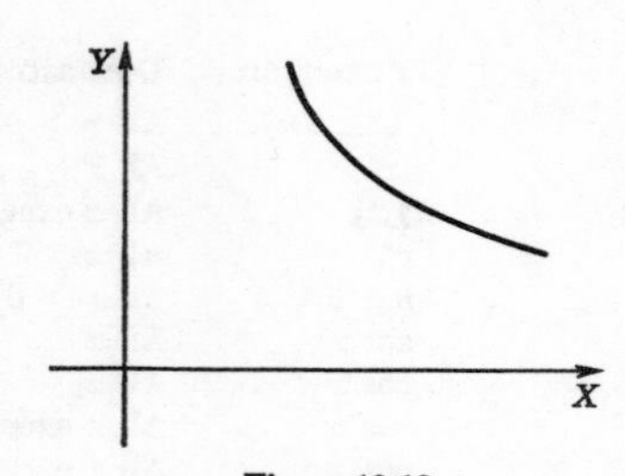

Figure 10.10

$$f(x) = \frac{1}{x} \qquad x > 0$$

Domain of definition: $0 < x < \infty$.
f is continuous.
f has neither maximum nor minimum.

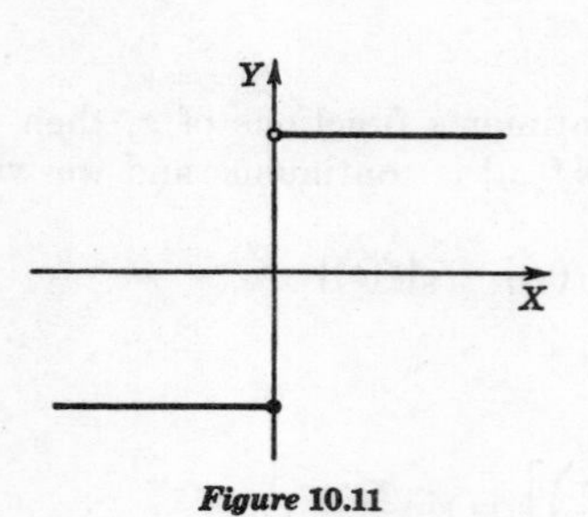

Figure 10.11

$$f(x) = \begin{cases} -1 & x \leq 0 \\ 1 & x > 0 \end{cases}$$

Domain of definition: $-\infty < x < \infty$.
Point of discontinuity: $x = 0$.
f is never zero.
f has both a maximum and a minimum.

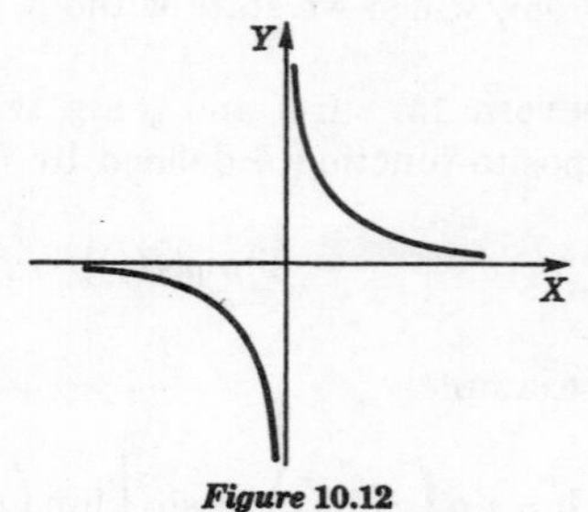

Figure 10.12

$$f(x) = \frac{1}{x} \qquad x \neq 0$$

Domain of definition: $-\infty < x < \infty$, $x \neq 0$.
Point of discontinuity: $x = 0$.
f is never zero.
f has neither maximum nor minimum.

Problems 10.10

1. Prove that $\lim_{x\to 0} \dfrac{ax^n + bx^{n-1} + \cdots + c}{Ax^n + Bx^{n-1} + \cdots + C} = \dfrac{c}{C}$, $C \neq 0$.
2. Prove that $\lim_{x\to \infty} \dfrac{ax^n + bx^{n-1} + \cdots + c}{Ax^n + Bx^{n-1} + \cdots + C} = \dfrac{a}{A}$, $A \neq 0$. HINT: Divide numerator and denominator by x^n; now let $x \to \infty$.

In Probs. 3 to 18 find the limit.

3. $\lim_{x\to 2} \dfrac{x^3 + 2x^2 - 3x - 4}{x + 2}$.
4. $\lim_{x\to -1} \dfrac{x^5 + 2x^4 + x^3 - 3x^2 - 2x}{x^7 + x^3 + 1}$.
5. $\lim_{x\to -2} \dfrac{x + 2}{x^2 - 4x - 12}$.
6. $\lim_{x\to 3} \dfrac{x - 3}{x^3 - 2x^2 - 10x + 21}$.
7. $\lim_{x\to (\pi/3)} \sin\left(x - \dfrac{\pi}{3}\right)$.
8. $\sin\left[\lim_{x\to (\pi/3)} \left(x - \dfrac{\pi}{3}\right)\right]$.
9. $\lim_{x\to (\pi/6)} \cos\left(x + \dfrac{\pi}{6}\right)$.
10. $\cos\left[\lim_{x\to (\pi/6)} \left(x + \dfrac{\pi}{6}\right)\right]$.
11. $\lim_{x\to \frac{1}{2}} \dfrac{\text{arc Sin } x}{x}$.
12. $\lim_{x\to \frac{1}{2}} x \text{ arc Cos } x$.
13. $\lim_{x\to 0} \log \dfrac{1 + x}{1 - x}$.
14. $\lim_{x\to 99} \log \sqrt{1 + x}$.
15. $\lim_{x\to (\pi/4)} \log \sin\left(x + \dfrac{\pi}{4}\right)$.
16. $\lim_{x\to (\pi/4)} \log |\cos 2x|$.
17. $\lim_{x\to 1} \log \text{arc Sin } x$.
18. $\lim_{x\to \frac{1}{2}} \log \text{arc Cos } x$.

In Probs. 19 to 22, examine for continuity at the point indicated.

19*. $f(x) = \begin{Bmatrix} \dfrac{x^2 - 1}{x(x - 1)}, & x \neq 1 \\ 2, & x = 1 \end{Bmatrix}$ at $x = 1$.

20*. $f(x) = \begin{Bmatrix} \dfrac{x(x + 1)}{x^2 - 1}, & x \neq 1 \\ \frac{1}{2} & x = 1 \end{Bmatrix}$ at $x = -1$.

21*. $f(x) = \begin{Bmatrix} \dfrac{x(x + 1)}{x^2 - 1}, & x \neq 1 \\ \frac{1}{2} & x = 1 \end{Bmatrix}$ at $x = 1$.

22*. $f(x) = \begin{Bmatrix} \dfrac{x(x + 2)}{x^2 - 4}, & x \neq -2 \\ \frac{1}{2}, & x = -2 \end{Bmatrix}$ at $x = -2$.

23*. $f(x) = \begin{Bmatrix} \dfrac{x(x + 2)}{x^2 - 4}, & x \neq -2 \\ \frac{1}{2}, & x = -2 \end{Bmatrix}$ at $x = 2$.

24*. $f(x) = \begin{Bmatrix} \dfrac{x^2 - 4}{x(x - 2)}, & x \neq 2 \\ \frac{1}{2}, & x = 2 \end{Bmatrix}$ at $x = 2$.

In Probs. 25 to 29 what are the points of discontinuity?

25. $\log |x - 1|$.

26. $\tan \left(x + \frac{\pi}{4}\right)$.

27. $\tan 3x$.

28. $\dfrac{x(x + 2)}{(x - 1)(x + 2)}$.

29. $\dfrac{x(x + 2)}{(x - 1)(x - 2)}$.

30. Under which of the following conditions is $\lim_{x \to 0} f(x) = c$?

(a) $|f(0)| = c$.
(b) $f(0) = c$.
(c) If, for every positive number ϵ, there exists a δ such that $|f(x)| - c < \epsilon$ when $|x| < \delta$.
(d) If, for every positive number ϵ, there exists a δ such that $|f(x) - c| < \epsilon$ when $0 < |x| < \delta$.
(e) If, for every positive number ϵ, there exists a δ such that $|f(x) - c| < \epsilon$ when $|x - c| < \delta$.

10.11 Area

In the introduction to this chapter we mentioned the Greeks' concern with limits and indicated Archimedes' method for finding the area of the parabola. In Book V of Euclid the following definition of the area of a circle is given:

Definition: The area of a circle is the limiting value of the area of an inscribed (or circumscribed) regular polygon of n sides as the number of sides n is increased indefinitely.

It was by means of inscribed (and circumscribed) polygons that Archimedes approximated π (Sec. 10.2, Prob. 4). But for more general curves we shall need another definition for area. To this end let us think of a continuous function f which is positive in the closed interval $[a,b]$, with $a < b$. We indicate its graph in Fig. 10.13, erecting the ordinates at the points a and b, which we agree to designate also as x_0 and x_n, respectively. It is for such a closed region, bounded by $y = f(x)$, $x = a$, $x = b$, $y = 0$, that we wish to formulate a definition of a quantity that will correspond to our intuitive notion of area. The interval $[a,b]$ is subdivided into n equal parts by the insertion of the $n - 1$ points $x_1, x_2, \ldots, x_{n-1}$, so that $x_0(= a) < x_1 < x_2 < \cdots < x_{n-1} < x_n(= b)$. Ordinates are erected at the new points of subdivision. Now, by Theorem 16, the function f has a maximum and a minimum value in each subinterval. Let these occur, in the general ith interval $[x_{i-1},x_i]$, $i = 1, 2, \ldots, n$, at the points x_{M_i} and x_{m_i}, respectively. In $[x_{i-1},x_i]$ the maximum value of f is $f(x_{M_i})$ and the minimum value is $f(x_{m_i})$, and further $f(x_{m_i}) \leq f(x_{M_i})$. Form the rectangles with these maximum and minimum values as heights and common width $(x_i - x_{i-1})$. Clearly, the areas of any two rectangles with the same

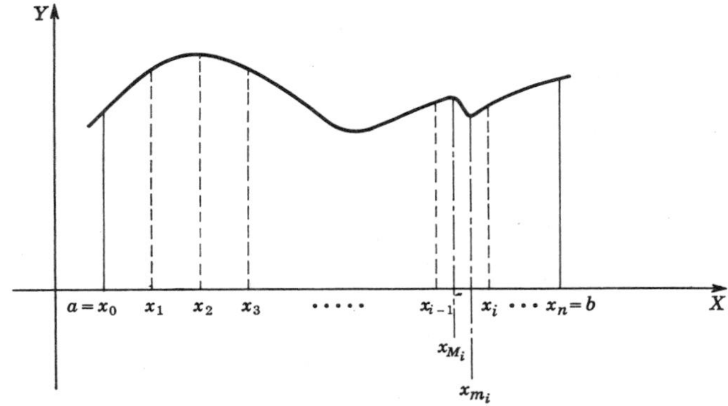

***Figure* 10.13**

base satisfy the inequality $f(x_{m_i})(x_i - x_{i-1}) \leq f(x_{M_i})(x_i - x_{i-1})$, and thus for the sum of all of the n such sets, we have

$$\begin{aligned} f(x_{m_1})(x_1 - x_0) + f(x_{m_2})(x_2 - x_1) + \cdots + f(x_{m_i})(x_i - x_{i-1}) + \cdots \\ + f(x_{m_n})(x_n - x_{n-1}) \leq f(x_{M_1})(x_1 - x_0) + f(x_{M_2})(x_2 - x_1) + \cdots \\ + f(x_{M_i})(x_i - x_{i-1}) + \cdots + f(x_{M_n})(x_n - x_{n-1}) \end{aligned}$$

In order to simplify the notation, we now introduce the Σ notation. The Greek Σ (corresponding to S) is used to indicate "the sum of". A dummy subscript, often i, is used with Σ. Thus $\sum_{i=1}^{3} x^i$ is read "the sum of x to the ith power, i running from 1 to 3". This is, when written out,

$$\sum_{i=1}^{3} x^i = x + x^2 + x^3$$

Again,

$$\sum_{i=0}^{n-1} ar^i = a + ar + ar^2 + \cdots + ar^{n-1}$$

is just the geometric series to n terms. The $n + 1$ terms in the arithmetic progression with a as first term and common difference d can be written, in summation form,

$$\sum_{i=0}^{n} (a + id) = a + (a + d) + (a + 2d) + \cdots + (a + nd)$$

A more complicated illustration is

$$\sin x = x - \frac{x^3}{3!} + \frac{x^5}{5!} - \cdots + (-1)^{n+1}\frac{x^{2n-1}}{(2n-1)!} + \cdots$$

$$= \sum_{i=1}^{\infty} (-1)^{i+1}\frac{x^{2i-1}}{(2i-1)!}$$

Exercise A. Write $e^x = 1 + x + \frac{x^2}{2!} + \cdots + \frac{x^n}{n!} + \cdots$ in summation notation.

Exercise B. Write $\cos x = 1 - \frac{x^2}{2!} + \frac{x^4}{4!} - \cdots + (-1)^{n+1}\frac{x^{2n-2}}{(2n-2)!} + \cdots$ in summation notation.

We now write the above inequality in Σ notation:

$$\sum_{i=1}^{n} f(x_{m_i})(x_i - x_{i-1}) \leq \sum_{i=1}^{n} f(x_{M_i})(x_i - x_{i-1}) \tag{13}$$

Exercise C. When (and only when) would the equality sign in (13) hold?

Here again, as so often before, we wish to appeal to our intuition. There is something that we should be able to express as a number A associated with the closed figure bounded by the curve $y = f(x)$ and by the three straight lines $x = a$, $x = b$, $y = 0$, and which we should like to call *area*. Our intuition tells us that the portion of this area above the segment $[x_{i-1}, x_i]$ should lie between the numbers $f(x_{m_i})(x_i - x_{i-1})$ and $f(x_{M_i})(x_i - x_{i-1})$. Hence we wish to have, for the whole area A,

$$\sum_{i=1}^{n} f(x_{m_i})(x_i - x_{i-1}) \leq A \leq \sum_{i=1}^{n} f(x_{M_i})(x_i - x_{i-1}) \tag{14}$$

As a matter of convenience in notation, let us designate by s_n the left-hand member of (14) and by S_n the right-hand member. Thus (14) becomes

$$s_n \leq A \leq S_n \tag{15}$$

Let us consider the effect on (15) of further subdividing each interval into two (equal) intervals. In the general interval $[x_{i-1}, x_i]$, let the midpoint be $\bar{x}_i$. Now the maximum value of f in $[x_{i-1}, \bar{x}_i]$ is equal to or less than the maximum in the whole interval $[x_{i-1}, x_i]$, and a like statement applies to the interval $[\bar{x}_i, x_i]$. Further, the minimum value of f in either of these two subintervals is at least as great as the minimum value in the whole interval. Hence

$$s_n \leq s_{2n} \leq A \leq S_{2n} \leq S_n \tag{16}$$

If this process of doubling the number of subdivisions is continued indefinitely, we arrive at two sequences of real numbers:

$$\left.\begin{array}{l}(1)\ s_n,\ s_{2n},\ s_{4n},\ \ldots,\ s_{pn},\ \ldots \\ (2)\ S_n,\ S_{2n},\ S_{4n},\ \ldots,\ S_{pn},\ \ldots\end{array}\right\} \text{where } p = 2^k$$

The sequence (1) is monotone increasing, and the sequence (2) is monotone decreasing. From theorems about sequences which we have not presented to you, we can conclude that:

(3) The limit $s = \lim_{k=\infty} s_{pn}$ exists, where $p = 2^k$.

(4) The limit $S = \lim_{k=\infty} S_{pn}$ exists, where $p = 2^k$.

(5) $s \leq S$.

Our intuition tells us that, for a well-behaved function,

$$\lim_{k \to \infty} (s_{pn} - S_{pn}) = 0$$

so that s should equal S, and that A should be defined to be the common value of s and S. Since it can be proved that s does equal S when $f(x)$ is continuous in $[a, b]$, we are confident that we are on the right track. It looks, however, as if this limit might depend upon the original number of subdivisions n, or upon the fact that these divide $[a, b]$ into equal parts. That this is not the case is difficult to prove. The fact is, one gets the same limit for any method of subdividing $[a, b]$ and for any method of refining this subdivision, provided that the lengths of the subintervals of the subdivision *all* tend to zero in the limiting process. Moreover, the same limit is obtained even though we go further and dispense with $f(x_{m_i})$ and $f(x_{M_i})$ in forming the heights of the rectangles but use instead the value $f(x_i^*)$, where x_i^* is any point whatsoever in the interval $[x_{i-1}, x_i]$. Thus we formulate the following general definition of area in terms of the notations just discussed.

Definition: The area enclosed by $y = f(x)$, $x = a$, $x = b$, $y = 0$, where f is continuous in $[a, b]$, is

$$A_a^b = \lim_{n \to \infty} \sum_{i=1}^{n} f(x_i^*)(x_i - x_{i-1}) \tag{17}$$

where the length of the greatest interval $[x_{k-1}, x_k]$ approaches zero and where $x_{i-1} \leq x_i^* \leq x_i$.

Of course, in a given problem, we may take equal intervals if we please and may also specialize x_i^* to be, say, the right-hand end point. Actually, at best, the direct evaluation of (17) is generally tedious or seemingly impossible, and therefore, in the next chapter, we shall develop a very powerful method of doing this by the calculus. At present we shall illustrate (17) in only one example, namely, that of Sec. 10.2, Prob. 1.

Illustration 1. Find the area enclosed by $y = x^3$, $x = 0$, $x = 1$, $y = 0$ (Fig. 10.14).

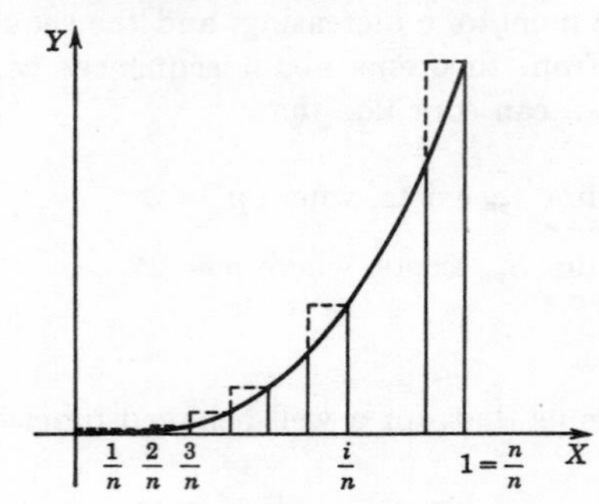

Figure 10.14

Solution: Form n equal intervals with the points

$$0 < \frac{1}{n} < \frac{2}{n} < \cdots < \frac{i}{n} < \cdots < \frac{n-1}{n} < 1$$

and form the product $f\left(\frac{i}{n}\right)(x_i - x_{i-1}) = \left(\frac{i}{n}\right)^3\left(\frac{1}{n}\right)$. We have, approximately,

$$A \approx \sum_{i=1}^{n} \frac{i^3}{n^4} = \frac{1}{n^4}[1^3 + 2^3 + \cdots + n^3]$$

Exactly, the area will be

$$\begin{aligned}
A &= \lim_{n\to\infty} \frac{1}{n^4}[1^3 + 2^3 + \cdots + n^3] \\
&= \lim_{n\to\infty} \frac{1}{n^4} \cdot \frac{n^2(n+1)^2}{4} \\
&= \lim_{n\to\infty} \frac{1}{4} \cdot \frac{(n+1)^2}{n^2} \\
&= \lim_{n\to\infty} \frac{1}{4} \cdot \frac{n^2 + 2n + 1}{n^2} \\
&= \lim_{n\to\infty} \frac{1}{4} \cdot \left[1 + \frac{2}{n} + \frac{1}{n^2}\right] \\
&= \frac{1}{4} \text{ square unit}
\end{aligned}$$

Conceivably, the definition of area given would not yield results for the triangle and rectangle, among other figures, consistent with previous definitions for these. In doing Probs. 8 and 9 below, you will find that the new definition does give the answers you have already accepted for these quantities. Thus we have a consistency proof of our generalized definition of area.

It should be recalled that the limit (17) in the definition always exists in the event the function is continuous, but that it fails to exist for some functions. We may extend our definition by saying that, for *any* given function, if the $\lim_{n \to \infty} \sum_{i=1}^{n} f(x_i^*)(x_i - x_{i-1})$ does exist, and has the same value regardless of the method of subdivision or choice of x_i^*, then we shall call its value the *area* of the associated region.

Problems 10.11

In Probs. 1 to 7 use the definition to find the following:

1. The area bounded by $y = x^2$, $y = 0$, $x = a$, $a > 0$.
2. The area bounded by $y = x^2$, $y = 0$, $x = a$, $x = b$, $b > a$.
3. The area bounded by $y = x^3$, $y = 0$, $x = a$, $a > 0$.
4. The area bounded by $y = x^3$, $y = 0$, $x = a$, $x = b$, $b > a$.
5. The area bounded by $y = x^4$, $y = 0$, $x = 1$, given

$$\sum_{k=1}^{n} k^4 = \frac{6n^5 + 15n^4 + 10n^3 - n}{30}$$

6. The area bounded by $y = x^4$, $y = 0$, $x = a$, $a > 0$ (see Prob. 5 above).
7. The area bounded by $y = x^4$, $y = 0$, $x = a$, $x = b$, $b > a$ (see Prob. 5 above).
8. Use the definition to verify that the area of the rectangle bounded by $y = a$, $y = 0$, $x = 0$, $x = b$ is ab, where $a > 0$, $b > 0$.
9. Use the definition to verify that the area of the triangle bounded by

$$y = (h/b)x,\ y = 0,\ x = b \qquad \text{is } \tfrac{1}{2}hb \qquad \text{where } h > 0,\ b > 0$$

10. Use the definition to verify that the area bounded by $y = |x|$, $y = 0$, $x = -2$, $x = 1$ is $\frac{1}{2}(2)(2) + \frac{1}{2}(1)(1)$.

10.12 Rates

When a particle moves, there are associated with the motion certain quantities such as time, distance, velocity, and acceleration. We shall restrict ourselves to the case where the motion takes place on a straight line, since we are unprepared at this time to consider general curvilinear motion.

Let $y = f(t)$ give the position of the particle on the Y-axis at any time t. The time variable is measured continuously by a clock and is usually thought of as positive or zero, although on occasion we may want to assign a negative value in order to describe a past event. The y-coordinate is a linear distance positive, negative, or zero from some fixed point on the line called the origin (Fig. 10.15). Suppose the particle to be at $y = f(t_1)$ and $y = f(t_1 + h)$, when t is t_1 and $t_1 + h$, respectively. Then the particle has moved $f(t_1 + h) - f(t_1)$ units of distance in $h > 0$ units of time.

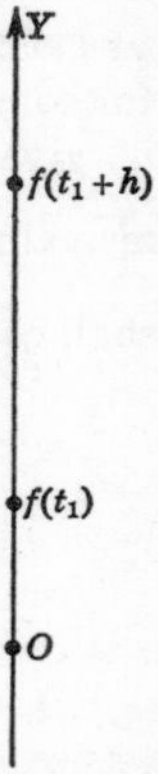

Y

$f(t_1+h)$

$f(t_1)$

O

Figure 10.15

Definition: If a particle moves a distance of

$$f(t_1 + h) - f(t_1)$$

in time h, then the ratio

$$\bar{v} = \frac{f(t_1 + h) - f(t_1)}{h} \tag{18}$$

is called the average velocity during the time interval h.

Average velocity is thus the change in distance per unit change in time. Units often encountered are miles per hour, centimeters per second, etc. These are abbreviated mi/hr, cm/sec, etc. Since distance may be negative, so also velocity may be negative. If only the absolute values of the distances are used, then average velocity is called average speed.

Now average velocity (and also average speed) is an interval property, since it describes what happens in an interval of time. Hence it cannot directly explain such a statement as "exactly at that instant the plane was traveling at 500 mi/hr", because there is no interval of time involved in this observation. And yet, intuitively, the statement does have some sense. It seems to say that, if the plane had continued at the same (constant) speed as it was traveling at that instant, then it would have covered 500 miles every hour thereafter. But this does not supply an answer to the inherent difficulty in the notion of traveling at 500 mi/hr *at* a certain (clock) value, say, t_1, of the time variable. But let us think of a small interval of time $[t_1, t_1 + h_1]$, where $h_1 > 0$, and the average velocity v_1 during this interval. Then consider the average velocity v_2 in the interval $[t_1, t_1 + h_2]$, where $h_1 > h_2 > 0$. In turn, consider intervals $[t_1, t_1 + h_3]$, $[t_1, t_1 + h_4]$, . . . , $[t_1, t_1 + h_n]$, . . . , where $h_1 > h_2 > h_3 > \ldots > h_n > 0$ and the corresponding average velocities $v_1, v_2, v_3, v_4, \ldots$. If this sequence of average velocities has a limit v, we may define this to be the velocity at t_1. This definition, indeed, corresponds to our intuitive idea of a velocity which cannot be measured directly but which must be approximated by a sequence of closer and closer average velocities.

This definition, however, does depend upon the particular sequence chosen, and possibly different sequences would lead to different answers. Hence it is preferable to require that we know y as a function of t and, in terms of this knowledge, define velocity in the following fashion:

Definition: Given distance y as a function f of t, velocity for a particular value of t, say, t_1, is defined to be

$$v(t_1) = \lim_{h \to 0} \frac{f(t_1 + h) - f(t_1)}{h} \tag{19}$$

provided this limit exists.

Remark. If we substitute $h = 0$ in the expression $\dfrac{f(t_1 + h) - f(t_1)}{h}$, it takes the meaningless form 0/0. As we have discussed before, however, the *limit* of this expression may still have meaning and be of great value.

The concept of acceleration is no more difficult to grasp mathematically than that of velocity. It is known to be the rate at which velocity is changing. To describe this precisely, let us compute the two values of instantaneous velocity $v(t_1)$ and $v(t_1 + h)$ corresponding to the two values of t, namely, t_1 and $t_1 + h$.

Definition: The ratio

$$\bar{a} = \frac{v(t_1 + h) - v(t_1)}{h} \tag{20}$$

is called the average acceleration during the interval h. It may be positive, negative, or zero.

Definition: The instantaneous acceleration, or simply, acceleration at t_1, is defined by

$$a(t_1) = \lim_{h \to 0} \frac{v(t_1 + h) - v(t_1)}{h} \tag{21}$$

provided this limit exists.

Average acceleration is an interval property. Instantaneous acceleration is a point property; it is a limit. The unit of acceleration is "units of velocity per unit of time", such as feet per second per second, miles per hour per minute, etc. These are abbreviated ft/sec/sec, or ft/sec^2; mi/hr/min; etc.

To summarize, velocity is rate of change of distance with respect to time. Acceleration is rate of change of velocity with respect to time.

Illustration 1. A particle moves vertically (up and down) in a straight line under the following law of motion: $y = 8t - t^2$, where t is in seconds and y is in feet. Find

(a) The velocity at any time t_1.
(b) The acceleration at any time t_1.
(c) The set of values of $t > 0$ for which velocity is positive.
(d) Maximum value of y.

Solution:

(a)
$$v = \lim_{h\to 0} \frac{f(t_1 + h) - f(t_1)}{h} = \lim_{h\to 0} \frac{[8(t_1 + h) - (t_1 + h)^2] - [8t_1 - t_1^2]}{h}$$
$$= \lim_{h\to 0} \frac{8t_1 + 8h - t_1^2 - 2t_1h - h^2 - 8t_1 + t_1^2}{h}$$
$$= \lim_{h\to 0} \frac{8h - 2t_1h - h^2}{h}$$
$$= \lim_{h\to 0} 8 - 2t_1 - h \qquad h \neq 0$$
$$= 8 - 2t_1 \qquad \text{ft/sec}$$

(b)
$$a = \lim_{h\to 0} \frac{v(t_1 + h) - v(t_1)}{h} = \lim_{h\to 0} \frac{[8 - 2(t_1 + h)] - [8 - 2t_1]}{h}$$
$$a = \lim_{h\to 0} \frac{8 - 2t_1 - 2h - 8 + 2t_1}{h}$$
$$= \lim_{h\to 0} \frac{-2h}{h} = \lim_{h\to 0} (-2) \qquad h \neq 0$$
$$= -2 \text{ ft/sec/sec}$$

(c) $v = 8 - 2t > 0$, or $t < 4$ sec. Since also $t > 0$, the answer is: $0 < t < 4$.

(d) The particle is at the origin when $t = 0$ and again when $t = 8$. Since it helps to sketch the graph of $y = 8t - t^2$, we do so in Fig. 10.16; however, this graph does

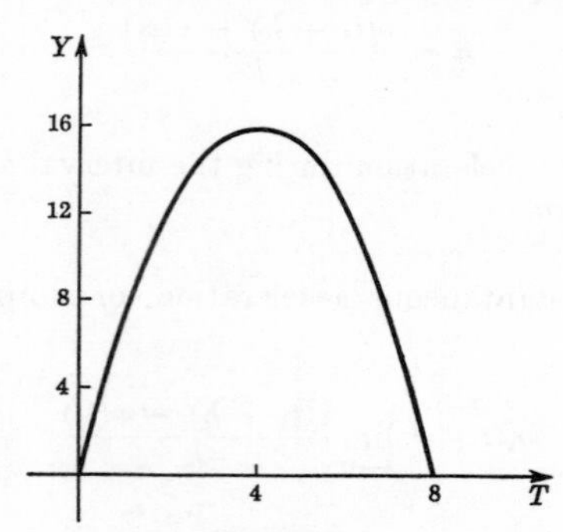

Figure 10.16

not represent the curve traversed by the particle—it simply shows more clearly how high the particle is at time t. The particle evidently rises to some maximum height, then falls back down again, reaching the "ground" in 8 sec. The velocity is positive ($0 < t < 4$) going up, negative ($4 < t < 8$) coming down. It therefore reached its maximum height at $t = 4$ when the velocity was zero. Its maximum height was $8 \cdot 4 - 4^2 = 16$ ft.

Problems 10.12

Given: Distance y, ft, from origin ($t = 0$) of a particle at time t, sec	Find: (a) Average velocity $\bar{v}$ during interval from:	(b) Average velocity $\bar{v}$ during interval from:	(c) Instantaneous velocity v at:
1. $y = 16t^2$	$t = 0$ to $t = 1$	$t = 1$ to $t = 2$	$t = 3$
2. $y = 16t^2 + 96t$	$t = 0$ to $t = 1$	$t = 0$ to $t = 2$	$t = 3$
3. $y = 16t^2 + 96t + 160$	$t = 0$ to $t = 1$	$t = 0$ to $t = 2$	$t = 2$
4. $y = 16t^2 - 96t + 160$	$t = 0$ to $t = 1$	$t = 0$ to $t = 2$	$t = 1$
5. $y = 6t - 3t^2$	$t = 0$ to $t = \frac{1}{2}$	$t = \frac{1}{2}$ to $t = 1$	$t = 1$
6. $y = 3t - 2t^2$	$t = 0$ to $t = 2$	$t = 1$ to $t = \frac{3}{2}$	$t = \frac{1}{3}$
7. $y = 16{,}000 - 16t^2$	$t = 0$ to $t = t_1$	$t = t_1$ to $t = t_2$	$t = t_1$
8. $y = 10{,}000 + 1{,}600t - 16t^2$	$t = 0$ to $t = t_1$	$t = t_1$ to $t = t_2$	$t = t_2$
9. $y = a + bt$	$t = 0$ to $t = t_1$	$t = t_1$ to $t = t_2$	$t = t_1$
10. $y = a + bt + ct^2$	$t = 0$ to $t = t_1$	$t = t_1$ to $t = t_2$	$t = t_2$

Given: Velocity v, ft/sec, at time t, sec	Find: (a) Velocity v at:	(b) Average acceleration $\bar{a}$ during interval from:	(c) Acceleration a at:
11. $v = 5 + 3t$	$t = 0$	$t = 0$ to $t = 1$	$t = 2$
12. $v = 6 - 3t$	$t = 2$	$t = 0$ to $t = 2$	$t = 2$
13. $v = 32t$	$t = 1$	$t = 0$ to $t = 1$	$t = 1$
14. $v = 32t + 100$	$t = 2$	$t = 0$ to $t = 2$	$t = 2$
15. $v = 100 - 32t$	$t = 3$	$t = 0$ to $t = 3$	$t = t_1$
16. $v = 32t - 32{,}000$	$t = 0$	$t = 0$ to $t = 2$	$t = t_1$
17. $v = 16{,}000 - 32t$	$t = 100$	$t = 0$ to $t = 100$	$t = t_1$
18. $v = t(t - 1)$	$t = 2$	$t = 0$ to $t = 100$	$t = t_1$
19. $v = a + bt$	$t = 0$	$t = 0$ to $t = 10$	$t = t_1$
20. $v = a + bt + ct^2$	$t = 1$	$t = 1$ to $t = 10$	$t = t_1$

Given: Distance y, ft, of a particle at time t, sec	Find: Distance at:	Velocity at:	Acceleration at:
21. $y = t^2 + t + 1$	$t = 1$	$t = t_1$	$t = 1$
22. $y = 3t^2 + 2t + 1$	$t = 1$	$t = t_1$	$t = 1$
23. $y = 4t^2 - 2t + 10$	$t = 1$	$t = t_1$	$t = t_1$
24. $y = 10t^2 - 5t - 100$	$t = t_1$	$t = t_1$	$t = t_1$
25. $y = a + bt + ct^2$	$t = t_1$	$t = t_1$	$t = t_1$
26. $y = a + bt$	$t = t_1$	$t = t_1$	$t = t_1$
27. $y = at^3$	$t = t_1$	$t = t_1$	$t = t_1$
28. $y = at^4$	$t = t_1$	$t = t_1$	$t = t_1$
29*. $y = t^n$, n a positive integer	$t = t_1$	$t = t_1$	$t = t_1$
30*. $y = t^{-n}$, n a positive integer	$t = t_1$	$t = t_1$	$t = t_1$

10.13 Tangent to a Curve

As a last example of how limits are used even in the very formulation of mathematical concepts we shall define what we mean by a line tangent to a curve at a point (tangent to a curve). From your study of plane geometry you may remember some definition of a tangent to a circle.

Exercise A. Write out what you believe to be the definition of a tangent to a circle that you came across in your study of plane geometry.

The notion is, of course, an intuitive one: we feel that a curve, though turning and bending, should have some sort of nearly constant direction in a very small interval. As it has so often in the past, our intuition gives us a clue.

Consider a curve C, such as is pictured in Fig. 10.17, and draw the line PQ, called a secant line. If Q were made to trace the curve until it approached the point P, the secant PQ would take different positions therewith.

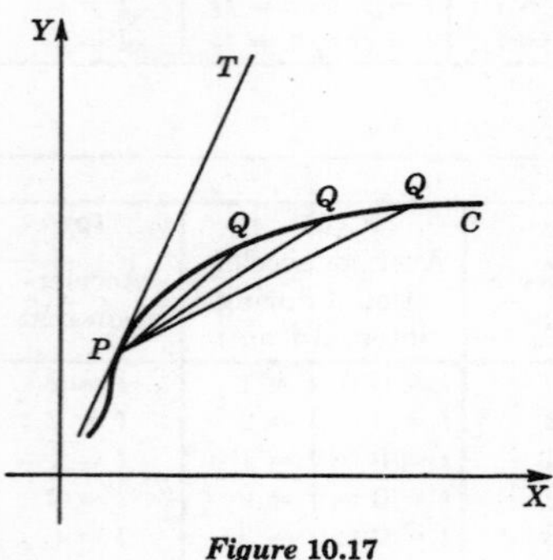

Figure 10.17

Definition: The line whose position is the limiting position PT of the secant line PQ, as $Q \to P$ along the curve, if one exists, is called the tangent line to the curve at the point P.

A function must be continuous at a given point if its graph is to have a tangent there. But continuity is not sufficient to ensure the existence of a tangent; there are curves continuous everywhere with a tangent at no point.

The definition and the discussion of a tangent so far have been geometric in nature. Now let us translate the geometric wording into an equation for this line. This is indeed quite essential. For our definition of the tangent to a curve, we used the phrase "limiting position of a secant". Although this makes intuitive sense, we have not defined the meaning of such a limit and hence cannot proceed deductively here. Instead, we must use our knowledge of analytic geometry to translate this intuitive idea into a sharp, clear one.

We know that a line is completely determined by a point and a slope, and in this case the point is given. Therefore we must seek the slope of the tangent. From our early discussion we might well infer that the slope of the tangent should equal the limit of the slopes of the secants. The slope of a secant which cuts the curve $y = f(x)$ in the points $(x_1, f(x_1))$,

$(x_1 + h, f(x_1 + h))$ is

$$\text{Slope of secant} = \frac{f(x_1 + h) - f(x_1)}{h}$$

Hence we state the following definition.

Definition: The tangent to the curve $y = f(x)$ at the point $(x_1, f(x_1))$ is the line passing through this point whose slope m is given by

$$m = \lim_{h \to 0} \frac{f(x_1 + h) - f(x_1)}{h} \tag{22}$$

provided this limit exists.

We recall that the equation of such a line is

$$y - y_1 = m(x - x_1) \tag{23}$$

Finally, therefore, we write down the equation of this tangent line by computing m and substituting in (23) above.

Illustration 1. Find the equation of the line tangent to the curve

$$y = x^2 - x + 1$$

at the point $x = 0$, $y = 1$.

Solution: We have

$$\begin{aligned}
f(x_1) &= f(0) = 1 \\
f(x_1 + h) &= (x_1 + h)^2 - (x_1 + h) + 1 \\
&= (0 + h)^2 - (0 + h) + 1 \\
&= h^2 - h + 1 \\
m &= \lim_{h \to 0} \frac{f(x_1 + h) - f(x_1)}{h} \\
&= \lim_{h \to 0} \frac{h^2 - h}{h} \\
&= \lim_{h \to 0} (h - 1) \qquad h \neq 0 \\
&= -1
\end{aligned}$$

The equation of the tangent is therefore

$$y - 1 = -1(x - 0) \text{ or } y - 1 = -x$$

Illustration 2. Find the equation of the tangent to $y = x^2$ at the point (x_1, y_1).

Solution:

$$\begin{aligned}
f(x_1) &= x_1^2 \\
f(x_1 + h) &= (x_1 + h)^2 = x_1^2 + 2x_1h + h^2 \\
m &= \lim_{h \to 0} \frac{f(x_1 + h) - f(x_1)}{h} \\
&= \lim_{h \to 0} \frac{x_1^2 + 2x_1h + h^2 - x_1^2}{h} \\
&= 2x_1
\end{aligned}$$

The equation of the tangent is therefore

$$y - y_1 = 2x_1(x - x_1)$$

Several of the seemingly independent aspects of limits which we have taken up in this chapter can, essentially, be united in one theory now known as *the calculus*. This we do in the next chapter.

Problems 10.13

Find the equation of the tangent line as indicated, and sketch.

1. $y = 2x + 5$, at (1,7).
2. $y = 3x - 5$, at (2,1).
3. $y = 2x^2 + 3x$, at (1,5).
4. $y = x^2 - 4x$, at $(-1,5)$.
5. $y = x^2 + x + 1$, at (1,3).
6. $y = x^2 - 3x + 2$, at (1,0).
7. $y = x^3 + 3$, at $(-1,2)$.
8. $y = x^3 - x$, at (0,0).
9. $y = \frac{1}{x}$, at $(3,\frac{1}{3})$.
10. $y = \frac{1}{x - 1}$, at (2,1).
11. $y = \frac{x}{x + 1}$, at $(1,\frac{1}{2})$.
12. $y = ax^2 + b$, at $(x_1, ax_1^2 + b)$.
13. $y = ax^3 + b$, at $(x_1, ax_1^3 + b)$.
14. $y = \frac{1}{x}$, at $\left(x_1, \frac{1}{x_1}\right)$.
15. $y = ax + b$, at the point where $x = x_1$.
16. $y = ax^2 + bx + c$, at the point where $x = x_1$.
17. $y = x^4$, at the point where $x = x_1$.
18. $y = x^5$, at the point where $x = x_1$.
19*. $y = x^n$ (n a positive integer), at (x_1, x_1^n).
20*. $y = x^{-n}$ (n a positive integer), at (x_1, x_1^{-n}).

References

Courant, Richard, and Herbert Robbins: "What Is Mathematics?", pp. 289–328, 399–401, Oxford, New York (1941).
Courant, Richard: "Differential and Integral Calculus", chap. 1, Interscience, New York (1937).
Heath, T. L.: "Works of Archimedes", Cambridge, London (1897).

Also consult the following articles in the *American Mathematical Monthly:*

Ballantine, J. P.: A Peculiar Function, vol. 37, p. 350 (1930).
Cajori, Florian: The History of Zeno's Arguments on Motion, vol. 22, pp. 1, 39, 77, 109, 143, 179, 215, 253, 292 (1915).
Club Topic: A Fibonacci Series, vol. 25, p. 235 (1918).
Knebelman, M. S.: An Elementary Limit, vol. 50, p. 507 (1943).
Moulton, E. J.: Two Teasers for Your Friends, vol. 55, p. 342 (1948).

The Calculus | 11

11.1 Integration

If f is a continuous function which is nonnegative in the interval $[a, b]$, we have seen (Sec. 10.11) that we can define the area bounded by $y = f(x)$, $y = 0$, $x = a$, and $x = b$ by the expression:

$$A_a^b = \lim_{n\to\infty} \sum_{i=1}^{n} f(x_i^*)(x_i - x_{i-1}) \tag{1}$$

We also saw that this can be simplified if we specify equal intervals $x_i - x_{i-1} = \Delta x$ and agree to set $x_i^* = x_i$, the right-hand end point. Then (1) becomes

$$A_a^b = \lim_{n\to\infty} \sum_{i=1}^{n} f(x_i)\,\Delta x \tag{2}$$

In order to simplify the notation, we introduce the new expression

$$\int_a^b f(x)\,dx$$

which, by definition, is equal to

$$\lim_{n\to\infty} \sum_{i=1}^{n} f(x_i)\,\Delta x$$

and is read "the integral of f, with respect to x, from a to b". (Some authors refer to this as the *definite integral.*) The function f is called the *integrand; a* and b are called the lower and upper limits† of integration,

† The word "limit" here is used in the sense of "bound" and has no connection with the limit of the sequence or of a function. This regrettable confusion is so well established in mathematics that the student will just have to keep alert to be sure of the sense in which "limit" is used.

respectively. The expression $\int_a^b f(x)\,dx$ is to be interpreted as a whole, namely, as a simpler way of writing (2). The individual symbols $\int$ and dx have no meanings of their own. Notice, however, that the integral notation is suggestive of the expression (2) from which it is obtained. Instead of $\lim \Sigma$, we write $\int$, an old form of the letter S, which suggests a sum; instead of Δx we write dx; and we include the function $f(x)$ in the middle. Remember that the integral is *not* a sum; it is the *limit* of a sequence of sums and is equal to a real number.

Several extensions of this definition are possible, which are departures from our usual concept of area.

(i) Although, usually, the interval notation $[a, b]$ implies that $a < b$, we need not here place this restriction on a and b. If $a > b$, then our points of division will be chosen to satisfy the inequalities

$$a = x_0 > x_1 > x_2 \cdots > x_{n-1} > x_n = b$$

The rest of the definition remains unchanged. In formula (1) the expressions $x_i - x_{i-1}$ are negative and the resulting "area" becomes negative if (as before) f is nonnegative in $[a, b]$.

(ii) f need not be nonnegative but may have any sign.

(iii) From (i) and (ii) it follows that A_a^b may be positive, negative, or even zero. We often need the notion of *signed* area just as we need other signed quantities.

For these reasons we need not consider the integral as an area; it is an operation on a continuous function f in a closed interval $[a, b]$. Further, there are many other interpretations that can be given to the integral, and we shall take some of these up in due course.

Therefore we may consider, abstractly, the operator $\left(\int_a^b \cdots dx\right)$ that operates on a given function f yielding (if the limit exists) a real number. We express this by writing $\int_a^b f(x)\,dx$, a real number, which depends upon f, a, and b, but not upon what we call the independent variable x. The letter x could be called t or z, etc.; for this reason the variable of integration is often referred to as a *dummy* variable. Hence we have:

Theorem 1. $\int_a^b f(x)\,dx = \int_a^b f(t)\,dt = \cdots = \int_a^b f(z)\,dz = \cdots$.

Earlier we said that we did not admit the case $a = b$. To include this case, we give the following intuitively reasonable definition.

Definition: $\int_a^a f(x)\,dx = 0.$

The following theorems are readily proved directly from the definition of integral.

Theorem 2. $\int_a^b k \cdot f(x)\,dx = k \cdot \int_a^b f(x)\,dx, \quad k \text{ constant.}$

Theorem 3. $\int_a^b f(x) \pm g(x)\,dx = \int_a^b f(x)\,dx \pm \int_a^b g(x)\,dx.$

Theorem 4. $\int_a^b f(x)\,dx = -\int_b^a f(x)\,dx.$

Theorem 5. $\int_a^c f(x)\,dx = \int_a^b f(x)\,dx + \int_b^c f(x)\,dx.$

Exercise A. Prove Theorems 2 to 5.

The integral $\int_a^b f(x)\,dx$ depends on f, a, and b. In order to study this dependence more fully, let us think of f as a given function, a as being fixed, and b as a variable. We should perhaps change the notation; therefore consider $y = f(u)$, a fixed and x, a variable, as the upper limit. We suppose, as usual, that $f(u)$ is continuous in an interval I which contains a and x.

Definition: The function F whose values are given by

$$F(x,a) = \int_a^x f(u)\,du \tag{3}$$

is called the integral of f, with respect to u, from a to x. It is a function of x and a and also depends upon f.

By changing a to some other number b in I, we get $F(x,b) = \int_b^x f(u)\,du$, another integral. How do $F(x,a)$ and $F(x,b)$ differ? To answer this, we consider the difference $F(x,a) - F(x,b)$. We have

$$\begin{aligned} F(x,a) - F(x,b) &= \int_a^x f(u)\,du - \int_b^x f(u)\,du \\ &= \int_a^x f(u)\,du + \int_x^b f(u)\,du \\ &= \int_a^b f(u)\,du = k \qquad \text{a constant} \end{aligned}$$

Hence any two integrals of one and the same function differ only by an additive constant; that is, $F(x,a) = F(x,b) + k$.

Exercise B. Give a geometric interpretation of $\int_a^x f(u)\,du$.

We are now ready to consider the question of finding an integral for a given function f. For an arbitrarily given function f, this becomes a highly complicated task, indeed, and would occupy the serious student for many years. However, we shall consider the simple examples of the nonnegative integral powers of x, namely, $1 = x^0, x, x^2, x^3, \ldots, x^n, \ldots$, some of which we have already worked with in finding area. Because we have been through some of the details, we omit some of the steps in the following analyses, but you should have no trouble in supplying the missing ones. We shall also set $a = 0$, since the integrals corresponding to other values of a differ from those below by only an additive constant.

$$\text{(4)} \qquad \int_0^x (1)\,du = \lim_{n\to\infty} \sum_{i=1}^{n} (1) \cdot \frac{x}{n}$$

$$= \lim_{n\to\infty} \frac{n}{n}\, x$$

$$= x$$

$$\text{(5)} \qquad \int_0^x u\,du = \lim_{n\to\infty} \sum_{i=1}^{n} \left(\frac{i}{n}\, x\right) \cdot \frac{x}{n}$$

$$= \lim_{n\to\infty} \frac{x^2}{n^2} [1 + 2 + \cdots + n]$$

$$= \lim_{n\to\infty} x^2 \left(\frac{n^2 + n}{2n^2}\right)$$

$$= \frac{x^2}{2}$$

$$\text{(6)} \qquad \int_0^x u^2\,du = \lim_{n\to\infty} \sum_{i=1}^{n} \left(\frac{i}{n}\, x\right)^2 \cdot \frac{x}{n}$$

$$= \lim_{n\to\infty} \frac{x^3}{n^3} [1^2 + 2^2 + \cdots + n^2]$$

$$= \lim_{n\to\infty} \frac{x^3}{n^3} \cdot \frac{n(n+1)(2n+1)}{6}$$

$$= \frac{x^3}{3}$$

$$\text{(7)} \qquad \int_0^x u^3\,du = \lim_{n\to\infty} \sum_{i=1}^{n} \left(\frac{i}{n}\, x\right)^3 \cdot \frac{x}{n}$$

$$= \lim_{n\to\infty} \frac{x^4}{n^4} [1^3 + 2^3 + \cdots + n^3]$$

$$= \lim_{n\to\infty} \frac{x^4}{n^4} \cdot \frac{n^2(n+1)^2}{4}$$

$$= \frac{x^4}{4}$$

(8)
$$\int_0^x u^4\,du = \lim_{n\to\infty} \sum_{i=1}^{n} \left(\frac{i}{n}x\right)^4 \cdot \frac{x}{n}$$
$$= \lim_{n\to\infty} \frac{x^5}{n^5}[1^4 + 2^4 + \cdots + n^4]$$
$$= \lim_{n\to\infty} \frac{x^5}{n^5} \cdot \frac{6n^5 + 15n^4 + 10n^3 - n}{30}$$
$$= \frac{x^5}{5}$$

When we discussed sequences, we warned about jumping to conclusions as to what the next term was in case no general term was given. But can you make a guess as to the value of $\int_0^x u^5\,du$? Guess what $\int_0^x u^n\,du$ is equal to. The answer is given by Theorem 6.

Theorem 6. $\int_0^x u^n\,du = x^{n+1}/(n+1)$, for any positive number n. This theorem is quite difficult to prove, and we shall have to take it on faith. And in general, for any real number $n \neq -1$,

$$\int_a^x u^n\,du = \frac{x^{n+1}}{n+1} - \frac{a^{n+1}}{n+1}$$

provided that u^n is continuous in the interval $[a, x]$. Note that when n is negative, u^n is discontinuous at $u = 0$, so that, in this case, 0 must not lie in $[a, x]$.

For your information we state Theorem 7 without proof.

Theorem 7. $\int_a^x \frac{1}{u}\,du = \log_e x - \log_e a \qquad a > 0, x > 0.$

$$\int_a^x e^u\,du = e^x - e^a.$$

These results are among the important reasons for introducing the number e.

Problems 11.1

In Probs. 1 to 16 plot the graph and find the signed area indicated.

1. $y = x(1 - x)$, $y = 0$, $x = 0$, $x = 1$.
2. $y = x(x + 2)(x - 2)$, $y = 0$, $x = -2$, $x = 0$.
3. $y = x(1 - x)$, $y = 0$, $x = 0$, $x = \frac{3}{2}$.
4. $y = x(x + 2)(x - 2)$, $y = 0$, $x = -2$, $x = 2$.
5. $y = x(1 - x)$, $y = 0$, $x = 1$, $x = \frac{3}{2}$.
6. $y = x(x + 2)(x - 2)$, $y = 0$, $x = 0$, $x = 2$.

7. $y = x(2 - x),\ y = x.$
8. $y = x(4 - x),\ y = x.$
9. $y = x(2 - x),\ y = -x.$
10. $y = x^4,\ y = x.$
11. $y = x^4,\ y = |x|.$
12. $y = 1/x^2,\ y = 0,\ x = 1,\ x = 2.$
13. $y = 1/x^2,\ y = 0,\ x = -2,\ x = -1.$
14. $y = 1/x^3,\ y = 0,\ x = 1,\ x = 2.$
15. $y = 1/x^3,\ y = 0,\ x = -2,\ x = -1.$
16. $y = x^{2n},\ y = 0,\ x = 0,\ x = 10,\ n \geq 0.$

In Probs. 17 to 26 perform the indicated operation.

17. $\int_0^1 x^{\frac{1}{2}}\,dx.$
18. $\int_4^9 x^{-\frac{1}{2}}\,dx.$
19. $\int_0^1 (1 + x - x^2 - 2x^3)\,dx.$
20. $\int_{-2}^{-1} \frac{3 - 2x}{x^3}\,dx.$
21. $\int_{-1}^{1} \frac{2x^2 + 3x - 20}{x + 4}\,dx.$
22. $\int_1^e \frac{1}{x}\,dx.$
23. $7 \int_0^1 e^x\,dx.$
24. $\int_0^{2\pi} \sin x\,dx.$ (BT)
25. (a) $\int_1^n \frac{1}{x^2}\,dx.$ (b) $\lim\limits_{n\to\infty} \int_1^n \frac{1}{x^2}\,dx.$
(c) Sketch $y = 1/x^2$. Interpret (a) and (b) as areas.
26. (a) $\int_1^n \frac{2}{x^3}\,dx.$ (b) $\lim\limits_{n\to\infty} \int_1^n \frac{2}{x^3}\,dx.$
(c) Sketch $y = 2/x^3$. Interpret (a) and (b) as areas.
27. Find the integral from 0 to x of the general polynomial of degree n. State each theorem used in obtaining this integral.
28. Is $\int_0^x u^n \cdot u^m\,du = \left(\int_0^x u^n\,du\right)\left(\int_0^x u^m\,du\right)$, n, m, positive integers? Prove your statement.
29. Is $\int_0^x \frac{u^n}{u^m}\,du = \frac{\int_0^x u^n\,du}{\int_0^x u^m\,du}$, n, m positive integers? Prove your statement.

11.2 Differentiation

In determining the slope m of a curve $y = f(x)$ at a point $x = x_1$, we were led to the formula (Sec. 10.13)

$$m(x_1) = \lim_{h\to 0} \frac{f(x_1 + h) - f(x_1)}{h} \tag{9}$$

Return to Sec. 10.12 and look at the formula for velocity at $t = t_1$. It is

$$v(t_1) = \lim_{h\to 0} \frac{f(t_1 + h) - f(t_1)}{h} \tag{10}$$

Again, in Sec. 10.12 the formula for acceleration at $t = t_1$ is

$$a(t_1) = \lim_{h\to 0} \frac{v(t_1 + h) - v(t_1)}{h} \tag{11}$$

It is a phenomenon worth recording that these three processes are abstractly identical: each is the same limit operation on a function. A name has been given to this operation.

Definition: The limit

$$\lim_{h \to 0} \frac{f(x_1 + h) - f(x_1)}{h} \tag{12}$$

if it exists, is called the derivative of f with respect to x at the point $x = x_1$.

The derivative is therefore a new function, which is the result of an operation on a function at a point. The process is called *differentiation.*

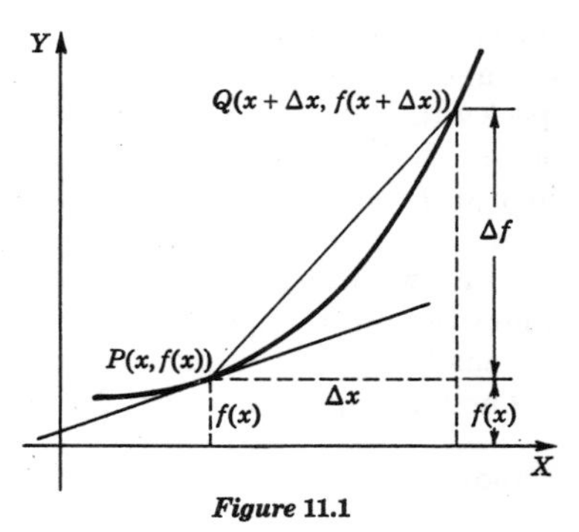

Figure 11.1

Exercise A. Are the domains of definition of a function and its derivative necessarily the same? Explain.

Often other notations are used. For instance, if we wish to think of this as the operation of determining the slope of a curve $y = f(x)$ at any point x, then P will be the point $P(x, f(x))$ (Fig. 11.1). Any other point Q will then have coordinates $(x + \Delta x, f(x + \Delta x))$, where Δx stands for the h previously used.

Let us agree to write Δf for the quantity $f(x + \Delta x) - f(x)$. The slope of the secant PQ will then be

$$\frac{\Delta f}{\Delta x} = \frac{f(x + \Delta x) - f(x)}{\Delta x}$$

and hence

$$m = \lim_{\Delta x \to 0} \frac{\Delta f}{\Delta x} = \lim_{\Delta x \to 0} \frac{f(x + \Delta x) - f(x)}{\Delta x} \tag{13}$$

is the derivative of f at the point x. This also follows immediately from (12) if you set $h = \Delta x$ and $x_1 = x$.

The symbols in (13), while standard, are still long to write down (except for the single symbol m), and thus we have devised other symbols which are also quite standard. Some of these, beginning with the fundamental one, are now given.

The derivative of f with respect to x at the point x is:

$$= \lim_{\Delta x \to 0} \frac{f(x + \Delta x) - f(x)}{\Delta x} \qquad \text{[definition]}$$

$$= \frac{df(x)}{dx} \qquad \text{[after Leibniz, 1646–1716]}$$

$$= f'_x(x) \text{ or } f'(x) \qquad \text{[after Lagrange, 1736–1813]}$$

$$= D_x f(x) \qquad \text{[after Cauchy, 1789–1857]}$$

When there can be no resulting confusion about what is meant, these may be simplified to read $df/dx = f'_x = D_x f$. These are still to be read "the derivative of f *with respect to* x at the point x", for f might be a function of x and several other variables, and then it would become imperative to specify the variable with respect to which the differentiation was to take place. If f is a function of a single letter only $[y = f(x)]$ and there is no chance of confusion, then we may write simply f' or Df or Dy for the derivative. Caution is urged where df/dx is used; it just stands for the derivative of f with respect to x at the point x; it is *one* whole symbol for this quantity even though it is made up of pieces, d and f and / and d and x. We shall not restrict ourselves exclusively to any one of these notations.

The following statements are easy to prove.

Theorem 8. $D_x cf = c \cdot D_x f$, c constant.

Theorem 9. $D_x(f \pm g) = D_x f \pm D_x g$.

Exercise B. Prove Theorems 8 and 9.

We now turn to the problem of differentiating the various nonnegative integral powers of x, namely, $1 = x^0, x, x^2, \ldots, x^n, \ldots$. Draw the associated figure, and remember that we are calling

$$\Delta f = f(x + \Delta x) - f(x)$$

Theorem 10. $D_x 1 = 0$.

Proof:

$$f(x) = 1$$
$$f(x + \Delta x) = 1$$
$$\Delta f = 0$$
$$\frac{\Delta f}{\Delta x} = 0 \qquad \Delta x \neq 0$$
$$D_x f = \lim_{\Delta x \to 0} \frac{\Delta f}{\Delta x} = 0$$

Exercise C. Prove that, for a constant C,

$$D_x C = 0$$

Theorem 11. $D_x x = 1$.

Proof:

$$\begin{aligned} f(x) &= x \\ f(x + \Delta x) &= x + \Delta x \\ \Delta f &= \Delta x \\ \frac{\Delta f}{\Delta x} &= \frac{\Delta x}{\Delta x} = 1 \qquad \Delta x \neq 0 \\ D_x f &= \lim_{\Delta x \to 0} \frac{\Delta f}{\Delta x} = 1 \end{aligned}$$

Theorem 12. $D_x x^2 = 2x$.

Proof:

$$\begin{aligned} f(x) &= x^2 \\ f(x + \Delta x) &= (x + \Delta x)^2 = x^2 + 2x\,\overline{\Delta x} + \overline{\Delta x}^2 \\ \Delta f &= 2x\,\overline{\Delta x} + \overline{\Delta x}^2 \\ \frac{\Delta f}{\Delta x} &= 2x + \Delta x \qquad \Delta x \neq 0 \\ D_x f &= \lim_{\Delta x \to 0} \frac{\Delta f}{\Delta x} = 2x \end{aligned}$$

Theorem 13. $D_x x^3 = 3x^2$.

Proof:

$$\begin{aligned} f(x) &= x^3 \\ f(x + \Delta x) &= (x + \Delta x)^3 = x^3 + 3x^2\,\overline{\Delta x} + 3x\,\overline{\Delta x}^2 + \overline{\Delta x}^3 \\ \Delta f &= 3x^2\,\overline{\Delta x} + 3x\,\overline{\Delta x}^2 + \overline{\Delta x}^3 \\ \frac{\Delta f}{\Delta x} &= 3x^2 + 3x\,\overline{\Delta x} + \overline{\Delta x}^2 \qquad \Delta x \neq 0 \\ D_x f &= \lim_{\Delta x \to 0} \frac{\Delta f}{\Delta x} = 3x^2 \end{aligned}$$

Theorem 14. $D_x x^4 = 4x^3$.

Proof:

$$\begin{aligned} f(x) &= x^4 \\ f(x + \Delta x) &= (x + \Delta x)^4 = x^4 + 4x^3\,\overline{\Delta x} + 6x^2\,\overline{\Delta x}^2 + 4x\,\overline{\Delta x}^3 + \overline{\Delta x}^4 \\ \Delta f &= 4x^3\,\overline{\Delta x} + 6x^2\,\overline{\Delta x}^2 + 4x\,\overline{\Delta x}^3 + \overline{\Delta x}^4 \\ \frac{\Delta f}{\Delta x} &= 4x^3 + 6x^2\,\overline{\Delta x} + 4x\,\overline{\Delta x}^2 + \overline{\Delta x}^3 \qquad \Delta x \neq 0 \\ D_x f &= \lim_{\Delta x \to 0} \frac{\Delta f}{\Delta x} = 4x^3 \end{aligned}$$

We have passed over such questions as the $\lim_{\Delta x \to 0} 6x^2 \overline{\Delta x}, 4x \overline{\Delta x}^2$, etc. By now it must be clear that these go to zero with Δx, for any given fixed x. Note carefully that $\Delta f \to 0$ as $\Delta x \to 0$, for otherwise the $\lim_{\Delta x \to 0} (\Delta f / \Delta x)$ would not exist. This says that f must be continuous at the point x if f is to be differentiable there.

Exercise D. Prove in detail: In order that f be differentiable at a point x, it is necessary that f be continuous there.

You might want to guess what the derivatives of the higher powers would be. A reasonable guess is that $D_x x^n = nx^{n-1}$ for n a positive integer. We shall prove this below.

Theorem 15. $D_x x^n = nx^{n-1}$, when n is a positive integer.

Proof: We proceed by induction. We know that for $n = 1$,

$$D_x x^1 = D_x x = 1$$

Thus the formula is verified. We must now prove that for all $k \geq 1$: if $D_x x^k = kx^{k-1}$, then $D_x x^{k+1} = (k + 1)x^k$. Now $x^{k+1} = x^k \cdot x$. To find its derivative, we consider

$$\begin{aligned}
\lim_{\Delta x \to 0} \frac{(x + \Delta x)^{k+1} - x^{k+1}}{\Delta x} &= \lim_{\Delta x \to 0} \frac{(x + \Delta x)^k (x + \Delta x) - x^k x}{\Delta x} \\
&= \lim_{\Delta x \to 0} \frac{[(x + \Delta x)^k - x^k]x}{\Delta x} + \lim_{\Delta x \to 0} \frac{(x + \Delta x)^k \cdot \Delta x}{\Delta x} \\
&= (D_x x^k)x + x^k \\
&= kx^{k-1} \cdot x + x^k \\
&= (k + 1)x^k
\end{aligned}$$

Hence the formula of the theorem is true for all positive integers n.

As a matter of fact, we state without proof that Theorem 15 holds for any real value of the exponent n.

Exercise E. Prove Theorem 15 by the Δ-process, using the Binomial Theorem (Sec. 12.13). Model your proof on that of Theorem 14.

For your information, we also state without proof the following two theorems.

Theorem 16. $D_x e^x = e^x$.

We have seen above that we can obtain the derivative of a polynomial by differentiating it term by term. This procedure is sometimes legitimate for a function whose values are given by a series. In the case of e^x we may differentiate term by term. Thus from

$$e^x = 1 + x + \frac{x^2}{2!} + \frac{x^3}{3!} + \cdots + \frac{x^n}{n!} + \cdots,$$

we find

$$\begin{aligned} D_x e^x &= 0 + 1 + x + \frac{x^2}{2!} + \cdots + \frac{x^{n-1}}{(n-1)!} + \cdots \\ &= e^x \end{aligned}$$

This result is one of the chief motivations for defining the number e. It is true that ce^x is the only function which is its own derivative.

Exercise F. Assuming that term-by-term differentiation is legitimate for the series for $\sin x$ and $\cos x$, prove that $D_x \sin x = \cos x$; $D_x \cos x = -\sin x$. Use the series in Sec. 10.7.

Theorem 17. $D_x \log_e x = 1/x$.

This result is another reason for considering e. $D_x \log_a x$ for any other base does not give such a simple result.

Exercise G. Assume Theorem 17, and show that $D_x \log_a x = (1/x) \log_a e$ (see Sec. 7.3, Exercise C).

Problems 11.2

In Probs. 1 to 20 find $D_x f$.

1. $f(x) = x^3 - 6x^2 + 3x - 5$.
2. $f(x) = x^4 + x^3 + 7x + \pi$.
3. $f(x) = x^0 + x + x^2$
4. $f(x) = x^{3/2} - 4x^{1/2}$.
5. $f(x) = x^{1/2} - 7x^0 - 3x^{-1/2}$.
6. $f(x) = \dfrac{4}{x^4} - \dfrac{2}{x^2} + 4x^4 - 2x^2$.
7. $f(x) = \dfrac{2 - x^3}{x}$.
8. $f(x) = \dfrac{(2 - x)^3}{x}$.
9. $f(x) = \dfrac{1 - x^{n+1}}{1 - x}$.
10. $f(x) = \dfrac{1 + x^{2n+1}}{1 + x}$.
11. $f(x) = \log_e 2x$.
12. $f(x) = \log_e x^2$.
13. $f(x) = \dfrac{1}{\log_x e}$.
14. $f(x) = \log_{10} x$.
15. $f(x) = \log_{10} 3x^2$.
16. $f(x) = \log_2 2^x$.
17. $f(x) = 5^{\log_5 x}$.
18. $f(x) = 5e^{x+3}$.
19. $f(x) = 3e^{x-5}$.
20. $f(x) = e^{ax}$.

In Probs. 21 to 30 find the equation of the tangent at the point indicated and sketch.

21. $y = \frac{1}{9}x^2$, (3,1).
22. $y = x^3 - x$, (1,0).
23. $y = -1/x$, (1,−1).
24. $y = x(1 - x)$, $(\frac{1}{2},\frac{1}{4})$.
25. $y = x(x + 2)(x - 2)$, (0,0).
26. $y = e^x$, (1,e).

27. $y = \log_e x$, (1,0).
28. $y = \sin x$, (0,0). (See Exercise F.)
29. $y = \cos x$, $(\pi/3,\frac{1}{2})$. (See Exercise F.)
30. $y = \cosh x$, (0,1). (See Prob. 20 above and Sec. 7.2.)

In Probs. 31 to 40 find the equation of the line *perpendicular* to the tangent line at the point indicated and sketch.

31. $y = \frac{1}{4}x^2$, (2,1).
32. $y = x^3 - 4x$, (0,0).
33. $y = 1/x$, $(-2,-\frac{1}{2})$.
34. $y = 1/x^2$, (1,1).
35. $y = \frac{1}{27}x^3$, (3,1).
36. $y = e^x$, (0,1).
37. $y = 3 \log_e x$, (1,0).
38. $y = \sin x$, $(\pi/6,\frac{1}{2})$.
39. $y = \cos x$, (0,1).
40. $y - 2 = 2(x - 2)^2$, (2,2).

41. From the definition of derivative find Dy for $y = x^{-2}$. Hence show that the general rule applies.
42. From the definition of derivative find Dy for $y = x^{-n}$, where n is a positive integer. Hence show that the same rule applies for either positive or negative powers of x. HINT: Use mathematical induction.
43. Illustrate with an example to show that, in general, $D(f \cdot g) \neq Df \cdot Dg$. (The derivative of a product is not the product of the derivatives.)
44. Illustrate with an example to show that, in general, $D(f/g) \neq Df/Dg$. (The derivative of a quotient is not the quotient of the derivatives.)
45. Draw a curve to show that continuity is not a sufficient condition for differentiability.

11.3 Comparison of Integration and Differentiation

As we have pointed out before, the problem of finding area (also length and volume) is several thousand years old, and the Greeks made some limited progress with the problem, which has been uppermost in the minds of mathematicians even up to the very present time. This problem led quite naturally to the integral calculus. On the other hand, the notions of velocity and acceleration are very modern by comparison; they led to the differential calculus.

And there seems to be no offhand reason under the sun why there should be a connection between these seemingly independent ideas:

$$\text{Integral} = \int_a^x f(u)\,du = \lim_{n\to\infty} \sum_{i=1}^{n} f(u_i^*)(u_i - u_{i-1}) \tag{14}$$

$$\text{Derivative} = D_x f = \lim_{\Delta x \to 0} \frac{\Delta f}{\Delta x} \tag{15}$$

But look at the following table for a good comparison:

The function f	The integral $\int_0^x f(u)\,du$	The derivative Df
x	$\frac{x^2}{2}$	1
x^2	$\frac{x^3}{3}$	$2x$
x^3	$\frac{x^4}{4}$	$3x^2$
x^4	$\frac{x^5}{5}$	$4x^3$
x^n, $n \geq 1$	$\frac{x^{n+1}}{n+1}$	nx^{n-1}

Exercise A. Apply the operation of differentiation to the second column, and show that the first column is recovered.

Exercise B. Integrate the functions in the third column from 0 to x, and show that the first column is recovered.

Thus it turns out that integration and differentiation are essentially inverse to one another, at least for the special functions used.

In order to explain this clearly, we consider two sets F^* and G^* of functions, which will be seen to be related. We suppose that f is given in advance and that it is continuous in some interval $I[a, b]$. Let x and a lie in this interval. We then define the following sets of functions.

First, the set F^* is the set of all functions such that, for each member F,

$$F(x) = \int_a^x f(u)\,du$$

for some a. If two functions in this set are distinct, their corresponding a's must be unequal. On the other hand, two distinct a's (say, a_1 and a_2) will yield the same F if

$$\int_{a_1}^{a_2} f(u)\,du = 0$$

F^* is called the set of *integrals of f*.

Second, the set G^* is the set of all functions such that, for each member G,

$$D_xG(x) = f(x)$$

for each x. This is called the set of *primitives of f*. (Some authors use the phrase *indefinite integral* instead of *primitive* in this connection.)

We saw in Sec. 11.1 that the set F^* of integrals has the important property that any two integrals $F_1(x)$ and $F_2(x)$ differ only by an additive constant; that is,

$$F_1(x) = F_2(x) + C$$

The set G^* of primitives has the same property. This follows from Theorems 18 and 19 below.

Theorem 18. If $G(x)$ has its derivative equal to zero at every point of an interval I, then $G(x) = C$ (a constant) in I.

We cannot give a proof of this theorem here, but we appeal to your intuition. If $D_xG(x) = 0$ at a point, the tangent to its graph has zero slope and hence is horizontal. Thus the graph of $G(x)$ has a horizontal tangent at each point of I. Therefore this graph must be itself a horizontal straight line; that is, $G(x)$ is a constant.

Theorem 19. If G_1 and G_2 are two functions whose derivatives are each equal to $f(x)$ in I, then $G_1(x) = G_2(x) + C$ in I.

Proof: Apply Theorem 18 to the function $G_1(x) - G_2(x)$. Its derivative is zero in I. Hence $G_1(x) - G_2(x) = C$.

Since the sets of functions F^* and G^* seem to have properties in common, you may wonder whether they are, in fact, identical sets. A partial answer is given in Theorem 20, which says in effect that *every integral of f is a primitive of f*, or that $F^* \subseteq G^*$.

Theorem 20. Every integral of a continuous function f is a primitive of f. In other words, if $F(x) = \int_a^x f(u)\,du$, then $D_xF(x) = f(x)$.

This theorem was first proved by Newton's teacher Isaac Barrow (1630–1677), in the year 1667. Barrow's theorem is now called the Fundamental Theorem of Integral Calculus.

Proof: We shall appeal to the interpretation of the integral as an area (Fig. 11.2) in order to make the proof seem less abstract; of course, logically, we do not need to do so. To find dF/dx, we must find the limit as $\Delta x \to 0$ of

$$\frac{F(x + \Delta x, a) - F(x,a)}{\Delta x} = \frac{1}{\Delta x}\left[\int_a^{x+\Delta x} f(u)\,du - \int_a^x f(u)\,du\right]$$
$$= \frac{1}{\Delta x}\int_x^{x+\Delta x} f(u)\,du$$

This last expression is $1/\Delta x$ times the area under $f(u)$ from $u = x$ to $u = x + \Delta x$. As before, call x_m and x_M the points in the interval

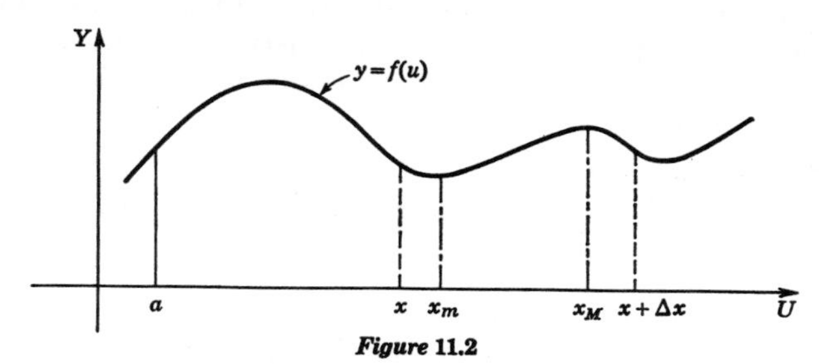

Figure 11.2

$[x, x + \Delta x]$, where $f(u)$ is a minimum and a maximum, respectively. Then the area from x to $x + \Delta x$, namely,

$$F(x + \Delta x, a) - F(x, a) = \int_x^{x+\Delta x} f(u)\, du$$

satisfies the inequalities

$$\Delta x f(x_m) \leq F(x + \Delta x, a) - F(x,a) \leq \Delta x f(x_M)$$

that is,
$$f(x_m) \leq \frac{F(x + \Delta x, a) - F(x,a)}{\Delta x} \leq f(x_M)$$

In the limit, as $\Delta x \to 0$, $f(x_m)$ and $f(x_M)$ must both tend to the limit $f(x)$ because of the continuity of the function f. Hence, by the domination principle (Sec. 10.9, Theorem 12), we can write

$$\lim_{\Delta x \to 0} \frac{F(x + \Delta x, a) - F(x,a)}{\Delta x} = f(x)$$

or
$$D_x F = f(x)$$

This completes the proof.

On the other hand, not every primitive is an integral. As a counterexample, let $f(x) = 0$. Then $G(x) = 1$ is a primitive, but $G(x)$ cannot be written in the form $1 = \int_a^x 0\, du$. This integral is zero for all values of a and hence cannot be equal to 1. Another and less trivial counterexample is the following. Let $f(x) = -1/x^2$. Then $G(x) = 1/x$ is a primitive, but $G(x)$ cannot be written in the form $\int_a^x \left(-\frac{1}{u^2}\right)$ du, since

$$\int_a^x \left(-\frac{1}{u^2}\right) du = \frac{1}{x} - \frac{1}{a}$$

and this is not equal to $\frac{1}{x}$ for any value of a.

The practical importance of Theorem 20 should now become clear. It is a difficult task to find integrals directly from the basic definition since limits of complicated sums must be evaluated. Often it is easier to find a primitive by looking for a function G whose derivative is the given function $f(x)$. We now show that *if a primitive can be found, then an integral can be obtained from it.*

We are seeking to find $F(x) = \int_a^x f(u)\, du$. Let G be any primitive of f. Since F is also a primitive of f (Theorem 20), it follows that

$$F(x) = G(x) + C \tag{16}$$

for a suitable constant C. To find the value of C, put $x = a$ in Eq. (16). Hence $F(a) = G(a) + C$. We note, however, that

$$F(a) = \int_a^a f(u)\, du = 0$$

Hence $$C = -G(a)$$

Therefore $$F(x) = G(x) - G(a)$$

Rule. This gives us a rule for evaluating any integral

$$\int_a^b f(u)\, du$$

First, find a primitive $G(x)$ such that $D_x G(x) = f(x)$ for all points in the interval $[a, b]$. Then

$$\int_a^b f(u)\, du = G(b) - G(a)$$

In computation we usually write this in the form:

$$\int_a^b f(u)\, du = G(x) \Big]_a^b = G(b) - G(a)$$

One final remark should be made about a primitive of f. Our definition says that G is a primitive of f if $D_x G(x) = f(x)$. A common notation for G is

$$G(x) = \int f(x)\, dx + C$$

where C is to be thought of as an arbitrary constant.

The fundamental problem of finding integrals and primitives of given functions can occupy the serious student for several years. The process is known simply as *integration.*

Exercise C. From Sec. 11.2, Theorems 16 and 17, show the following:

(a) $\int e^x \, dx = e^x + C$.

(b) $\int \frac{1}{x} \, dx = \log_e x + C$.

Illustration 1. Find the value of $\int_1^2 (x^3 - \frac{1}{2}x^4 + x - 3) \, dx$.

Solution:

$$\begin{aligned}\int_1^2 \left(x^3 - \frac{1}{2}x^4 + x - 3\right) dx &= \left[\frac{x^4}{4} - \frac{1}{10}x^5 + \frac{x^2}{2} - 3x\right]_1^2 \\ &= (\tfrac{16}{4} - \tfrac{32}{10} + \tfrac{4}{2} - 6) - (\tfrac{1}{4} - \tfrac{1}{10} + \tfrac{1}{2} - 3) \\ &= -\tfrac{17}{20}\end{aligned}$$

Illustration 2. Find a primitive of $6x^2 + \frac{7}{x} - 2$.

Solution:

$$\begin{aligned}G(x) &= \int \left(6x^2 + \frac{7}{x} - 2\right) dx \\ &= 2x^3 - 7 \log_e x - 2x + C\end{aligned}$$

We can apply this method of integration to solve problems involving distance and velocity.

Illustration 3. The velocity of a particle moving on the X-axis is given by $v = 2t - 3t^2 + 1$. At $t = 0$, the particle is at the origin. Where is it when $t = 1$?

Solution: We know that $D_t x = v = 2t - 3t^2 + 1$. Thus x is some primitive of $2t - 3t^2 + 1$, or

$$x = t^2 - t^3 + t + C$$

The value of C is obtained by putting $x = 0$, $t = 0$, in this equation and solving for C. We find that $C = 0$; thus

$$x = t^2 - t^3 + t$$

At $t = 1$, $x = 1 - 1 + 1 = 1$ unit from the origin.

Illustration 4. Find the area under the curve $y = e^x$ from $x = 0$ to $x = 2$.

Solution:

$$\begin{aligned}A_0^2 &= \int_0^2 e^x \, dx \\ &= e^x \Big]_0^2 \\ &= e^2 - e^0 \\ &= e^2 - 1 \text{ sq unit}\end{aligned}$$

Problems 11.3

In Probs. 1 to 10 find (*a*) a primitive of f, and (*b*) an integral of f where f is given by:

1. $f(x) = x^2 + 3x - 5$.
2. $f(x) = 4x^3 - 3x^2 + 2x - 1$.
3. $f(x) = x^{-2} + x^{-1}$.
4. $f(x) = \frac{2}{3}x^{1/2} - 2x^{-1/2}$.
5. $f(x) = x^{2/3} + 2^{2/3}$.
6. $f(x) = (4 - x)^3$.
7. $f(x) = \dfrac{(4 - x)^2}{x}$.
8. $f(x) = x^n$, $n \neq -1$.
9. $f(x) = 7e^x$.
10. $f(x) = \dfrac{e^{2x} - 3e^x}{e^x}$.

In Probs. 11 to 20 find $f(x)$ where a primitive G of f is given by:

11. $G(x) = 4$.
12. $G(x) = x - 4$.
13. $G(x) = x^3 - x + C$, C constant.
14. $G(x) = x^4 + x^3 + x + C$, C constant.
15. $G(x) = x^{-2} - x^{-1}$.
16. $G(x) = (3 - x)^3$.
17. $G(x) = \dfrac{(3 - x)^3}{x} + C$, C constant.
18. $G(x) = x^n$.
19. $G(x) = \dfrac{e^{2x} + 4e^x}{e^x} + C$, C constant.
20. $G(x) = \sin^2 \frac{1}{2}x$.

21. The distance from the origin at time t of a particle is $y = t^4 + t^2 - 2t + 1$.

(a) Find the velocity at any time t. (b) Find the velocity when $t = 0$.

22. The velocity at any time t is given by $v = t^3 + 2t + 5$.

(a) Find the acceleration at any time t. (b) Find the acceleration when $t = 0$.

23. The distance from the origin at time t is given by $y = t^3 + 2t^2 - 3t + 4$. Find:

(a) The velocity at any time t. (b) The acceleration at any time t.

24. The velocity at any time t is given by $v = 3t^2 + t + 3$. If the particle was at the origin when $t = 0$, find the distance traveled between $t = 0$ and $t = 1$.

25. Prove: The rate of change of area [enclosed by $y = f(x)$, $x = a$, $x = x_0$, $y = 0$] per unit change in x, at $x = x_0$, is $f(x_0)$.

26. If the graph of $y = f(x)$ is a curve passing through the point (1,0) and if $D_xf = x - 1$, what is the exact expression for $f(x)$?

27. The area A bounded by the curve $y = f(x)$, the X-axis, and the lines $x = 0$ and $x = x_1$ is given by $A = x_1^2$, for all x_1. Find $f(x)$.

11.4 Rules of Differentiation

By means of the fundamental theorem, once we have found an integral, $\int f(x)\,dx = F(x) + C$, then we know how to differentiate the function F. And vice versa, where we have discovered the derivative of a given function, $D_xF(x) = f(x)$, then we immediately know an integral of the function f. Therefore we can use either process as the basic one to build up tables of integrals and derivatives. It happens that the process of differentiating is by far the simpler one; therefore let us turn to the development of some rules of differentiation.

At present we know the formulas

$$\int x^n\,dx = \frac{x^{n+1}}{n+1} + C \qquad (n \neq -1) \qquad \int \frac{1}{x}\,dx = \log_e x + C$$

$$\int e^x\,dx = e^x + C \qquad \int \sin x\,dx = -\cos x + C \qquad \int \cos x\,dx = \sin x + C$$

$$D_x x^n = nx^{n-1} \qquad D_x \log_e x = \frac{1}{x} \qquad D_x e^x = e^x$$

$$D_x \sin x = \cos x \qquad D_x \cos x = -\sin x$$

We should like to be able to talk about differentiating and integrating other elementary functions, but it will not be possible to do so in this brief treatment. Practically all our work will be with polynomials, rational functions, and a few other functions involving roots and powers.

We now know how to integrate and to differentiate any polynomial and monomial term of the form x^n, where n is any real number. In order to help us with the problem of differentiating more complicated functions, we first prove the following two theorems. Theorem 21 deals with the derivative of a product $f \cdot g$ of two functions f and g. Theorem 22, the so-called "chain rule", deals with the derivative of a composite function $f[u(x)]$.

Theorem 21. $D_x[f(x) \cdot g(x)] = f(x) \cdot D_x g(x) + g(x) \cdot D_x f(x)$.

Exercise A. Write this theorem out in words.

Proof: The derivative of the product of two functions is, directly from the definition,

$$D_x[f(x) \cdot g(x)] = \lim_{\Delta x \to 0} \frac{f(x+\Delta x)g(x+\Delta x) - f(x)g(x)}{\Delta x}$$

This limit can readily be evaluated by the trick of adding to and subtracting from the numerator the quantity $f(x + \Delta x)g(x)$. That is,

$$\begin{aligned} D_x[f(x) \cdot g(x)] \\ &= \lim_{\Delta x \to 0} \frac{f(x+\Delta x)[g(x+\Delta x) - g(x)] + g(x)[f(x+\Delta x) - f(x)]}{\Delta x} \\ &= \lim_{\Delta x \to 0} f(x+\Delta x)\frac{g(x+\Delta x) - g(x)}{\Delta x} + \lim_{\Delta x \to 0} g(x)\frac{f(x+\Delta x) - f(x)}{\Delta x} \\ &= f(x)D_x g(x) + g(x)D_x f(x) \end{aligned}$$

Theorem 22. $D_x f(u(x)) = D_{u(x)} f(u(x)) \cdot D_x u(x)$.

Exercise B. Write this theorem out in words.

Proof: Directly from the definition of derivative we have

$$\begin{aligned} D_x f(u(x)) &= \lim_{\Delta x \to 0} \frac{f[u(x + \Delta x)] - f[u(x)]}{\Delta x} \\ &= \lim_{\Delta x \to 0} \frac{f[u(x) + \Delta u(x)] - f[u(x)]}{\Delta u(x)} \cdot \frac{\Delta u(x)}{\Delta x} \qquad \Delta u(x) \neq 0 \\ &= D_{u(x)} f[u(x)] \cdot D_x u(x) \end{aligned}$$

This general rule takes on a rather nice special form in case f is a power of u. Here $f(u(x)) = [u(x)]^n$. The derivative becomes

$$D_x[u(x)]^n = n[u(x)]^{n-1} D_x u(x) \tag{17}$$

This should be memorized.

Illustration 1. Find the derivative of y with respect to x, given

$$y = x^3 \cdot x^7$$

Solution: Of course, $y = x^3 \cdot x^7 = x^{10}$ and $Dy = 10x^9$. But we wish to use the formula for the derivative of a product in a very simple case first. In Illustrations 4 and 5 below, we take up more complicated examples.

$$\begin{aligned} y &= x^3 \cdot x^7 \\ Dy &= x^3 Dx^7 + x^7 Dx^3 \\ &= x^3 \cdot 7x^6 + x^7 \cdot 3x^2 \\ &= 10x^9 \end{aligned}$$

Illustration 2. Find the derivative of $y = u^3 + u - 5$, where

$$u = x^2 + 6x$$

with respect to x.

Solution: Theorem 22 becomes, in y and u notation,

$$\frac{dy}{dx} = \frac{dy}{du} \cdot \frac{du}{dx} = (3u^2 + 1) \cdot (2x + 6)$$

Illustration 3. Find the derivative of $y = (x^3 + 6x - 1)^{17}$ with respect to x.

Solution: Think of $x^3 + 6x - 1$ as being u; that is, $y = u^{17}$, $u = x^3 + 6x - 1$. Then

$$\begin{aligned} \frac{dy}{dx} &= nu^{n-1} \frac{du}{dx} \\ &= 17(x^3 + 6x - 1)^{16}(3x^2 + 6) \end{aligned}$$

Illustration 4. Differentiate $y = x^3 \sqrt{x^2 + 1}$.

Solution: First of all we think of it as a product; when we get to the point of differentiating the second factor, we find we have a u^n problem also.

$$y = x^3(x^2 + 1)^{1/2}$$
$$Dy = x^3 \cdot \tfrac{1}{2}(x^2 + 1)^{-1/2}(2x) + (x^2 + 1)^{1/2}3x^2$$

This is the answer; we can simplify algebraically or not.

$$Dy = \frac{x^4}{(x^2 + 1)^{1/2}} + 3x^2(x^2 + 1)^{1/2}$$
$$= \frac{x^4 + 3x^2(x^2 + 1)}{(x^2 + 1)^{1/2}} = \frac{4x^4 + 3x^2}{(x^2 + 1)^{1/2}}$$

Illustration 5. Differentiate $x^{1/2}/(1 + x^3)$ with respect to x.

Solution:

$$\frac{x^{1/2}}{1 + x^3} = x^{1/2}(1 + x^3)^{-1}$$
$$D_x\left(\frac{x^{1/2}}{1 + x^3}\right) = x^{1/2}[-1(1 + x^3)^{-2}3x^2] + (1 + x^3)^{-1}\left[\tfrac{1}{2}x^{-1/2}\right]$$
$$= \frac{-3x^{1/2}x^2}{(1 + x^3)^2} + \frac{1}{2x^{1/2}(1 + x^3)}$$
$$= \frac{-6x^3 + 1 + x^3}{2x^{1/2}(1 + x^3)^2} = \frac{1 - 5x^3}{2x^{1/2}(1 + x^3)^2}$$

Illustration 6. Find the equation of the tangent to the hyperbola $(x^2/a^2) - (y^2/b^2) = 1$ at the point (x_1,y_1). Note that this implies that $|x_1| \geq a$.

Solution: This equation defines a relation, and not a function. In order to differentiate, we must consider one of the functions which can be derived from this relation. We may choose either

$$y = +\frac{b}{a}\sqrt{x^2 - a^2} \qquad \text{or} \qquad y = -\frac{b}{a}\sqrt{x^2 - a^2}$$

The domain of definition of each function is $|x| \geq a$. The given point on the hyperbola (x_1,y_1) will satisfy exactly one of these equations, and we then select this one as the definition of a function:

$$y = f(x)$$

With this definition of $f(x)$, the function

$$F(x) = \frac{x^2}{a^2} - \frac{[f(x)]^2}{b^2} - 1$$

has the value zero for all x such that $|x| \geq a$. Its derivative, $F'(x)$, must also be zero. Hence

$$F'(x) = \frac{2x}{a^2} - \frac{2f(x)f'(x)}{b^2} = 0$$

Solving, we find

$$f'(x) = \frac{b^2}{a^2} \frac{x}{f(x)} = \frac{b^2}{a^2} \frac{x}{y}$$

Hence the slope of the hyperbola at (x_1,y_1) is

$$m = \frac{b^2}{a^2} \frac{x_1}{y_1}$$

Hence the equation of the tangent is

$$y - y_1 = \frac{b^2}{a^2} \frac{x_1}{y_1} (x - x_1)$$

or simplifying,

$$a^2yy_1 - a^2y_1^2 = b^2xx_1 - b^2x_1^2$$
$$b^2xx_1 - a^2yy_1 = b^2x_1^2 - a^2y_1^2$$

Since the right-hand member is a^2b^2 [the point (x_1,y_1) is on the hyperbola, and therefore the coordinates satisfy its equation], we have

$$b^2xx_1 - a^2yy_1 = a^2b^2$$

or finally,

$$\frac{xx_1}{a^2} - \frac{yy_1}{b^2} = 1$$

Illustration 7. Differentiate $y = e^{ax}$.

Solution: Put $u = ax$, so that $y = e^u$. From Theorem 22,

$$D_xy = e^u \cdot \frac{du}{dx} = ae^u = ae^{ax}$$

Exercise C

(a) Show that $a^x = e^{x \log_e a}$.
(b) Show that $D_xa^x = a^x \log_e a$.

Exercise D. Show that:

(a) $\int e^{ax}\,dx = \frac{1}{a} e^{ax} + C$.

(b) $\int a^x\,dx = a^x/(\log_e a) + C$.

Exercise E. Find a primitive of each of the following functions as indicated:

(a) $\frac{1}{2} \int \frac{1 - 5x^3}{x^{1/2}(1 + x^3)^2}\,dx$ (see Illustration 5).

(b) $\int 17(x^3 + 6x - 1)^{16}(3x^2 + 6)\,dx$ (see Illustration 3).

(c) $\int \frac{4x^4 + 3x^2}{(x^2 + 1)^{\frac{1}{2}}}\,dx$ (see Illustration 4).

(d) $\int x^7(x^2 - 1)\,dx$.

There are some general rules (methods) of integration, but we shall not take them up except for the one that is the inverse of rule (17) for differentiation. Rule (17) says that if $y = u^n$, where u is a function of x, then $D_x y = nu^{n-1}D_x u$. If we immediately turn around and integrate $D_x y$, we recover the function y. Hence

$$\int (nu^{n-1}D_x u)\,dx = u^n + C$$

We should rather consider the integral of u^n than of u^{n-1}, however. We have only to modify the above to read

$$(18) \qquad \int u^n D_x u\,dx = \frac{u^{n+1}}{n+1} + C \qquad n \neq -1$$

Illustration 8. Integrate $y = x^2\sqrt{x^3 - 1}$.

Solution: Write

$$\begin{aligned} y &= \tfrac{1}{3}(x^3 - 1)^{1/2}(3x^2) \\ &= \tfrac{1}{3}u^{1/2}D_x u \end{aligned}$$

Hence

$$\begin{aligned} \int x^2\sqrt{x^3 - 1}\,dx &= \frac{1}{3}\int u^{1/2}D_x u\,dx = \frac{\frac{1}{3}u^{3/2}}{\frac{3}{2}} + C \\ &= \tfrac{2}{9}(x^3 - 1)^{3/2} + C \end{aligned}$$

Problems 11.4

In Probs. 1 to 10 differentiate with respect to the independent variable indicated.

1. $y = x^2(1 - x^2)^5$.
2. $y = \dfrac{x(1 - x^3)^4}{2 + x}$.
3. $y = \dfrac{t^4 - t^2}{3 - t}$.
4. $y = \dfrac{1}{\sqrt{1 - t^3}}$.
5. $f(z) = \sqrt[3]{z^2 + 2z}$.
6. $f(z) = (4 - z^2)^{3/2}$.
7. $\rho = (1 - 2\theta)^n$.
8. $\rho = \dfrac{\theta}{(4 - 3\theta)^2}$.
9. $g(x) = e^x \sin x$.
10. $g(x) = \log_e \sin x$.

In Probs. 11 to 20 integrate as indicated.

11. $\int x^2(x^3 + 1)^4\,dx$.
12. $\int [3x - 7x(1 - x^2)^5]\,dx$.
13. $\int (1 - x - x^2)^7(1 + 2x)\,dx$.
14. $\int (x^2 + x + 1)^{3/4}(2x + 1)\,dx$.
15. $\int \sqrt[3]{1 + x}\,dx$.
16. $\int \dfrac{x}{\sqrt{x^2 - 1}}\,dx$.

17. $\int (2e^{7x} + \cos x)\, dx.$

18. $\int \frac{1}{1 - x^2} (-2x\, dx).$ HINT: See Theorem 7.

19. $\int \{3x^2(1 + x)^4 + x^3[4(1 + x)^3]\}\, dx.$ HINT: See Theorem 21.

20. $\int \frac{1}{\sin x} (\cos x\, dx).$ HINT: See Theorem 7.

In Probs. 21 to 26 find the derivative of the derivative.

21. $y = ax^2 + bx + c.$
22. $y = (a/x^2) + (b/x) + c.$
23. $y = \sqrt{1 - x}.$
24. $y = \sin x + e^{-2x}.$
25. $y = \log_e x.$
26. $y = x/(1 - x).$

In Probs. 27 to 30 find a primitive of a primitive.

27. $y = ax^2 + bx + c.$
28. $y = (a/x^3) + (b/x^2) + c.$
29. $y = \sqrt{4 - x}.$
30. $y = 3e^{2x}.$

31. Derive the formula for the derivative of a quotient of two functions f/g:

$$D\left(\frac{f}{g}\right) = \frac{gDf - fDg}{g^2}$$

32. Find the equation of the tangent to the parabola $y^2 = 4px$ at (x_1,y_1) on the curve.
33. Find the equation of the tangent to the ellipse $(x^2/a^2) + (y^2/b^2) = 1$ at (x_1,y_1) on the curve.

11.5 Second Derivatives

Since $D_xf(x) = f'(x)$ is itself a function f' of x it has a derivative, namely,

$$D_x(D_xf(x)) = \lim_{\Delta x \to 0} \frac{f'(x + \Delta x) - f'(x)}{\Delta x}$$

(provided this limit exists). We write

$$D_x^2f(x) = \lim_{\Delta x \to 0} \frac{f'(x + \Delta x) - f'(x)}{\Delta x} = f''(x)$$

and call this the second derivative of f with respect to x at the point x. The superscript 2 on D is not a square; it stands for the *second* derivative. Where $y = f(x)$, we may write (d^2y/dx^2) for $f''(x)$. Still higher derivatives could be written

$$D_x^3f, \quad . \ . \ . \ , D_x^nf, \text{ or } \frac{d^3f}{dx^3}, \quad . \ . \ . \ , \frac{d^nf}{dx^n}, \text{ or } f'''(x), \quad . \ . \ . \ , f^{(n)}(x)$$

We have already seen that, for motion in a straight line, velocity is the derivative of distance with respect to time and that acceleration is the derivative of velocity with respect to time. Therefore acceleration is the second derivative of distance with respect to time. Thus, if $y = f(t)$ is the distance from the origin at any time t,

$$\text{Distance:} \qquad y = f(t)$$

$$\text{Velocity:} \qquad v(t) = \frac{dy}{dt} = f'(t)$$

$$\text{Acceleration:} \quad a(t) = \frac{d^2y}{dt^2} = f''(t)$$

Illustration 1. If the distance y from the origin at time t is given by $y = -16t^2 + 3{,}000t + 50{,}000$, find:

(a) The initial distance; i.e., the value of y when $t = 0$.
(b) The velocity at any time t and the initial velocity.
(c) The acceleration at any time t and the initial acceleration.

Solution:

(a) $y(0) = 50{,}000$.
(b) $v(t) = -32t + 3{,}000$.
$v(0) = 3{,}000$.
(c) $a(t) = -32$.
$a(0) = -32$.

Illustration 2. For a particle moving vertically, up or down, the acceleration is known from experiment to be constant and equal to -32 ft/sec/sec. If initially (that is, at $t = 0$), the particle is y_0 ft high and moving with velocity v_0, find an expression for:

(a) The velocity at any time t.
(b) The height (distance from origin) at any time t.

Solution: Here we are given

$$a(t) = -32$$

from which, by integration, we find

$$v(t) = -32t + C_1$$

where C_1 is an arbitrary constant of integration. But in the given problem C_1 must be determined so that $v(0) = v_0$; that is,

$$v(0) = v_0 = -32(0) + C_1$$

whence $C_1 = v_0$. Therefore the velocity at any time t for this problem is

$$v(t) = -32t + v_0$$

After integration this relation will yield the height y.

$$y = -16t^2 + v_0 t + C_2$$

where C_2 is an arbitrary constant. This C_2 must be determined so that $y(0) = y_0$. That is,

$$y(0) = y_0 = -16(0)^2 + v_0(0) + C_2$$

Finally, we have

$$y(t) = -16t^2 + v_0 t + y_0$$

Look back at Illustration 1 now and note the connection between these two examples. Illustration 2 is the general problem of the falling body.

We have also seen that, for the graph of $y = f(x)$, the slope of the curve at any point x is given by the derivative $f'(x)$. The second derivative $f''(x)$ is therefore the rate of change of slope per unit change in x. These ideas are useful in plotting.

Exercise A. Consider that industrial production y is a differentiable function of t. With regard to the statement, "Industrial production is declining at a diminishing rate", discuss the shape of the curve $y = f(t)$. Make use of $D_t y$ and $D_t^2 y$.

11.6 Maxima and Minima

In this section we apply the ideas of the calculus to help us draw the graphs of certain functions.

Definition: A function f is said to be increasing at the point x_0 if, for all $|\Delta x|$ sufficiently small,

$$\begin{aligned} &f(x_0 + \Delta x) < f(x_0) \qquad \text{when } \Delta x < 0 \\ &f(x_0 + \Delta x) > f(x_0) \qquad \text{when } \Delta x > 0 \end{aligned} \tag{19}$$

A function is increasing in an interval if it is increasing at each point of the interval. As x traces such an interval in the positive direction, the graph of $y = f(x)$ rises.

Theorem 23. If $f'(x_0) > 0$, then f is increasing at x_0.

Proof: Given

$$f'(x_0) = \lim_{\Delta x \to 0} \frac{f(x_0 + \Delta x) - f(x_0)}{\Delta x} > 0$$

If in the limit the ratio $\dfrac{f(x_0 + \Delta x) - f(x_0)}{\Delta x}$ is positive for $\Delta x < 0$, then, for sufficiently small $|\Delta x|$ it must be true that $f(x_0 + \Delta x) - f(x_0) < 0$. That is, $f(x_0 + \Delta x) < f(x_0)$, which is the first part of condition (19). Again if $\Delta x > 0$ and is small, it must be true that

$$f(x_0 + \Delta x) - f(x_0) > 0$$

and the second condition of (19), $f(x_0 + \Delta x) > f(x_0)$, is satisfied. Hence the theorem is proved.

Exercise A. State and prove the converse of Theorem 23 for a differentiable function f.

Exercise B. Write out a definition of decreasing function and a theorem (and its converse) corresponding to Theorem 23.

Consider the curve $y = f(x)$, where f is a differentiable function (Fig. 11.3). The value $f(x_1)$ is the largest that the function f assumes in a small

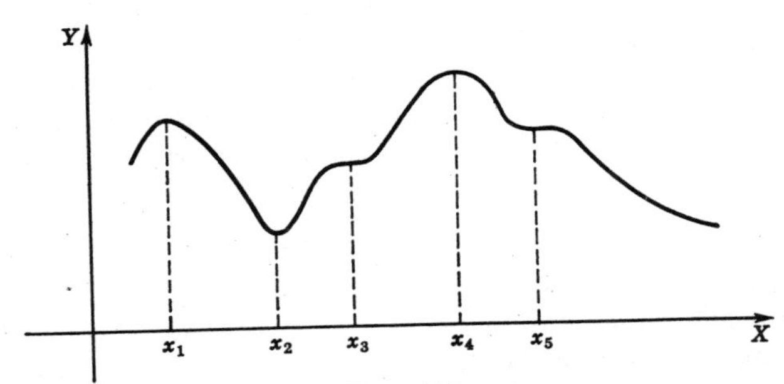

Figure 11.3

interval containing x_1. Such a value of the function f is called a *relative maximum* of f. Similarly, $f(x_2)$ is called a *relative minimum* of f. We often omit the adjective "relative", but it will still be understood. (Of course, at times we may be interested in the *absolute* maximum or *absolute* minimum of a function—usually in a given interval—in which case we shall so state.) At each point of a suitably small interval to the left of x_1, the derivative $f'(x) > 0$ (Exercise A). At x_1, the derivative $f'(x_1) = 0$. At each point of a small interval to the right of x_1, the derivative $f'(x) < 0$ (Exercise B).

Exercise C. What are the corresponding facts for small intervals to the left and to the right of x_2?

Let us put this into systematic form:

Definitions: The point $x = x_0$ is a *relative maximum point* of f if and only if $f(x_0 \pm \Delta x) < f(x_0)$ for all sufficiently small positive numbers Δx. It is a *relative minimum point* of f if and only if $f(x_0 \pm \Delta x) > f(x_0)$ for all such Δx.

As an aid to finding relative maxima and minima we prove the next theorem.

Theorem 24. If x_0 is a relative maximum or minimum of f, and if f has a derivative at x_0, then $f'(x_0) = 0$.

Proof: We must have $f'(x_0) > 0, = 0$, or <0. If $f'(x_0) > 0$, f is increasing at x_0, and hence x_0 cannot be a maximum. Similarly, we must exclude $f'(x_0) < 0$. Therefore $f'(x_0) = 0$.

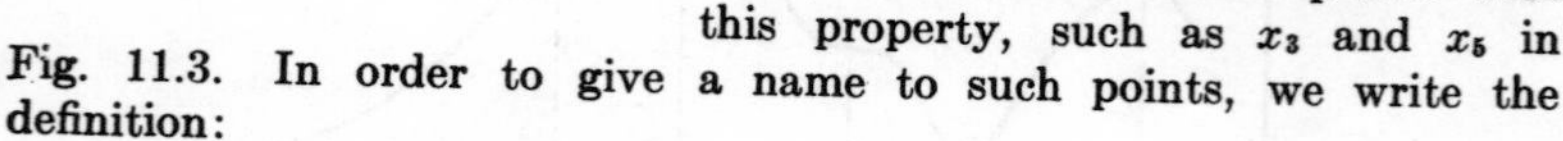

Figure 11.4

Exercise D. Why does the theorem not apply to the relative maxima and minima exhibited in the graph in Fig. 11.4?

The condition $f'(x_0) = 0$ is not sufficient for x_0 to be a maximum or minimum, for there are other points with this property, such as x_3 and x_5 in Fig. 11.3. In order to give a name to such points, we write the definition:

Definition: The point $x = x_0$ is a *stationary point* of f if and only if $f'(x_0) = 0$ and either $f'(x_0 \pm \Delta x) > 0$ or $f'(x_0 \pm \Delta x) < 0$ for all sufficiently small positive numbers Δx.

An overall term including maximum, minimum, and stationary points is that of a *critical point:*

Definition: The point $x = x_0$ is a *critical* point of f if and only if $f'(x_0) = 0$.

We summarize with a rule as follows:

Rule for Finding the Relative Maximum (Minimum) Value of a Function. First, find the function to be maximized! This function may be given. Again the statement of the original problem may be in words, and you will then have to translate these into the appropriate mathematical expressions. You may have to differentiate some given function several times, or you may have to perform other operations on given quantities, but regardless of what the operations are, you must first find the function whose maximum (minimum) is sought. Call this function f.

Second, find $f'(x)$. The solutions of $f'(x) = 0$ are the critical values, and they must be tested in order to determine whether a certain one yields a maximum value of f, a minimum value of f, or a stationary value of f.

In the table below the test may be made by using $f(x)$ or by using $f'(x)$ as indicated. In this table α and β are used to designate certain positive constants; each plays the role of a Δx to be chosen so as to simplify the test.

Testing for maxima and minima

x	$f(x)$	$f'(x)$	Comments
x_0	$f(x_0)$	Given $f'(x_0) = 0$	Testing $f(x_0)$ for a maximum;
$x_0 - \alpha$	$f(x_0 - \alpha) < f(x_0)$	$f'(x_0 - \alpha) > 0$	$\therefore$ $f(x_0)$ is a relative maximum
$x_0 - \beta$	$f(x_0 + \beta) < f(x_0)$	$f'(x_0 + \beta) < 0$	[because of the inequalities in
			either the $f(x)$ or the $f'(x)$ column]
x_0	$f(x_0)$	Given $f'(x_0) = 0$	Testing $f(x_0)$ for a minimum;
$x_0 - \alpha$	$f(x_0 - \alpha) > f(x_0)$	$f'(x_0 - \alpha) < 0$	$\therefore$ $f(x_0)$ is a relative minimum
$x_0 + \beta$	$f(x_0 + \beta) > f(x_0)$	$f'(x_0 + \beta) > 0$	[because of the inequalities in
			either the $f(x)$ or the $f'(x)$ column]

CAUTION: The interval $[x_0 - \alpha, x_0 + \beta]$ must not be so large as to include other critical values or points of discontinuity of f or of f'.

Exercise E. Make out a similar table for a stationary point.

When we have gained information about the points where f is stationary and about the maximum and minimum values of f, we are in a better position to plot the curve. Hence the calculus is a powerful tool indeed in curve tracing.

Illustration 1. Sketch the graph of $y = 2x^3 + 3x^2 - 12x$.

Solution: The zeros of the polynomial $2x^3 + 3x^2 - 12x$ are

$$x = 0 \qquad x = -\tfrac{3}{4} \pm \frac{\sqrt{105}}{4}$$

The domain of definition is $-\infty < x < \infty$. There is no symmetry, and the function is everywhere continuous. We find $f'(x)$ and set $f'(x) = 0$.

$$D_x y = 6x^2 + 6x - 12 = 0$$

that is

$$x^2 + x - 2 = (x - 1)(x + 2) = 0$$

$$x = 1, \ -2$$

These are the critical points which must be tested.

x	$f(x)$	$f'(x)$	Comments
1	$f(1) = -7$	$f'(1) = 0$	Testing $f(1)$;
Set $x_0 - \alpha = 1 - 1 = 0$	$f(0) = 0 > -7$	$f'(0) = -12 < 0$	$\therefore f(1) = -7$ is a
Set $x_0 + \beta = 1 + 1 = 2$	$f(2) = 4 > -7$	$f'(2) = 24 > 0$	relative minimum
-2	$f(-2) = 20$	$f'(-2) = 0$	Testing $f(-2)$;
Set $x_0 - \alpha = -2 - 1 = -3$	$f(-3) = 9 < 20$	$f'(-3) = 24 > 0$	$\therefore f(-2) = 20$ is a
Set $x_0 + \beta = -2 + 2 = 0$	$f(0) = 0 < 20$	$f'(0) = -12 < 0$	relative maximum

We compute a few more values of the function and sketch in Fig. 11.5.

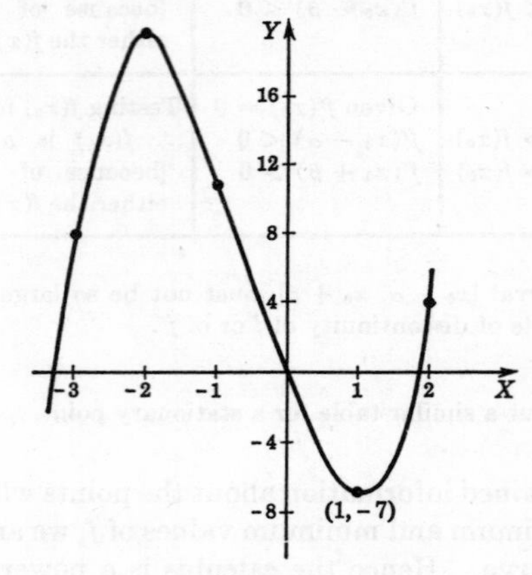

Figure 11.5

Illustration 2. Prove that, among all rectangles with fixed perimeter P, the square is the one with maximum area.

Solution: Call the sides of the general rectangle x and y. Then

$$P = 2x + 2y \tag{20}$$

The quantity to be maximized is the area A, where

$$A = xy$$

We cannot yet proceed to differentiate A, however, because there are two variables momentarily present, namely, x and y. But, using relation (20), we may eliminate either x or y. From (20),

$$y = \frac{P}{2} - x$$

so that $A = xy$ becomes

$$A = x\left(\frac{P}{2} - x\right)$$

Now we may proceed.

$$\frac{dA}{dx} = \frac{P}{2} - 2x$$

The critical value is obtained by setting

$$\frac{dA}{dx} = \frac{P}{2} - 2x = 0$$

and solving; hence $x = P/4$. Using this, we find, from (20), that $y = P/4$; also that the rectangle has equal sides, i.e., is a square.

We still do not know (except intuitively) that these values correspond to a maximum; we must test.

x	$A(x)$	$A'(x)$	Comments
$\frac{P}{4}$	This test will not be used in this problem	$A'\left(\frac{P}{4}\right) = 0$	Testing $A\left(\frac{P}{4}\right)$;
$\frac{P}{4} - \alpha = 0$		$A'(0) = \frac{P}{2} > 0$	$A\left(\frac{P}{4}\right) = \frac{P^2}{16}$ is a relative maximum given by $x = y = \frac{P}{4}$;
$\frac{P}{4} + \beta = \frac{P}{2}$		$A'\left(\frac{P}{2}\right) = -\frac{P}{2} < 0$	∴ rectangle of maximum area is a square

Problems 11.6

In Probs. 1 to 17 find all critical points and the relative maxima and minima, and plot the curve.

1. $y = x^2 + 6x + 8$.
2. $y = x^2 + 5x + 6$.
3. $y = x^2 - x - 2$.
4. $y = x^2 - 10x + 24$.
5. $y = x^3 - x^2 - 2x$.
6. $y = x^3 - 2x^2 + x$.
7. $y = x^3 + 6x^2 + 8x$.
8. $y = 6x^3 - 11x^2 - 1$.
9. $y = x^2 - x^4$.
10. $y = (x^4/4) - (x^3/3)$.
11. $y = (x^5/5) - (x^3/3)$.
12. $y = \dfrac{x-1}{x(x+2)}$.
13. $y = \dfrac{x-1}{x^2(x+2)}$.
14. $y = 2\sqrt{1-(x-3)^2}$.
15. $y = -\frac{5}{3}\sqrt{-x^2 - 6x}$.
16. $y = xe^{-2x}$.
17. $y = x^2e^x$.

18. If the velocity of a particle at any time t is given by $v = t(t + 2)$, find the minimum velocity.

19. If the velocity of a particle at any time t is given by $v = t^3 - 3t^2 + t - 5$, find the minimum acceleration.

20. If the distance of a particle from the origin at any time t is given by $s = (t^4/24) - (t^3/3) + 1$, find time when:

(a) Distance was minimum. (b) Velocity was maximum.
(c) Acceleration was minimum.

21. A particle starts at the origin and moves out along the positive X-axis for a while, then stops and moves back toward the origin, and then stops again and moves away from the origin. The distance of the particle from the origin is given by $x = 2t^3 - 9t^2 + 12t$. Find:

(a) The time t_1 when the particle stopped for the first time.
(b) The time t_2 when the particle stopped for the second time.
(c) The velocity at t_1 and t_2. (d) The acceleration at t_1 and t_2.
(e) The time when the velocity was a minimum.

22. A bomb is dropped from a stationary flying saucer 6,400 ft high. [Use $a(t) = -32$ ft/sec/sec.]

(a) When did it strike the ground?
(b) With what velocity did it strike the ground?

23. From a height of 16,000 ft a particle is hurled vertically upward with an initial velocity of 1,440 ft/sec. [Use $a(t) = -32$ ft/sec/sec.]

(a) What was the maximum height? (b) When did it strike the ground?
(c) With what velocity did it strike the ground?

24. A man has P running feet of chicken wire and with it wishes to form a rectangular pen, making use of an existing stone wall as one side. Find the dimensions so that the pen will have maximum area.

25. Prove that, among all rectangles with fixed area A, the square is the one with minimum perimeter.

26. (a) Find the relative dimensions of a closed tin can (cylindrical) to be made from a given amount of metal (without losses in cutting, etc.) that will have maximum volume.
(b) Same for an open tin cup.

27. A watermelon grower wishes to ship as early as possible in the season to catch the higher prices. He can ship now 6 tons at a profit of \$4 per ton. By waiting he estimates he can add 3 tons per week to his shipment, but that the profit will be reduced \$$\frac{1}{3}$ per ton per week. How long should he wait for maximum profit?

28. A man in a boat offshore 3 mi from the nearest point P wishes to reach a point Q down the shore 6 mi from P. On water he can travel 4 mi/hr, on land 5 mi/hr. Where should he land in order to minimize his total travel time? What if Q were 4 mi down shore? Discuss the problem if Q were 3 mi down shore.

29. What is the absolute maximum value assumed by the function given by $y = 1 - |x - 1|$?

30. (a) What is the absolute minimum value assumed by the function given by $y = 1 + \sqrt[3]{(x - 2)^2}$ in the interval $0 \leq x \leq 3$?
(b) The absolute maximum?

11.7 Related Rates

Theorem 22 gave the formula for $D_x f[u(x)]$. Since this has many applications in rate problems involving time as independent variable, we shall call x by the letter t and write

$$D_t f[u(t)] = D_{u(t)} f[u(t)] \cdot D_t u(t) \tag{21}$$

Usually the variable t does not enter explicitly: we are given some relation such as $y = f(u)$, it being understood that u is a function of t. Most often we are given enough information to compute dy/dt or to compute du/dt.

Illustration 1. The radius of a circle is increasing at the rate of 2 ft/min. How fast is the area increasing when $r = r$ ft? when $r = 3$ ft?

Solution: Evidently we have

$$A = \pi r^2$$

where r is such a function of t that dr/dt (given) $= 2$ ft/min. We are asked to find dA/dt; we therefore differentiate A with respect to t, getting

$$\begin{aligned} \frac{dA}{dt} &= 2\pi r \frac{dr}{dt} \\ &= 4\pi r \qquad \text{ft}^2/\text{min} \end{aligned} \tag{22}$$

which is the first part of the answer. For the second part we substitute $r = 3$ in (22), getting

$$\frac{dA}{dt} = 12\pi \qquad \text{ft}^2/\text{min}$$

when $r = 3$ ft.

Illustration 2. A man, 100 ft away from the base of a flagpole, starts walking toward the base at 10 ft/sec just as a flag at the top of the pole is lowered at the rate of 5 ft/sec. If the pole is 70 ft tall, find how the distance between the man and the flag is changing per unit of time at the end of 2 sec.

Solution: Call x the distance the man is from the base, y the height of flag, and z the distance between man and flag at any time t (Fig. 11.6). Then we are given

$$\frac{dx}{dt} = -10 \text{ ft/sec}$$
$$\frac{dy}{dt} = -5 \text{ ft/sec}$$

(The minus sign is present because x and y are decreasing.) Hence, integrating, we get

$$x = -10t + C_1$$
$$y = -5t + C_2$$

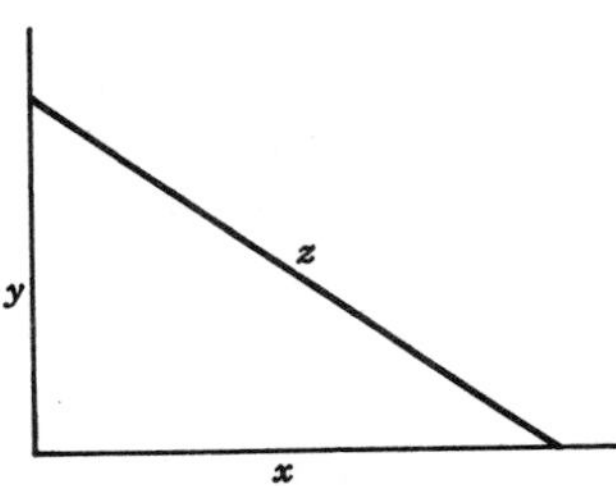

Figure 11.6

where $C_1 = 100$ ft (the initial distance) and $C_2 = 70$ ft (the initial height). Thus

$$x = -10t + 100 \qquad y = -5t + 70 \tag{23}$$

Now, always, we have

$$z^2 = x^2 + y^2 \tag{24}$$

and differentiating with respect to time gives

$$2z\frac{dz}{dt} = 2x\frac{dx}{dt} + 2y\frac{dy}{dt} \tag{25}$$

[z^2, x^2, and y^2 each is a u^n problem, Sec. 11.4, formula (17)]. From (23) we compute $x = 80$ ft, $y = 60$ ft, when $t = 2$; and then, from (24),

$$z = \sqrt{x^2 + y^2} = \sqrt{(80)^2 + (60)^2} = 100 \text{ ft}$$

Hence (25) becomes, canceling the multiplicative factor 2,

$$100\frac{dz}{dt} = 80(-10) + 60(-5) = -1{,}100$$

or $dz/dt = -11$ ft/sec. The minus sign says that the distance z is decreasing at $t = 2$.

Problems 11.7

1. A baseball diamond is a square 90 ft to the side. If a batter runs down the first base line at 30 ft/sec, how fast is his distance from third base changing as he passes first base?
2. Gas is being pumped into a spherical balloon at the rate of 2,000 ft³/min. When $r = 50$ ft, find:

   (a) The rate at which r increases.
   (b) The rate at which the surface area ($= 4\pi r^2$) increases.

3. A man 6 ft tall walks at 6 ft/sec directly away from a light that is 18 ft above the ground.

   (a) How fast does his shadow lengthen?
   (b) At what rate is the head of his shadow moving away from the base of the light?

4. A boat is being pulled toward a pier which is 10 ft above the water. The rope is pulled in at the rate of 4 ft/sec. How fast is the boat approaching the base (water line) of the pier when 30 ft of rope remains to be pulled in?
5. The edge of a cube increases at the rate of 0.0002 cm/min. When the edge is 2 cm, find:

   (a) How fast the volume is changing. (b) How the surface area is changing.

6. The volume of a cube is increasing at the rate of 0.001 cc/min. How fast is the surface area changing at time t?

7. An airplane, flying due north at an elevation of 1 mi and a speed of 600 mi/hr, passes directly over a ship traveling due east at 20 mi/hr. How fast are they separating 6 min later?
8. Under certain physical conditions, the height of a right circular cylinder increases at a rate of 0.0002 mm/min while the radius decreases at the rate of 0.0001 mm/min. Find the rate at which the volume changes at any time t.
9. The radius of a conical filter is 2 in., and the height is 3 in. Liquid passes through the filter at the constant rate of 3 cc/min. How fast is the level of the liquid falling when the depth of the liquid is 2 in.? (BT)
10. For what value of x is the rate of change of x^3 forty-eight times that of x?

References

Courant, Richard, and Herbert Robbins: "What Is Mathematics?" chap. 8, Oxford, New York (1941).

Further details are given in the many standard texts on calculus.

12 | Probability

12.1 Random Experiment

The mathematical problems we have discussed thus far in this book have led to definite answers such as

$(p \wedge q) \to p$ is a tautology.
$\{1, 2, 3\} \cap \{2, 3, 4, 5, 6\} = \{2, 3\}$.
The integers, mod 3, form a field.
No real number x satisfies the equation $x^2 + 1 = 0$.

We now wish to consider some problems for which there are no definite answers but which are, nevertheless, subject to logical analysis. A large body of such material is treated in the branch of mathematics called the *theory of probability*, the basic idea being that of a *random experiment.*

The term random experiment, or random phenomenon, is not usually defined in mathematics, but is used in reference to any and all physical experiments or other situations where nonidentical data result from what appear to be essentially identical processes. Repeated measurements on some physical object, such as measurements of length, weight, velocity, etc., generally lead to a set of readings not all of which are alike regardless of the care we exercise in making them. Random variations, sometimes called random errors, creep in quite beyond our precautions. A stamping machine may on occasion, and for some unknown reason, produce a faulty part. As yet we cannot control the sex of an embryo, which may therefore be considered a random phenomenon. A name taken from a list of registered voters may turn up a Democrat or Republican. A fair toss of a coin or die results in one of the outcomes head (H), tail (T), or 1, 2, 3, 4, 5, 6, entirely at random.

These notions are applicable to a wide variety of problems in games of chance and in the physical, biological, and social sciences and to many other areas of human endeavor.

12.2 The Sample Space

A random experiment may lead not only to many outcomes, but also to different categories, or sets, of outcomes, and our first task is to decide what we mean by a permissible set of outcomes. Let us suppose that we draw a card from a standard deck and find that it is the five of clubs. We may focus our attention upon one or more of the following *outcomes* of this draw: (1) the card is a club; (2) the card is black; (3) the card is a five; or possibly (4) the card has a cigarette burn on it. All these are perfectly acceptable *outcomes*, but we must decide which of them is of interest to us at the moment. In making the draw we are seeking the answer to one or more of the questions: (1) What suit is the card? (2) What color is the card? (3) What is the denomination of the card? (4) What is the physical condition of the card? The possible answers to these questions form sets of outcomes which we shall call *permissible.*

Definition: A set of outcomes of a random experiment is called *permissible* if and only if:

(1) It is exhaustive; i.e., we can assign at least one element of this set to every performance of the experiment.
(2) The elements of the set are mutually exclusive; i.e., we can assign no more than one element of this set to a single performance of the experiment.

Illustration 1

(a) The following are permissible sets of outcomes of the experiment of drawing one card from a deck (let c, d, h, s refer to club, diamond, heart, spade, respectively): $\{$red, black$\}$; $\{c, d, h, s\}$; $\{$burned, unburned$\}$; $\{2, 3, 4, 5, 6, 7, 8, 9, 10, J, Q, K, A\}$; $\{2c, 3c, \ldots, Ac, 2d, 3d, \ldots, Ad, 2h, 3h, \ldots, Ah, 2s, 3s, \ldots, As\}$.
(b) The set $\{2, 3, 5\}$ is not permissible since it is not exhaustive.
(c) The set $\{$red, black, club$\}$ is not permissible since its elements are not mutually exclusive. For instance, the three of clubs would be assigned to two elements of this set.

The technical words used to describe this situation are given below:

Definitions: A *sample space* is a permissible set of outcomes of a random experiment. A *simple event* is an element of a sample space.

Notation. We shall write e_1, e_2, e_3, etc., for simple events, and $U = \{e_1, e_2, e_3, \ldots, e_n\}$ for a sample space. Thus, for a deck of cards, we might write $U = \{e_1, e_2, \ldots, e_{52}\}$, where each e_i refers to a specific card.

Illustration 2. In a single toss of a coin, where we are interested only in knowing whether head or tail shows, we use as sample space $U = \{H, T\}$.

Illustration 3. In a single toss of an ordinary die, we take as sample space $U = \{1, 2, 3, 4, 5, 6\}$.

Illustration 4. An experiment consists of drawing a card from a deck and tossing a coin. $U = \{e_1H, e_2H, \ldots, e_{52}H, e_1T, e_2T, \ldots, e_{52}T\}$ is a permissible sample space that contains all the possible outcomes which are usually considered pertinent.

The situation in Illustration 4, where two distinct experiments have been combined into one, might be described as a joint experiment having a joint sample space. The importance of this point of view is explained in Sec. 12.6.

Illustration 5. An experiment consists of making random drawings from a hat containing a nickel (N), a dime (D), and a quarter (Q).

(**a**) One coin is drawn. $U_1 = \{N, D, Q\}$.
(**b**) Two coins are drawn simultaneously. $U_2 = \{ND, NQ, DQ\}$.
(**c**) One coin is drawn and examined. Then another coin is drawn. $U_3 = \{ND, DN, NQ, QN, DQ, QD\}$. Here we are interested in the *order* in which the coins are drawn.
(**d**) One coin is drawn and examined, and then it is returned to the hat. A second drawing is made. An appropriate sample space is $U_4 = \{ND, DN, NQ, QN, DQ\ QD, NN, DD, QQ\}$.

Suppose that, in our card-drawing experiment, we have chosen the sample space: $U = \{2, 3, 4, 5, 6, 7, 8, 9, 10, J, Q, K, A\}$ and that our interest now centers on whether the draw is below an 8. The simple events satisfying this condition form a subset of U, namely: $\{2, 3, 4, 5, 6, 7\}$. We shall call such a subset an *event*. In general, we have the definitions:

Definitions: An *event* is a subset of a sample space U. Two events A and B are called *complementary events* if A and B are complementary subsets of U.

The subsets of $U = \{e_1, e_2, \ldots, e_n\}$ are:

$\emptyset$, the null set: this is an impossible event.

$\{e_1\}, \{e_2\}, \ldots, \{e_n\}$; each of these is a simple event.*

$\{e_1, e_2\} = \{e_1\} \cup \{e_2\}$, $\{e_1, e_3\} = \{e_1\} \cup \{e_3\}$, $\ldots$, $\{e_{n-1}, e_n\} = \{e_{n-1}\} \cup \{e_n\}$; each is the union of two simple events.

$\{e_1, e_2, e_3\} = \{e_1, e_2\} \cup \{e_3\} = \{e_1\} \cup \{e_2\} \cup \{e_3\}$, $\ldots$; $\{e_{n-2}, e_{n-1}, e_n\} = \{e_{n-2}\} \cup \{e_{n-1}\} \cup \{e_n\}$; union of three simple events.

. . . .

$U = \{e_1, e_2, \ldots, e_n\} = \{e_1\} \cup \{e_2\} \cup \cdots \cup \{e_n\}$; the whole sample space. Our interest will be in finite sample spaces, namely, those which contain only a finite number of simple events.

* Usually, we shall omit the braces in the case of a simple event, writing e_1 instead of $\{e_1\}$, etc.

All the five regular solids make beautiful *dice*. Inasmuch as many of the problems in this chapter are connected with them, they are shown in Figs. 12.1 to 12.5, along with ways of constructing them.

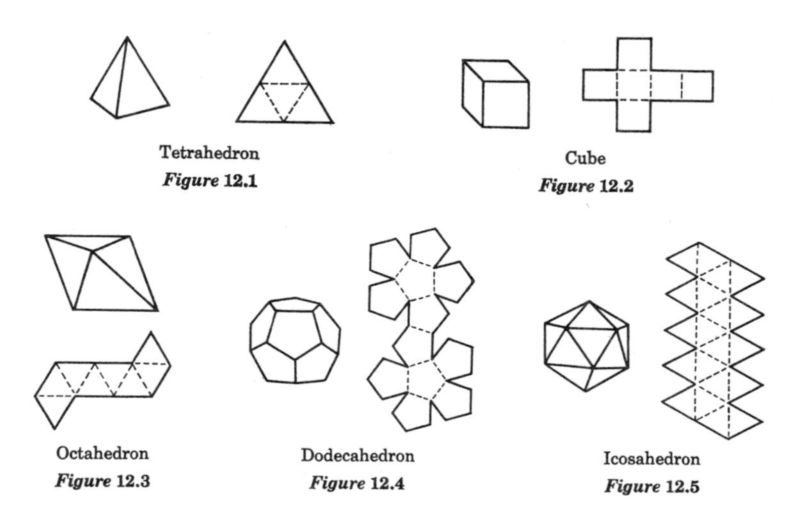

Tetrahedron
Figure 12.1

Cube
Figure 12.2

Octahedron
Figure 12.3

Dodecahedron
Figure 12.4

Icosahedron
Figure 12.5

Problems 12.2

In Probs. 1 to 16 set up a sample space.

1. Two coins are tossed once.
2. Three coins are tossed once.
3. A coin and a die are tossed once.
4. Two ordinary dice are tossed once.
5. A tetrahedral (regular) die is tossed once. Count the *down* face.
6. A tetrahedral die is tossed once. Count the sum of the visible faces.
7. A regular homogeneous octahedron (8 faces) is tossed once. Count the *up* face.
8. A regular dodecahedron (12 faces) is tossed once. Count the *up* face.
9. A regular icosahedron (20 faces) is tossed once. Count the *up* face.
10. One toss of a coin with two heads.
11. One toss of a die marked 1, 1, 2, 3, 4, 5.
12. One toss of a die marked 1, 1, 1, 2, 2, 3.
13. A certain radio tube consists of just three parts, a grid, a filament, and a plate, each one of which (when first made) might test defective or not.
14. A random sample of two people is drawn from a population of 1,000 Democrats and 500 Republicans. (The numbers are extraneous.)
15. Assume that, in a certain town, the population of all three-children families in which only one is a boy shows the following distribution according to the order of the male birth: male first child, 25 cases; second child, 35 cases; third child, 40 cases. (The numbers are extraneous.)
16. Our present calendar repeats in cycles of 400 years. In each cycle the 13th of the month falls on the days of the week with the frequencies $S(687)$, $M(685)$, $T(685)$, $W(687)$, $T(684)$, $F(688)$, $S(684)$. (The numbers are extraneous.)

In Probs. 17 to 20 which are permissible sets of outcomes, i.e., sample spaces? Explain.

17. Of the 10 digits one is selected at random.

(a) $U = \{0, 1, 2, 3, 4, 5, 6, 7, 8, 9\}$.
(b) $U = \{0, 2, 4, 6, 8\}$.
(c) $U = \{\text{even, odd}\}$.
(d) $U = \{\text{less than 2, 2, greater than 2}\}$.
(e) $U = \{0, 1, 2, 3, \text{less than 9}, 9\}$.

18. Two coins and a die are tossed.

(a) $U = \{(\text{coins alike, die shows 1}), (\text{alike}, 2), \ldots, (\text{alike}, 6)\}$.
(b) $U = \{(\text{coins alike, die is even number}), (\text{alike, odd number})\}$.
(c) $U = \{(\text{coins alike, die is 1}), \ldots, (\text{alike}, 6), (\text{unlike}, 1), \ldots, (\text{unlike}, 6)\}$.
(d) $U = \{(H, H, \text{die shows number} < 6), (H, T, <6), (T, H, <6), (T, T, <6)\}$.
(e) $U = \{HH, HT, TH, TT\}$.

19. Three dice are tossed.

(a) $U = \{(a, b, c) \mid 1 \leq a \leq 6, 1 \leq b \leq 6, 1 \leq c \leq 6\}$.
(b) $U = \{(3 \leq a + b + c \leq 18) \mid 1 \leq a \leq 6, 1 \leq b \leq 6, 1 \leq c \leq 6\}$.
(c) $U = \{\text{sum even, sum odd}\}$.
(d) $U = \{\text{number on first die even, number on first die odd}\}$.

20. Two icosahedra (one is red, one is green) are tossed.

(a) $U_1 = \{(a, b) \mid 1 \leq a \leq 20, 1 \leq b \leq 20\}$.
(b) $U_2 = \{\text{sum even, sum odd}\}$.
(c) $U_3 = \{\text{neither number is prime, exactly one is prime, two primes}\}$.
(d) $U_4 = \{(\text{red is even, green is odd})\}$.
(e) $U_5 = \{\text{red is even, green is odd}\}$.
(f) $U_6 = \{(\text{red is even, green is odd}), (\text{red is odd, green is even})\}$.
(g) $U_7 = \{(\text{red is even, green is even}), (\text{red is odd, green is odd})\}$.
(h) $U_8 = U_6 \cup U_7$ (Union of U_6 and U_7; either U_6, or U_7, or both).
(i) $U_9 = U_6 \cap U_7$ (Intersection; both U_6 and U_7).
(j) $U_{10} = U_1 \cup U_2$.
(k) $U_{11} = U_1 \cup U_2'$ (U' is complement of U).

12.3 Frequency Definition of Probability

Consider that an experiment is designed and that the category of outcomes is stated; i.e., the sample space $U = \{e_1, e_2, \ldots, e_n\}$ is specified. The experiment is now repeated N times (N large), and the number of times an event A occurs is found to be N_A. N_A is the *absolute frequency* of A, and $f_1(A) = N_A/N$ is the *relative frequency* of A in N trials. The experiment is again repeated, say, M times (M large). The relative frequency of A in M trials is $f_2(A) = M_A/M$. A series of such repetitions produces a series of relative frequencies $f_1(A), f_2(A), \ldots, f_i(A), \ldots,$ and it is man's experience that these relative frequencies do not differ greatly from one set of repetitions to another, but seem to cluster around a fixed but unknown number $P(A)$. Because of this experience, we lay down the following hypothesis, the validity of which is essential if the theory of probability is to be applicable to the physical world.

Hypothesis. There exists a constant $P(A)$ around which the experimental frequencies $f_i(A)$ cluster.

Definition: The number $P(A)$ is called the *probability* of the event A.

Thus the relative frequencies $f_i(A)$ are experimental approximations to $P(A)$. If we have only one such, say, $f(A)$, then the best we can do on an experimental basis is to use it for $P(A)$.

Illustration 1. The faces of a hexagonal pencil are marked 1 to 6, and some of the wooden edges are shaved off a bit here and there. This makes a biased object, a roll of which on a smooth floor constitutes a random experiment. The sample space is $U = \{1, 2, 3, 4, 5, 6\}$. If we wish to estimate $P(4)$, the probability that face 4 will turn up, we must make a long sequence of rolls and compute $f(4)$, which can then be taken as an approximation to $P(4)$. Essentially, there is no other way of getting an estimate of $P(4)$.

Illustration 2. What is the probability that the next child born in the United States will be a girl? This type of problem cannot be treated without some sensible information as to how we should assign probabilities to the two sexes. Biologically, we do not know too much about the way in which sex is determined. It is only natural that we turn to the records of some past time interval. United States vital statistics indicate that in 1957 the total number of live births was 4,254,784, of which 2,074,824 were girls. Counting these cases as a long run of a random experiment, we find that the answer to the question is about $2{,}074{,}824/4{,}254{,}784 = 0.4876$. For a number of years there have been fewer female than male births.

Exercise A. What is the sample space in Illustration 2?

Illustration 3. If today is your twentieth birthday, what is the probability that you will be alive on your twenty-fifth birthday? To get some kind of meaningful answer, we think of dying as a random process and search the records for a long run of such events. Mortality-experience tables used by insurance companies indicate that, of 95,991 persons (United States) alive on their twentieth birthday, 95,400 can be expected to be alive on their twenty-fifth. Your probability is therefore about $95{,}400/95{,}991 = 0.9938$. But drive safely.

Exercise B. What is the sample space in Illustration 3?

Illustration 4. A taw, or die, is made from some homogeneous material in the form of a right prism with equilateral triangles for base and top and with square sides (Fig. 12.6). The base and top are marked 1, 2, and the sides 3, 4, 5. A sample space is $U = \{1, 2, 3, 4, 5\}$, and a toss is a random experiment. Undoubtedly, the die could come to rest with any number 1 to 5 *down*, and, because of the symmetries involved, we are willing to believe that $f(1) \approx f(2)$ and $f(3) \approx f(4) \approx f(5)$ in a long sequence of tosses. Even before the experiment, we should consider it appropriate to assign $P(1) = P(2)$ and $P(3) = P(4) = P(5)$, but a sensible assignment could be made only after experimentation. Another interesting sample space for this die is $U = \{\text{triangle down, square down}\}$.

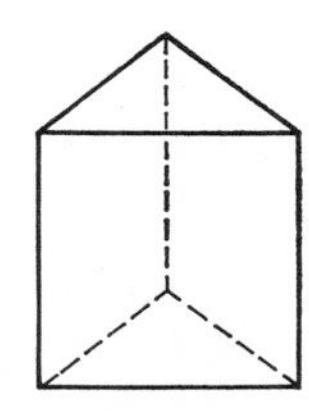

Figure **12.6**

Most of our work will be with conceptual rather than actual experiments, but it will be highly instructive for you to carry out one or more of the random experiments in the problems below.

Problems 12.3

1. Tossing a thumbtack, which may fall point up or point down, is a random experiment. Estimate the probability of point up. What is the sample space? To simplify the work, put 10 like thumbtacks in a box, shake, and examine for the number showing point up. Do this 10 times. This will give a reasonable approximation to the problem of making 100 tosses with a single tack.
2. Repeat as in Prob. 1 with 10 new bottle caps to estimate the probability of top up. What is the sample space?
3. Stack eight pennies head to tail and smoothly scotch-tape. This is now a three-sided die. Set up the sample space, toss 50 times, and estimate the probability of "cylindrical side".
4. Take four paper book matches (unused) and toss them about a foot above your head and let fall on a hard, smooth floor, such as asphalt tile. You may very well find some instances where a book comes to rest on an edge. Take $U = \{$scratch side up, scratch side down, edge$\}$ as sample space and estimate the probability of edge from a sequence of 25 tosses.
5. Glue together a dime and a quarter, tail to tail. Toss 25 times and estimate the probability of dime up.
6. Cover a sheet of paper with parallel lines drawn 1 in. apart (or use graph paper so ruled). Cut a square $\frac{1}{2}$ in. on a side from some reasonably stiff paper or cardboard. Drop the square from about 1 ft onto the rulings $n = 30$ times, counting the number of times it intersects a ruling. Estimate the probability P of an intersection. (Here there is a theoretical answer: $P \to 2/\pi$ as $n \to \infty$.)

12.4 Idealized Probability

Casual examination of a coin reveals that it is far from being a symmetric cylinder. Cylinders, cubes, and the like do not exist in the real world, but only in our minds. Like points and lines in geometry, they are abstractions of things we see in nature and, as such, have application to physical objects that are *almost* cylinders or *almost* cubes.

In the theory of probability it is highly desirable to idealize a random experiment—*by making suitable hypotheses*—so that we may shift attention from relative frequency, $f(A)$, directly to probability, $P(A)$. This affords economy of time and speeds applicability, whenever it can be done, and is especially helpful in problems involving games of chance.

Illustration 1. What is the probability of tossing a head with a penny? We take as sample space $U = \{H, T\}$, make a long sequence of fair tosses, and compute $f(H)$. Or we may repeat this many times, average the f's, and use the average $f(H)$ as an approximation for $P(H)$. If a sequence consisted in as many as 9^9 tosses, we have no doubt that $f(H)$ would be near $\frac{1}{2}$, but that actually, because a coin *is* biased, $f(H)$ would differ from $\frac{1}{2}$ in such a way as to make the assumption that $P(H) = \frac{1}{2}$ quite untenable. On the other hand, in a short game of tossing, the *assumption* that

$P(H) = \frac{1}{2}$ is reasonable. We therefore idealize the coin, thinking of it as a perfect cylinder, reason that in a long run we should expect to find $f(H)$ just about $\frac{1}{2}$, dispense with the testing, and *arbitrarily* assign $P(H) = \frac{1}{2}$. We also assign $P(T) = \frac{1}{2}$ for similar reasons. Note that $P(H) + P(T) = 1$. This is the common basis for action in games of chance using honest coins.

Illustration 2. What is the probability of tossing a 6 with an ordinary die which is made of homogeneous material and marked in the usual way with 1, 2, 3, 4, 5, 6 dots on the six faces, respectively? To make a dot, a small amount of material, roughly hemispherical, is removed and the hole painted. This method of manufacture obviously produces bias, and the face with only one dot, being the heaviest, should turn down most often, placing uppermost the opposite face, with six dots. Thus, calling $f(i)$ the relative frequency of the face with exactly i dots, we should expect to find, in a long sequence of tosses, $f(6) > f(1)$. With some 315,672 tosses, the English biologist Weldon verified this conjecture; his experiments showed that we are unjustified in making the hypothesis that $f(6)$ is about $\frac{1}{6}$. [Actually, Weldon tossed 12 dice 26,306 times, counting a five or a six as a success. Ideally, $P(5 \text{ or } 6)$ would be assumed to be $\frac{1}{3}$ for a symmetric die. He found a relative frequency of 0.3377, which is significantly too high.] However, in games of chance involving dice, this bias is usually disregarded, complete symmetry is assumed, and, arbitrarily, to each face is assigned the same probability $\frac{1}{6}$ so that $\sum_{i=1}^{6} P(i) = P(1) + P(2) + P(3) + P(4) + P(5) + P(6) = 1$.

Illustration 3. A gadget resembling a clock face is marked uniformly around the perimeter with the integers 1 to 12. A central hand is rotated at high uniform angular velocity, and when a button is pressed, an instrument will record the next number which the sweeping hand passes. What is the probability that the number will be 5? Here again is an example of a random process. It seems reasonable to assign equal probabilities to each number so that the answer to the question should be $\frac{1}{12}$, because, no doubt, the experimental method would yield a relative frequency $f(5)$ very close to $\frac{1}{12}$ for large N. This means, of course, that we are assuming that $\sum_{i=1}^{12} P(i) = 1$.

Problems 12.4

Set up a sample space, idealize the random experiment, and make suitable probability assignments assuming regular solids, regular markings (or as indicated), and one toss. For a sample space $U = \{e_1, e_2, \ldots, e_n\}$ be sure that $\sum_{i=1}^{n} P(e_i) = 1$.

1. Regulation coin.
2. Coin with two heads.
3. Tetrahedron.
4. Tetrahedron marked 1, 1, 2, 3.
5. Cube marked 1, 2, 2, 3, 3, 3.
6. Cube marked 1, 2, 3, 4, 5, blank.
7. Octahedron.
8. Octahedron marked 1, 1, 1, 1, 2, 2, 2, 3.
9. Dodecahedron.
10. Dodecahedron marked with eight 1's, three 2's, and one 3.
11. Icosahedron.
12. Icosahedron marked with seventeen H's, three T's.

12.5 The Probability Distribution

These somewhat vague and intuitive notions must be formalized. Associated with every random experiment there is a sample space $U = \{e_1, e_2, \ldots, e_n\}$. To each simple event e_i there must be assigned a probability $P(e_i)$, and this must be done solely on the basis of experimentation or hypothesis. In a long sequence of N repetitions of the experiment there are certain relations that hold for relative frequencies, and these shall serve as guides in assigning probabilities. These relations are:

(I) Let each simple event e_i occur with absolute frequency N_i. Then, since $f(e_i) = N_i/N$, the relative frequency $f(e_i)$ satisfies the double inequality $0 \leq f(e_i) \leq 1$. This suggests that the assignment $P(e_i)$ be made in such a way that $0 \leq P(e_i) \leq 1$.

(II) The fact that

$$\begin{aligned} f(e_1) + f(e_2) + \cdots + f(e_n) &= \frac{N_1}{N} + \frac{N_2}{N} + \cdots + \frac{N_n}{N} \\ &= \frac{N_1 + N_2 + \cdots + N_n}{N} = \frac{N}{N} \\ &= 1 \end{aligned}$$

suggests that the assignments be made in such a way that

$$P(e_1) + P(e_2) + \cdots + P(e_n) = 1$$

(III) Let s simple events, say, $e_1, e_2, \ldots, e_s$, occur with absolute frequencies $N_1, N_2, \ldots, N_s$, respectively, and let the event A be the union of these s simple events. (There is no loss in generality in using the first s simple events since the order of writing down the simple events is unimportant.) Thus $A = e_1 \cup e_2 \cup \cdots \cup e_s$. Then

$$\begin{aligned} f(A) = \frac{N_1 + N_2 + \cdots + N_s}{N} &= \frac{N_1}{N} + \frac{N_2}{N} + \cdots + \frac{N_s}{N} \\ &= f(e_1) + f(e_2) + \cdots + f(e_s) \end{aligned}$$

This suggests that we take $P(A) = P(e_1) + P(e_2) + \cdots + P(e_s)$.

(IV) Finally, the relative frequency of an impossible event $\emptyset$ is $f(\emptyset) = 0/N = 0$. This suggests the assignment $P(\emptyset) = 0$.

These intuitive considerations supply the motivation for a formal definition.

Definition: A *probability distribution* is an assignment of a real number to each event of a sample space, $U = \{e_1, e_2, \ldots, e_n\}$, such that the following four conditions are satisfied:

Condition I. For each simple event e_i, $0 \leq P(e_1) \leq 1$.

Condition II. $\sum_{i=1}^{n} P(e_i) = 1$.

Condition III. If $A = e_1 \cup e_2 \cup \cdots \cup e_s$, then $P(A) = \sum_{i=1}^{s} P(e_i)$.

Condition IV. $P(\emptyset) = 0$.

When probabilities are assigned in this manner to, or *distributed over*, the simple events of U, we say that a probability model, or *probability distribution over* U, has been established for the experiment.

Definition: The probability distribution in which there has been assigned the same real number to each simple event is called the *uniform probability distribution*, or simply the uniform distribution.

Conditions I and II are the important ones in the definition because they must be followed when the probabilities are initially assigned. Fortunately, they are easily checked in most instances. Condition III simply states what shall be regarded as the probability of event A (any subset of U). Condition IV only rarely enters into numerical work, but is useful in proving theorems.

Theorem 1. $P(U) = 1$.

This is an immediate consequence of Conditions II and III, since $U = e_1 \cup e_2 \cup \cdots \cup e_n$, $P(U) = \sum_{i=1}^{n} P(e_i) = 1$.

Theorem 2. For any event A, $P(A)$ satisfies the double inequality $0 \leq P(A) \leq 1$. This follows since A is a subset of U.

Exercise A. Write out the details of the proof of Theorem 2.

Theorem 3. If A and B are any two events such that

$$A \cap B = e_1 \cup e_2 \cup \cdots \cup e_t$$

then $P(A \cap B) = \sum_{i=1}^{t} P(e_i)$.

This follows from Condition III.

Exercise B. By considering relative frequencies, prove that $f(A \cup B) = f(A) + f(B) - f(A \cap B)$.

The result in Exercise B suggests the next theorem.

Theorem 4. $P(A \cup B) = P(A) + P(B) - P(A \cap B)$.

Proof: A is a subset of U and is the union of some simple events. If we add the probabilities of these simple events, the result is $P(A)$ by Condition III. Likewise, B is the union of simple events some of which (say, $e_j, \ldots, e_k$) overlap with those in A. If we add the probabilities of the simple events in B, the result is $P(B)$. But in $P(A) + P(B)$ we have included $P(e_j) + \cdots + P(e_k)$ twice, once in $P(A)$ and once in $P(B)$. We should therefore subtract $P(e_j \cup \cdots \cup e_k)$ from $P(A) + P(B)$ in order to get $P(A \cup B)$. But, since $e_j \cup \cdots \cup e_k = A \cap B$, it follows that $P(e_j \cup \cdots \cup e_k) = P(A \cap B)$. Hence the proof is complete.

Stated in words, Theorem 4 says that the probability that at least one of two events occurs is the sum of the probabilities of the separate events minus the probability that both events occur. This is often referred to as the *addition rule of probability*.

Definition: Two events A and B are *mutually exclusive* if and only if $A \cap B = \emptyset$.

Theorem 5. If A and B are mutually exclusive events, then

$$P(A \cup B) = P(A) + P(B)$$

Proof: Since $A \cap B = \emptyset$, therefore $P(A \cap B) = P(\emptyset) = 0$, by Condition IV.

Exercise C. Complete the proof of Theorem 5.

Exercise D. State and prove the converse of Theorem 5.

Theorem 6. If every pair A_i, A_j of r events $A_1, A_2, \ldots, A_r$ are mutually exclusive, then $P(A_1 \cup A_2 \cup \cdots \cup A_r) = \sum_{i=1}^{r} P(A_i)$.

Exercise E. Prove Theorem 6 by induction. HINT: $P(A_1 \cup A_2 \cup \ldots \cup A_k) \cup A_{k+1}) = P(A_1 \cup A_2 \cup \cdots \cup A_k) + P(A_{k+1})$.

Theorem 7. If A and A' are complementary events, $P(A) + P(A') = 1$.

This follows from $A \cup A' = U$, $A \cap A' = \emptyset$, Condition IV, Theorem 1, and Theorem 4.

Exercise F. Write out the details of the proof of Theorem 7.

In later sections we shall make use of these theorems.

Problems 12.5

In Probs. 1 to 12 set up the indicated probability distribution for the random experiment of tossing the given regular solid.

1. Cube. $U = \{\text{odd, even}\}$.
2. Cube. $U = \{\text{number} < 4, 4, >4\}$.
3. Tetrahedron. $U = \{\text{odd, even}\}$.
4. Tetrahedron. $U = \{\text{number} < 3, 3, >3\}$.
5. Octahedron. $U = \{\text{odd, even}\}$.
6. Octahedron. $U = \{\text{multiple of 4, otherwise}\}$.
7. Dodecahedron. $U = \{1, \text{prime, composite}\}$.
8. Dodecahedron. $U = \{\text{number} < 3, 3 \leq \text{number} < 7, 7, >7\}$.
9. Icosahedron. $U = \{\text{multiple of 3, otherwise}\}$.
10. Icosahedron. $U = \{\text{multiple of 2, multiple of 11, otherwise}\}$.
11. Icosahedron. $U = \{\text{multiple of 6, multiple of 7, otherwise}\}$.
12. Icosahedron. $U = \{\text{number} < 2, 2 < \text{number} < 18, \text{otherwise}\}$.

In Probs. 13 to 16 set up the indicated probability distribution for the random experiment of drawing one card from a full deck where:

13. $U = \{c, d, h, s\}$.
14. $U = \{\text{red, black}\}$.
15. $U = \{2, 3, 4, 5, 6, 7, 8, 9, 10, J, Q, K, A\}$.
16. $U = \{e_1, e_2, \ldots, e_{52}\}$.

Problems 17 to 20 refer to drawing one ball from a bag containing five red balls marked 1, 2, 3, 4, 5, four green balls marked 1, 2, 3, 4, and three black balls marked 1, 2, 3. Set up the probability distribution where:

17. $U = \{\text{red, green, black}\}$.
18. $U = \{\text{red} \cup \text{green, black}\}$.
19. $U = \{\text{odd, even}\}$.
20. $U = \{\text{red and odd, green and even, otherwise}\}$.

12.6 Joint Distributions

We shall have immediate need of a certain argument known as the *multiplication principle.*

Multiplication principle. If there are n ways of doing one operation and, after that, m ways of doing a second operation, then there are $n \times m$ ways of doing the two operations together.

Illustration 1. If there are five airlines flying from San Francisco to New York and eight airlines flying from New York to London, then there are $5 \times 8 = 40$ air routes from San Francisco to London.

Exercise A. Generalize this principle for n operations.

Think of the toss of a symmetric die as one experiment and the toss of a symmetric coin as another. The two uniform distributions are $U_1 = \{1, 2, 3, 4, 5, 6\}, P(1) = P(2) = \cdots = P(6) = \frac{1}{6}$ and $U_2 = \{H, T\}$, $P(H) = P(T) = \frac{1}{2}$. If the two experiments are performed jointly, there is a joint sample space $U = U_1 \times U_2$ which is the cartesian product $\{1, 2, 3, 4, 5, 6\} \times \{H, T\} = \{1H, 2H, 3H, 4H, 5H, 6H, 1T, 2T, 3T, 4T, 5T, 6T\}$. It is helpful to display this joint sample space in an array of rows and columns (something like a coordinate system) as in Fig. 12.7. Because of this representation, simple events in a sample space are sometimes called *sample points.*

A joint sample space is thus determined, but as yet there is no probability distribution because no probabilities have been assigned to the sample points of the joint sample space. While any assignment consistent with the requirements for a probability distribution is possible, there is one which, in this case, seems more reasonable than any of the others. To any simple event such as $\{1H\}$, say, the assignment $P(1H) = P(1) \cdot P(H)$ is supported by the following argument. Let $n_1, n_2, n_3, n_4, n_5, n_6$ be the number of times 1, 2, 3, 4, 5, 6 turn up, respectively, in a large number N of tosses of the die. By reasons of symmetry we *assume* that $n_1 \approx n_2 \approx n_3 \approx n_4 \approx n_5 \approx n_6 \approx N/6$, where $\sum_{i=1}^{6} n_i = N$. Similarly, let m_1, m_2 be the number of times H and T show on the die in a large number M of tosses. Again we assume that $m_1 \approx m_2 \approx M/2$, where $m_1 + m_2 = M$.

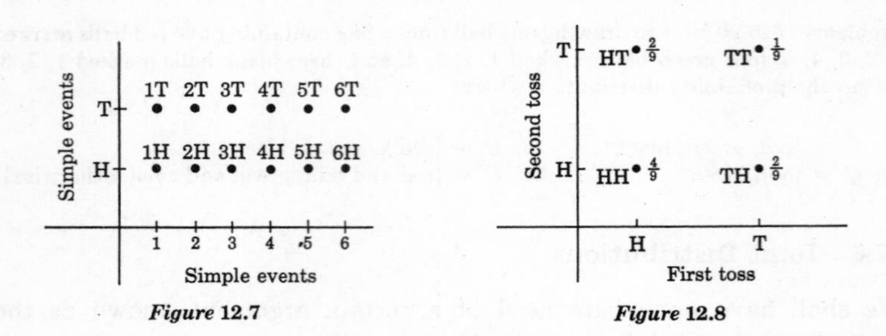

Figure 12.7 *Figure* 12.8

Then, by the Multiplication Principle, $n_1 \times m_1$ gives the number of times $1H$ occur together in a joint toss of coin and die. The relative frequency of $1H$ is n_1m_1/NM; that is,

$$f(1H) = \frac{n_1m_1}{NM} = \frac{n_1}{N} \cdot \frac{m_1}{M} \approx \frac{N/6}{N} \cdot \frac{M/2}{M} = \frac{1}{6} \cdot \frac{1}{2} = \frac{1}{12}$$

We therefore arbitrarily assign $\frac{1}{12}$ to each simple event in the joint sample space following the rule that $P(1H) = P(1) \cdot P(H)$, etc.

Exercise B. Carry out the above argument, and assign probabilities in the joint sample space of two tosses of a symmetric coin.

Symmetry is not an essential feature of the above considerations. Two tosses are made with a coin so biased that $P(H) = \frac{2}{3}, P(T) = \frac{1}{3}$. A joint probability distribution over the joint sample space

$$U = \{HH, HT, TH, TT\}$$

is to be made. $U = U_1 \times U_1 = \{H, T\} \times \{H, T\}$ (Fig. 12.8). What is a sensible way of assigning the probabilities? First, we must understand what the statement "$P(H) = \frac{2}{3}$" means. This implies that, if n_1 and n_2 are the number of times H and T turn up, respectively, in a large number N of tosses, then $n_1 \approx \frac{2}{3}N$, $n_2 \approx \frac{1}{3}N$, $n_1 + n_2 = N$. Similarly, if m_1 and m_2 are the number of times H and T show in a large number M of tosses, then $m_1 \approx \frac{2}{3}M$, $m_2 \approx \frac{1}{3}M$, $m_1 + m_2 = M$. By the Multiplication Principle, $n_1 \times m_1$ gives the number of times HH occurs in one toss followed by another. (Or we might have two such coins and toss them once simultaneously.) The relative frequency of HH is n_1m_1/NM; that is,

$$f(HH) = \frac{n_1m_1}{NM} = \frac{n_1}{N} \cdot \frac{m_1}{M} \approx \frac{\frac{2}{3}N}{N} \cdot \frac{\frac{2}{3}M}{M} = \frac{2}{3} \cdot \frac{2}{3} = \frac{4}{9}$$

We therefore arbitrarily assign $\frac{4}{9}$ to the simple event HH in the joint sample space following the rule $P(HH) = P(H) \cdot P(H)$. In like manner $P(HT) = P(H) \cdot P(T) = \frac{2}{3} \cdot \frac{1}{3} = \frac{2}{9}$, etc. Note that the distributions are not uniform.

Exercise C. Show that Condition II is satisfied by this assignment.

In general, if two probability distributions are given,*

$$U_1 = \{e_1, e_2, \ldots, e_n\}, \qquad P(e_1), \qquad P(e_2), \ldots, P(e_n)$$

and $$U_2 = \{e_1', e_2', \ldots, e_m'\}, \qquad P(e_1'), \qquad P(e_2'), \ldots, P(e_m'),$$

then the joint distribution is

$$U = U_1 \times U_2, \qquad P(e_ie_j'), \qquad i = 1, 2, \ldots, n, \qquad j = 1, 2, \ldots, m$$

Our only concern will be with distributions where we assign to the simple events e_ie_j' in the joint distribution the probabilities $P(e_ie_j')$ in accordance

* Here the primes are used merely to indicate that a second set of simple events e_j' are being considered and are not associated with complements.

with the rule $P(e_ie'_j) = P(e_i) \cdot P(e'_j)$. This is called the case of *independent experiments.*

That this assignment satisfies Condition II is clear from the following computations. The cartesian product $U_1 \times U_2$ may be written in rows and columns, thus exhibiting the simple events of the joint sample space.

$$U = U_1 \times U_2 = \{e_1, e_2, \ldots, e_n\} \times \{e'_1, e'_2, \ldots, e'_m\}$$

$$\begin{array}{cccc} e_1e'_1 & e_1e'_2 & \cdots & e_1e'_m \\ e_2e'_1 & e_2e'_2 & \cdots & e_2e'_m \\ \cdots & \cdots & \cdots & \cdots \\ e_ne'_1 & e_ne'_2 & \cdots & e_ne'_m \end{array}$$

Now consider a *fixed* simple event in U_1, say, e_i, and add the probabilities assigned to the elements in the ith row of the joint distribution.

$$\sum_{j=1}^{m} P(e_ie'_j) = \sum_{j=1}^{m} P(e_i) \cdot P(e'_j) = P(e_i)$$

since, for U_2, $\sum_{j=1}^{m} P(e'_j) = 1$. Now if we take the sum of the probabilities of all the rows, we have $\sum_{i=1}^{n} P(e_i)$, which shows that Condition II is satisfied.

Exercise D. Write out the above proof using $\sum_{i=1}^{n}\sum_{j=1}^{m} P(e_ie'_j)$ notation. (See Sec. 10.11.)

Exercise E. Argue that the proof follows from $\sum_{i=1}^{n}\sum_{j=1}^{m} P(e_i) \cdot P(e'_j) = [P(e_1) + P(e_2) + \cdots + \cdots + P(e_n)] \cdot [P(e'_1) + P(e'_2) + \cdots + P(e'_m)]$.

Generalization to more than two basic probability distributions is immediate. Thus, for U_1, U_2, U_3, we have $U_1 \times U_2 \times U_3$ and assume that

$$P(e_ie'_je''_k) = P(e_i) \cdot P(e'_j) \cdot P(e''_k)$$
$$i = 1, \ldots, n, j = 1, \ldots, m, k = 1, \ldots, p$$

Exercise F. Show that Condition II is satisfied by this assignment.

Illustration 2. Three symmetric coins are tossed. We take $P(HHT) = P(H) \cdot P(H) \cdot P(T) = \frac{1}{2} \cdot \frac{1}{2} \cdot \frac{1}{2} = \frac{1}{8}$, etc. The joint distribution is uniform.

Illustration 3. A biased coin with $P(H) = \frac{2}{3}$ is tossed, a symmetric die marked 1, 1, 1, 2, 2, 3 is tossed, and a card is drawn from a full deck. To the simple event $(H,1,K$ spades) in the joint sample space we assign $P(H,1,K$ spades$) = \frac{2}{3} \cdot \frac{1}{2} \cdot \frac{1}{52} = \frac{1}{156}$, etc. The joint distribution is not uniform.

This disposes of the assignment of probabilities to the simple events in a joint probability distribution. Theorems 3 and 4 must be used in computing $P(A \cap B)$ and $P(A \cup B)$, where A and B are any events in either a simple or joint distribution. We turn to these tasks in the next two sections.

Problems 12.6

In Probs. 1 to 20 set up a joint probability distribution assuming symmetry, ordinary markings (or as indicated), and one toss.

1. Two tetrahedra.
2. Two octahedra.
3. Two dodecahedra.
4. Two icosahedra.
5. One tetrahedron, one octahedron.
6. One tetrahedron, one dodecahedron.
7. One tetrahedron, one icosahedron.
8. One octahedron, one dodecahedron.
9. One octahedron, one icosahedron.
10. One dodecahedron, one icosahedron.
11. Two tetrahedra marked 1, 1, 2, 3 and 1, 2, 2, 3.
12. Two octahedra each marked 1, 2, 2, 3, 3, 3, 3, 4.
13. Two dodecahedra each marked with eleven 1's and one 2.
14. Two icosahedra each marked with eighteen H's and two T's.
15. Two "regular" n-gons. Devise a gadget simulating one.
16. One "regular" n-gon and one "regular" m-gon. Devise gadgets simulating them.
17. Three coins.
18. Three tetrahedra.
19. Three tetrahedra marked 1, 1, 1, 2.
20. Four coins.

In Probs. 21 to 26 set up a probability distribution for the experiment of drawing:

21. A card from a deck, examining it, replacing it, and drawing a second card.
22. A card from a deck, not examining it at the moment, not replacing it, drawing a second card, and examining both, without knowing which was drawn first.
23. A ball from a bag containing four red balls, five black balls, examining it, replacing it, and drawing a second ball.
24. A ball from a bag containing four red balls, five black balls, not examining it at the moment, not replacing it, drawing a second ball, and examining both, without knowing which was drawn first.
25. A ball from a bag containing four red balls, five black balls, not ever examining it, not replacing it, and drawing a second ball.
26. Four balls from a bag containing four red balls, five black balls, throwing them away *without examining them*, and drawing one more ball *and examining it.*

27. A College Entrance Examination Board test has 50 questions with five choices for each answer, one and only one being correct. If an answer is chosen at random for each question, what is the probability that all questions will be answered correctly? How many simple events are there in the sample space? Is the distribution uniform?

28. Of 100 people, 60 are males, of whom 30 smoke; 15 of the females smoke. One person is selected at random. Set up a probability distribution. Is it uniform?
29. If 3 per 1,000 males are stillborn and 1 per 1,000 females is stillborn. what is the probability that the next child born is female and alive? Assume the ratio male female = 51:49.
30. The following fictitious data refer to a population of 100,000 people and polio shots.

	Number of polio shots				
	0	1	2	3	Total
Contracted polio	5	1	1	0	7
Did not contract polio	293	700	1,000	98,000	99,993
Total	298	701	1,001	98,000	100,000

From the 100,000, one is selected at random. For this experiment, set up a probability distribution.
31. Each of three identical boxes has two identical compartments each of which has one coin as follows: box I, silver, silver; box II, silver, gold; box III, gold, gold. A box is selected, then one of its two compartments. What is the probability distribution?

12.7 Probability of Both A and B. $P(A \cap B)$

In order to find the probability that both A and B will occur, we must turn to the sample space, determine the sample points in $A \cap B$, and compute $P(A \cap B)$ directly. This is the essence of Theorem 3, Sec. 12.5.

Illustration 1. A symmetric coin and a symmetric die are tossed. Let A be the event "H on coin", and let B be the event "3 on die". Find $P(A \cap B)$. The uniform distribution is assumed. $U = \{H1, H2, H3, H4, H5, H6, T1, T2, T3, T4, T5, T6\}$, and each of the simple events has probability $\frac{1}{12}$.

$$A = \{H1, H2, H3, H4, H5, H6\}$$
$$B = \{H3, T3\}$$
$$A \cap B = \{H3\}$$
$$P(A \cap B) = \tfrac{1}{12}$$

Exercise A. Show that $P(A \cap B) = P(A) \cdot P(B)$ in this case.

Illustration 2. A symmetric penny and a symmetric nickel are tossed. Let A be the event "coins match", and let B be "penny falls head". Find $P(A \cap B)$. We assume the uniform distribution over $U = \{HH, HT, TH, TT\}$, where HT means H on penny, T on nickel, etc.

$$A = \{HH, TT\}$$
$$B = \{HH, HT\}$$
$$A \cap B = \{HH\}$$
$$P(A \cap B) = \tfrac{1}{4}$$

Exercise B. Show that $P(A \cap B) = P(A) \cdot P(B)$ in this case.

Illustration 3. A penny and a nickel are tossed. Let A be the event "coins match", and let B be "at least one head shows". Find $P(A \cap B)$. Again we use the uniform distribution over $U = \{HH, HT, TH, TT\}$, with the following results:

$$\begin{aligned} A &= \{HH, TT\} \\ B &= \{HH, HT, TH\} \\ A \cap B &= \{HH\} \\ P(A \cap B) &= \tfrac{1}{4} \end{aligned}$$

Exercise C. Show that $P(A \cap B) \neq P(A) \cdot P(B)$ in this case.

Illustration 4. Data and events are the same as in Illustration 2, except that penny and nickel are similarly biased with $P(H) = \frac{2}{3}$. Find $P(A \cap B)$. Here we have a nonuniform joint distribution. $P(HH) = \frac{4}{9}$, $P(HT) = \frac{2}{9}$, $P(TH) = \frac{2}{9}$, $P(TT) = \frac{1}{9}$.

$$\begin{aligned} A \cap B &= \{HH\} \\ P(A \cap B) &= \tfrac{4}{9} \end{aligned}$$

Exercise D. Show that $P(A \cap B) \neq P(A) \cdot P(B)$ in this case.

Illustration 5. A die marked 1, 2, 3, 4 in red, 5, 6 in green, is tossed. Let A be "number is even" and B be "number is red". Find $P(A \cap B)$. The uniform distribution is over $U = \{1R, 2R, 3R, 4R, 5G, 6G\}$.

$$\begin{aligned} A &= \{2R, 4R, 6G\} \\ B &= \{1R, 2R, 3R, 4R\} \\ A \cap B &= \{2R, 4R\} \\ P(A \cap B) &= \tfrac{1}{3} \end{aligned}$$

Exercise E. Show that $P(A \cap B) = P(A) \cdot P(B)$ in this case.

Illustration 6. A die marked 1, 2, 3 in red, 4, 5, 6 in green, is tossed. Let A be "number is even" and B be "number is red". Find $P(A \cap B)$. The uniform distribution is over $U = \{1R, 2R, 3R, 4G, 5G, 6G\}$.

$$\begin{aligned} A &= \{2R, 4G, 6G\} \\ B &= \{1R, 2R, 3R\} \\ A \cap B &= \{2R\} \\ P(A \cap B) &= \tfrac{1}{6} \end{aligned}$$

Exercise F. Show that $P(A \cap B) \neq P(A) \cdot P(B)$ in this case.

Exercises A to F indicate that in some cases $P(A \cap B) = P(A) \cdot P(B)$, and in some cases, not.

Definition: Two events A and B are said to be *independent* if and only if $P(A \cap B) = P(A) \cdot P(B)$.

The notion of independent events is important in advanced probability theory.

Problems 12.7

In Probs. 1 to 14 find $P(A \cap B)$, where A, B refer to either solid if alike; otherwise A refers to first. One toss is made.

1. Two tetrahedra. A, odd number; B, even.
2. Two octahedra. A, odd number; B, perfect square.
3. Two dodecahedra. A, 1 or prime; B, perfect cube.
4. Two icosahedra. A, number exceeds 17; B, power of 3.
5. One tetrahedron, one octahedron. A, odd; B, even.
6. One tetrahedron, one dodecahedron. A, 3; B, prime or even.
7. One tetrahedron, one icosahedron. A, number ≥ 1; B, perfect cube.
8. One octahedron, one dodecahedron. A, number ≥ 7; B, number ≤ 3.
9. One octahedron, one icosahedron. $A = \{x \mid (x-1)(x-2) = 0\}$; B, 1 or prime.
10. One dodecahedron, one icosahedron. $A = \{x \mid (x-4)(x-7) = 0\}$; B, power of 4.
11. Two tetrahedra, marked 1, 1, 2, 3 and 1, 2, 2, 3. A, 1; B, 2.
12. Two octahedra, each marked 1, 2, 2, 3, 3, 3, 3, 4. A, 2; B, 3.
13. Two dodecahedra, each marked with eleven 1's, one 2. A, 2; B, 1.
14. Two icosahedra, each marked with eighteen H's, two T's. A, H; B, T.

In Probs. 15 to 20 three coins are tossed once. Find $P(A \cap B)$.

15. A, coins match; B, at least one H.
16. A, coins match; B, at least two H.
17. A, coins match; B, at least three H.
18. A, coins match; B, not more than one H.
19. A, coins match; B, not more than two H.
20. A, coins match; B, not more than three H.

21. An octahedron, marked 1, 2, 3, 4, 5, 6 in red, 7, 8 in green, is tossed. Find $P(A \cap B)$, where A is "number is even", B is "number is green".
22. A dodecahedron, marked 1, 2, 3, 4, 5, 6, 7, 8 in red, 9, 10, 11, 12 in green, is tossed. Find $P(A \cap B)$, where A is "number is even", B is "number is green".
23. In bag I there are three red and four black balls; in bag II there are four red and five black balls. One ball is drawn from each bag. Find $P(R,R)$.
24. In bag I there are three red and four black balls; in bag II there are four red and five black balls. One ball is drawn from bag I, and if (and only if) it is red, it is put in bag II. Then a ball is drawn from bag II. Find $P(R,R)$.
25. Four random strikes on the letter keys of a typewriter are made. What is the probability of typing the word "good"?
26. The letter keys of a typewriter are repeatedly struck at random 26 times. What is the probability of typing "idontunderstandthisproblem"?
27. Find the probability that three people chosen at random have the same birthday (February 29 is to be omitted.)
28. Prove that if A and B are independent events, then A and B' are independent. (So are A' and B, A' and B'.)

In Probs. 29 to 32 find $P(1,1,1)$ in one toss of the three given dice.

29. Three tetrahedra.
30. Three octahedra.
31. Three dodecahedra.
32. Three icosahedra.

33. A bag contains five red balls and two black balls; one is drawn. A tetrahedron marked 1, 1, 2, 3 is tossed. And a card is drawn from a full deck. Find P(black, 1, king).

34. A coin, a die, and a tetrahedron are tossed. Find $P(H,3,2)$.

35. What is the probability of H on the first of:

(a) Two tosses of a coin? (b) Three tosses? (c) n tosses?

36. Three bags contain red, blue, and white balls as follows: $I(7,4,2)$, $II(3,1,5)$, and $III(6,1,4)$. If a bag is chosen at random and a ball drawn:

(a) What is the probability of getting a white ball?
(b) If the ball is replaced after each draw, what is the probability of drawing a white ball 2 times in succession? n times?

37. A bag contains 10 counters marked with the integers 1 to 10. A counter is drawn and replaced. A counter is again drawn. Find the probability that:

(a) The same counter was drawn the second time.
(b) Each counter drawn was an odd number.

38. Four persons in turn each draw a card from one and the same full deck without replacements. Find the probability that:

(a) Each suit is represented. (b) All are of the same suit.
(c) No two are of the same value.

12.8 Probability of Either A or B or Both. $P(A \cup B)$

Theorem 4, Sec. 12.5, furnishes the method of computing the probability that of two events at least one occurs:

$$P(A \cup B) = P(A) + P(B) - P(A \cap B)$$

The first step is to find $P(A \cap B)$.

Illustration 1. Find the probability of at most two tails or at least two heads on a toss of three coins. The uniform distribution is over $U = \{HHH, HHT, HTH, THH, HTT, THT, TTH, TTT\}$. Let A be the event "at most two tails", B be "at least two heads".

$$\begin{aligned}
A &= \{HHH, HHT, HTH, THH, HTT, THT, TTH\} \\
P(A) &= \tfrac{7}{8} \\
B &= \{HHH, HHT, HTH, THH\} \\
P(B) &= \tfrac{4}{8} \\
A \cap B &= B \\
P(A \cap B) &= \tfrac{4}{8} \\
A \cup B &= A \\
P(A \cup B) &= \tfrac{7}{8} + \tfrac{4}{8} - \tfrac{4}{8} \\
&= \tfrac{7}{8}
\end{aligned}$$

Illustration 2. A coin is tossed three times. Find the probability of $A = \{HHH\}$ or $B = \{HHT\}$ or $C = \{HHT\}$. In this case the events are mutually exclusive and Theorem 4, Sec. 12.5, applies.

$$P(A \cup B \cup C) = \tfrac{3}{8}$$

Illustration 3. If A is "H on coin" and if B is "3 on die", find $P(A \cup B)$.

$$\begin{aligned} A &= \{H1, H2, H3, H4, H5, H6\} \\ P(A) &= \tfrac{6}{12} \\ B &= \{H3, T3\} \\ P(B) &= \tfrac{2}{12} \\ A \cap B &= \{H3\} \\ P(A \cap B) &= \tfrac{1}{12} \\ P(A \cup B) &= \tfrac{6}{12} + \tfrac{2}{12} - \tfrac{1}{12} \\ &= \tfrac{7}{12} \end{aligned}$$

Illustration 4. A penny and a nickel are tossed. If A is "coins match" and if B is "at least one head shows", find $P(A \cup B)$.

$$\begin{aligned} A &= \{HH, TT\} \\ B &= \{HH, HT, TH\} \\ A \cap B &= \{HH\} \\ P(A \cap B) &= \tfrac{1}{4} \\ P(A \cup B) &= \tfrac{2}{4} + \tfrac{3}{4} - \tfrac{1}{4} \\ &= 1 \end{aligned}$$

Illustration 5. A coin is tossed five times. Find the probability of getting at least one head. The answer is $P(\text{one } H) + P(\text{two } H\text{'s}) + P(\text{three } H\text{'s}) + P(\text{four } H\text{'s}) + P(\text{five } H\text{'s})$, but this involves many calculations. The better approach is to use Theorem 7, Sec. 12.5. The event complementary to "at least one head" is "all tails", the probability of which is clearly $(\tfrac{1}{2})^5$. The answer to the problem is simply $1 - (\tfrac{1}{2})^5 = \tfrac{31}{32}$.

Problems 12.8

In Probs. 1 to 8 one toss is made. Compute the probability indicated.

1. Two coins. A, (HH); B, (TT). $P(A \cup B)$.
2. Two coins. A, (HT); B, (TH). $P(A \cup B)$.
3. Two coins. A, (at least one H); B(no T). $P(A \cup B)$.
4. Two coins. A, (at most one H); B, (no T). $P(A \cup B)$.
5. Three coins. A, (at least two H); B, (no T). $P(A \cup B)$.
6. Three coins. A, (at least two H); B, (exactly one T). $P(A \cup B)$.
7. Three coins. A, (HHH); B, (TTT); C, (exactly one H). Compute $P(A \cup B \cup C)$.
8. Three coins. A, (HHH); B, (TTT); C, (exactly two H). Compute $P(A \cup B \cup C)$.

9*. Prove: $P(A \cup B \cup C) = P(A) + P(B) + P(C) - P(A \cap B) - P(A \cap C) - P(B \cap C) + P(A \cap B \cap C)$. HINT: Consider $P((A \cup B) \cup C)$; apply Theorem 4 of this section and the distributive law for $(A \cup B) \cap C$.

In Probs. 10 to 25 one toss is made. Compute the probability indicated.

10. Three coins. A, (exactly one H); B, (at least one T); C, (HTH). $P(A \cup B \cup C)$. Use result of Prob. 9.
11. Three coins. A, (HHH); B, (HHT,THH); C, (at least two T). $P(A \cup B \cup C)$. Use result of Prob. 9.
12. Two dice. A, (1,2); B, (6,3). $P(A \cup B)$.
13. Two dice. A, $\{(1,2), (1,3)\}$; B, (3,1). $P(A \cup B)$.
14. Two dice. A, (sum $= 6$). $P(A)$.
15. Two dice. A, (sum ≥ 4). $P(A)$.
16. Two dice. A, (sum < 4); B, (sum > 5). $P(A \cup B)$.
17. Two dice. A, (sum ≤ 4); B, $\{(2,1),(3,1),(4,1)\}$. [Here (3,1) is to mean 3 on die 1 and 1 on die 2.] $P(A \cup B)$.
18. Two dice. A, $\{(1,2), (2,1), (3,4)\}$; B, $\{(1,6), (2,1)\}$; C, $\{(1,2)\}$. Use Prob. 9.
19. Two dice. A, (sum ≤ 2); B, (sum ≤ 3); (sum ≤ 4). $P(A \cup B \cup C)$. Use Prob. 9.
20. One coin, one die. A, (H); B, $\{3,4\}$. $P(A \cap B)$, $P(A \cup B)$.
21. One coin, one die. A, (H); B, (even). $P(A \cap B)$, $P(A \cup B)$.
22. Two dice. A, (sum ≤ 10). $P(A)$. HINT: Use A'.
23. Two dice. A, (sum ≤ 11). $P(A)$. HINT: Use A'.
24. Two dice. A, (sum ≤ 10); B, (sum even). $P(A \cup B)$. HINT: Use $(A \cup B)' = A' \cap B'$.
25. Two dice. A, (sum < 10); B, (sum odd). $P(A \cup B)$. HINT: Use $(A \cup B)' = A' \cap B'$.

26. A special deck of cards consists of one A, two K's, three Q's, and four J's. A die is marked 1, 1, 1, 2, 3, 3. A card is drawn, and the die tossed. A, $\{K3, Q2\}$; B, $\{2\}$. $P(A \cup B)$.
27. A bag contains four dimes and four nickels. If three coins are drawn at random, what is the probability that they will make change for a quarter?
28. In a single throw of two dice, what is the probability that there will be:

(**a**) No doublet. (**b**) Neither doublet nor a six. (**c**) No even number?

29. In three throws of two dice, what is the probability of throwing doublets not more than two times?
30. Bag A contains five red and three black balls. Bag B contains four red and six black balls. If a bag is chosen at random and a ball drawn, what is the probability that it will be black?
31. Bag A has four dimes and three nickels; bag B has three dimes and two nickels. A bag has been selected, and a dime has been drawn. What is the probability that it comes from bag B?
32. On the average, a trackman can pole-vault 12 ft two times in five trials and broad-jump 24 ft one time in three.

(**a**) On a given day what is the probability that he will succeed in both if given but one trial in each event?
(**b**) If he is to enter but one event, in which he will have two trials, and chooses this event by tossing a coin, what is his probability of success?

33. A narrow stream has one high bank and one low bank. A boy can (safely) jump from high to low four times out of five, from low to high two times in five. If he chooses to start from a bank by tossing a coin and makes two jumps:

(a) What is the probability that he ends up on the opposite bank? (If he misses a jump, he crawls out on the bank from which this jump was made.)
(b) What is the probability of his landing in the drink at least once?

34. The following data come from the *Monthly Weather Review*, vol. 81, no. 3, p. 54 (1953).

Probabilities of storm occurrences per month

Month	July	Aug.	Sept.	Oct.
At least one storm	0.39	0.75	0.92	0.83

What is the probability of the following:

(a) At least one storm in September and at least one in October?
(b) At least one in September or at least one in October, but not both?
(c) At least one in September or at least one in October or both?

35. A small college has 100 students each of whom plays at least one sport as follows:

49 play football
38 play basketball
35 play track
7 play football and basketball
9 play football and track
8 play basketball and track
2 play all three

One is picked at random. What is the probability that he plays:

(a) At least one sport? (b) Only one sport?

36. Of 100 readers:

60 read the *Times*
40 read the *Guardian*
37 read the *Sun*
20 read the *Times* and the *Guardian*
15 read the *Times* and the *Sun*
7 read the *Guardian* and the *Sun*
5 read all three

One is picked at random. What is the probability that he reads:

(a) At least two? (b) Not more than two?

12.9 Conditional Probability of A, Given B. $P(A \mid B)$

Thus far the probability questions have been directly related to events (subsets) of a sample space. For example, an honest die is tossed, and an unqualified question is asked: What is the probability of 6, namely, $P(6)$? The only information we have is that the die is tossed; the question is unconditional. But there is a whole class of problems where the question is a conditional one based on additional information. For example, the die is tossed, and we are told that an even number shows. What is the probability of 6 now that we have the additional information that the number showing is even? Surely this changes things since we know that 1, 3, 5 are ruled out and that only 2, 4, 6 are possibilities.

In searching for an answer to the question we return to an examination of relative frequencies in a long sequence N of tosses. If n_i is the number of times the face i turns up in N tosses, then $n_i \approx N/6$ and the absolute frequency of the event *both* 6 *and even* is about $N/6$. The absolute frequency of the event *even* is about $3N/6$, and, consequently, the relative frequency of *both* 6 *and even* when *even* is about $(N/6)/(3N/6) = \frac{1}{3}$. Therefore a plausible answer is $\frac{1}{3}$. This also seems reasonable upon examining, first, the uniform distribution over $U = \{1, 2, 3, 4, 5, 6\}$, $P(i) = \frac{1}{6}$, $i = 1, 2, \ldots, 6$, and then, when we are told that the die shows even, upon examining the uniform distribution over the (*reduced*) sample space $U(\text{reduced}) = \{2, 4, 6\}$, $P(i) = \frac{1}{3}$, $i = 1, 2, 3$.

In general, for a long sequence of N repetitions of an experiment, let $N_{A \cap B}$ and N_B be the absolute frequencies of the events $A \cap B$ and B, respectively. The relative frequencies are $f(A \cap B) = N_{A \cap B}/N$ and $f(B) = N_B/N$. Now the relative frequency with which $A \cap B$ occurs when B occurs is

$$\frac{N_{A \cap B}}{N_B} = \frac{N_{A \cap B}/N}{N_B/N} = \frac{f(A \cap B)}{f(B)} \qquad f(B) \neq 0$$

This makes the following definition a reasonable one.

Definition: The probability of A, *given* B, is $P(A \cap B)/P(B)$, $P(B) \neq 0$.

Notation. Another symbol for the probability of A, given B, is $P(A \mid B)$, so that

$$P(A \mid B) = \frac{P(A \cap B)}{P(B)} \qquad P(B) \neq 0$$

which is also called the *conditional* probability of event A subject to the condition that event B has happened.

Illustration 1. A tetrahedron is tossed twice. Find $P(A \mid B)$ if A is "down faces match", B is "sum of the down faces exceeds 5". That is, given that the sum of the down faces (on the two throws) exceeds 5, find the probability that the down faces match. The sample space is $U = \{1, 2, 3, 4\} \times \{1, 2, 3, 4\}$, which is listed in rows and columns in Fig. 12.9. Simple events making up A are down the main diagonal; those in B are enclosed in the triangle.

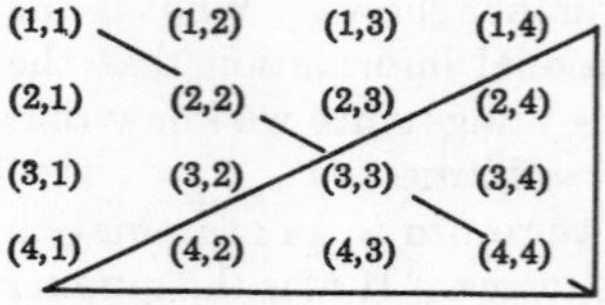

(1,1)	(1,2)	(1,3)	(1,4)
(2,1)	(2,2)	(2,3)	(2,4)
(3,1)	(3,2)	(3,3)	(3,4)
(4,1)	(4,2)	(4,3)	(4,4)

Figure 12.9

$$
\begin{aligned}
A &= \{(1,1), (2,2), (3,3), (4,4)\} \\
B &= \{(2,4), (3,3), (3,4), (4,2), (4,3), (4,4)\} \\
P(B) &= \tfrac{6}{16} \\
A \cap B &= \{(3,3), (4,4)\} \\
P(A \cap B) &= \tfrac{2}{16} \\
P(A \mid B) &= \frac{P(A \cap B)}{P(B)} = \frac{\frac{2}{16}}{\frac{6}{16}} \\
&= \tfrac{2}{6}
\end{aligned}
$$

(On occasion we may not reduce certain fractions so that we may the better see the principle involved.)

This problem may be worked from another point of view. It is given that the event B has occurred; consider B as a *reduced* sample space. Since the original distribution was uniform, it seems reasonable to assign a uniform distribution to B, which contains six simple events, each of which will now have probability $\frac{1}{6}$. Since there are just two simple events in B answering the description "faces match", it follows that $P(A \mid B) = \frac{2}{6}$. This checks with the previous result.

It is easy to prove that the reduced-sample-space method always works for uniform distributions and, when modified as in Exercise A below, for nonuniform distributions as well. Let the original distribution be uniform over U, containing n points, each with probability $1/n$. Let B contain n_1 of these, and let $A \cap B$ contain n_2. Then

$$P(A \mid B) = P(A \cap B)/P(B) = (n_2/n)(n_1/n) = n_2/n_1$$

Now set up a uniform distribution over B, considered as a reduced sample space. Each point in B now has probability $1/n_1$; also $P(A \cap B) = n_2/n_1$, and $P(B) = 1$. Hence $P(A \mid B) = (n_2/n_1)/1$, and this agrees with our previous result.

Exercise A. Let the distribution over U be nonuniform. Prove that, if in the reduced sample space B the assigned probabilities are proportional to the original ones in B (satisfying Condition II), then the reduced-sample-space method works.

Exercise B. In Illustration 1 above, assume bias; let $P(1) = P(2) = \frac{1}{8}$, $P(3) = \frac{2}{8}$, $P(4) = \frac{4}{8}$. Find $P(A \mid B) = \frac{20}{64}/\frac{44}{64}$ directly from the original nonuniform sample space. Now find $P(A \mid B) = \frac{20}{44}/1$ from the results of Exercise A.

Illustration 2. A tetrahedron is tossed twice. Find $P(A \mid B)$ if A is "the number on second toss is 2", B is "the number on the first toss is 4". By the reduced-sample-space method, we find $P(A \mid B) = \frac{1}{4}$. From the original sample space,

$$A = \{(1,2), (2,2), (3,2), (4,2)\}$$
$$B = \{(4,1), (4,2), (4,3), (4,4)\}$$
$$P(B) = \tfrac{4}{16}$$
$$A \cap B = \{(4,2)\}$$
$$P(A \cap B) = \tfrac{1}{16}$$
$$P(A \mid B) = \frac{P(A \cap B)}{P(B)} = \frac{\frac{1}{16}}{\frac{4}{16}}$$

Although $P(A)$ is not involved, note, however, that $P(A \mid B) = P(A)$. This means that the information furnished by knowing that the number 4 has occurred on the first toss did not affect the probability of 2 on the second toss. Whenever $P(A \mid B) = P(A)$, it follows that $P(A \cap B) = P(A) \cdot P(B)$ and events A and B are independent by definition (Sec. 12.8).

Exercise C. Prove that, if A and B are independent and if $P(A) \neq 0$, $P(B) \neq 0$, then $P(A \mid B) = P(A)$, and $P(B \mid A) = P(B)$.

Problems 12.9

In Probs. 1 to 20 the event B is given (has happened). Find $P(A \mid B)$.

1. A coin is tossed three times. B, HH on first two tosses; A, H on third toss.
2. A die is tossed three times. B, (6,5) on first two tosses; A, 4 on third toss.
3. A coin is tossed three times. B, at most two H; A, at least two H.
4. A coin is tossed three times. B, at least one H; A, at most two H.
5. A tetrahedron is tossed once. B, sum of faces *showing* exceeds 5; A, sum of faces *showing* exceeds 6.
6. A tetrahedron is tossed once. B, sum of the down face and the smallest face *showing* exceeds 3; A, sum of the down face and the smallest face *showing* exceeds 4.
7. A tetrahedron is tossed twice. B, first down face exceeds 2; A, second down face is 2.
8. A tetrahedron is tossed twice. B, sum of the down faces exceeds 5; A, first down face is even.
9. Two coins are tossed once. B, one coin is H; A, one coin is T.
10. Two coins are tossed once. B, no H; A, no T.
11. Two coins and two tetrahedra are tossed once. B, at least one T and sum 4; A, at least one H and sum ≤ 4.
12. Two coins and two tetrahedra are tossed once. B, at least one T and sum ≤ 3; A, two T and sum ≥ 3.
13. Two bags contain red and black balls as follows: $I(4,4)$, $II(2,6)$. B, a ball has been drawn from one of the bags at random and found to be black; A, black ball came from bag I.
14. Three bags contain red and black balls as follows: $I(4,4)$, $II(2,6)$, $III(7,1)$. B, a ball has been drawn from one of the bags at random and found to be black; A, black ball came from bag I.
15. Each of three identical boxes has two identical compartments each of which has one coin as follows: box I, silver, silver; box II, silver, gold; box III, gold, gold. A box is selected, then one of its two compartments. B, coin is gold; A, box II.

16. A person is picked at random. B, his birthday is the 13th of the month; A, Friday. (See Sec. 6.2, Prob. 16.)
17. Mother, father, and son line up at random for a family picture. B, son on one end; A, father in middle.
18. Two boys and two girls line up at random. B, girl on an end; A, girls are separated.
19. A bridge club is made up of three married couples, two single men, four single women. One member is chosen at random. B, man; A, married.
20. A committee is composed of eight Democrats, of whom two are women, and six Republicans, of whom five are men. A member is chosen at random. B, man; A, Republican.

21. Prove: $P(A' \mid B) = 1 - P(A \mid B)$.
22. Prove: $P(A \mid B') \cdot (1 - P(B)) = P(A) - P(A \cap B)$.

12.10 Counting Procedures

In the work thus far in this chapter the emphasis has been on theory, the illustrative material requiring but little computation. This was by design. The numerical details were held to a minimum in order to eliminate any possible confusion arising from arithmetic difficulties. Thus the sample spaces have had only a few simple events and have otherwise been elementary. However, in more realistic situations, for example, in the applications of probability to problems arising in the sciences, it may well be that sample spaces involve many elements, and it then becomes a real problem in itself to make an accurate count of these. We therefore turn to a study of some of the standard counting procedures so that more complicated examples may be examined.

12.11 Permutations

In Sec. 4.4 we introduced the notion of a permutation. In that instance it was a rearrangement of the numbers (1,2,3) into some other order such as (2,3,1). In general, a permutation is defined as follows.

Definition: A *permutation* of n objects is an arrangement of these objects into a particular order.

The question before us here is, How many permutations can be formed with a set of n objects? First, we consider an illustration.

Illustration 1. A mail carrier has six letters in his hand and has six mailboxes in front of him. In how many ways can he distribute the six letters into the six boxes so that just one letter goes into each box?

Solution: He can put any one of the six letters in the first box and any one of the remaining five letters in the second box. By the Multiplication Principle he has

6×5 ways of placing letters in the first two boxes. By a repetition of the argument it follows that he can distribute the six letters in $6 \times 5 \times 4 \times 2 \times 1 = 720$ ways.

In order to state Theorem 1 below, we introduce two new symbols.

Definition: The symbol $P_{n,n}$ stands for the number of permutations of n distinct elements taken all at a time.

Definition: The symbol $n!$ (for n an integer ≥ 1), read "n factorial", stands for the product

$$n! = 1 \times 2 \times 3 \times \cdots \times n$$

Further, $0!$ is defined to be 1. Factorials are not defined for negative integers or for other real numbers.

Exercise A. Compute: $5!$, $6!$, $7!$.

Exercise B. Show that $n!/n = (n - 1)!$.

Exercise C. Compute $8!/5!$; $n!/(n - 1)!$; $n!/r!$, where $r < n$.

Theorem 1. $P_{n,n} = n!$.

Proof: The first choice can be made in n ways, the second in $(n - 1)$ ways, etc. Applying the Multiplication Principle, we get at once

$$P_{n,n} = n(n - 1)(n - 2) \cdots 3 \times 2 \times 1 = n!$$

We may be interested in arranging only $r < n$ of our n elements and forgetting about those left over. In Illustration 1, the mail carrier might have had only four boxes for his six letters. Then he would have had six choices for the first box, five for the second, four for the third, and three for the fourth. In this case we speak of the "permutations of n elements taken r at a time".

Definition: The symbol $P_{n,r}$ $(r \leq n)$ stands for the number of permutations of n distinct elements taken r at a time.

Theorem 2. $P_{n,r} = \dfrac{n!}{(n - r)!} \qquad (r \leq n)$.

Proof: The proof is similar to that of Theorem 1. From the Multiplication Principle,

$$P_{n,r} = \underbrace{n(n - 1) \cdots (n - r + 1)}_{r \text{ factors}}$$

This can be written

$$P_{n,r} = \frac{n(n-1)(n-r+1)(n-r)\cdots 2\cdot 1}{(n-r)\cdots 2\cdot 1}$$

or

$$P_{n,r} = \frac{n!}{(n-r)!}$$

In the foregoing we have assumed that all n elements were distinct. It may happen that some of them are identical, and in this case the above reasoning will not hold.

Illustration 2. In how many ways can three red flags, one white flag, and one blue flag be arranged on a staff?

Solution: There are five flags present. If all five were distinct, the answer would be $P_{5,5} = 5!$. Let us label the red flags: R_1, R_2, R_3. In our 5! arrangements we have counted the permutations $R_1R_2R_3WB$, $R_2R_1R_3WB$, $R_3R_1R_2WB$, etc., as distinct arrangements. According to the statement of the problem, this is an error. There are, in fact, 3! ways of arranging $R_1R_2R_3$, all of which are to be counted as identical. Therefore the correct answer is $5!/3! = 20$.

Definition: The symbol $P_{n,n}^{r_1,r_2,\cdots}$ (for $r_1 + r_2 + \cdots = n$) stands for the number of permutations of n elements taken all at a time where there are r_1 identical elements of one kind, r_2 identical elements of another kind, etc.

Theorem 3. $P_{n,n}^{r_1,r_2,\cdots} = \dfrac{n!}{r_1!r_2!\cdots}$.

Proof: Suppose that we list all the permutations thus described. Then let us think of the r_1 identical elements as actually different and carry out the $r_1!$ arrangements in each permutation in our list, leaving all the other elements fixed. Then do the same for the r_2 identical elements, etc. When we have finished, we shall have obtained all the permutations of n distinct elements. By the Multiplication Principle,

$$P_{n,n}^{r_1,r_2,\cdots} \times r_1! \times r_2! \times \cdots = P_{n,n} = n!$$

From this the theorem follows.

Exercise D. Write out the permutations of the elements $AABC$, and verify Theorem 3.

Problems 12.11

1. If there are 20 steamers plying between New York and London, in how many ways could the round trip from New York be made if the return was made on:

   (a) The same ship? (b) A different ship?

2. A company has 15 planes to fly its executives between office and construction site. In how many ways could the round trip from the office be made if the return was made on:

   (a) The same plane? (b) A different plane?

3. Seven travelers stop overnight in a town where there are seven hotels. If no two persons are to stay at the same hotel, in how many ways may they choose hotels?
4. Seven travelers stop overnight in a town where there are six hotels. In how many ways can the travelers choose hotels?
5. How many six-letter nonsense words can be made with the letters b, c, d, t, v, and z:

   (a) With no repetitions? (b) With repetitions?

6. A cake mix is to be made with eight ingredients. In how many orders can these be put together?
7. A semaphore has three arms, and each arm has five distinct positions. What is the total number of signals that can be made?
8. Of a boat crew of eight men, three can row only on bow side and two others can row only on stroke side. In how many ways can the crew be assigned to rowing positions (four on a side)?
9. Each of 10 socks is a different color. How many color combinations can a person wear if left and right foot arrangements are:

   (a) Considered distinct? (b) Unimportant?

10. In how many of the distinct permutations of the letters in "syzygy" do the three y's come together?
11. How many permutations of nine letters can be made out of the letters of the word "Tennessee"?
12. How many permutations of 11 letters can be made out of the letters of the word "Mississippi"?
13. In how many of the distinct permutations of the letters in "Mississippi" do the four i's not come together?
14. At a political rally there are five speakers. In how many orders may they speak if:

    (a) A is to precede B? (b) A is to precede B immediately?

15. In how many ways can the letters of "abstemiously" be rearranged without changing the order (with respect to each other) of the vowels?
16. How many football teams could be formed using just 11 players (switching positions)?
17. How many football teams could be formed using 15 players (switching positions)?
18. How many six-place automobile tags could be made up using one and only one letter of the alphabet (omitting I and O) and the 10 digits, allowing a letter and five digits? [Assume that 0(zero) may not be used in the first (left-hand) position.]

19*. How many identification tags, each in the form of a matrix with two rows and three columns, can be made using two letters (omitting I and O) and four digits (allowing repetitions)?

20*. How many integers x, such that $1,111 \leq x \leq 4,321$, can be formed with the digits 1, 2, 3, 4 (allowing repetitions)?

21*. How many natural numbers not exceeding 4,321 can be formed with the digits 1, 2, 3, 4 (allowing repetitions)?

22. A circular conference table seats 10 persons. In how many ways may 10 persons seat themselves? (Consider only positions relative to each other, not with respect to the table.)

23. In how many ways can n keys be put on a key ring?

24. If $21P_{n,4} = 7P_{n,5}$, find n.

25. In how many ways can 10 different things be divided equally among 5 persons?

26. Each face of a regular tetrahedron is to be painted a solid and different color. How many tetrahedra can thus be painted with:

(a) Four colors? (b) Five colors?

27. Each face of a cube is to be painted a solid and different color. How many cubes can thus be painted with six colors?

28*. A daring young lady has three shades of nail polish with which to paint her fingernails. How many ways can she do this (each nail is to be one solid color) if:

(a) There are no other restrictions?
(b) There are not more than two shades on one hand?

29. In how many ways can n identifiable letters be stuffed into m linearly arranged pigeonholes?

30. In how many ways can n unidentifiable letters be stuffed into m linearly arranged pigeonholes?

12.12 Combinations

In permutations the *order* of the elements is important. If we ignore the order of the elements, we speak of "combinations". A combination of a set of elements is a subset of all or a part of these elements taken without regard to the order of the elements. We wish to know how many subsets consisting of r elements each can be formed from the elements of a set containing n elements. This is called the number of combinations of n elements taken r at a time.

Illustration 1. How many committees of five can be chosen from a group of eight persons?

Solution: This is a *combination* problem since *order* of the members of the committee is unimportant. Suppose that we call this number $C_{8,5}$. For each combination we can form 5! permutations by arranging the members of the committee in different orders. This gives us $P_{8,5}$ permutations. From the Multiplication Principle,

$$C_{8,5} \times 5! = P_{8,5} = \frac{8!}{3!}$$

Therefore

$$C_{8,5} = \frac{8!}{5!3!} = 56$$

Definition: The symbol $C_{n,r}$ $(0 \leq r \leq n)$ stands for the number of combinations of n distinct elements taken r at a time, and $C_{n,0} = 1$. The simpler symbol $\binom{n}{r}$ is widely used for $C_{n,r}$.

Theorem 4. $\binom{n}{r} = \dfrac{n!}{r!(n-r)!}$.

Proof: By reasoning like that of Illustration 1,

$$\binom{n}{r} \times r! = P_{n,r} = \frac{n!}{(n-r)!}$$

Problems 12.12

1. Show that $\binom{n}{r} = \binom{n}{n-r}$.
2. Show that $\binom{n}{r} = P_{n,n}^{r,n-r}$. Can you prove this directly from the definitions?
3. If $\binom{n}{9} = \binom{n}{8}$, find $\binom{n}{17}$.
4. If $P_{n,r} = 12$ and $\binom{n}{r} = 6$, find n and r.
5. An airline had only 7 seats left on a given flight when a party of 10 applied for tickets.
   (a) In how many ways could the 7 seats be filled?
   (b) Suppose a second flight had just 3 vacant seats. In how many ways could the 10 passengers then be accommodated?
6. A man buys five new tires and tubes. In how many ways could the tires and tubes be paired?
7. A gardener has 20 different plants, but places for only 17. In how many ways may the planting be made?
8. A woman has seven dresses, three of which she proposes to wear daily (one for breakfast, one for lunch, one for dinner). For how many days can she do this before repeating some daily selection?
9. A bookstore stocks works A, B, C, D bound singly, in pairs, in triplets, and all together. In how many editions can A be bought?
10. If four straight lines are drawn in a plane, no two parallel, no three concurrent, how many triangles will be formed?
11. How many triangles can be formed if each side is to be 3 or 4 or 5 or 6 in. long?

12. Each night the Sultan selects some 4 of his 14 storytellers to meet with him as a group, each group of 4 differing in at least one storyteller. On how many nights can they meet without repeating a group?

13. Two identical pieces of chalk and one eraser are to be issued to four pupils. In how many ways can this be done if:

(a) No pupil is to receive more than one object?
(b) No pupil is to receive both pieces of chalk?

14. At a luncheon for six there is a choice of three beverages, coffee, tea, or milk. How many different selections are possible (such as two coffees, one tea, three milks)? Assume that each person takes one and only one beverage.

15. An automobile dealer has just four automobiles. How many different selections (of at least one) can be made?

16. A post office keeps four kinds of stamps. How many different selections (of at least one but not more than three) can be made?

17. Two opposite faces of a cube are to be left unpainted. The other four faces are to be painted each a solid color. How many ways can the cube be painted if there are four colors to choose from but no color is to be used for more than two faces?

18. Two opposite faces of a regular dodecahedron are to be left unpainted. The other 10 faces are to be painted each a solid and a different color. How many ways can the dodecahedron be painted if there are 15 colors to choose from?

19. A travel agency advertises round trips to 30 countries. In how many ways could a tourist make two such trips (nonrepeating) per month per year?

20. How many different sums of money can be formed from a cent, a nickel, a dime, a quarter, a half dollar, and a dollar?

21. In how many ways can a party of 10 persons be divided into:

(a) Two groups of 7 and 3 each?
(b) Two named groups (A and B) of 5 each?
(c) Two unnamed groups of 5 each?

22. In how many ways can at least four, but not more than six, persons be chosen from nine?

23. We wish to select six persons from eight, but if A is chosen, then B must also be chosen. In how many ways can the selection be made?

24. From a group consisting of 15 Republicans and 10 Democrats, how many committees can be formed consisting of either 4 Republicans and 6 Democrats or 6 Republicans and 4 Democrats?

25. A committee of five is to be selected from a club consisting of 60 men and 40 women. In how many ways can this be done so that the committee women outnumber the committee men?

26. A plane is divided into how many regions by n lines, no two parallel and no three concurrent? HINT: Use induction.

27. A super checkerboard has $2n$ rows and $2n$ columns. In how many ways can two squares of different color be selected?

28. How many natural numbers less than 1,000 do not contain the digit 9?

29*. In how many ways can three distinct natural numbers, each less than 100, be chosen so that their sum is divisible by 3? HINT: Consider residue classes, mod 3.

30*. A room has seven lights, each controlled by a separate switch. In how many ways can three be turned on so that no two are turned on together more than once?

12.13 Binomial Theorem

By direct multiplication, we can easily establish the following formulas:

$$
\begin{aligned}
(a+b)^2 &= a^2 + 2ab + b^2 \\
(a+b)^3 &= a^3 + 3a^2b + 3ab^2 + b^3 \\
(a+b)^4 &= a^4 + 4a^3b + 6a^2b^2 + 4ab^3 + b^4 \\
(a+b)^5 &= a^5 + 5a^4b + 10a^3b^2 + 10a^2b^3 + 5ab^4 + b^5
\end{aligned}
$$

The coefficients in these products have a pattern which is illustrated by the following scheme, known as *Pascal's Triangle.*

$$
\begin{array}{lccccccccccc}
(a+b)^0 & & & & & & 1 & & & & & \\
(a+b)^1 & & & & & 1 & & 1 & & & & \\
(a+b)^2 & & & & 1 & & 2 & & 1 & & & \\
(a+b)^3 & & & 1 & & 3 & & 3 & & 1 & & \\
(a+b)^4 & & 1 & & 4 & & 6 & & 4 & & 1 & \\
(a+b)^5 & 1 & & 5 & & 10 & & 10 & & 5 & & 1 \\
\cdots & \cdots & \cdots & \cdots & \cdots & \cdots & \cdots & \cdots & \cdots & \cdots & \cdots & \cdots
\end{array}
$$

In this array each horizontal line begins and ends with a 1, and each other entry is the sum of the two numbers to its left and right in the horizontal row above.

Exercise A. By direct multiplication, verify the above formulas for $(a + b)^2$, $(a + b)^3$, $(a + b)^4$, $(a + b)^5$.

Exercise B. From Pascal's Triangle determine the coefficients of the terms in the expansion of $(a + b)^6$, and verify your result by direct multiplication. ANSWER: 1, 6, 15, 20, 15, 6, 1.

Whenever we discover a pattern like this, we suspect that there must be some general way of describing it, and so we are led to ask whether there is some general formula for $(a + b)^n$, where n is any positive integer. There is indeed such a formula, known as the *Binomial Formula,* and we shall now proceed to develop it.

Consider

$$
\begin{aligned}
(a+b)^n &= \underbrace{(a+b)(a+b) \cdots (a+b)}_{n \text{ factors}} \\
&= a^n + \cdots + Ca^{n-r}b^r + \cdots + b^r
\end{aligned}
$$

where C is the coefficient of a typical term $a^{n-r}b^r$ $(0 \leq r \leq n)$. Now, in multiplying out the product $(a + b)(a + b) \cdots (a + b)$, $a^{n-r}b^r$ will be obtained by selecting b from each of some r factors (and a from the remaining $n - r$ factors). Since a selection of r things from n can be made in $\binom{n}{r}$ ways, it appears that $C = C_{n,r} = \binom{n}{r}$.

Exercise C. Verify that Pascal's Triangle can be written in the form:

$$
\begin{array}{ll}
(a+b)^0 & \binom{0}{0} \\
(a+b)^1 & \binom{1}{0} \quad \binom{1}{1} \\
(a+b)^2 & \binom{2}{0} \quad \binom{2}{1} \quad \binom{2}{2} \\
(a+b)^3 & \binom{3}{0} \quad \binom{3}{1} \quad \binom{3}{2} \quad \binom{3}{3} \\
(a+b)^4 & \binom{4}{0} \quad \binom{4}{1} \quad \binom{4}{2} \quad \binom{4}{3} \quad \binom{4}{4} \\
(a+b)^5 & \binom{5}{0} \quad \binom{5}{1} \quad \binom{5}{2} \quad \binom{5}{3} \quad \binom{5}{4} \quad \binom{5}{5} \\
\cdots\cdots & \cdots\cdots\cdots\cdots\cdots\cdots\cdots\cdots\cdots\cdots
\end{array}
$$

We have derived (and essentially proved) the Binomial Formula, or Theorem.

Theorem 5. Binomial Theorem. Let n be a positive integer. Then

$$
(a + b)^n = a^n + \binom{n}{1} a^{n-1}b + \binom{n}{2} a^{n-2}b^2 + \cdots
$$
$$
+ \binom{n}{r} a^{n-r}b^r + \cdots + \binom{n}{n-2} a^2b^{n-2} + \binom{n}{n-1} ab^{n-1} + b^n
$$

Or, in the expansion of $(a + b)^n$, the coefficient of $a^{n-r}b^r$ is $\binom{n}{r}$.

But let us give a more formal proof by mathematical induction. The formula is trivially verified for $n = 1$, and indeed we have verified it for $n = 1, 2, 3, 4, 5$. We therefore prove: If the formula is true for $n = k$, then it is true for $n = k + 1$. To do so, we wish to show that the coefficient of $a^{k+1-r}b^r$ in the expansion of $(a + b)^{k+1}$ is $\binom{k+1}{r}$. By hypothesis

$$
(a + b)^k = a^k + \cdots + \binom{k}{r-1} a^{k+1-r}b^{r-1} + \binom{k}{r} a^{k-r}b^r + \cdots + b^k
$$

Then $(a + b)^{k+1}$ is given by the product:

$$
\begin{array}{l}
a^k + \cdots + \binom{k}{r-1} a^{k+1-r}b^{r-1} + \binom{k}{r} a^{k-r}b^r + \cdots + b^k \\
a + b \\
\hline
a^{k+1} + \cdots + \binom{k}{r-1} a^{k+2-r}b^{r-1} + \binom{k}{r} a^{k+1-r}b^r + \cdots + ab^k \\
\qquad a^k b + \cdots \qquad\qquad + \binom{k}{r-1} a^{k+1-r}b^r + \binom{k}{r} a^{k-r}b^{r+1} \\
\qquad\qquad\qquad\qquad\qquad\qquad\qquad\qquad\qquad + \cdots + b^{k+1} \\
\hline
a^{k+1} + \cdots \qquad\qquad + \left[\binom{k}{r} + \binom{k}{r-1}\right] a^{k+1-r}b^r + \cdots + b^{k+1}
\end{array}
$$

Therefore the coefficient of $a^{k+1-r}b^r$ is $\binom{k}{r} + \binom{k}{r-1}$. We must now simplify this.

$$
\begin{aligned}
\binom{k}{r} + \binom{k}{r-1} &= \frac{k!}{r!(k-r)!} + \frac{k!}{(r-1)!(k-r+1)!} \\
&= \frac{k!(k-r+1) + k!r}{r!(k-r+1)!} \\
&= \frac{k!(k+1)}{r!(k+1-r)!} = \frac{(k+1)!}{r!(k+1-r)!} = \binom{k+1}{r}
\end{aligned}
$$

Therefore the formula is verified for $n = k + 1$, and by the axiom of induction (Sec. 3.2), the theorem is proved.

Exercise D. Relate the last computation in the proof above to the method of constructing Pascal's Triangle.

The Binomial Theorem permits us to write down rather quickly the expansions of powers of binomials which are tedious to compute by repeated multiplication.

Illustrations

1. Expand $(2x + 5y)^3$.

$$
\begin{aligned}
(2x + 5y)^3 &= (2x)^3 + 3(2x)^2(5y) + 3(2x)(5y)^2 + (5y)^3 \\
&= 8x^3 + 60x^2y + 150xy^2 + 125y^3
\end{aligned}
$$

2. Expand $(3x - 2y)^4$.

$$
\begin{aligned}
(3x - 2y)^4 &= (3x)^4 + 4(3x)^3(-2y) + 6(3x)^2(-2y)^2 + 4(3x)(-2y)^3 + (-2y)^4 \\
&= 81x^3 - 216x^3y + 216x^2y^2 - 96xy^3 + 16y^4
\end{aligned}
$$

3. Compute the term involving x^4y^3 in the expansion of $(3x - 5y)^7$. This term will involve $(3x)^4(-5y)^3$ with an appropriate coefficient. The theorem tells us that this coefficient is $\binom{7}{3} = 35$. So the term is

$$35(3x)^4(-5y)^3 = -354{,}375x^4y^3$$

Problems 12.13

1. Compute: $\binom{6}{i}$, $i = 0, 1, \ldots, 6$.

2. Compute: $\binom{7}{i}$, $i = 0, 1, \ldots, 7$.

In Probs. 3 to 6 verify the given formulas by direct computation.

3. $\binom{6}{2} + \binom{6}{3} = \binom{7}{3}$.

4. $\binom{5}{3} + \binom{5}{4} + \binom{6}{4}$.

5. $1 + \sum_{i=1}^{4} \binom{4}{i} = 2^4$.

6. $1 + \sum_{i=1}^{5} \binom{5}{i} = 2^5$.

7. Use the Binomial Theorem to show that $1 + \sum_{i=1}^{n} \binom{n}{i} = 2^n$.

In Probs. 8 to 16 expand by the binomial theorem.

8. $(2x + y)^6$.

9. $(a - 2b)^5$.

10. $\left(\frac{1}{2s} + t\right)^5$.

11. $(x + \overline{\Delta x})^6$.

12. $(x - 2\overline{\Delta x})^6$.

13. $[(2/x) - x^2]^5$.

14. $\left(\frac{1}{x} + \frac{1}{y}\right)^4$.

15. $(1.02)^4 = (1 + 0.02)^4$.

16. $(1.99)^4 = (2 - 0.01)^4$.

17. Find the coefficient of $a^{10}b^3$ in the expansion of $(a - b)^{13}$.
18. Find the coefficient of r^4s^5 in the expansion of $(2r - 3s)^9$.
19. Find the coefficient of x^2y^3z in the expansion of $(x + y + z)^6$.
20. What is the coefficient of $x^ry^sz^t$ in the expansion of $(x + y - z)^n$, where $n = r + s + t$?

12.14 Binomial Probability Distribution

The binomial theorem is most useful in probability theory. Consider an experiment in which our interest centers on an event A and its complement A'. Since one of these will be thought of as the successful event and the other as the event of failure, we might as well call them S and F. An appropriate probability distribution is $U_1 = \{F, S\}; P(F), P(S)$. For simplicity, let $P(S) = p$ and $P(F) = q(= 1 - p)$. Further, let $P(0)$ be the probability of exactly 0 successes and $P(1)$ the probability of exactly 1 success in one performance (or trial) of the experiment. Thus

$$P(0) = q \qquad P(1) = p$$

Note that

$$(q + p)^1 = q + p$$

If the experiment is performed a second time (two trials), the joint distribution is over

$$U_2 = \{F, S\} \times \{F, S\} = \{FF, FS, SF, SS\}$$

with respective probabilities

$$P(FF) = qq \qquad P(FS) = qp \qquad P(SF) = pq \qquad P(SS) = pp$$

From these it follows that the probabilities $P(0)$, $P(1)$, $P(2)$ of exactly 0, 1, 2 successes in two trials are, respectively,

$$P(0) = q^2 \qquad \begin{aligned} P(1) &= qp + pq \\ &= 2qp \end{aligned} \qquad P(2) = p^2$$

Note that

$$(q + p)^2 = q^2 + 2qp + p^2$$

The reason for adding qp and pq to obtain $2qp$ is that we have no interest in whether the first or the second trial resulted in the one success.

If the experiment is performed a third time, the distribution is over

$$\begin{aligned} U_3 &= \{F, S\} \times \{F, S\} \times \{F, S\} \\ &= \{FFF, FFS, FSF, SFF, FSS, SFS, SSF, SSS\} \end{aligned}$$

with probabilities

$$qqq,\ qqp,\ qpq,\ pqq,\ qpp,\ pqp,\ ppq,\ ppp$$

Therefore $P(0)$, $P(1)$, $P(2)$, $P(3)$ of exactly 0, 1, 2, 3 successes in three trials are

$$P(0) = q^3$$

$$\begin{aligned} P(1) &= qqp + qpq + pqq \\ &= 3q^2p \end{aligned} \qquad \begin{aligned} P(2) &= qpp + pqp + ppq \\ &= 3qp^2 \end{aligned}$$

$$P(3) = p^3$$

Note that

$$(q + p)^3 = q^3 + 3q^2p + 3qp^2 + p^3$$

Again, like terms q^2p have been added and like terms qp^2 have been added because the order of the successes and failures is considered unimportant.

These three cases suggest that there is a general rule which gives, in n trials, $P(r)$ the probability of exactly r successes (and, of course, $n - r$ failures) and that this number $P(r)$ is exactly the $(r + 1)$st term in the binomial expansion of $(q + p)^n$. The reasoning is precisely that of the initial argument used in proving Theorem 5 in the previous section. Consequently,

$$P(r) = \binom{n}{r} q^{n-r}p^r$$

$$(q + p)^n = q^n + \binom{n}{1} q^{n-1}p + \cdots + \binom{n}{r} q^{n-r}p^r + \cdots + p^n$$
$$= P(0) + P(1) + \cdots + P(r) + \cdots + P(n)$$

Exercise A. Why is $\sum_{i=0}^{n} P(i) = 1$?

Exercise B. Is the distribution over $\{F, S\}$ necessarily uniform?

Where the order of S and F is unimportant, we write S^aF^b for the event of a successes and b failures. Also, we have agreed to write $\binom{n}{0} = 1$. This leads to the following useful definition:

Definition: The distribution over $U = \{F^n, F^{n-1}S, \ldots, F^{n-r}S^r, \ldots, S^n\}$ with probabilities $P(r) = \binom{n}{r} q^{n-r}p^r$, $r = 0, 1, \ldots, n$, is called the *Binomial Probability Distribution.*

Illustration 1. A coin is tossed 100 times. What is the probability of getting exactly 47 heads? ANSWER: $P(47\ H\text{'s}) = \binom{100}{47} (\frac{1}{2})^{100-47}(\frac{1}{2})^{47} = \binom{100}{47}/2^{100}$.

Illustration 2. A die is tossed 100 times. What is the probability of getting exactly 47 3's? Here $p = \frac{1}{6}$, $q = \frac{5}{6}$. ANSWER: $P(47\ 3\text{'s}) = \binom{100}{47} (\frac{5}{6})^{100-47}(\frac{1}{6})^{47} = 5^{53}\binom{100}{47}/6^{100}$.

Illustration 3. A biased coin with $p = P(H) = \frac{2}{3}$ is tossed 100 times. What is the probability of 47 or 48 heads? ANSWER: $P(47 \text{ or } 48\ H\text{'s}) = \binom{100}{47}(\frac{1}{3})^{53}(\frac{2}{3})^{47} + \binom{100}{48}(\frac{1}{3})^{52}(\frac{2}{3})^{48}$.

Illustration 4. An icosahedron is tossed 100 times. What is the probability of not more than 47 16's? Here $p = \frac{1}{20}$. ANSWER:

$$\begin{aligned} P(\leq 47\ 16\text{'s}) &= \binom{100}{0}(\tfrac{19}{20})^{100} + \binom{100}{1}(\tfrac{19}{20})^{99}(\tfrac{1}{20}) + \cdots \\ &\quad + \binom{100}{47}(\tfrac{19}{20})^{53}(\tfrac{1}{20})^{47} \\ &= \sum_{r=0}^{47}\binom{100}{r}(\tfrac{19}{20})^{100-r}(\tfrac{1}{20})^{r} \end{aligned}$$

Illustration 5. Two dice are tossed 100 times. What is the probability that the sum on a given toss will not exceed 7 more than 97 times?

Solution: For a single toss of two dice the probability of getting a sum of 2 or 3 or 4 or 5 or 6 or 7 is

$$p = \tfrac{1}{36} + \tfrac{2}{36} + \tfrac{3}{36} + \tfrac{4}{36} + \tfrac{5}{36} + \tfrac{6}{36} = \tfrac{21}{36} = \tfrac{7}{12} \qquad q = \tfrac{15}{36} = \tfrac{5}{12}$$

The answer to the problem is therefore

$$\begin{aligned} P(\text{sum} \leq 7 \text{ not more than 97 times}) &= \sum_{r=0}^{97}\binom{100}{r} q^{100-r}p^{r} \\ &= \sum_{r=0}^{97}\binom{100}{r}(\tfrac{5}{12})^{100-r}(\tfrac{7}{12})^{r} \\ &= 1 - \sum_{r=98}^{100}\binom{100}{r}(\tfrac{5}{12})^{100-r}(\tfrac{7}{12})^{r} \end{aligned}$$

Exercise C. Explain the last line in the answer above.

In each of the above five illustrations there would be great numerical difficulties in reducing the answer directly to simple decimal form. There are easy methods for getting good approximations, but these will not be considered in this book. (The idea is to use the normal probability distribution as an approximation to the binomial distribution.)

Illustration 6. From a bag containing m red balls and n black balls, k balls are drawn. Given that they are all of one color, what is the probability that they are black?

Solution: First an adequate sample space must be chosen. A simple event will be any combination of k balls taken from the $(m + n)$ balls. The total number of such is $\binom{m+n}{k}$, and, for a uniform distribution, the probability of each simple event is $1\Big/\binom{m+n}{k}$. The events of special interests are A(k black), B(k black or k red), and $(A \cap B) = A$. The number of simple events in A is $\binom{n}{k}$ and in B is $\binom{n}{k} + \binom{m}{k}$. We are asked to find $P(A \mid B)$.

$$P(A \mid B) = \frac{P(A \cap B)}{P(B)} = \frac{P(A)}{P(B)}$$

$$= \frac{\dfrac{1}{\binom{m+n}{k}} \cdot \binom{n}{k}}{\dfrac{1}{\binom{m+n}{k}} \cdot \left[\binom{n}{k} + \binom{m}{k}\right]}$$

$$= \frac{\binom{n}{k}}{\binom{n}{k} + \binom{m}{k}}$$

The final form of the answer gives a short cut to the solution.

Problems 12.14

1. A batter is now batting 250. What is the probability that he will make:

   (a) A hit next time at bat?
   (b) Exactly two hits out of his next three trips to bat?
   (c) At least two hits?

2. The probability that a marksman will hit a target is given as $\frac{1}{10}$. What is his probability of at least 1 hit in 10 shots?
3. The probability of winning a certain contest is $\frac{1}{5}$. What is the contestant's probability of winning at least four out of six such contests?
4. What is the probability of obtaining:

   (a) Exactly one head in tossing a coin six times?
   (b) Two heads? (c) Three heads? (d) At least three heads?

5. If four dice are tossed, what is the probability that:

   (a) Exactly two will turn up aces? (b) At least two?

6. What is the probability of obtaining the sum 7 at least once in three throws of two dice?
7. A die is marked 0, 0, 0, 0, 2, 3. What is the probability of obtaining the sum 12 in five tosses?
8. A tetrahedron is tossed 10 times. Find $P(\text{sum} = 34)$.

In Probs. 9 to 12 a coin is tossed 500 times. Find:

9. P(at least $2T$).
10. P(at most $2T$).
11. P(at least $2T$ but not more than $5T$).
12. P(at least $200T$ but not more than $300T$).

In Probs. 13 to 16 a die marked x, x, y, y, z, z is tossed 5 times. Find:

13. P(at least $4y$'s).
14. P(at most $4y$'s).
15. P(at least 1 but not more than $3y$'s).
16. P($2y$'s or $3z$'s).

In Probs. 17 to 20 a die marked x, x, x, y, y, z is tossed five times. Find:

17. P(at least $4x$'s).
18. P(at most $4z$'s).
19. P(at least $1x$ and not more than $1z$).
20. P(at least $1y$ and not more than $1z$).

21. In one toss of two dice, consider a success to be either a 7 or an 11 or a 12. What is the probability of success?
22. From a deck consisting of the cards 2, 3, 4, 5, 6, 7, a single draw is made, success being either a 3 or a 5. Find the probability of x successes in seven draws (with replacements).
23. A batter has a probability of 0.25 of making a hit in a single trip to the plate. Find the probability of x hits in 25 times at bat.
24. Suppose that the probability of catching a trout on a single cast is 0.01. Find the probability of catching x trout in 250 casts.
25. Two letters fall out of the word "little". If they are replaced at random, what is the probability that it still reads *little?* HINT: Write down the fall-out sample space $\{l_1i, l_1t_1, l_1t_2, \ldots, l_2e\}$; treat l_1l_2, t_1t_2 separately.
26. A red card is removed from a pack. Thirteen cards are then drawn and found to be all of one color. What is the probability that the color is black? (See Illustration 6 above.)
27. What is the probability that two squares marked at random on a chessboard have contact only at a corner?
28. What is the probability that a hand of five cards contains:

(a) No ace? (b) Exactly one ace?

29. What is the probability that a hand of five cards contains:

(a) No pair? (b) At least two aces?

30. What is the probability of throwing a doublet with:

(a) Two true dice? (b) One true and one biased die?

31. Find the probability of throwing a sum of 14 with:

(a) Three dice. (b) Eleven dice.

32. The integers 1, 2, . . . , 20 are written down in a row so that no two of the integers 1, 2, 3, 4 come together. What is the probability that 2 is next to 7?

33. The integers 1, 2, . . . , 20 are written down in a ring so that no two of the integers 1, 2, 3, 4 come together. What is the probability that 1 is next to 8?
34. Ten people stand in a line. What is the probability that A stands next to B?
35. Ten people sit at a round table. What is the probability that A sits next to B?
36. Five men speak at a political rally. If it is known that A speaks after B, what is the probability that it is immediately after?

12.15 Testing Hypotheses

Suppose we have tossed a coin 10 times and have obtained 10 heads. What may we conclude as to whether or not the coin is biased? The procedure is to make a hypothesis about the coin and to establish a means of accepting or rejecting this hypothesis. We may make any hypothesis we please and then carry out the test, but we must make a specific hypothesis.

To be definite, we shall make the hypothesis that the coin is true, i.e., that the probability of a head on a single toss is $p = \frac{1}{2}$. On the basis of this hypothesis, the probability of 10 heads in 10 tosses is

$$\left(\frac{1}{2}\right)^{10} = \frac{1}{1{,}024} = 0.00098$$

Hence, if the coin were true, it is very unlikely that the observed result would have taken place. We are therefore inclined to reject the hypothesis that the coin is true.

Suppose the coin is biased with $p = P(\text{head}) = \frac{2}{3}$. On the basis of this hypothesis, the probability of 10 heads in 10 tosses is

$$\left(\frac{2}{3}\right)^{10} = \frac{1{,}024}{59{,}049} = 0.017$$

This is still not very likely, but is certainly more likely than in the hypothesis that $p = \frac{1}{2}$. We may be in doubt as to acceptance or rejection of this hypothesis.

Finally, if we have a badly warped coin and make the hypothesis that $p = 0.9$, the probability of 10 heads in 10 tosses is $(0.9)^{10} = 0.34$. This is so likely that we should certainly accept this hypothesis. We should similarly accept the hypotheses that $p = 0.99$ and $p = 1.0$.

We need to understand clearly what is meant by the acceptance of a hypothesis. We have chosen this hypothesis because of evidence not connected with our particular experiment, and we wish to know whether the experiment confirms or denies the hypothesis. If we reject the hypothesis, we certainly must abandon it. If we accept the hypothesis, we are essentially saying that we have no valid reason for rejecting it. We cannot be at all certain that an accepted hypothesis is true. Thus we may accept a number of hypotheses on the basis of the same experiment. The decision as to which of these is correct (if indeed any is) cannot be made with certainty, and we proceed on the assumption that any of them may be correct.

In order to have a definite rule for deciding whether to accept or reject a hypothesis, we must decide in advance on some level of probability which will separate these two situations. In common practice this value is taken to be 0.05 or 0.01 or sometimes 0.005. For consistency in this book we arbitrarily use the value of 0.01. We now adopt the following procedure:

Rule for Testing Hypotheses (Connected with a Binomial Distribution)

(1) Make the hypothesis that the probability of success in a single trial of the event under consideration is equal to some definite value p.
(2) Using this value of p, in a binomial distribution compute the probability $\bar{p}$ of the observed number of successes in the stated number of trials n.
(3) If $\bar{p} < 0.01$, reject the hypothesis (at the 0.01 level).
If $\bar{p} \geq 0.01$, accept the hypothesis (at the 0.01 level).

In these terms we then state the conclusions to the questions raised at the beginning of this section and we use the following language:

At the 0.01 level we reject the hypothesis that $p = \frac{1}{2}$.
At the 0.01 level we accept the hypotheses that p equals $\frac{2}{3}$, 0.9, 0.99, and 1.0.

Note that we cannot conclude from our observations what the correct value of p actually is. The best that we can do is to reject certain values of p and to accept others. Even when we follow this rule, we may make errors. We may reject a true hypothesis, but this will happen in only a few cases. We may also accept a false hypothesis. We must run the risk of these errors if we are to come to any conclusions at all. Statistical inference may therefore be called "probable inference" in distinction to the exact inferences we draw in other types of reasoning.

Problems 12.15

1. A coin is tossed five times, and three heads are obtained. Let p be the probability of obtaining a head on a single toss. At the 0.01 level, test for acceptance or rejection the hypothesis:

(a) $p = \frac{1}{10}$. (b) $p = \frac{1}{3}$. (c) $p = \frac{1}{2}$.
(d) $p = \frac{2}{3}$. (e) $p = \frac{3}{4}$.

2. Toss a coin 10 times and count the number of heads. Assuming the coin to be true, compute the probability of obtaining this number of heads. Do you accept or reject (at the 0.01 level) the hypothesis that your coin is true?

3. A coin is tossed 10 times and gives 8 heads and 2 tails. Test the hypothesis that the coin is true (using the 0.01 level).

4. A die is rolled six times, in which the two-spot turns up twice. Test the hypothesis (at the 0.01 level) that the die is true.

5. A pair of dice are rolled five times, and on each roll the total number of spots is seven. In case the dice belong to your opponent (who is a stranger), decide whether or not you should shoot him, taking into account the laws of the land and of probability.

6. Six thumbtacks are tossed simultaneously onto the floor, and exactly three land with their points straight up. Test (at the 0.01 level) the hypothesis that the probability that a tack will land point up is:

(a) $\frac{1}{3}$. (b) $\frac{1}{2}$.

7. An automatic bolt machine has been making 3 faulty bolts in 100 on the average. An inspector takes five bolts at random and finds one defective among them. Test (at the 0.01 level) the hypothesis that the machine is out of adjustment.

8. A random sample of five people reveals that one is left-handed and four are right-handed. At the 0.01 level test the hypothesis that:

(a) One person in eight is left-handed.
(b) One person in nine is left-handed.

9. On the average 5 persons in 100 still get disease X even though they have been inoculated against it. The use of a new drug with a sample of 100 persons reduces the number who contract the disease to just three. Test (at the 0.01 level) the hypothesis that the new drug is better. (Use logarithms for the calculations.)

10. A seed house sells seed in packets of 100. Experience indicates that, without a special treatment, only 90 per cent of the seed will germinate, on the average. After treatment a random packet tested 95 per cent viable. At the 0.01 level, test the hypothesis that the treatment is effective. (Use logarithms for the calculations.)

References

Feller, William: "Probability Theory and Its Applications", Wiley, New York (1950).
Mosteller, Frederick, Robert E. K. Rourke, and George B. Thomas, Jr.: "Probability with Statistical Applications", Addison-Wesley, Reading, Mass. (1961).

Boolean Algebra | 13

13.1 The Algebra of Sets

In Chap. 1 you became acquainted with the concept of a set and learned about the operations on subsets of a universal set, namely, intersection, union, and complementation. Our purpose here is to study this subject in more detail as an illustration of an abstract mathematical system whose properties are quite different from those of a field or those of any algebra with which you are familiar. We shall see that this abstract system, called *Boolean Algebra,* includes not only the algebra of sets, but also the algebra of propositions developed in Chap. 1. Moreover, it has a surprising application to the design of certain types of electric circuits, and in this way has become an important tool for the electrical engineer of today.

Let us begin by reviewing what we learned about the algebra of sets in Chap. 1. We shall be dealing with a universal set U and the collection of its subsets: A, B, C, etc., together with the empty set $\emptyset$, which is a subset of U and also of every A, B, C, etc., which is a subset of U. We have defined the *intersection* $A \cap B$ as the subset of U whose elements belong to both A and B; the *union* $A \cup B$ as the subset of U whose elements belong either to A or to B or to both A and B; and the *complement* A' as the subset of U whose elements do not belong to A.

We have also noted a number of important properties of these operations, such as:

(a) Intersection and union are commutative; that is, $A \cap B = B \cap A$ and $A \cup B = B \cup A$.

(b) Intersection and union are associative; that is,

$$(A \cap B) \cap C = A \cap (B \cap C)$$

and

$$(A \cup B) \cup C = A \cup (B \cup C)$$

(c) There are two distributive laws:

$$A \cap (B \cup C) = (A \cap B) \cup (A \cap C)$$
$$A \cup (B \cap C) = (A \cup B) \cap (A \cup C)$$

There are other identities of this kind, but before giving you a basic list of them, let us see how they can be derived.

Our first method of derivation was given in Chap. 1. The method was to prove the parallel tautology for propositions by means of a truth table and then to use Theorem 17 to translate this into an identity for subsets of U. A second method in common use is that of Venn diagrams. Let us illustrate this by using this method to demonstrate the first distributive law stated above.

Illustration 1. Use Venn diagrams to establish the identity:

$$A \cap (B \cup C) = (A \cap B) \cup (A \cap C)$$

Solution: Suppose that A, B, and C have the relative positions shown in Fig. 13.1. Working on the left side of the equality, we first shade $B \cup C$ and obtain Fig. 13.2.

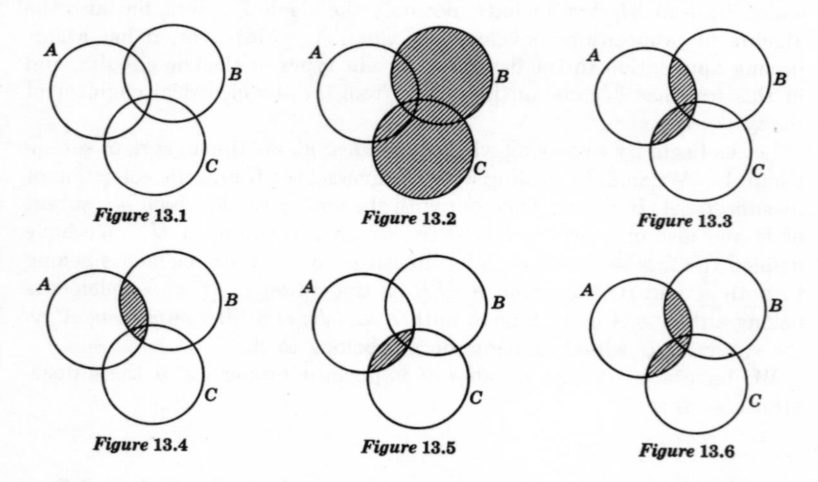

Figure 13.1 *Figure* 13.2 *Figure* 13.3

Figure 13.4 *Figure* 13.5 *Figure* 13.6

We then reconsider and shade $A \cap (B \cup C)$ in Fig. 13.3. Now working on the right side of the equality, we shade $A \cap B$ (Fig. 13.4) and $(A \cap C)$ (Fig. 13.5). Finally, from Figs. 13.4 and 13.5, we shade $(A \cap B) \cup (A \cap C)$ as in Fig. 13.6. The desired result is verified by noting that Figs. 13.3 and 13.6 are identical.

The method of Illustration 1, though helpful to the intuition, has certain logical deficiencies. We had to assume a particular position of the sets A, B, and C in Fig. 13.1 and have proved the identity only for this special case. We should complete the proof by considering all possible cases, but this is very tedious and is hardly ever done.

There is, however, a straightforward and rigorous method for proving such identities which is essentially the same as the method of truth tables for proving tautologies. We adopt a notation which will also be helpful later in this chapter. Consider a subset A of U and an arbitrary element x of U. If x is in A, we write the symbol 1; if x is not in A, we write the symbol 0. Our symbols 1 and 0 in fact correspond to the truth values T and F of Chap. 1. If there are two subsets A and B, all possible positions of an arbitrary element x are included in the table.

A	B	**Interpretation**
1	1	x in both A and B
1	0	x in A, not in B
0	1	x not in A, in B
0	0	x not in either A or B

When there are three sets, A, B, C, the corresponding table requires eight horizontal lines.

Exercise A. Write the table for three subsets: A, B, C.

In this notation our three basic operations are then defined by the tables:

Intersection

A	B	$A \cap B$
1	1	1
1	0	0
0	1	0
0	0	0

Union

A	B	$A \cup B$
1	1	1
1	0	1
0	1	1
0	0	0

Complementation

A	A'
1	0
0	1

We can now proceed just as with truth tables to prove the identities for subsets of U.

Illustration 2. By the above method prove that $A \cap (B \cup C) = (A \cap B) \cup (A \cap C)$.

Solution: Form the table:

A	B	C	$B \cup C$	$\overset{*}{A \cap (B \cup C)}$	$A \cap B$	$A \cap C$	$\overset{*}{(A \cap B) \cup (A \cap C)}$
1	1	1	1	1	1	1	1
1	0	1	1	1	0	1	1
0	1	1	1	0	0	0	0
0	0	1	1	0	0	0	0
1	1	0	1	1	1	0	1
1	0	0	0	0	0	0	0
0	1	0	1	0	0	0	0
0	0	0	0	0	0	0	0

The identity follows from the equality of the two starred columns.

In this way we can prove all the true relationships among subsets of U. From among these we shall extract a basic collection which will constitute the axioms of our abstract Boolean Algebra.

13.2 Definition of a Boolean Algebra

In this section we shall give our definition of a Boolean Algebra. Although our axioms are motivated by the algebra of sets, and although we use the symbols A, B, C, etc., and $\cap$, $\cup$, $'$, remember that now we are dealing with an abstract system, and not with the algebra of sets.

Definition: A Boolean Algebra is an abstract system given by:

Undefined Terms. A set S of undefined elements, A, B, C, etc., containing at least two distinct elements.

Undefined Operations. $\cap, \cup, '$, called intersection, union, and complementation, respectively.

Axioms

B1. CLOSURE. If A and B are elements of S, then $A \cap B$, $A \cup B$, and A' are uniquely defined elements of S.

B2. ASSOCIATIVITY. For any three elements of S, A, B, and C:

(a) $(A \cap B) \cap C = A \cap (B \cap C)$; and
(b) $(A \cup B) \cup C = A \cup (B \cup C)$.

B3. COMMUTATIVITY. For any two elements of S, A, and B:

(a) $A \cap B = B \cap A$; and
(b) $A \cup B = B \cup A$.

B4. EXISTENCE OF IDENTITY ELEMENTS. There exist distinct elements of S, O, and I such that, for any A in S:

(**a**) $I \cap A = A \cap I = A$; and
(**b**) $O \cup A = A \cup O = A$.
We call I the *universal element*, or the *identity for intersection*, and O the *null element*, or the *identity for union.*

B5. EXISTENCE OF COMPLEMENTS. Corresponding to each element A of S there exists a unique element A', called the complement of A, with the properties:

(**a**) $A \cap A' = A' \cap A = O$; and
(**b**) $A \cup A' = A' \cup A = I$.

B6. DISTRIBUTIVE LAWS. For any three elements of S, A, B, and C:

(**a**) $A \cap (B \cup C) = (A \cap B) \cup (A \cap C)$; and
(**b**) $A \cup (B \cap C) = (A \cup B) \cap (A \cup C)$.

Remarks

1. Let us compare *these axioms* with those of a field (Sec. 2.9). For this purpose let $\cup$ correspond to $+$, $\cap$ to $\times$, O to 0, and I to 1. Then we see that Axioms B1 to B4 are the same as the corresponding field axioms. Our B5 is quite different from the field axioms concerning inverses, and in B6 we now have two distributive laws instead of the single distributive law for fields.

2. There is complete duality in the axioms of Boolean Algebra in the sense that, if $\cap$ and I are interchanged with $\cup$ and O, respectively, the axioms are left unchanged. For this reason, if we can prove a theorem in this algebra, we can automatically conclude the truth of the dual theorem, which is obtained from the proved theorem by the above interchange of symbols.

3. These axioms are redundant, since some of them can be proved from others. We have chosen the above set for convenience in our exposition and because of its analogy with the axioms of a field.

4. If: A, B, and C represent subsets of a universal set U,
O represents the empty set $\emptyset$,
I represents U,
$\cap$, $\cup$, and $'$ have their customary meanings in terms of sets,
then these axioms are true relationships in the algebra of sets. (See Probs. 13.2.)

5. If: A, B, and C are replaced by propositional variables p, q, r,
O by the (constant) proposition f (which is false),
I by the (constant) proposition t (which is true),
$\cap$ by conjunction $\wedge$,
$\cup$ by disjunction $\vee$,
$'$ by negation $\sim$,
$=$ by equivalence $\leftrightarrow$,
then these axioms become tautologies in the algebra of propositions. (See Probs. 13.2.)

Problems 13.2

In Probs. 1 to 10 verify the truth of the given axiom as a theorem in the algebra of sets

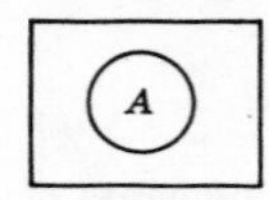

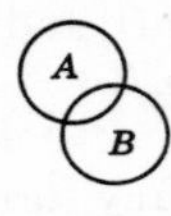

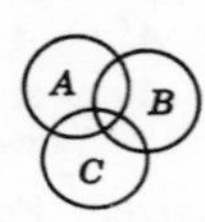

using Venn diagrams. Begin your series of diagrams with whichever of the above figures is relevant.

1. B2*a*.	**2.** B2*b*.
3. B3*a*.	**4.** B3*b*.
5. B4*a*.	**6.** B4*b*.
7. B5*a*.	**8.** B5*b*.
9. B6*a*.	**10.** B6*b*.

In Probs. 11 to 20 prove the truth of the given axioms as a theorem in the algebra of sets using the method of Illustration 2, Sec. 13.1.

11. B2*a*.	**12.** B2*b*.
13. B3*a*.	**14.** B3*b*.
15. B4*a*.	**16.** B4*b*.
17. B5*a*.	**18.** B5*b*.
19. B6*a*.	**20.** B6*b*.

In Probs. 21 to 30 reinterpret the given axiom as a tautology in the algebra of propositions (see Remark 5 above), and prove that this is a tautology by the use of truth tables.

21. B2*a*.	**22.** B2*b*.
23. B3*a*.	**24.** B3*b*.
25. B4*a*.	**26.** B4*b*.
27. B5*a*.	**28.** B5*b*.
29. B6*a*.	**30.** B6*b*.

13.3 Theorems in a Boolean Algebra

As noted in Remark 2 of Sec. 13.2, theorems in Boolean Algebra come in dual pairs. In Theorem 1 we shall give dual proofs, and after that you are expected to be able to construct the proof of the dual by yourselves.

Theorem 1. $A \cap O = O$; $A \cup I = I$.

Proof:

$$\begin{aligned} O &= A \cap A' && [\text{B5}a] \\ &= A \cap (A' \cup O) && [\text{B4}b] \\ &= (A \cap A') \cup (A \cap O) && [\text{B6}a] \\ &= O \cup (A \cap O) && [\text{B5}a] \\ &= A \cap O && [\text{B4}b] \end{aligned} \qquad \begin{aligned} I &= A \cup A' && [\text{B5}b] \\ &= A \cup (A' \cap I) && [\text{B4}a] \\ &= (A \cup A') \cap (A \cup I) && [\text{B6}b] \\ &= I \cap (A \cup I) && [\text{B5}b] \\ &= A \cup I && [\text{B4}a] \end{aligned}$$

The first part of this theorem reminds us of our earlier theorem in a field to the effect that $a \times 0 = 0$. Note that the proof is quite different. The field analog of the dual is $a + 1 = 1$, which is false!

Theorem 2. Idempotent laws:

$$A \cap A = A \qquad A \cup A = A$$

Proof:

$$\begin{aligned} A &= A \cap I && [\text{B4}a] \\ &= A \cap (A \cup A') && [\text{B5}b] \\ &= (A \cap A) \cup (A \cap A') && [\text{B6}a] \\ &= (A \cap A) \cup O && [\text{B5}a] \\ &= A \cap A && [\text{B4}b] \end{aligned}$$

Exercise A. Prove that $A \cup A = A$. HINT: Dualize the above proof.

Theorem 3. De Morgan's laws:

$$(A \cap B)' = A' \cup B' \qquad (A \cup B)' = A' \cap B'$$

Proof: We construct the proof of the first of these by showing that $A' \cup B'$ is the complement of $A \cap B$. It must be shown that both parts of B5 hold. As for the first:

$$\begin{aligned} (A \cap B) \cap (A' \cup B') &= (A \cap B \cap A') \cup (A \cap B \cap B') \\ &= (O \cap B) \cup (A \cap O) \\ &= O \cup O \\ &= O \end{aligned}$$

as is required. As for the second,

$$\begin{aligned} (A \cap B) \cup (A' \cup B') &= (A' \cup B') \cup (A \cap B) \\ &= (A' \cup B' \cup A) \cap (A' \cup B' \cup B) \\ &= (I \cup B') \cap (I \cup A') \\ &= I \cap I \\ &= I \end{aligned}$$

Since the complement of $A \cap B$ is unique, we must therefore have:

$$(A \cap B)' = A' \cup B'$$

Exercise B. Give the reasons for the steps in the proof above.

Exercise C. Prove the dual theorem.

Theorem 4. Law of involution: $(A')' = A$.
Proof left to the reader.

Theorem 5. Laws of absorption:

$$A \cap (A \cup B) = A \qquad A \cup (A \cap B) = A$$

Proof:

$$\begin{aligned} A \cap (A \cup B) &= (A \cup O) \cap (A \cup B) \\ &= A \cup (O \cap B) \qquad [\text{"Factor": use B6}b \text{ right to left}] \\ &= A \cup O \\ &= A \end{aligned}$$

Exercise D. Prove the dual theorem.

Theorem 6

$$I' = O \qquad O' = I$$

Proof left to the reader.

It is frequently convenient to introduce a relation into a Boolean Algebra which is somewhat analogous to inequality. We have already met this as the inclusion relation for two sets (Sec. 1.4) where we wrote $A \subseteq B$ and read "A is included in B". In Exercise A, Sec. 1.13, we saw that, for sets, $A \subseteq B$ is equivalent to the statement $A \cap B' = \emptyset$, and we use this to motivate our abstract definition.

Definition: The relation $A \subseteq B$ is defined to be equivalent to the proposition $A \cap B' = O$.

The next few theorems give us the properties of the inclusion relation.

Theorem 7. For every A, $O \subseteq A \subseteq I$.

Proof: From the definition, $O \subseteq A$ is equivalent to $O \cap A' = O$, which is true by Theorem 1. Similarly, $A \subseteq I$ is equivalent to $A \cap I' = O$, or to $A \cap O = O$, which is true by Theorem 1.

Theorem 8. If $A \subseteq B$ and $B \subseteq C$, then $A \subseteq C$.

Proof: $A \subseteq B$ is equivalent to $A \cap B' = O$, and $B \subseteq C$ to $B \cap C' = O$. We must show that $A \cap C' = O$. Now

$$\begin{aligned} A \cap C' &= (A \cap C') \cap (B \cup B') \\ &= (A \cap C' \cap B) \cup (A \cap C' \cap B') \\ &= O \cup O \\ &= O \end{aligned}$$

Theorem 9. If $A \subseteq B$ and $B \subseteq A$, then $A = B$.

Proof: The hypotheses are equivalent to $A \cap B' = O$ and $B \cap A' = O$. From Theorems 3 and 6, $B \cap A' = O$ is equivalent to $A \cup B' = I$. Hence, from B5, B' is the complement of A, or $B' = A'$. Therefore $B = A$.

Problems 13.3

1. Prove Theorem 4.
2. Prove Theorem 6.
3. Prove that $A \subseteq B$ if and only if $A \cap B = A$.
4. Prove that $A \subseteq B$ if and only if $A \cup B = B$.
5. Prove that $A \subseteq B$ if and only if $A' \cup B = I$.
6. Prove that, if $A \subseteq B$ and if $A \subseteq C$, then $A \subseteq (B \cap C)$.
7. Prove that, if $A \subseteq B$, then $A \subseteq (B \cup C)$ for any C.
8. Prove that $A \subseteq B$ if and only if $B' \subseteq A'$.
9. Prove that $A \cup (A' \cap B) = A \cup B$.
10. Prove that $A \cap (A' \cup B) = A \cap B$.
11. Prove that $(A \cup B) \cap (C \cup D) = (A \cap C) \cup (A \cap D) \cup (B \cap C) \cup (B \cap D)$.
12. Prove that $(A \cap B) \cup (C \cap D) = (A \cup C) \cap (A \cup D) \cap (B \cup C) \cap (B \cup D)$.
13. Prove that $(A \cup B) \cap (A \cup B') = A$.
14. Prove that $(A \cap B) \cup (A \cap B') = A$.
15. Prove that, if $A \cup C = B \cup C$ and if $A \cup C' = B \cup C'$, then $A = B$. HINT: Write $(A \cup C) \cap (A \cup C') = (B \cup C) \cap (B \cup C')$, and use Prob. 13.
16. Prove that $(A \cup C) \cap (A' \cup B) \cap (B \cup C) = (A \cap B) \cup (A' \cap C)$. HINT: Expand the right-hand side.
17. Prove that $(A \cap B) \cup (A' \cap C) = (A \cup C) \cap (A' \cup B)$. HINT: Write $A \cap B = A \cap (A' \cup B)$; $A' \cap C = A' \cap (A \cup C)$.
18. From Probs. 16 and 17, or otherwise, prove that $(A \cup C) \cap (A' \cup B) \cap (B \cup C) = (A \cup C) \cap (A' \cup B)$.
19. Prove that $(A \cap C) \cup (A' \cap B) \cup (B \cap C) = (A \cap C) \cup (A' \cap B)$.

20*. Prove that Axiom B2 can be proved from the other axioms of a Boolean Algebra.

13.4 Boolean Algebra as an Algebra of Sets

Our axioms of Boolean Algebra were abstracted from the algebra of sets, and so we may wonder whether there are Boolean Algebras which are in fact more general than algebras of sets. The answer to this question was proved to be "no" by M. H. Stone.* What he proved can be described briefly as follows. Suppose we are given any abstract Boolean Algebra; then we can find a universal set such that an algebra of its subsets is isomorphic (Sec. 4.6) with the given abstract Boolean Algebra.

One consequence of this theorem is that we can now use the methods of Sec. 13.1 to prove identities in Boolean Algebra. The method of Illustration 2 of that section, which is analogous to that of truth tables, works automatically, although it is tedious if there are many variables in the identity under consideration.

This method unfortunately does not apply to statements involving *inclusion*, $A \subseteq B$. The difficulty is that $A \subseteq B$ is a statement about the sets A and B which is true or false; hence it is a proposition in the sense of Chap. 1 rather than an identity in Boolean Algebra.

13.5 The Boolean Algebra (0,1)

The most elementary Boolean Algebra consists of the elements 0, 1 with the operations $+$, $\times$, and $'$. In the following treatment we use the letters a, b, c, etc., to stand for arbitrary elements of the set $\{0, 1\}$. We shall identify 0 with the null element O in the axioms, 1 with I, $+$ with union $\cup$, and $\times$ with intersection $\cap$. The corresponding tables are:

Multiplication

a	b	$a \times b$
1	1	1
1	0	0
0	1	0
0	0	0

Addition

a	b	$a + b$
1	1	1
1	0	1
0	1	1
0	0	0

Complementation

a	a'
1	0
0	1

Doubtless you have noticed that these tables are identical with the truth tables of Chap. 1 if we set up the correspondence:

Logic	**Boolean algebra (0,1)**
T	1
F	0
$\wedge$	$\times$
$\vee$	$+$
$\sim$	$'$
$\leftrightarrow$	$=$
$p, q, r, \ldots$	$a, b, c, \ldots$

* In a paper entitled The Theory of Representations for Boolean Algebras, *Transactions of the American Mathematical Society*, vol. 40, pp. 37–111 (1936).

Note that implication ($\rightarrow$) does not appear in this table. In Prob.1, Sec. 1.11, we showed that

$$(p \rightarrow q) \leftrightarrow (\sim p) \vee q$$

Hence the Boolean Algebra equivalent of ($\rightarrow$) is $a' + b$. This suggests the following table, which is seen to correspond to the truth table for implication:

a	b	a'	$a' + b$
1	1	0	1
1	0	0	0
0	1	1	1
0	0	1	1

This correspondence between elementary logic and the Boolean Algebra (0,1) is of great importance. It means that any statement which is true in one system becomes a true statement in the other if we carry through the stated change in notation. For example, the distributive laws $B6$ were presented to you in logical notation as Probs. 20 and 21, Sec. 1.11:

$$p \wedge (q \vee r) \leftrightarrow (p \wedge q) \vee (p \wedge r)$$
$$p \vee (q \wedge r) \leftrightarrow (p \vee q) \wedge (p \vee r)$$

We thus have nothing new, but have our old friend "logic" dressed up in new clothes. It is for this reason that Boolean Algebra is often called the "Algebra of Logic". It was in this form, indeed, that George Boole invented it.

13.6 Electrical Networks

An important application of this binary system is to electrical networks. Two switches are said to be connected in "parallel" when current will flow through the system when either or both of these is closed. This is illustrated in the diagram below, where S stands for a switch. Similarly, the switches are connected in "series" when current will flow only when both of these are closed.

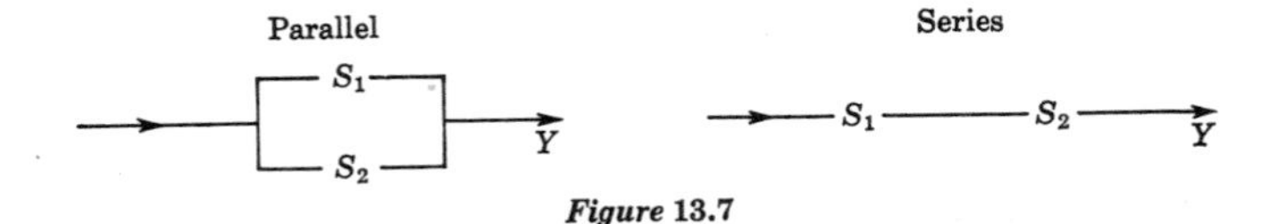

Figure 13.7

Give each switch S the value 1 if it is closed and the value 0 if it is open. Give the combined switch Y the value 1 if it is closed and the value 0 if it is open. Then we can describe this situation by the equations:

Parallel: $S_1 + S_2 = Y$

Series: $S_1 \times S_2 = Y$

where $+$ and $\times$ have the meanings in Sec. 13.5.

The axioms of Boolean Algebra can now be given concrete representations in terms of circuits involving parallel and series connections.

Illustration 1. Consider the associative law:

$$S_1 + (S_2 + S_3) = (S_1 + S_2) + S_3$$

In circuit language this may be represented by Fig. 13.8. By tracing through these

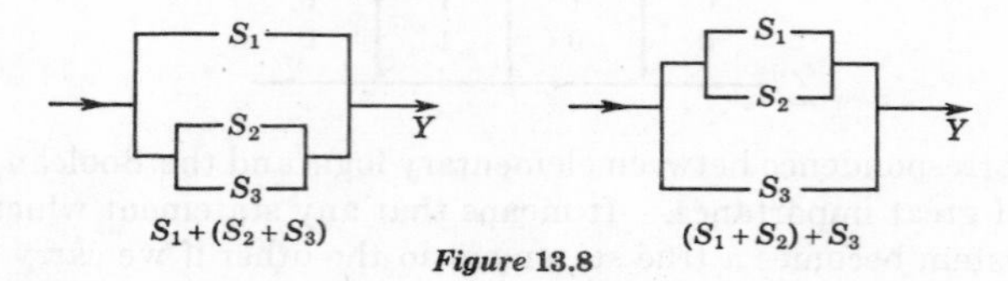

Figure 13.8

diagrams you should verify that they have the same electrical properties, and hence are equivalent circuits.

Illustration 2. As in Illustration 1, we show the representation in circuits of the

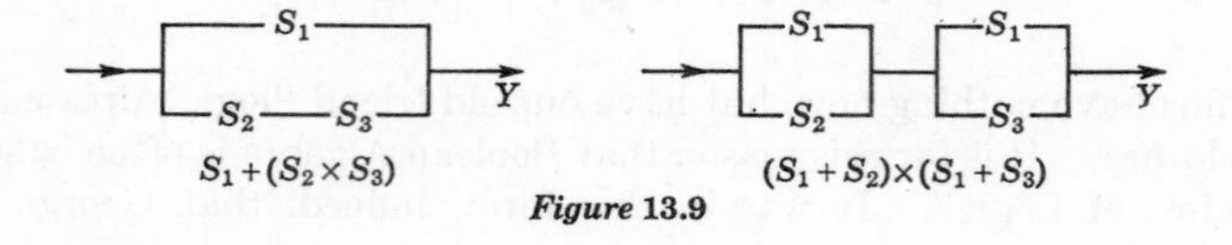

Figure 13.9

distributive law:

$$S_1 + (S_2 \times S_3) = (S_1 + S_2) \times (S_1 + S_3)$$

In terms of circuits, $S_1 + (S_2 \times S_3)$ is the simpler of the two. This suggests that we may be able to simplify complicated circuits through the use of Boolean Algebra.

Illustration 3. Write the Boolean equation of the circuit (Fig. 13.10).

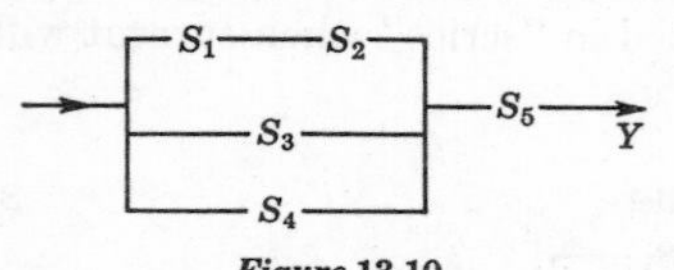

Figure 13.10

ANSWER: $[(S_1 \times S_2) + S_3 + S_4] \times S_5 = Y.$

Problems 13.6

In Probs. 1 to 20 draw circuits to illustrate the given Boolean identity.

1. $(S_1 \times S_2) \times S_3 = S_1 \times (S_2 \times S_3)$.

2. $S_1 \times S_2 = S_2 \times S_1$.

3. $S_1 + S_2 = S_2 + S_1$.

4. $1 \times S = S$. (1 is a circuit which is always closed.)

5. $0 + S = S$. (0 is a circuit which is always open.)

6. $S + S' = 1$. (S' is open if S is closed and is closed when S is open.)

7. $S \times S' = 0$.

8. $S_1 \times (S_2 + S_3) = (S_1 \times S_2) + (S_1 \times S_3)$.

9. $S \times 0 = 0$.

10. $S + 1 = 1$.

11. $S \times S = S$.

12. $S + S = S$.

13. $(S_1 \times S_2)' = S_1' + S_2'$.

14. $(S_1 + S_2)' = S_1' \times S_2'$.

15. $S_1 + (S_1' \times S_2) = S_1 + S_2$.

16. $S_1 \times (S_1' + S_2) = S_1 \times S_2$.

17. $(S_1 + S_2) \times (S_1 + S_2') = S_1$.

18. $(S_1 \times S_2) + (S_1 \times S_2') = S_1$.

19. $(S_1 + S_3) \times (S_1' + S_2) \times (S_2 + S_3) = (S_1 + S_3) \times (S_1' + S_2)$.

20. $(S_1 \times S_3) + (S_1' \times S_2) + (S_2 \times S_3) = (S_1 \times S_3) + (S_1' \times S_2)$.

In Probs. 21 to 26 write the Boolean equation of each circuit, and hence, by Boolean algebra, show them to be equivalent. You may use the identities in Probs. 1 to 20.

21.

S_1 — S_2 — S_3'
S_1' — S_2 → Y
S_2 — S_3
and → S_2 → Y

22.

S_1 — S_2 — S_3
S_1 — S_2' — S_3 → Y
S_1' — S_2' — S_3
and → S_3 — [S_1 / S_2'] → Y

23.

[S_1 / S_2] — [S_1 / S_2'] — [S_1' / S_2'] → Y and → S_1 — S_2' → Y

24.

S_1
S_2 — S_3
→ Y
S_1' — S_2
S_1
and → [S_1 / S_2] → Y

25.

S_1 — S_2
S_1' — S_3 → Y
S_2 — S_3
and → [S_1 — S_2 / S_1' — S_3] → Y

26.

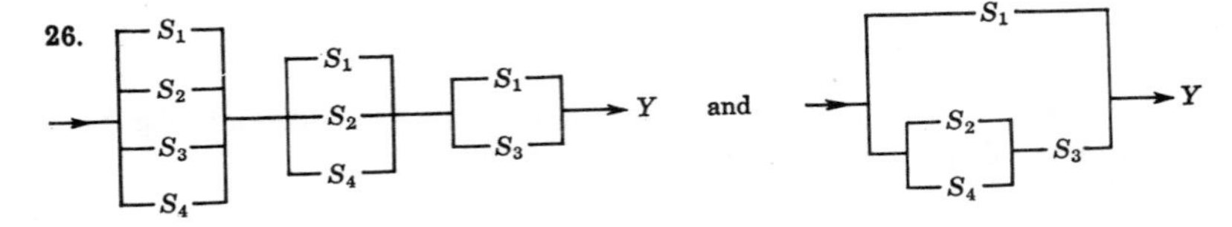

13.7 Design of Circuits

We can use the ideas of Sec. 13.6 to help us design a circuit which will have desired properties. Since the general theory is a little (but not very) complicated, we proceed by an illustration.

Illustration 1. Suppose that we wish to control a hall light by two switches, one upstairs and one downstairs. By throwing either switch we wish to be able to change the light from "off" to "on" or from "on" to "off".

Let us set up a table of what we want, using 0 and 1 as above to indicate the states of our switches and currents. Let D represent the downstairs switch, U the upstairs switch, and Y the light. Then we have the following table:

D	U	Y
1	1	1
1	0	0
0	1	0
0	0	1

We wish to find an equation which fits this table. One equation which works is

$$(D \times U) + (D' \times U') = Y$$

Another is

$$(D + U') \times (D' + U) = Y$$

Exercise A. Show that these two equations are in fact satisfied by the entries in the table of Illustration 1.

Exercise B. From the axioms of Boolean Algebra show that

$$(D \times U) + (D' \times U') = (D + U') \times (D' + U)$$

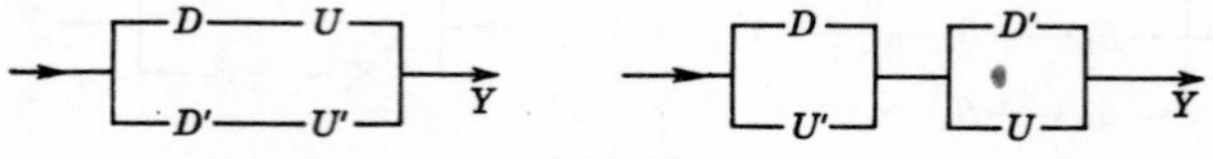

Figure 13.11

The desired circuit may then be drawn in either of the forms shown in Fig. 13.11. The corresponding wiring diagrams are drawn in Fig. 13.12.

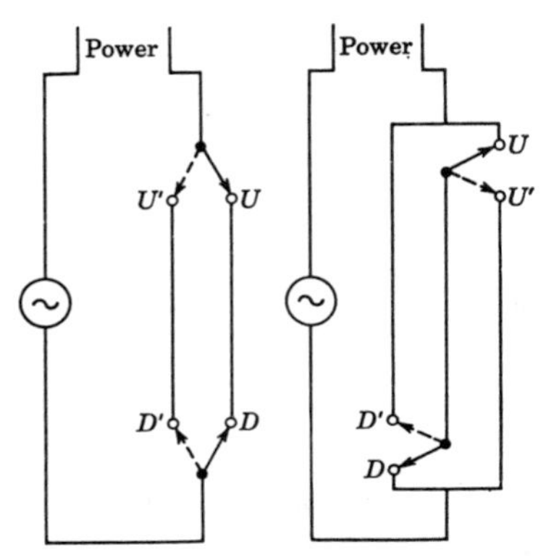

Figure 13.12

You may wonder how we wrote down these two equations. There is a general theorem which tells us what to do:

Theorem 10. Let $Y = f(A,B)$ represent a function in which A, B, and Y are elements of the Boolean Algebra (0,1). The function is given completely by the following table:

A	B	$f(A,B)$
1	1	$f(1,1)$
1	0	$f(1,0)$
0	1	$f(0,1)$
0	0	$f(0,0)$

where the entries in the right-hand column are arbitrarily chosen numbers in the set $\{0,1\}$. Then $f(A,B)$ is explicitly given by the formulas:

$$\begin{aligned} f(A,B) &= [f(1,1) \times A \times B] + [f(1,0) \times A \times B'] + [f(0,1) \times A' \times B] \\ &\qquad + [f(0,0) \times A' \times B'] \\ &= [f(1,1) + A' + B'] \times [f(1,0) + A' + B] \times [f(0,1) + A + B'] \\ &\qquad \times [f(0,0) + A + B] \end{aligned}$$

The proof is obtained by substituting for A and B and observing that the results are correct. For example, in the first formula,

$$\begin{aligned} f(1,0) &= [f(1,1) \times 0] + [f(1,0) \times 1] + [f(0,1) \times 0] + [f(0,0) \times 0] \\ &= 0 + f(1,0) + 0 + 0 \\ &= f(1,0) \end{aligned}$$

In the second formula, for instance,

$$\begin{aligned} f(0,0) &= [f(1,1) + 1] \times [f(1,0) + 1] \times [f(0,1) + 1] \times [f(0,0) + 0] \\ &= 1 \times 1 \times 1 \times f(0,0) \\ &= f(0,0) \end{aligned}$$

If we apply these formulas to the table of Illustration 1 with $D = A$ and $U = B$, we obtain the results given there.

Illustration 2. This idea can also be used to solve logical problems. Consider the following situation: Suppose that A is a man who is always a liar or always a truthteller, but that we do not know which he is. In a room there are two chairs which look alike, but one of which is lethal. Assume that A knows which chair is which. A man B wishes to sit in one of the chairs, but does not wish to be killed. He may ask A a single question, "Is r true?" What should the proposition r be?

Solution: Set up the propositions:

p: A is a truthteller.
q: The left-hand chair is lethal.
r: The statement in B's question: "Is r true?"

Then form the following table:

p	q	A's answer to "Is r true?"	r
T	T	T	T
T	F	F	F
F	T	T	F
F	F	F	T

In the third column we chose values of T and F to agree with those of q. Thus A's answer will tell B the status of the chair. The "r" column is obtained from the third column by considering whether or not A is a truthteller.

We wish, therefore, to have a combination of p and q equivalent to r. We change the problem into Boolean Algebra notation, as described in Sec. 13.5. When we do so, we find that this problem is abstractly the same as the upstairs-downstairs switch problem of Illustration 1.

Hence the proposition r must be

$$(p \wedge q) \vee (\sim p \wedge \sim q)$$

or

$$(p \vee \sim q) \wedge (\sim p \vee q)$$

A suitable question for B is: "Are you a truthteller, and is the left-hand chair lethal, or are you a liar, and is the left-hand chair safe?"

Exercise C. Show that $p \leftrightarrow q$ is another correct answer.

Problems 13.7

1. Design a circuit so that any one of three switches A, B, C will control the state of a light.
2. Design a circuit for an automatic voting machine in which each of three voters records his vote by throwing a switch to "yes" or "no". A light is to show if a majority votes "yes".
3. A missile is to be fired when the commanding officer (CO) *and* either of his two aides (A_1 and A_2) throw their switches to "fire" position. Design a suitable circuit.
4. A missile is to be fired if the commanding officer (CO) *and* any two of his three aides (A_1, A_2, A_3) throw their switches to "fire" position. Design a suitable circuit.
5. A committee consists of two men and two women. According to the rules, a measure will pass if both men favor it or if both women together with either man favor it. They vote by throwing switches. Construct a circuit which will sound a buzzer if a measure is to pass.
6. In a corporation, A has 20 shares, B has 45 shares, C has 10 shares, and D has 25 shares. Each votes his shares in a block by throwing a switch. Construct a circuit which will sound a buzzer if a majority of the shares has been voted in favor of a measure.
7. A victim is confronted with four vials of clear liquid R, R', L, L'. He is told that one of the pair R, R' contains water and the other contains poison. The same holds for the pair L, L'. Moreover, the two poisons neutralize each other, so that if both are drunk, no harm ensues.

   The victim is required to choose one vial from R, R' and one from L, L' and to drink their contents. In the interests of good sportsmanship he is allowed to ask one question of an attendant who knows all the facts and who tells the truth. The attendant will answer "yes" or "no" only.

   What question should the victim ask so that he can be certain of survival?
8. The situation is as in Prob. 7 except that it is not known whether the attendant is a consistent truthteller or a consistent liar. Now what question should the victim ask?
9. A telephone network connecting A to B is constructed as in Fig. 13.13, where R_1, R_2, R_3, and R_4 are repeaters. The probability that repeater R_i is functioning is p_i. Find the probability that a message can be sent from A to B.

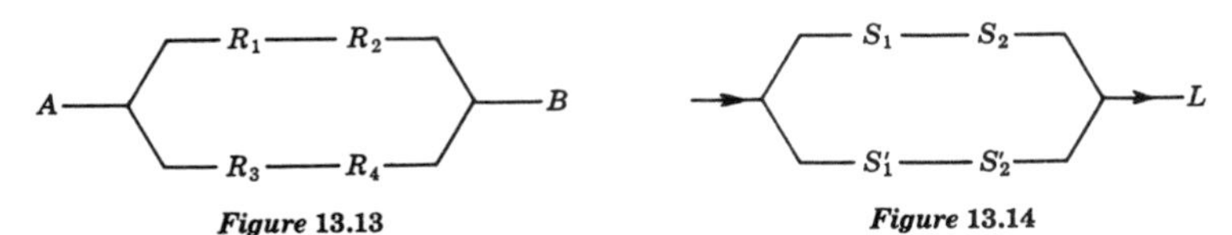

Figure 13.13

Figure 13.14

10. Two switches control a light as in Fig. 13.14. The probabilities that S_1 and S_2 are closed are p_1 and p_2, respectively. Find the probability that the light is on.

References

Hohn, Franz E.: "Applied Boolean Algebra", Macmillan, New York (1960).

Whitesitt, J. Eldon: "Boolean Algebra", Addison-Wesley, Reading, Mass. (1961).

Appendix

Table I Values of e^x and e^{-x}

x	e^x	e^{-x}	x	e^x	e^{-x}
0.00	1.0000	1.00000	2.10	8.1662	0.12246
0.01	1.0101	0.99005	2.20	9.0250	0.11080
0.02	1.0202	0.98020	2.30	9.9742	0.10026
0.03	1.0305	0.97045	2.40	11.023	0.09072
0.04	1.0408	0.96079	2.50	12.182	0.08208
0.05	1.0513	0.95123	2.60	13.464	0.07427
0.06	1.0618	0.94176	2.70	14.880	0.06721
0.07	1.0725	0.93239	2.80	16.445	0.06081
0.08	1.0833	0.92312	2.90	18.174	0.05502
0.09	1.0942	0.91393	3.00	20.086	0.04979
0.10	1.1052	0.90484	3.10	22.198	0.04505
0.20	1.2214	0.81873	3.20	24.533	0.04076
0.30	1.3499	0.74082	3.30	27.113	0.03688
0.40	1.4918	0.67032	3.40	29.964	0.03337
0.50	1.6487	0.60653	3.50	33.115	0.03020
0.60	1.8221	0.54881	3.60	36.598	0.02732
0.70	2.0138	0.49659	3.70	40.447	0.02472
0.80	2.2255	0.44933	3.80	44.701	0.02237
0.90	2.4596	0.40657	3.90	49.402	0.02024
1.00	2.7183	0.36788	4.00	54.598	0.01832
1.10	3.0042	0.33287	4.10	60.340	0.01657
1.20	3.3201	0.30119	4.20	66.686	0.01500
1.30	3.6693	0.27253	4.30	73.700	0.01357
1.40	4.0552	0.24660	4.40	81.451	0.01228
1.50	4.4817	0.22313	4.50	90.017	0.01111
1.60	4.9530	0.20190	4.60	99.484	0.01005
1.70	5.4739	0.18268	4.70	109.95	0.00910
1.80	6.0496	0.16530	4.80	121.51	0.00823
1.90	6.6859	0.14957	4.90	134.29	0.00745
2.00	7.3891	0.13534	5.00	148.41	0.00674

Table II Common Logarithms (**Base 10**)

N.	0	1	2	3	4	5	6	7	8	9
10	0000	0043	0086	0128	0170	0212	0253	0294	0334	0374
11	0414	0453	0492	0531	0569	0607	0645	0682	0719	0755
12	0792	0828	0864	0899	0934	0969	1004	1038	1072	1106
13	1139	1173	1206	1239	1271	1303	1335	1367	1399	1430
14	1461	1492	1523	1553	1584	1614	1644	1673	1703	1732
15	1761	1790	1818	1847	1875	1903	1931	1959	1987	2014
16	2041	2068	2095	2122	2148	2175	2201	2227	2253	2279
17	2304	2330	2355	2380	2405	2430	2455	2480	2504	2529
18	2553	2577	2601	2625	2648	2672	2695	2718	2742	2765
19	2788	2810	2833	2856	2878	2900	2923	2945	2967	2989
20	3010	3032	3054	3075	3096	3118	3139	3160	3181	3201
21	3222	3243	3263	3284	3304	3324	3345	3365	3385	3404
22	3424	3444	3464	3483	3502	3522	3541	3560	3579	3598
23	3617	3636	3655	3674	3692	3711	3729	3747	3766	3784
24	3802	3820	3838	3856	3874	3892	3909	3927	3945	3962
25	3979	3997	4014	4031	4048	4065	4082	4099	4116	4133
26	4150	4166	4183	4200	4216	4232	4249	4265	4281	4298
27	4314	4330	4346	4362	4378	4393	4409	4425	4440	4456
28	4472	4487	4502	4518	4533	4548	4564	4579	4594	4609
29	4624	4639	4654	4669	4683	4698	4713	4728	4742	4757
30	4771	4786	4800	4814	4829	4843	4857	4871	4886	4900
31	4914	4928	4942	4955	4969	4983	4997	5011	5024	5038
32	5051	5065	5079	5092	5105	5119	5132	5145	5159	5172
33	5185	5198	5211	5224	5237	5250	5263	5276	5289	5302
34	5315	5328	5340	5353	5366	5378	5391	5403	5416	5428
35	5441	5453	5465	5478	5490	5502	5514	5527	5539	5551
36	5563	5575	5587	5599	5611	5623	5635	5647	5658	5670
37	5682	5694	5705	5717	5729	5740	5752	5763	5775	5786
38	5798	5809	5821	5832	5843	5855	5866	5877	5888	5899
39	5911	5922	5933	5944	5955	5966	5977	5988	5999	6010
40	6021	6031	6042	6053	6064	6075	6085	6096	6107	6117
41	6128	6138	6149	6160	6170	6180	6191	6201	6212	6222
42	6232	6243	6253	6263	6274	6284	6294	6304	6314	6325
43	6335	6345	6355	6365	6375	6385	6395	6405	6415	6425
44	6435	6444	6454	6464	6474	6484	6493	6503	6513	6522
45	6532	6542	6551	6561	6571	6580	6590	6599	6609	6618
46	6628	6637	6646	6656	6665	6675	6684	6693	6702	6712
47	6721	6730	6739	6749	6758	6767	6776	6785	6794	6803
48	6812	6821	6830	6839	6848	6857	6866	6875	6884	6893
49	6902	6911	6920	6928	6937	6946	6955	6964	6972	6981
50	6990	6998	7007	7016	7024	7033	7042	7050	7059	7067
51	7076	7084	7093	7101	7110	7118	7126	7135	7143	7152
52	7160	7168	7177	7185	7193	7202	7210	7218	7226	7235
53	7243	7251	7259	7267	7275	7284	7292	7300	7308	7316
54	7324	7332	7340	7348	7356	7364	7372	7380	7388	7396
N.	0	1	2	3	4	5	6	7	8	9

Table II Common Logarithms (Continued)

N.	0	1	2	3	4	5	6	7	8	9
55	7404	7412	7419	7427	7435	7443	7451	7459	7466	7474
56	7482	7490	7497	7505	7513	7520	7528	7536	7543	7551
57	7559	7566	7574	7582	7589	7597	7604	7612	7619	7627
58	7634	7642	7649	7657	7664	7672	7679	7686	7694	7701
59	7709	7716	7723	7731	7738	7745	7752	7760	7767	7774
60	7782	7789	7796	7803	7810	7818	7825	7832	7839	7846
61	7853	7860	7868	7875	7882	7889	7896	7903	7910	7917
62	7924	7931	7938	7945	7952	7959	7966	7973	7980	7987
63	7993	8000	8007	8014	8021	8028	8035	8041	8048	8055
64	8062	8069	8075	8082	8089	8096	8102	8109	8116	8122
65	8129	8136	8142	8149	8156	8162	8169	8176	8182	8189
66	8195	8202	8209	8215	8222	8228	8235	8241	8248	8254
67	8261	8267	8274	8280	8287	8293	8299	8306	8312	8319
68	8325	8331	8338	8344	8351	8357	8363	8370	8376	8382
69	8388	8395	8401	8407	8414	8420	8426	8432	8439	8445
70	8451	8457	8463	8470	8476	8482	8488	8494	8500	8506
71	8513	8519	8525	8531	8537	8543	8549	8555	8561	8567
72	8573	8579	8585	8591	8597	8603	8609	8615	8621	8627
73	8633	8639	8645	8651	8657	8663	8669	8675	8681	8686
74	8692	8698	8704	8710	8716	8722	8727	8733	8739	8745
75	8751	8756	8762	8768	8774	8779	8785	8791	8797	8802
76	8808	8814	8820	8825	8831	8837	8842	8848	8854	8859
77	8865	8871	8876	8882	8887	8893	8899	8904	8910	8915
78	8921	8927	8932	8938	8943	8949	8954	8960	8965	8971
79	8976	8982	8987	8993	8998	9004	9009	9015	9020	9025
80	9031	9036	9042	9047	9053	9058	9063	9069	9074	9079
81	9085	9090	9096	9101	9106	9112	9117	9122	9128	9133
82	9138	9143	9149	9154	9159	9165	9170	9175	9180	9186
83	9191	9196	9201	9206	9212	9217	9222	9227	9232	9238
84	9243	9248	9253	9258	9263	9269	9274	9279	9284	9289
85	9294	9299	9304	9309	9315	9320	9325	9330	9335	9340
86	9345	9350	9355	9360	9365	9370	9375	9380	9385	9390
87	9395	9400	9405	9410	9415	9420	9425	9430	9435	9440
88	9445	9450	9455	9460	9465	9469	9474	9479	9484	9489
89	9494	9499	9504	9509	9513	9518	9523	9528	9533	9538
90	9542	9547	9552	9557	9562	9566	9571	9576	9581	9586
91	9590	9595	9600	9605	9609	9614	9619	9624	9628	9633
92	9638	9643	9647	9652	9657	9661	9666	9671	9675	9680
93	9685	9689	9694	9699	9703	9708	9713	9717	9722	9727
94	9731	9736	9741	9745	9750	9754	9759	9763	9768	9773
95	9777	9782	9786	9791	9795	9800	9805	9809	9814	9818
96	9823	9827	9832	9836	9841	9845	9850	9854	9859	9863
97	9868	9872	9877	9881	9886	9890	9894	9899	9903	9908
98	9912	9917	9921	9926	9930	9934	9939	9943	9948	9952
99	9956	9961	9965	9969	9974	9978	9983	9987	9991	9996
N.	0	1	2	3	4	5	6	7	8	9

Table III Natural Logarithms (Base *e*)

	.00	.01	.02	.03	.04	.05	.06	.07	.08	.09
1.0	0.0000	0.0100	0.0198	0.0296	0.0392	0.0488	0.0583	0.0677	0.0770	0.0862
1.1	0.0953	0.1044	0.1133	0.1222	0.1310	0.1398	0.1484	0.1570	0.1655	0.1740
1.2	0.1823	0.1906	0.1989	0.2070	0.2151	0.2231	0.2311	0.2390	0.2469	0.2546
1.3	0.2624	0.2700	0.2776	0.2852	0.2927	0.3001	0.3075	0.3148	0.3221	0.3293
1.4	0.3365	0.3436	0.3507	0.3577	0.3646	0.3716	0.3784	0.3853	0.3920	0.3988
1.5	0.4055	0.4121	0.4187	0.4253	0.4318	0.4383	0.4447	0.4511	0.4574	0.4637
1.6	0.4700	0.4762	0.4824	0.4886	0.4947	0.5008	0.5068	0.5128	0.5188	0.5247
1.7	0.5306	0.5365	0.5423	0.5481	0.5539	0.5596	0.5653	0.5710	0.5766	0.5822
1.8	0.5878	0.5933	0.5988	0.6043	0.6098	0.6152	0.6206	0.6259	0.6313	0.6366
1.9	0.6419	0.6471	0.6523	0.6575	0.6627	0.6678	0.6729	0.6780	0.6831	0.6881
2.0	0.6932	0.6981	0.7031	0.7080	0.7129	0.7178	0.7227	0.7275	0.7324	0.7372
2.1	0.7419	0.7467	0.7514	0.7561	0.7608	0.7655	0.7701	0.7747	0.7793	0.7839
2.2	0.7885	0.7930	0.7975	0.8020	0.8065	0.8109	0.8154	0.8198	0.8242	0.8286
2.3	0.8329	0.8373	0.8416	0.8459	0.8502	0.8544	0.8587	0.8629	0.8671	0.8713
2.4	0.8755	0.8796	0.8838	0.8879	0.8920	0.8961	0.9002	0.9042	0.9083	0.9123
2.5	0.9163	0.9203	0.9243	0.9282	0.9322	0.9361	0.9400	0.9439	0.9478	0.9517
2.6	0.9555	0.9594	0.9632	0.9670	0.9708	0.9746	0.9783	0.9821	0.9858	0.9895
2.7	0.9933	0.9969	1.0006	1.0043	1.0080	1.0116	1.0152	1.0188	1.0225	1.0260
2.8	1.0296	1.0332	1.0367	1.0403	1.0438	1.0473	1.0508	1.0543	1.0578	0.0613
2.9	1.0647	1.0682	1.0716	1.0750	1.0784	1.0818	1.0852	1.0886	1.0919	1.0953
3.0	1.0986	1.1019	1.1053	1.1086	1.1119	1.1151	1.1184	1.1217	1.1249	1.1282
3.1	1.1314	1.1346	1.1378	1.1410	1.1442	1.1474	1.1506	1.1537	1.1569	1.1600
3.2	1.1632	1.1663	1.1694	1.1725	1.1756	1.1787	1.1817	1.1848	1.1878	1.1909
3.3	1.1939	1.1969	1.2000	1.2030	1.2060	1.2090	1.2119	1.2149	1.2179	1.2208
3.4	1.2238	1.2267	1.2296	1.2326	1.2355	1.2384	1.2413	1.2442	1.2470	1.2499
3.5	1.2528	1.2556	1.2585	1.2613	1.2641	1.2669	1.2698	1.2726	1.2754	1.2782
3.6	1.2809	1.2837	1.2865	1.2892	1.2920	1.2947	1.2975	1.3002	1.3029	1.3056
3.7	1.3083	1.3110	1.3137	1.3164	1.3191	1.3218	1.3244	1.3271	1.3297	1.3324
3.8	1.3350	1.3376	1.3403	1.3429	1.3455	1.3481	1.3507	1.3533	1.3558	1.3584
3.9	1.3610	1.3635	1.3661	1.3686	1.3712	1.3737	1.3762	1.3788	1.3813	1.3838
4.0	1.3863	1.3888	1.3913	1.3938	1.3962	1.3987	1.4012	1.4036	1.4061	1.4085
4.1	1.4110	1.4134	1.4159	1.4183	1.4207	1.4231	1.4255	1.4279	1.4303	1.4327
4.2	1.4351	1.4375	1.4398	1.4422	1.4446	1.4469	1.4493	1.4516	1.4540	1.4563
4.3	1.4586	1.4609	1.4633	1.4656	1.4679	1.4702	1.4725	1.4748	1.4771	1.4793
4.4	1.4816	1.4839	1.4861	1.4884	1.4907	1.4929	1.4951	1.4974	1.4996	1.5019
4.5	1.5041	1.5063	1.5085	1.5107	1.5129	1.5151	1.5173	1.5195	1.5217	1.5239
4.6	1.5261	1.5282	1.5304	1.5326	1.5347	1.5369	1.5390	1.5412	1.5433	1.5454
4.7	1.5476	1.5497	1.5518	1.5539	1.5560	1.5581	1.5602	1.5623	1.5644	1.5665
4.8	1.5686	1.5707	1.5728	1.5748	1.5769	1.5790	1.5810	1.5831	1.5851	1.5872
4.9	1.5892	1.5913	1.5933	1.5953	1.5974	1.5994	1.6014	1.6034	1.6054	1.6074
5.0	1.6094	1.6114	1.6134	1.6154	1.6174	1.6194	1.6214	1.6233	1.6253	1.6273
5.1	1.6292	1.6312	1.6332	1.6351	1.6371	1.6390	1.6409	1.6429	1.6448	1.6467
5.2	1.6487	1.6506	1.6525	1.6544	1.6563	1.6582	1.6601	1.6620	1.6639	1.6658
5.3	1.6677	1.6696	1.6715	1.6734	1.6752	1.6771	1.6790	1.6808	1.6827	1.6845
5.4	1.6864	1.6882	1.6901	1.6919	1.6938	1.6956	1.6974	1.6993	1.7011	1.7029

Table III Natural Logarithms (Continued)

	.00	.01	.02	.03	.04	.05	.06	.07	.08	.09
5.5	1.7047	1.7066	1.7084	1.7102	1.7120	1.7138	1.7156	1.7174	1.7192	1.7210
5.6	1.7228	1.7246	1.7263	1.7281	1.7299	1.7317	1.7334	1.7352	1.7370	1.7387
5.7	1.7405	1.7422	1.7440	1.7457	1.7475	1.7492	1.7509	1.7527	1.7544	1.7561
5.8	1.7579	1.7596	1.7613	1.7630	1.7647	1.7664	1.7681	1.7699	1.7716	1.7733
5.9	1.7750	1.7766	1.7783	1.7800	1.7817	1.7834	1.7851	1.7868	1.7884	1.7901
6.0	1.7918	1.7934	1.7951	1.7967	1.7984	1.8001	1.8017	1.8034	1.8050	1.8066
6.1	1.8083	1.8099	1.8116	1.8132	1.8148	1.8165	1.8181	1.8197	1.8213	1.8229
6.2	1.8245	1.8262	1.8278	1.8294	1.8310	1.8326	1.8342	1.8358	1.8374	1.8390
6.3	1.8405	1.8421	1.8437	1.8453	1.8469	1.8485	1.8500	1.8516	1.8532	1.8547
6.4	1.8563	1.8579	1.8594	1.8610	1.8625	1.8641	1.8656	1.8672	1.8687	1.8703
6.5	1.8718	1.8733	1.8749	1.8764	1.8779	1.8795	1.8810	1.8825	1.8840	1.8856
6.6	1.8871	1.8886	1.8901	1.8916	1.8931	1.8946	1.8961	1.8976	1.8991	1.9006
6.7	1.9021	1.9036	1.9051	1.9066	1.9081	1.9095	1.9110	1.9125	1.9140	1.9155
6.8	1.9169	1.9184	1.9199	1.9213	1.9228	1.9242	1.9257	1.9272	1.9286	1.9301
6.9	1.9315	1.9330	1.9344	1.9359	1.9373	1.9387	1.9402	1.9416	1.9430	1.9445
7.0	1.9459	1.9473	1.9488	1.9502	1.9516	1.9530	1.9544	1.9559	1.9573	1.9587
7.1	1.9601	1.9615	1.9629	1.9643	1.9657	1.9671	1.9685	1.9699	1.9713	1.9727
7.2	1.9741	1.9755	1.9769	1.9782	1.9796	1.9810	1.9824	1.9838	1.9851	1.9865
7.3	1.9879	1.9892	1.9906	1.9920	1.9933	1.9947	1.9961	1.9974	1.9988	2.0001
7.4	2.0015	2.0028	2.0042	2.0055	2.0069	2.0082	2.0096	2.0109	2.0122	2.0136
7.5	2.0149	2.0162	2.0176	2.0189	2.0202	2.0215	2.0229	2.0242	2.0255	2.0268
7.6	2.0281	2.0295	2.0308	2.0321	2.0334	2.0347	0.0360	2.0373	2.0386	2.0399
7.7	2.0412	2.0425	2.0438	2.0451	2.0464	2.0477	2.0490	2.0503	2.0516	2.0528
7.8	2.0541	2.0554	2.0567	2.0580	2.0592	2.0605	2.0618	2.0631	2.0643	2.0656
7.9	2.0669	2.0681	2.0694	2.0707	2.0719	2.0732	2.0744	2.0757	2.0769	2.0782
8.0	2.0794	2.0807	2.0819	2.0832	2.0844	2.0857	2.0869	2.0882	2.0894	2.0906
8.1	2.0919	2.0931	2.0943	2.0956	2.0968	2.0980	2.0992	2.1005	2.1017	2.1029
8.2	2.1041	2.1054	2.1066	2.1078	2.1090	2.1102	2.1114	2.1126	2.1138	2.1150
8.3	2.1163	2.1175	2.1187	2.1199	2.1211	2.1223	2.1235	2.1247	2.1259	2.1270
8.4	2.1282	2.1294	2.1306	2.1318	2.1330	2.1342	2.1353	2.1365	2.1377	2.1389
8.5	2.1401	2.1412	2.1424	2.1436	2.1448	2.1459	2.1471	2.1483	2.1494	2.1506
8.6	2.1518	2.1529	2.1541	2.1552	2.1564	2.1576	2.1587	2.1599	2.1610	2.1622
8.7	2.1633	2.1645	2.1656	2.1668	2.1679	2.1691	2.1702	2.1713	2.1725	2.1736
8.8	2.1748	2.1759	2.1770	2.1782	2.1793	2.1804	2.1815	2.1827	2.1838	2.1849
8.9	2.1861	2.1872	2.1883	2.1894	2.1905	2.1917	2.1928	2.1939	2.1950	2.1961
9.0	2.1972	2.1983	2.1994	2.2006	2.2017	2.2028	2.2039	2.2050	2.2061	2.2072
9.1	2.2083	2.2094	2.2105	2.2116	2.2127	2.2138	2.2148	2.2159	2.2170	2.2181
9.2	2.2192	2.2203	2.2214	2.2225	2.2235	2.2246	2.2257	2.2268	2.2279	2.2289
9.3	2.2300	2.2311	2.2322	2.2332	2.2343	2.2354	2.2364	2.2375	2.2386	2.2396
9.4	2.2407	2.2418	2.2428	2.2439	2.2450	2.2460	2.2471	2.2481	2.2492	2.2502
9.5	2.2513	2.2523	2.2534	2.2544	2.2555	2.2565	2.2576	2.2586	2.2597	2.2607
9.6	2.2618	2.2628	2.2638	2.2649	2.2659	2.2670	2.2680	2.2690	2.2701	2.2711
9.7	2.2721	2.2732	2.2742	2.2752	2.2762	2.2773	2.2783	2.2793	2.2803	2.2814
9.8	2.2824	2.2834	2.2844	2.2854	2.2865	2.2875	2.2885	2.2895	2.2905	2.2915
9.9	2.2925	2.2935	2.2946	2.2956	2.2966	2.2976	2.2986	2.2996	2.3006	2.3016

Table III Natural Logarithms (Continued)

N	Nat Log	N	Nat Log	N	Nat Log	N	Nat Log	N	Nat Log
0	— ∞	40	3.68 888	80	4.38 203	120	4.78 749	160	5.07 517
1	0.00 000	41	3.71 357	81	4.39 445	121	4.79 579	161	5.08 140
2	0.69 315	42	3.73 767	82	4.40 672	122	4.80 402	162	5.08 760
3	1.09 861	43	3.76 120	83	4.41 884	123	4.81 218	163	5.09 375
4	1.38 629	44	3.78 419	84	4.43 082	124	4.82 028	164	5.09 987
5	1.60 944	45	3.80 666	85	4.44 265	125	4.82 831	165	5.10 595
6	1.79 176	46	3.82 864	86	4.45 435	126	4.83 628	166	5.11 199
7	1.94 591	47	3.85 015	87	4.46 591	127	4.84 419	167	5.11 799
8	2.07 944	48	3.87 120	88	4.47 734	128	4.85 203	168	5.12 396
9	2.19 722	49	3.89 182	89	4.48 864	129	4.85 981	169	5.12 990
10	2.30 259	50	3.91 202	90	4.49 981	130	4.86 753	170	5.13 580
11	2.39 790	51	3.93 183	91	4.51 086	131	4.87 520	171	5.14 166
12	2.48 491	52	3.95 124	92	4.52 179	132	4.88 280	172	5.14 749
13	2.56 495	53	3.97 029	93	4.53 260	133	4.89 035	173	5.15 329
14	2.63 906	54	3.98 898	94	4.54 329	134	4.89 784	174	5.15 906
15	2.70 805	55	4.00 733	95	4.55 388	135	4.90 527	175	5.16 479
16	2.77 259	56	4.02 535	96	4.56 435	136	4.91 265	176	5.17 048
17	2.83 321	57	4.04 305	97	4.57 471	137	4.91 998	177	5.17 615
18	2.89 037	58	4.06 044	98	4.58 497	138	4.92 725	178	5.18 178
19	2.94 444	59	4.07 754	99	4.59 512	139	4.93 447	179	5.18 739
20	2.99 573	60	4.09 434	100	4.60 517	140	4.94 164	180	5.19 296
21	3.04 452	61	4.11 087	101	4.61 512	141	4.94 876	181	5.19 850
22	3.09 104	62	4.12 713	102	4.62 497	142	4.95 583	182	5.20 401
23	3.13 549	63	4.14 313	103	4.63 473	143	4.96 284	183	5.20 949
24	3.17 805	64	4.15 888	104	4.64 439	144	4.96 981	184	5.21 494
25	3.21 888	65	4.17 439	105	4.65 396	145	4.97 673	185	5.22 036
26	3.25 810	66	4.18 965	106	4.66 344	146	4.98 361	186	5.22 575
27	3.29 584	67	4.20 469	107	4.67 283	147	4.99 043	187	5.23 111
28	3.33 220	68	4.21 951	108	4.68 213	148	4.99 721	188	5.23 644
29	3.36 730	69	4.23 411	109	4.69 135	149	5.00 395	189	5.24 175
30	3.40 120	70	4.24 850	110	4.70 048	150	5.01 064	190	5.24 702
31	3.43 399	71	4.26 268	111	4.70 953	151	5.01 728	191	5.25 227
32	3.46 574	72	4.27 667	112	4.71 850	152	5.02 388	192	5.25 750
33	3.49 651	73	4.29 046	113	4.72 739	153	5.03 044	193	5.26 269
34	2.52 636	74	4.30 407	114	4.73 620	154	5.03 695	194	5.26 786
35	3.55 535	75	4.31 749	115	4.74 493	155	5.04 343	195	5.27 300
36	3.58 352	76	4.33 073	116	4.75 359	156	5.04 986	196	5.27 811
37	3.61 092	77	4.34 381	117	4.76 217	157	5.05 625	197	5.28 320
38	3.63 759	78	4.35 671	118	4.77 068	158	5.06 260	198	5.28 827
39	3.66 356	79	4.36 945	119	4.77 912	159	5.06 890	199	5.29 330
40	3.68 888	80	4.38 203	120	4.78 749	160	5.07 517	200	5.29 832

Table IV Trigonometric Functions of Real Numbers

x	Sin x	Tan x	Cot x	Cos x	x	Sin x	Tan x	Cot x	Cos x
.00	.00000	.00000	∞	1.00000	**.50**	.47943	.54630	1.8305	.87758
.01	.01000	.01000	99.997	0.99995	.51	.48818	.55936	1.7878	.87274
.02	.02000	.02000	49.993	.99980	.52	.49688	.57256	1.7465	.86782
.03	.03000	.03001	33.323	.99955	.53	.50553	.58592	1.7067	.86281
.04	.03999	.04002	24.987	.99920	.54	.51414	.59943	1.6683	.85771
.05	.04998	.05004	19.983	.99875	.55	.52269	.61311	1.6310	.85252
.06	.05996	.06007	16.647	.99820	.56	.53119	.62695	1.5950	.84726
.07	.06994	.07011	14.262	.99755	.57	.53963	.64097	1.5601	.84190
.08	.07991	.08017	12.473	.99680	.58	.54802	.65517	1.5263	.83646
.09	.08988	.09024	11.081	.99595	.59	.55636	.66956	1.4935	.83094
.10	.09983	.10033	9.9666	.99500	**.60**	.56464	.68414	1.4617	.82534
.11	.10978	.11045	9.0542	.99396	.61	.57287	.69892	1.4308	.81965
.12	.11971	.12058	8.2933	.99281	.62	.58104	.71391	1.4007	.81388
.13	.12963	.13074	7.6489	.99156	.63	.58914	.72911	1.3715	.80803
.14	.13954	.14092	7.0961	.99022	.64	.59720	.74454	1.3431	.80210
.15	.14944	.15144	6.6166	.98877	.65	.60519	.76020	1.3154	.79608
.16	.15932	.16138	6.1966	.98723	.66	.61312	.77610	1.2885	.78999
.17	.16918	.17166	5.8256	.98558	.67	.62099	.79225	1.2622	.78382
.18	.17903	.18197	5.4954	.98384	.68	.62879	.80866	1.2366	.77757
.19	.18886	.19232	5.1997	.98200	.69	.63654	.82534	1.2116	.77125
.20	.19867	.20271	4.9332	.98007	**.70**	.64422	.84229	1.1872	.76484
.21	.20846	.21314	4.6917	.97803	.71	.65183	.85953	1.1634	.75836
.22	.21823	.22362	4.4719	.97590	.72	.65938	.87707	1.1402	.75181
.23	.22798	.23414	4.2709	.97367	.73	.66687	.89492	1.1174	.74517
.24	.23770	.24472	4.0864	.97134	.74	.67429	.91309	1.0952	.73847
.25	.24740	.25534	3.9163	.96891	.75	.68164	.93160	1.0734	.73169
.26	.25708	.26602	3.7591	.96639	.76	.68892	.95045	1.0521	.72484
.27	.26673	.27676	3.6133	.96377	.77	.69614	.96967	1.0313	.71791
.28	.27636	.28755	3.4776	.96106	.78	.70328	.98926	1.0109	.71091
.29	.28595	.29841	3.3511	.95824	.79	.71035	1.0092	.99084	.70385
.30	.29552	.30934	3.2327	.95534	**.80**	.71736	1.0296	.97121	.69671
.31	.30506	.32033	3.1218	.95233	.81	.72429	1.0505	.95197	.68950
.32	.31457	.33139	3.0176	.94924	.82	.73115	1.0717	.93309	.68222
.33	.32404	.34252	2.9195	.94604	.83	.73793	1.0934	.91455	.67488
.34	.33349	.35374	2.8270	.94275	.84	.74464	1.1156	.89635	.66746
.35	.34290	.36503	2.7395	.93937	.85	.75128	1.1383	.87848	.65998
.36	.35227	.37640	2.6567	.93590	.86	.75784	1.1616	.86091	.65244
.37	.36162	.38786	2.5782	.93233	.87	.76433	1.1853	.84365	.64483
.38	.37092	.39941	2.5037	.92866	.88	.77074	1.2097	.82668	.63715
.39	.38019	.41105	2.4328	.92491	.89	.77707	1.2346	.80998	.62941
.40	.38942	.42279	2.3652	.92106	**.90**	.78333	1.2602	.79355	.62161
.41	.39861	.43463	2.3008	.91712	.91	.78950	1.2864	.77738	.61375
.42	.40776	.44657	2.2393	.91309	.92	.79560	1.3133	.76146	.60582
.43	.41687	.45862	2.1804	.90897	.93	.80162	1.3409	.74578	.59783
.44	.42594	.47078	2.1241	.90475	.94	.80756	1.3692	.73034	.58979
.45	.43497	.48306	2.0702	.90045	.95	.81342	1.3984	.71511	.58168
.46	.44395	.49545	2.1084	.89605	.96	.81919	1.4284	.70010	.57352
.47	.45289	.50797	1.9686	.89157	.97	.82489	1.4592	.68531	.56530
.48	.46178	.52061	1.9208	.88699	.98	.83050	1.4910	.67071	.55702
.49	.47063	.53339	1.8748	.88233	.99	.83603	1.5237	.65631	.54869
.50	.47943	.54630	1.8305	.87758	**1.00**	.84174	1.5574	.64209	.54030
x	Sin x	Tan x	Cot x	Cos x	x	Sin x	Tan x	Cot x	Cos x

Table IV Trigonometric Functions of Real Numbers (Continued)

x	Sin x	Tan x	Cot x	Cos x	x	Sin x	Tan x	Cot x	Cos x
1.00	.84147	1.5574	.64209	.54030	**1.50**	.99749	14.101	.07091	.07074
1.01	.84683	1.5922	.62806	.53186	1.51	.99815	16.428	.06087	.06076
1.02	.85211	1.6281	.61420	.52337	1.52	.99871	19.670	.05084	.05077
1.03	.85730	1.6652	.60051	.51482	1.53	.99917	24.498	.04082	.04079
1.04	.86240	1.7036	.58699	.50622	1.54	.99953	32.461	.03081	.03079
1.05	.86742	1.7433	.57362	.49757	1.55	.99978	48.078	.02080	.02079
1.06	.87236	1.7844	.56040	.48887	1.56	.99994	92.621	.01080	.01080
1.07	.87720	1.8270	.54734	.48012	1.57	1.00000	1255.8	.00080	.00080
1.08	.88196	1.8712	.53441	.47133	1.58	.99996	-108.65	-.00920	-.00920
1.09	.88663	1.9171	.52162	.46249	1.59	.99982	-52.067	-.01921	-.01920
1.10	.89121	1.9648	.50897	.45360	**1.60**	.99957	-34.233	-.02921	-.02920
1.11	.89570	2.0143	.49644	.44466	1.61	.99923	-25.495	-.03922	-.03919
1.12	.90010	2.0660	.48404	.43568	1.62	.99879	-20.307	-.04924	=.04918
1.13	.90441	2.1198	.47175	.42666	1.63	.99825	-16.871	-.05927	-.05917
1.14	.90863	2.1759	.45959	.41759	1.64	.99761	-14.427	-.06931	-.06915
1.15	.91276	2.2345	.44753	.40849	1.65	.99687	-12.599	-.07937	-.07912
1.16	.91680	2.2958	.43558	.39934	1.66	.99602	-11.181	-.08944	-.08909
1.17	.92075	2.3600	.42373	.39015	1.67	.99508	-10.047	-.09953	-.09904
1.18	.92461	2.4273	.41199	.38092	1.68	.99404	- 9.1208	-.10964	-.10899
1.19	.92837	2.4979	.40034	.37166	1.69	.99290	- 8.3492	-.11977	-.11892
1.20	.93204	2.5722	.38878	.36236	**1.70**	.99166	- 7.6966	-.12993	-.12884
1.21	.93562	2.6503	.37731	.35302	1.71	.99033	- 7.1373	-.14011	-.13875
1.22	.93910	2.7328	.36593	.34365	1.72	.98889	- 6.6524	-.15032	-.14865
1.23	.94249	2.8198	.35463	.33424	1.73	.98735	- 6.2281	-.16056	-.15853
1.24	.94578	2.9119	.34341	.32480	1.74	.98572	- 5.8535	-.17084	-.16840
1.25	.94898	3.0096	.33227	.31532	1.75	.98399	- 5.5204	-.18115	-.17825
1.26	.95209	3.1133	.32121	.30582	1.76	.98215	- 5.2221	-.19149	-.18808
1.27	.95510	3.2236	.31021	.29628	1.77	.98022	- 4.9534	-.20188	-.19789
1.28	.95802	3.3413	.29928	.28672	1.78	.97820	- 4.7101	-.21231	-.20768
1.29	.96084	3.4672	.28842	.27712	1.79	.97607	- 4.4887	-.22278	-.21745
1.30	.96356	3.6021	.27762	.26750	**1.80**	.97385	- 4.2863	-.23330	-.22720
1.31	.96618	3.7471	.26687	.25785	1.81	.97153	- 4.1005	-.24387	-.23693
1.32	.96872	3.9033	.25619	.24818	1.82	.96911	- 3.9294	-.25449	-.24663
1.33	.97115	4.0723	.24556	.23848	1.83	.96659	- 3.7712	-.26517	-.25631
1.34	.97348	4.2556	.23498	.22875	1.84	.96398	- 3.6245	-.27590	-.26596
1.35	.97572	4.4552	.22446	.21901	1.85	.96128	- 3.4881	-.28669	-.27559
1.36	.97786	4.6734	.21398	.20924	1.86	.95847	- 3.3608	-.29755	-.28519
1.37	.97991	4.9131	.20354	.19945	1.87	.95557	- 3.2419	-.30846	-.29476
1.38	.98185	5.1774	.19315	.18964	1.88	.95258	- 3.1304	-.31945	-.30430
1.39	.98370	5.4707	.18279	.17981	1.89	.94949	- 3.0257	-.33051	-.31381
1.40	.98545	5.7979	.17248	.16997	**1.90**	.94630	- 2.9271	-.34164	-.32329
1.41	.98710	6.1654	.16220	.16010	1.91	.94302	- 2.8341	-.35284	-.33274
1.42	.98865	6.5811	.15195	.15023	1.92	.93965	- 2.7463	-.36413	-.34215
1.43	.99010	7.0555	.14173	.14033	1.93	.93618	- 2.6632	-.37549	-.35153
1.44	.99146	7.6018	.13155	.13042	1.94	.93262	- 2.5843	-.38695	-.36087
1.45	.99271	8.2381	.12139	.12050	1.95	.92896	- 2.5095	-.39849	-.37018
1.46	.99387	8.9886	.11125	.11057	1.96	.92521	- 2.4383	-.41012	-.37945
1.47	.99492	9.8874	.10114	.10063	1.97	.92137	- 2.3705	-.42185	-.38868
1.48	.99588	10.983	.09105	.09067	1.98	.91744	- 2.3058	-.43368	-.39788
1.49	.99674	12.350	.08097	.08071	1.99	.91341	- 2.2441	-.44562	-.40703
1.50	.99749	14.101	.07091	.07074	**2.00**	.90930	- 2.1850	-.45766	-.41615
x	Sin x	Tan x	Cot x	Cos x	x	Sin x	Tan x	Cot x	Cos x

Table V Trigonometric Functions of Angles

Degrees	Radians	Sin	Cos	Tan	Cot	Sec	Csc		
0	.0000	.0000	1.0000	.0000	....	1.0000	....	1.5708	90
1	.0175	.0175	.9998	.0175	57.2900	1.0002	57.299	1.5533	89
2	.0349	.0349	.9994	.0349	28.6363	1.0006	28.654	1.5359	88
3	.0524	.0523	.9986	.0524	19.0811	1.0014	19.107	1.5184	87
4	.0698	.0698	.9976	.0699	14.3007	1.0024	14.336	1.5010	86
5	.0873	.0872	.9962	.0875	11.4301	1.0038	11.474	1.4835	85
6	.1047	.1045	.9945	.1051	9.5144	1.0055	9.5668	1.4661	84
7	.1222	.1219	.9925	.1228	8.1443	1.0075	8.2055	1.4486	83
8	.1396	.1392	.9903	.1405	7.1154	1.0098	7.1853	1.4312	82
9	.1571	.1564	.9877	.1584	6.3138	1.0125	6.3925	1.4137	81
10	.1745	.1736	.9848	.1763	5.6713	1.0154	5.7588	1.3963	80
11	.1920	.1908	.9816	.1944	5.1446	1.0187	5.2408	1.3788	79
12	.2094	.2079	.9781	.2126	4.7046	1.0223	4.8097	1.3614	78
13	.2269	.2250	.9744	.2309	4.3315	1.0263	4.4454	1.3439	77
14	.2443	.2419	.9703	.2493	4.0108	1.0306	4.1336	1.3265	76
15	.2618	.2588	.9659	.2679	3.7321	1.0353	3.8637	1.3090	75
16	.2793	.2756	.9613	.2867	3.4874	1.0403	3.6280	1.2915	74
17	.2967	.2924	.9563	.3057	3.2709	1.0457	3.4203	1.2741	73
18	.3142	.3090	.9511	.3249	3.0777	1.0515	3.2361	1.2566	72
19	.3316	.3256	.9455	.3443	2.9042	1.0576	3.0716	1.2392	71
20	.3491	.3420	.9397	.3640	2.7475	1.0642	2.9238	1.2217	70
21	.3665	.3584	.9336	.3839	2.6051	1.0711	2.7904	1.2043	69
22	.3840	.3746	.9272	.4040	2.4751	1.0785	2.6695	1.1868	68
23	.4014	.3907	.9205	.4245	2.3559	1.0864	2.5593	1.1694	67
24	.4189	.4067	.9135	.4452	2.2460	1.0946	2.4586	1.1519	66
25	.4363	.4226	.9063	.4663	2.1445	1.1034	2.3662	1.1345	65
26	.4538	.4384	.8988	.4877	2.0503	1.1126	2.2812	1.1170	64
27	.4712	.4540	.8910	.5095	1.9626	1.1223	2.2027	1.0996	63
28	.4887	.4695	.8829	.5317	1.8807	1.1326	2.1301	1.0821	62
29	.5061	.4848	.8746	.5543	1.8040	1.1434	2.0627	1.0647	61
30	.5236	.5000	.8660	.5774	1.7321	1.1547	2.0000	1.0472	60
31	.5411	.5150	.8572	.6009	1.6643	1.1666	1.9416	1.0297	59
32	.5585	.5299	.8480	.6249	1.6003	1.1792	1.8871	1.0123	58
33	.5760	.5446	.8387	.6494	1.5399	1.1924	1.8361	.9948	57
34	.5934	.5592	.8290	.6745	1.4826	1.2062	1.7883	.9774	56
35	.6109	.5736	.8192	.7002	1.4281	1.2208	1.7434	.9599	55
36	.6283	.5878	.8090	.7265	1.3764	1.2361	1.7013	.9425	54
37	.6458	.6018	.7986	.7536	1.3270	1.2521	1.6616	.9250	53
38	.6632	.6157	.7880	.7813	1.2799	1.2690	1.6243	.9076	52
39	.6807	.6293	.7771	.8098	1.2349	1.2868	1.5890	.8901	51
40	.6981	.6428	.7660	.8391	1.1918	1.3054	1.5557	.8727	50
41	.7156	.6561	.7547	.8693	1.1504	1.3250	1.5243	.8552	49
42	.7330	.6691	.7431	.9004	1.1106	1.3456	1.4945	.8378	48
43	.7505	.6820	.7314	.9325	1.0724	1.3673	1.4663	.8203	47
44	.7679	.6947	.7193	.9657	1.0355	1.3902	1.4396	.8029	46
45	.7854	.7071	.7071	1.0000	1.0000	1.4142	1.4142	.7854	45
		Cos	Sin	Cot	Tan	Csc	Sec	Radians	Degrees

Table VI Squares, Cubes, Roots

n	n^2	$\sqrt{n}$	n^3	$\sqrt[3]{n}$	n	n^2	$\sqrt{n}$	n^3	$\sqrt[3]{n}$
1	1	1.000	1	1.000	51	2,601	7.141	132,651	3.708
2	4	1.414	8	1.260	52	2,704	7.211	140,608	3.733
3	9	1.732	27	1.442	53	2,809	7.280	148,877	3.756
4	16	2.000	64	1.587	54	2,916	7.348	157,464	3.780
5	25	2.236	125	1.710	55	3,025	7.416	166,375	3.803
6	36	2.449	216	1.817	56	3,136	7.483	175,616	3.826
7	49	2.646	343	1.913	57	3,249	7.550	185,193	3.849
8	64	2.828	512	2.000	58	3,364	7.616	195,112	3.871
9	81	3.000	729	2.080	59	3,481	7.681	205,379	3.893
10	100	3.162	1,000	2.154	60	3,600	7.746	216,000	3.915
11	121	3.317	1,331	2.224	61	3,721	7.810	226,981	3.936
12	144	3.464	1,728	2.289	62	3,844	7.874	238,328	3.958
13	169	3.606	2,197	2.351	63	3,969	7.937	250,047	3.979
14	196	3.742	2,744	2.410	64	4,096	8.000	262,144	4.000
15	225	3.873	3,375	2.466	65	4,225	8.062	274,625	4.021
16	256	4.000	4,096	2.520	66	4,356	8.124	287,496	4.041
17	289	4.123	4,913	2.571	67	4,489	8.185	300,763	4.062
18	324	4.243	5,832	2.621	68	4,624	8.246	314,432	4.082
19	361	4.359	6,859	2.668	69	4,761	8.307	328,509	4.102
20	400	4.472	8,000	2.714	70	4,900	8.367	343,000	4.121
21	441	4.583	9,261	2.759	71	5,041	8.426	357,911	4.141
22	484	4.690	10,648	2.802	72	5,184	8.485	373,248	4.160
23	529	4.796	12,167	2.844	73	5,329	8.544	389,017	4.179
24	576	4.899	13,824	2.884	74	5,476	8.602	405,224	4.198
25	625	5.000	15,625	2.924	75	5,625	8.660	421,875	4.217
26	676	5.099	17,576	2.962	76	5,776	8.718	438,976	4.236
27	729	5.196	19,683	3.000	77	5,929	8.775	456,533	4.254
28	784	5.292	21,952	3.037	78	6,084	8.832	474,552	4.273
29	841	5.385	24,389	3.072	79	6,241	8.888	493,039	4.291
30	900	5.477	27,000	3.107	80	6,400	8.944	512,000	4.309
31	961	5.568	29,791	3.141	81	6,561	9.000	531,441	4.327
32	1,024	5.657	32,768	3.175	82	6,724	9.055	551,368	4.344
33	1,089	5.745	35,937	3.208	83	6,889	9.110	571,787	4.362
34	1,156	5.831	39,304	3.240	84	7,056	9.165	592,704	4.380
35	1,225	5.916	42,875	3.271	85	7,225	9.220	614,125	4.397
36	1,296	6.000	46,656	3.302	86	7,396	9.274	636,056	4.414
37	1,369	6.083	50,653	3.332	87	7,569	9.327	658,503	4.431
38	1,444	6.164	54,872	3.362	88	7,744	9.381	681,472	4.448
39	1,521	6.245	59,319	3.391	89	7,921	9.434	704,969	4.465
40	1,600	6.325	64,000	3.420	90	8,100	9.487	729,000	4.481
41	1,681	6.403	68,921	3.448	91	8,281	9.539	753,571	4.498
42	1,764	6.481	74,088	3.476	92	8,464	9.592	778,688	4.514
43	1,849	6.557	79,507	3.503	93	8,649	9.644	804,357	4.531
44	1,936	6.633	85,184	3.530	94	8,836	9.695	830,584	4.547
45	2,025	6.708	91,125	3.557	95	9,025	9.747	857,375	4.563
46	2,116	6.782	97,336	3.583	96	9,216	9.798	884,736	4.579
47	2,209	6.856	103,823	3.609	97	9,409	9.849	912,673	4.595
48	2,304	6.928	110,592	3.634	98	9,604	9.899	941,192	4.610
49	2,401	7.000	117,649	3.659	99	9,801	9.950	970,299	4.626
50	2,500	7.071	125,000	3.684	100	10,000	10.000	1,000,000	4.642
n	n^2	$\sqrt{n}$	n^3	$\sqrt[3]{n}$	n	n^2	$\sqrt{n}$	n^3	$\sqrt[3]{n}$

Answers to Odd-numbered Problems

Problems 1.4

1. Unsatisfactory. **3.** Satisfactory. **5.** Unsatisfactory.
7. Unsatisfactory. **9.** Unsatisfactory. **11.** No.
13. Yes. **15.** No. **17.** Yes.
19. Yes. **21.** $\{2\}$. **23.** $\{-1\}$.
25. $\emptyset$. **27.** $\{0\}$. **29.** $\{2\}$.
31. $n \leftrightarrow -n$. **33.** Not possible because of bigamy.
35. Not possible. **37.** State fielding position of each batter.

39.

0	1	−1	2	−2	3	−3	. . .
1	2	3	4	5	6	7	

45. $\emptyset$, $\{2\}$, $\{6\}$, $\{2, 6\}$; all but $\{2, 6\}$.
47. $\emptyset$, $\{3\}$, $\{5\}$, $\{7\}$, $\{3, 5\}$, $\{3, 7\}$, $\{5, 7\}$, $\{3, 5, 7\}$; all but $\{3, 5, 7\}$.
49. $\emptyset$, $\{a\}$, $\{b\}$, $\{c\}$, $\{d\}$, $\{a, b\}$, $\{a, c\}$, $\{a, d\}$, $\{b, c\}$, $\{b, d\}$, $\{c, d\}$, $\{a, b, c\}$, $\{a, b, d\}$, $\{a, c, d\}$, $\{b, c, d\}$, $\{a, b, c, d\}$; all but $\{a, b, c, d\}$.

Problems 1.7

1. $\forall_x$. **3.** None. **5.** $\forall_x$.
7. $\forall_x$. **9.** $\exists_x$.

11. John is a student, and Mary is beautiful. John is a student, or Mary is beautiful.
13. All lines are straight, and all circles are round. All lines are straight, or all circles are round.

	Conjunction	*Disjunction*
15.	$\{1\}$	$\{-1, 1\}$
17.	$\{-2\}$	$\{-2\}$
19.	$\{x \mid x > 5\}$	$\{x \mid x > 3\}$
21.	$\{7\}$	$\{x \mid x > 5\}$
23.	X	X

25. $\{3\}$

27. $\{1, 2, 4, 5, 8, 9, 11, 23, 60\}$. **29.** $\{x \mid x \text{ is a positive integer}\}$.

31. $\{x \mid x \neq 0\}$.

Problems 1.10

1. Jones is not a carpenter. **3.** 7 is not a prime number.
5. These two lines are not parallel. **7** $\{1, 2, 3, 5, 7\}$.

9. $\{2, 4, 6, 8, \ldots\}$. **11.** $\{x \mid x \neq -1 \text{ or } 1\}$. **13.** $\{x \mid x \leq 3\}$.
15. X. **17.** $\{x \mid x \neq 1\}$. **19.** $\{x \mid x \neq 8\}$.
21. $\{x \mid x \neq -2 \text{ or } 2\}$. **23.** X. **25.** $\{x \mid -4 < x < 4\}$.

27. If $3 + 7 = 10$, then $4 + 9 = 13$, true. If $4 + 9 = 13$, then $3 + 7 = 10$, true.
29. If $2 + 5 = 7$, then $3 + 3 = 8$, false. If $3 + 3 = 8$, then $2 + 5 = 7$, true.
31. If this circle of radius 5 in. has an area of 25π in.^{2}, it has a circumference of 10π in., true. If this circle of radius 5 in. has a circumference of 10π in., it has an area of 25π in.^{2}, true.
33. If 24 is a perfect square, 13 is a composite number, true. If 13 is a composite number, 24 is a perfect square, true.
35. If Philadelphia is in New Jersey, then Atlanta is in Georgia, true. If Atlanta is in Georgia, then Philadelphia is in New Jersey, false.

37. X. **39.** X. **41.** $\{x \mid x \neq 3\}$.
43. X. **45.** $\emptyset$. **47.** Equivalent.
49. Not equivalent. **51.** Not equivalent. **53.** Equivalent.
55. $\{x \mid x \neq 1 \text{ or } -1\}$. **57.** X. **59.** X.
61. $\{x \mid x \neq -1\}$. **63.** $\emptyset$.

Problems 1.11

27. $P \cup (Q \cup R)$. **29.** $P \cap (Q \cup R)$. **31.** $(P' \cup Q) \cap (Q' \cup P)$.

Problems 1.12

1. $p \wedge q$. **3.** $[(\sim p) \vee (\sim q)] \wedge (\sim r)$.
5. $[p \wedge (\sim r)] \wedge (\sim q)$. **7.** $[(\sim p) \wedge (\sim r)] \wedge (q)$.
9. $(p \vee q) \wedge [q \wedge (\sim r)]$.

11. All numbers are not even; all numbers are odd; no numbers are even.

13. Some rooms are not taken.
15. $\forall_x(x^2 \geq 0)$. **17.** $\exists_x(x^2 < 0)$.

19. For some triangles, the sum of the measures of the interior angles is not 180°
21. Some people are rich, and all people are not poor. Some people are rich, and no people are poor.
23. Some roads do not lead to Rome, or all roads are passable.
25. Some windows are not open, and I am not too hot.
27. There exists a pair of congruent angles whose measures are not equal.
29. There exists an integer x such that x^2 is even and x is odd.
31. There exists a pair of congruent triangles whose corresponding altitudes are not equal.

Problems 1.14

1. $(P \cap Q)' = (P' \cup Q')$. **3.** $(P')' = P$.
5. $P \cap (Q \cap R) = (P \cap Q) \cap R$. **7.** $P \cup P = P$.
9. $P \cap U = P$. **11.** (Fig. 1.14.11). **13.** (Fig. 1.14.13).

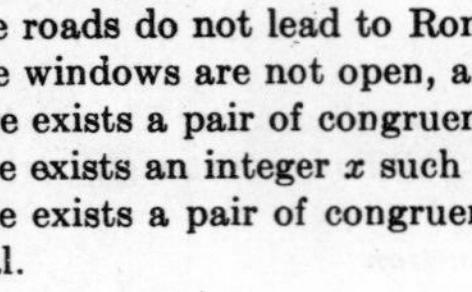

Similar
Congruent

Figure 1.14.11

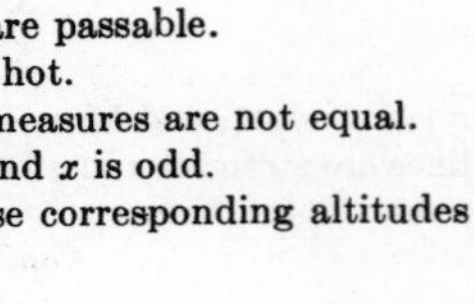

Rectangles
Squares

Figure 1.14.13

15. Converse: If $2a$ is divisible by 8, then a is divisible by 4. Contrapositive: If $2a$ is not divisible by 8, then a is not divisible by 4.
17. Converse: If the diagonals of a quadrilateral bisect each other, then it is a parallelogram. Contrapositive: If the diagonals of a quadrilateral do not bisect each other, then it is not a parallelogram.
19. Converse: If $a - c > b - c$ then $a > b$. Contrapositive: If $a - c \leq b - c$, then $a \leq b$.
23. $\forall_x[(\sim p_x) \rightarrow (\sim q_x)]$.
25. A sufficient condition that a triangle be isosceles is that its base angles be equal.
27. A sufficient condition that two lines be parallel is that they be perpendicular to the same line.
29. A sufficient condition that $x = 1$ is that $3x + 2 = x + 4$.
31. A necessary condition that a triangle be inscribed in a semicircle is that it be a right triangle.
33. A necessary condition that a body be in static equilibrium is that the vector sum of all forces acting on it is zero.
35. A necessary condition that two forces be in equilibrium is that they be equal, opposite, and collinear.
37. A triangle is inscribed in a semicircle only if it is a right triangle.
39. A body is in static equilibrium only if the vector sum of all forces acting on it is zero.
41. Two forces are in equilibrium only if they are equal, opposite, and collinear.
43. A necessary condition that a triangle be isosceles is that its base angles be equal. A sufficient condition that the base angles of a triangle be equal is that the triangle be isosceles.
45. A necessary condition that two lines be parallel is that they be perpendicular to the same line. A sufficient condition that two lines be perpendicular to the same line is that they be parallel.
47. A necessary condition that $x = 1$ is that $3x + 2 = x + 4$. A sufficient condition that $3x + 2 = x + 4$ is that $x = 1$.
49. A triangle is a right triangle only if it is inscribed in a semicircle.
51. The vector sum of all forces acting on a body is zero only if it is in static equilibrium.
53. Two forces are equal, opposite, and collinear only if they are in equilibrium.
55. A necessary and sufficient condition that two lines be parallel is that they be equidistant.
57. A necessary and sufficient condition that three concurrent forces be in equilibrium is that their vector sum be zero.
59. No. Promise is converse of that required for her to win.

Problems 1.17

1. True.
3. True.
5. True, $x = 4$.
7. False. The sum of the roots is -5.
9. False. $x = 1$ is a counterexample.
11. False. This is the "ambiguous case" for triangles. If a, b, and B are given and if b is neither too long nor too short (see Fig. 1.17.11), there are two noncongruent triangles each of which has these three given parts.

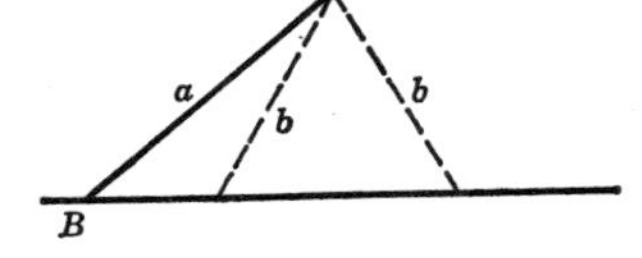

Figure 1.17.11

13. True. The roots are real if $b^2 - 4ac \geq 0$. Hence $b^2 - 4ac > 5$ is a sufficient condition.

15. True: $x = \dfrac{c - b}{a}$ is such a solution.

17. False. From Fig. 1.17.17.

$$\begin{aligned} \angle 1 + \angle 4 &= 180^\circ \\ \angle 2 + \angle 5 &= 180^\circ \\ \angle 3 + \angle 6 &= 180^\circ \\ \hline (\angle 1 + \angle 2 + \angle 3) + (\angle 4 + \angle 5 + \angle 6) &= 540^\circ \end{aligned}$$

But $\angle 4 + \angle 5 + \angle 6 = 180^\circ$;
so $\angle 1 + \angle 2 + \angle 3 = 360^\circ$.

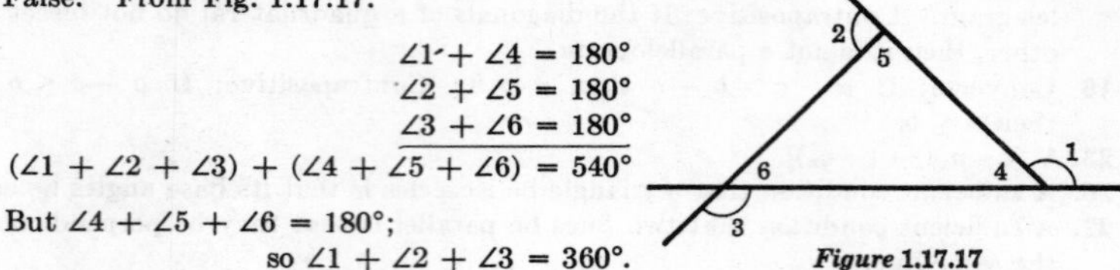

Figure 1.17.17

19. Assume the negation of the conclusion, namely: Some two distinct lines meet in more than one point. Choose any two such points A and B. Then there are two distinct lines through A and B. This contradicts the axiom.

21. Assume the negation of the conclusion: Some equation of the form $a + x = b$ has more than one solution. Let x_1 and x_2 be two such solutions with $x_1 \neq x_2$. Then

$$a + x_1 = b$$
$$a + x_2 = b$$

Subtracting, we have: $x_1 - x_2 = 0$, which is a contradiction.

Problems 2.5

5. $(a + b + c) + d$.
7. $-4, 2, -\pi, 0, \sqrt{5}$.
9. $7, 3, 0, \frac{2}{3}, \frac{1}{2}$.
11. $(a \times b) \times c$.

17. $3, -\frac{1}{4}, -\frac{4}{3}, 1$, the inverse of 0 is not defined

19. No.
21. No.
23. No.

25. $4/0$ is meaningless, $0/4 = 0$, $4/4 = 1$, $0/\frac{1}{4} = 0$, $0/0$ is indeterminate.
27. $x = -1, x = 0, x = 2$, none, $x = 1, 4$.
29. $x = 0, x = -2$, none, $x = 0, x = 3$.

31. $-2, -3$.
33. $\pm\sqrt{6}$.
35. True.
37. False.
39. True.

41. It is closed, commutative, associative. 0 is identity. $a(\neq -1)$ has addiplicative inverse $-a/(1 + a)$.

Problems 2.7

1. $13 + i$.
3. $-5 - 16i$.
5. $-3 - 12i$.
7. $2 + 7i$.
9. $5 - 7i$.
11. $2 + 7i$.
13. $5 + 2i$.
15. $-29 - 29i$.
17. $41 + i$.
19. 4.
21. 89.
23. $21 - 15i$.
25. $-45 + 9i$.
27. -21.
29. i.
31. $(49 - 10i)/61$.
33. $(11 + 52i)/25$.
35. $2 + i$.
37. $(45 + 30i)/117$.
39. $(8 - 3i)/6$.
41. $x = -\frac{23}{26}, y = \frac{15}{26}$.
43. $x = \frac{29}{17}, y = \frac{31}{17}$.
45. $x = 0, y = -\frac{3}{2}$.
55. $(5, -1)$.
57. $(-10, 11)$.
59. $(\frac{10}{13}, \frac{11}{13})$.

Problems 2.8

1.

+	0	1
0	0	1
1	1	0

×	0	1
0	0	0
1	0	1

3.

+	0	1	2	3	4	5
0	0	1	2	3	4	5
1	1	2	3	4	5	0
2	2	3	4	5	0	1
3	3	4	5	0	1	2
4	4	5	0	1	2	3
5	5	0	1	2	3	4

×	0	1	2	3	4	5
0	0	0	0	0	0	0
1	0	1	2	3	4	5
2	0	2	4	0	2	4
3	0	3	0	3	0	3
4	0	4	2	0	4	2
5	0	5	4	3	2	1

5. $-5, 3, 11, 19, 27$.

7. $-9, -2, 5, 12, 19$.

13. Because $(x - kb)$ is an integer and a is an integer $\neq \pm 1$, so $a(x - kb) \neq 1$.

Problems 2.12

9. The implication is false when a is negative and b is negative or zero.

13. When a and b have same sign.

15. Equality holds when $a = b$.

Problems 2.14

1. $0.\overline{571428}$.

3. $0.\overline{1}$.

5. $3.\overline{7}$.

7. $\frac{5}{9}$.

9. $\dfrac{1{,}508}{99}$.

11. $\dfrac{27{,}327}{9{,}990} = \dfrac{9{,}109}{3{,}330}$.

15. If a is divisible by 2, then a^2 is divisible by 2.

21. 1.

23. $\{\sqrt{7}\}$; many other answers possible.

Problems 3.2

29. I is false.

31. II is false.

33. I is false.

35. I is false.

37. I is false.

41. If $b = a$, there are no elements of A less than b, so I is verified. We must prove that if there are only a finite number of elements of A less than k, there are only a finite number of these less than $k + 1$. This follows from Prob. 40.

43. Using indirect proof, we assume that there are elements of A which are not in S. From the Well-ordering Principle there is a least of these, say, q. Of course, $q \neq a$, since a is in S by I. Then $q - 1$ is in A. But $q - 1$ cannot be in S, for if it were in S, then by II q would be in S. Hence q is not the least element of A which is not in S as we had supposed. This contradiction establishes the result.

Problems 3.4

1. $x = 1, y = -2$.

3. $x = -1, y = 2$.

5. $x = -14, y = 27$.

7. $x = 1, y = -2$.

9. $x = 1, y = -1$.

13. We assume that m is a prime and wish to show that every residue class, $a(\neq 0)$, has a multiplicative inverse. Hence we wish to solve $ax = 1 + km$ for the integers x and k. Writing this in the form $ax - km = 1$, it follows from Theorem 2 that such x and k exist. The class corresponding to x is the desired inverse.

Problems 3.6

11. No. **13.** No. **15.** Yes.

Problems 3.8

9. $\begin{pmatrix} 4 & -3 \\ 3 & 4 \end{pmatrix}$. **11.** $\begin{pmatrix} 10 & -9 \\ 4 & -1 \end{pmatrix}$. **13.** $\begin{pmatrix} -1 & -6 \\ 6 & 10 \end{pmatrix}$.

15. $\begin{pmatrix} 0 & 0 \\ 0 & 0 \end{pmatrix}$. **17.** $\begin{pmatrix} -8 & 22 \\ 9 & 25 \end{pmatrix}$. **19.** $\begin{pmatrix} \frac{2}{7} & -\frac{1}{7} \\ -\frac{1}{7} & \frac{4}{7} \end{pmatrix}$.

21. $\begin{pmatrix} -\frac{3}{2} & \frac{5}{4} \\ \frac{1}{2} & -\frac{1}{4} \end{pmatrix}$. **23.** The inverse does not exist.

Problems 4.3

3. (**c**) 0. (**d**) 0, 3, 2, 1.

5. (**a**)

+	0	1	2	3	4
0	0	1	2	3	4
1	1	2	3	4	0
2	2	3	4	0	1
3	3	4	0	1	2
4	4	0	1	2	3

(**c**) Yes. 0.

(**d**) 0, 4, 3, 2, 1. (**f**) Yes.

9.

×	a	b
a	a	b
b	b	a

a is the identity.
$a' = a$.
$b' = b$.

Problems 4.4

5. (**a**) The rotation through an angle equal to 0° is the identity.
(**b**) The inverse of a rotation through $n \times 60°$ is the rotation through $(-n) \times 60°$.

13. e is the identity.

$$p' = q$$
$$q' = p$$
$$r' = r$$
$$s' = s$$
$$t' = t$$

15. (**a**)

∘	e	m	n	p	q	r	t	u
e	e	m	n	p	q	r	t	u
m	m	n	p	e	u	t	q	r
n	n	p	e	m	r	q	u	t
p	p	e	m	n	t	u	r	q
q	q	t	r	u	e	n	m	p
r	r	u	q	t	n	e	p	m
t	t	r	u	q	p	m	e	n
u	u	q	t	r	m	p	n	e

(**b**) No.

(**c**) $m' = p$, $n' = n$, $p' = m$, $q' = q$, $r' = r$, $t' = t$, $u' = u$.

17.

$\circ$	e	a	b
e	e	a	b
a	a	b	e
b	b	e	a

No.

21. The order of a, of b, of c is 2.

23. (**b**) A.

25. There are 15 additional semigroups. For a complete list see Edwin Hewitt and Herbert S. Zuckerman, Finite Dimensional Convolution Algebras, *Acta Mathematica*, vol. 93, p. 114 (1955).

Problems 4.7

1. $Se = Sp = Sq = \{e, p, q\}$
$Sr = Ss = St = \{r, s, t\}$

3. Let $S = \{e, m, n, p\}$
$eS = mS = nS = pS = \{e, m, n, p\}$
$qS = nS = tS = uS = \{q, t, r, u\}$

7. Use the correspondance $a \leftrightarrow a'$.

Problems 5.3

3. Positive; negative.

5. $(2, -4)$ lies in quadrant IV.
$(3,7)$ lies in quadrant I.
$(-5, -6)$ lies in quadrant III.
$(-9,11)$ lies in quadrant II.
$(-4,7)$ lies in quadrant II.

7. 3; 2; 11; 11; 3.

15. $K = 273° + C$.

19. $I = 12\,F$.

23. $F = \frac{9}{5}C + 32°$; $-40°$.

Problems 5.5

1. $A = B$.

3. $A \supset B$.

5. $A = B$.

7. $A \subset B$.

9. $A \supset B$.

Problems 5.7

1. $x \equiv 5$, mod 7.

3. $x \equiv 4$, mod 5.

5. $x \equiv 7$, mod 11.

11. (Fig. 5.7.11).

13. (Fig. 5.7.13).

15. (Fig. 5.7.15).

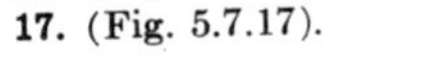

17. (Fig. 5.7.17).

19. (Fig. 5.7.19).

21. (Fig. 5.7.21).

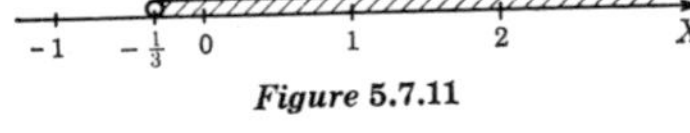

Figure 5.7.11

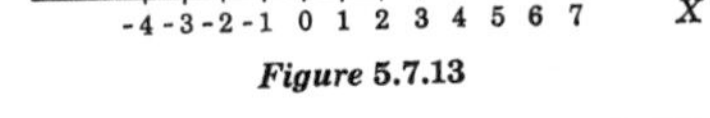

Figure 5.7.13

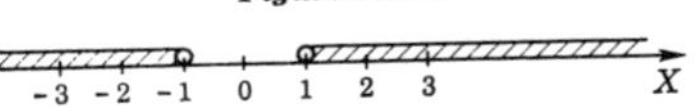

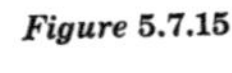

Figure 5.7.15

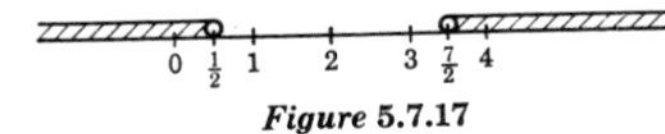

Figure 5.7.17

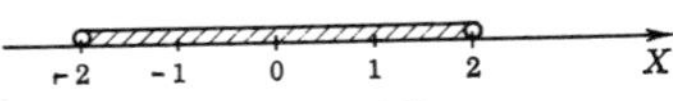

Figure 5.7.19

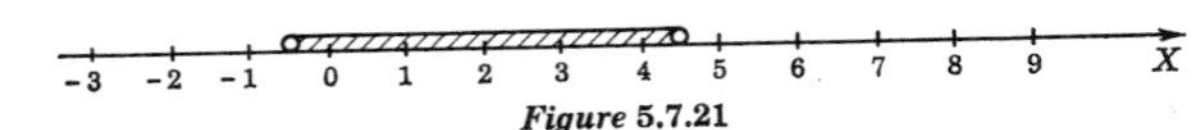

Figure 5.7.21

23. (Fig. 5.7.23). **25.** (Fig. 5.7.25). **27.** (Fig. 5.7.27).

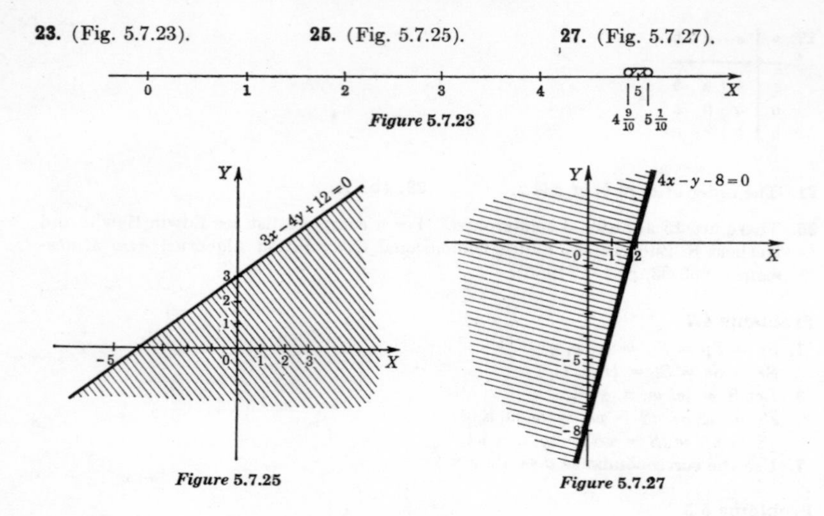

Figure 5.7.25

Figure 5.7.27

29. (Fig. 5.7.29). **31.** (Fig. 5.7.31). **33.** (Fig. 5.7.33).
35. (Fig. 5.7.35). **37.** (Fig. 5.7.37).

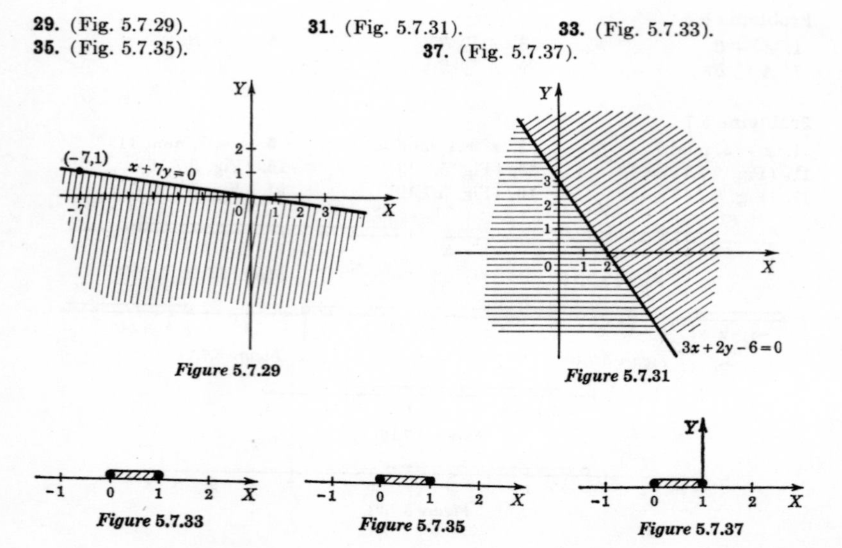

Figure 5.7.33 *Figure* 5.7.35 *Figure* 5.7.37

Problems 5.9

1. (2,2). (Fig. 5.9.1) **3.** (0,3). (Fig. 5.9.3) **5.** (3,1). (Fig. 5.9.5)

7. No solution. (Fig. 5.9.7) **9.** Infinitely many solutions. (Fig. 5.9.9)

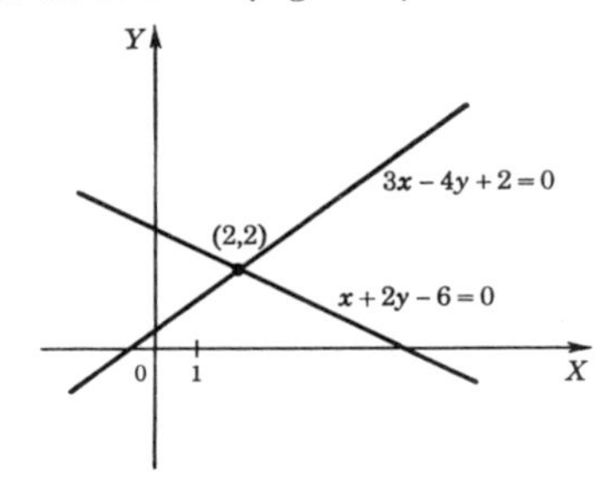

Figure 5.9.1

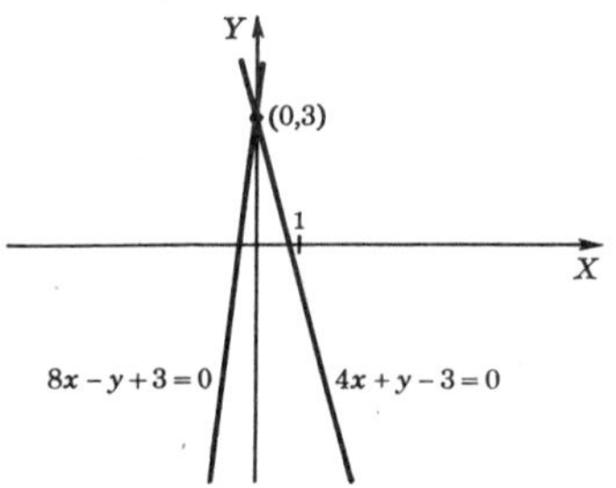

Figure 5.9.3

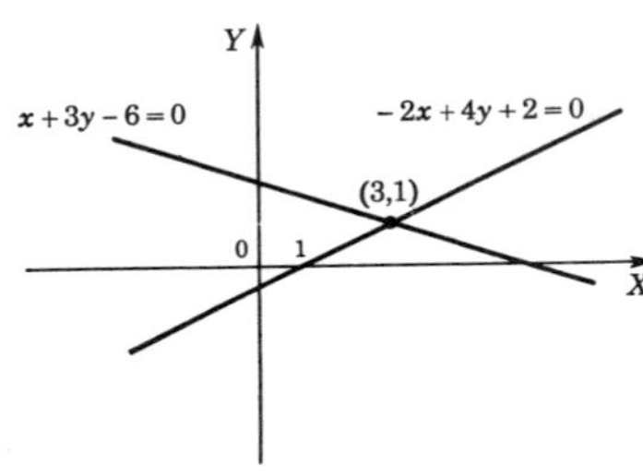

Figure 5.9.5

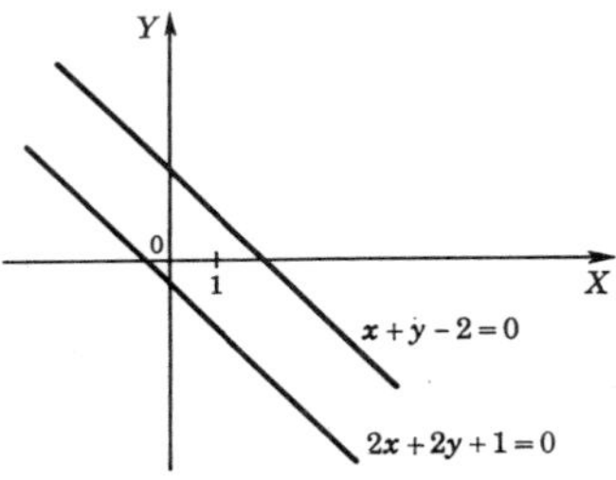

Figure 5.9.7

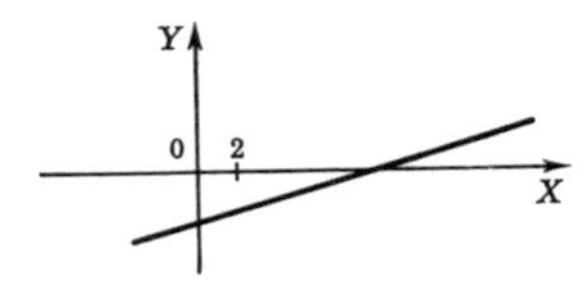

Figure 5.9.9

11. (1,4). **13.** (3,1).

15. (0,3), (1,4), (2,0), (3,1), (4,2).

17. No solution. **19.** (Fig. 5.9.19). **21.** (Fig. 5.9.21).

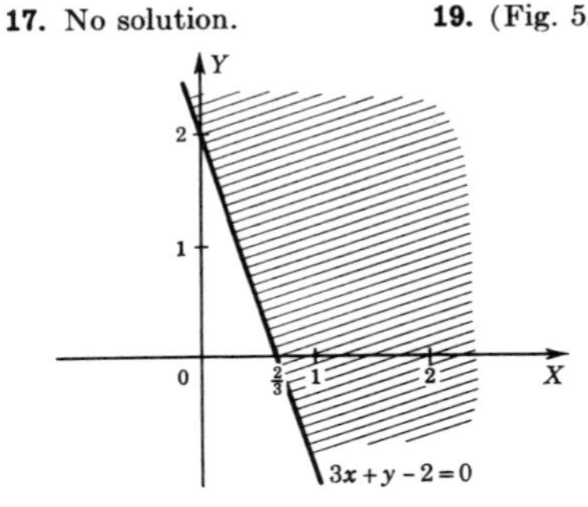

Figure 5.9.19

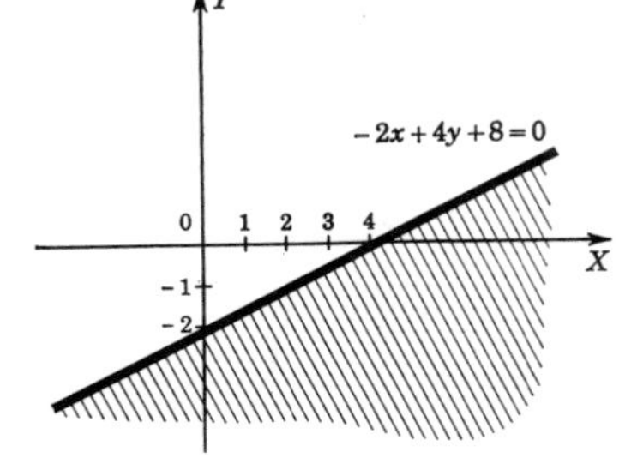

Figure 5.9.21

23. (Fig. 5.9.23). **25.** (Fig. 5.9.25). **27.** (Fig. 5.9.27).
29. (Fig. 5.9.29). **31.** (Fig. 5.9.31). **33.** (Fig. 5.9.33).

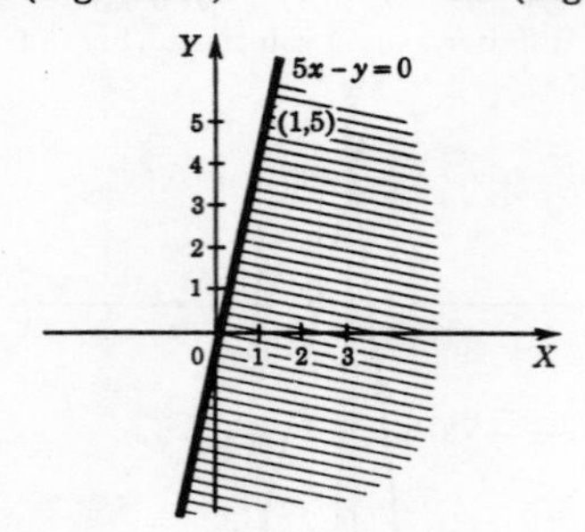

Figure **5.9.23**

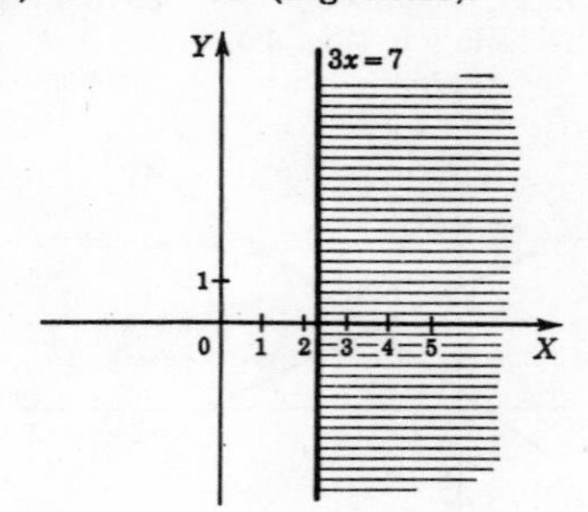

Figure **5.9.25**

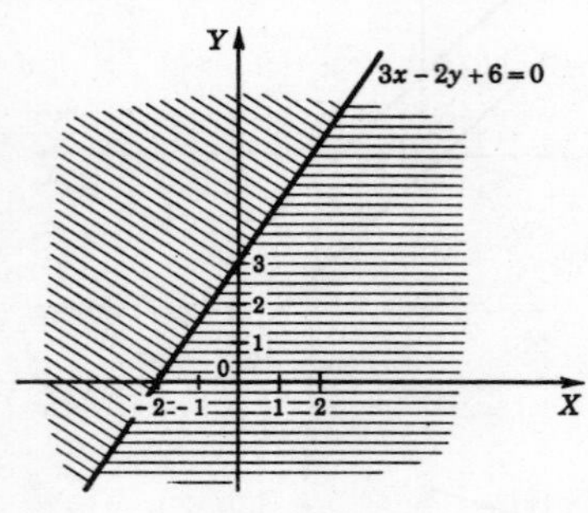

Figure **5.9.27**

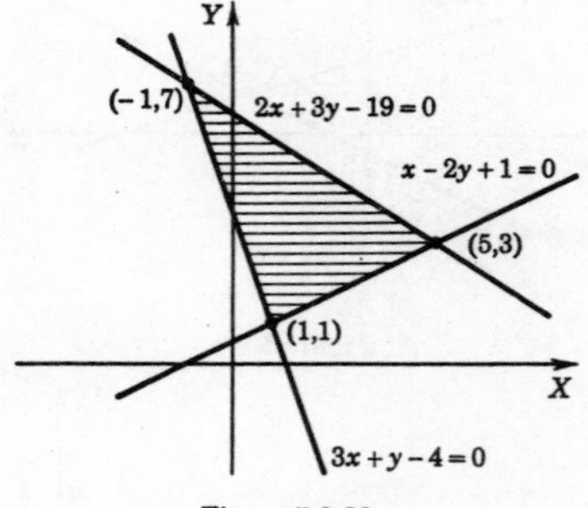

Figure **5.9.29**

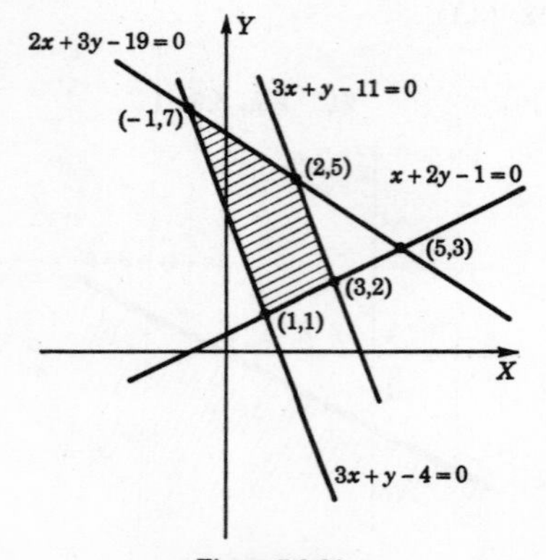

Figure **5.9.31**

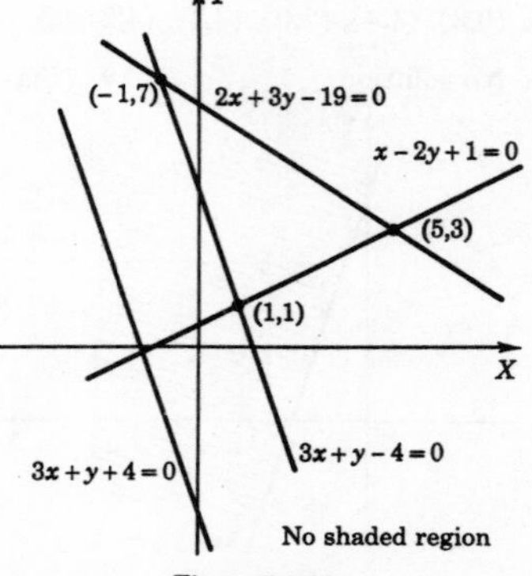

Figure **5.9.33**

35.

r	$0 \cdots 0$	$1 \cdots 1$	$2 \cdots 2$	3
p	$10 \cdots 18$	$10 \cdots 15$	$9 \cdots 12$	9

when 180 sec left to play; impossible if 90 sec left to play.

37. $x' = 2x + 3$.

39. $x' = \frac{3}{4}x + 25$; 100.

Problems 5.11

1. $-2 \pm i$.

3. $-\frac{1}{3}$ twice.

5. $-\frac{5}{4}, \frac{7}{9}$.

7. $1 + 3i, 2 - i$.

9. $3 + i, 2 + i$.

11. $(x + \frac{5}{2})^2 = \frac{21}{4}$.

13. $(x - 4)^2 = 14$.

15. $(2x - \frac{5}{2})^2 = \frac{13}{4}$.

17. 3, 10.

19. $\frac{7}{4}$, 3.

21. $-\dfrac{4 + 3i}{2 - i}, \dfrac{-1 + i}{2 - i}$.

23. $k = \frac{9}{4}$.

25. $k = \frac{49}{8}$.

27. $\pm\frac{4}{3}$.

29. $x^2 - 3x + 4 = 0$.

31. $3x^2 - x + 2 = 0$.

33. $x^2 - (3 - i)x + (2 + 5i) = 0$.

35. (Fig. 5.11.35).

37. (Fig. 5.11.37).

39. (Fig. 5.11.39).

41. $\{x \mid -1 < x < \frac{1}{4}\}$.

43. All real x.

45. No real x.

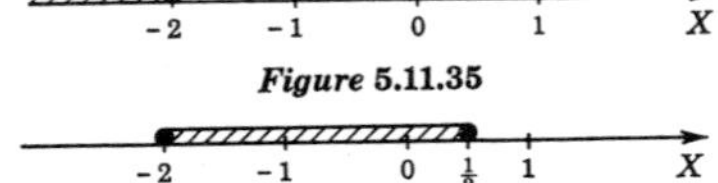

Figure 5.11.35

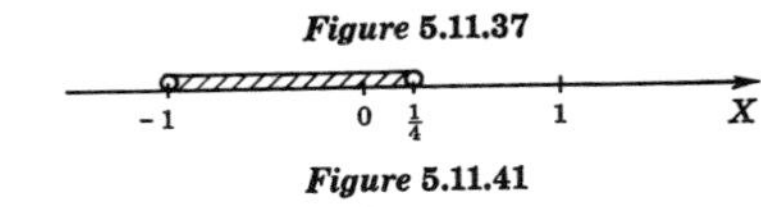

Figure 5.11.37

Figure 5.11.39

Figure 5.11.41

47. (Fig. 5.11.47).

49. (Fig. 5.11.49).

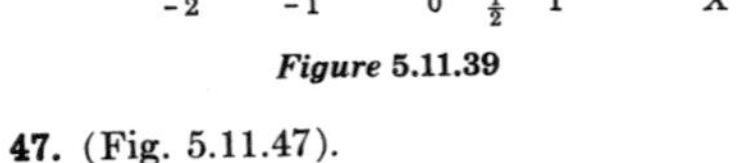

Figure 5.11.47

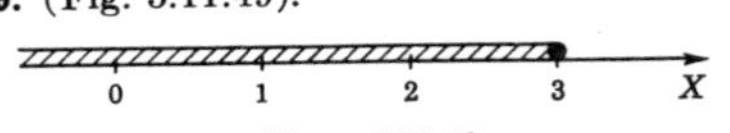

Figure 5.11.49

Problems 5.12

1. (Fig. 5.12.1).

3. (Fig. 5.12.3).

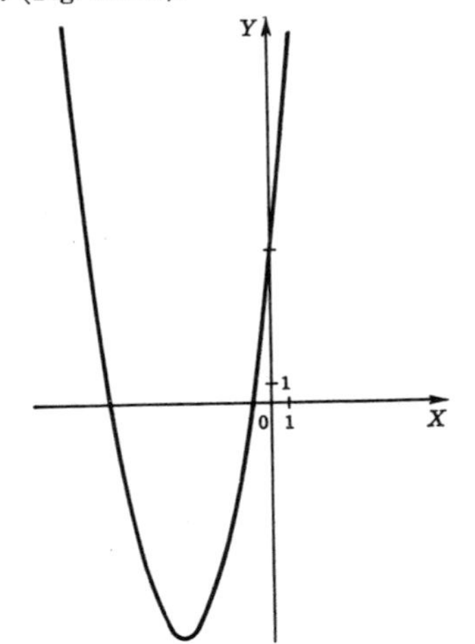

Figure 5.12.1

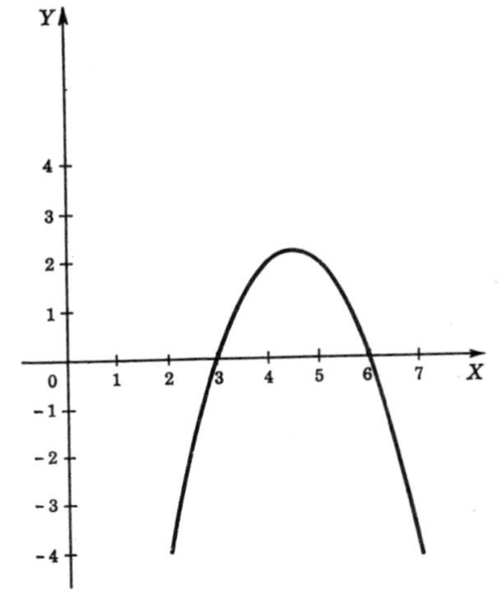

Figure 5.12.3

5. (Fig. 5.12.5).

7. (Fig. 5.12.7).

9. (Fig. 5.12.9).

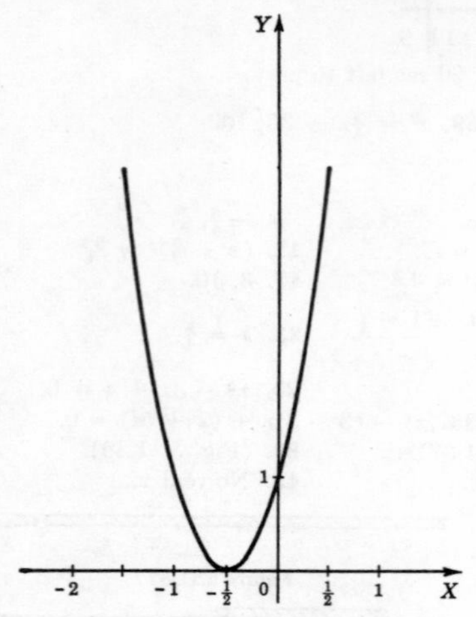

Figure **5.12.5**

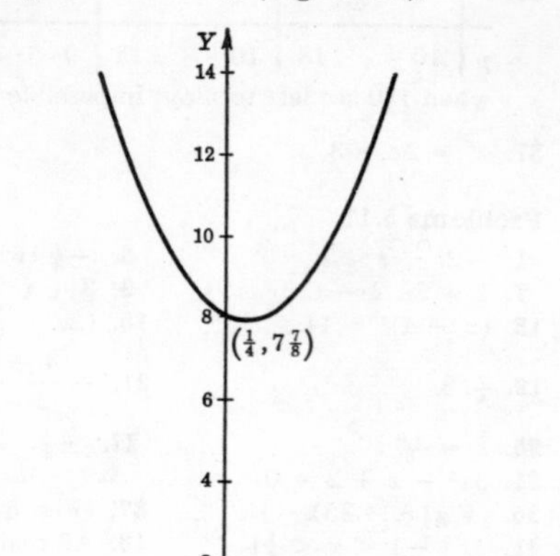

Figure **5.12.7**

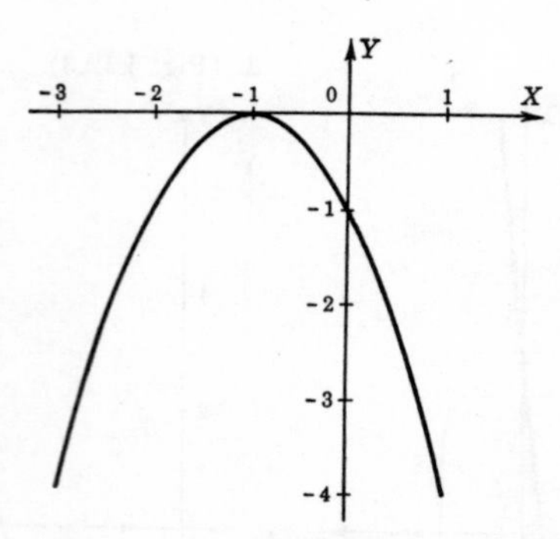

Figure **5.12.9**

11. (Fig. 5.12.11).

13. (Fig. 5.12.13).

15. (Fig. 5.12.15).

17. (Fig. 5.12.17).

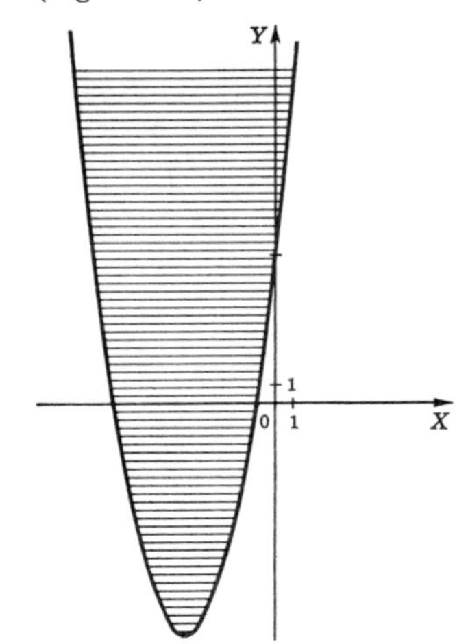

Figure **5.12.11**

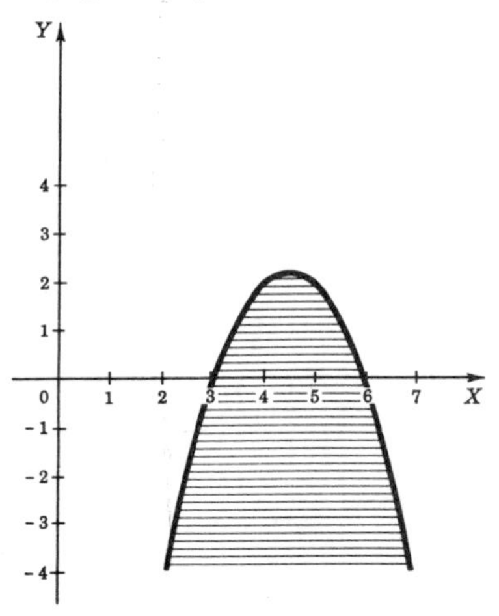

Figure **5.12.13**

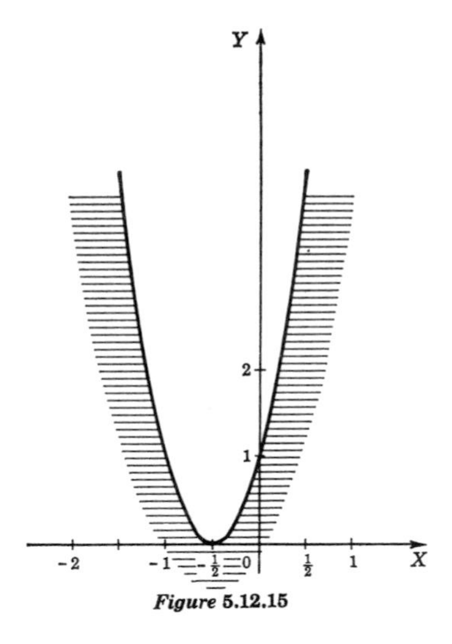

Figure **5.12.15**

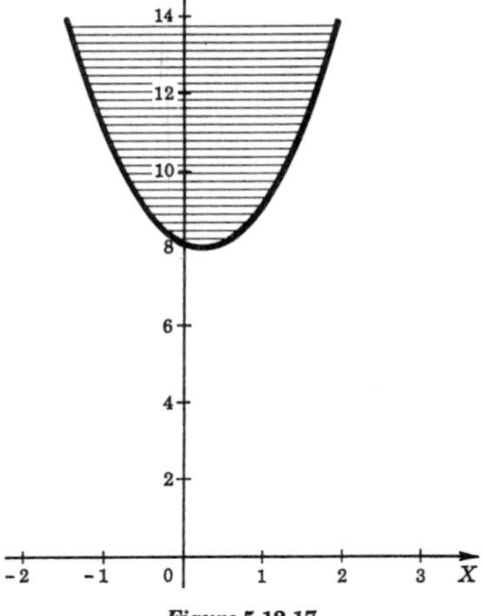

Figure **5.12.17**

19. Entire plane. **21.** (Fig. 5.12.21). **23.** (Fig. 5.12.23).

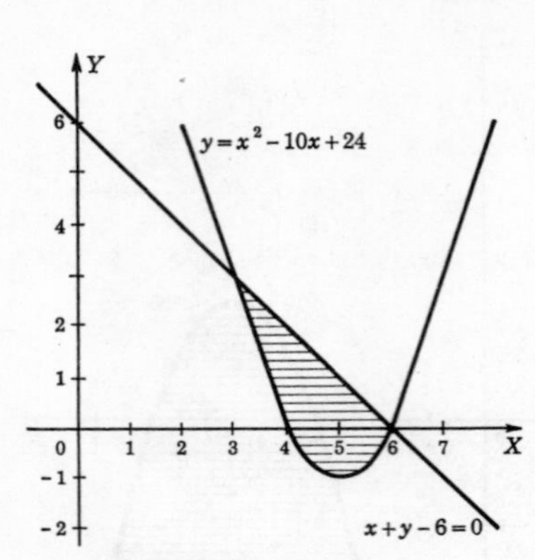

Figure **5.12.21**

$y=x^2+x+1$

$y=-1$

No graph

Figure **5.12.23**

25. y increases as x approaches $\pm\infty$. **27.** $y = 0$ when $x = r_1$ or r_2.

29. Since there are no real roots, y is never zero.

31. The highest or lowest point occurs when $\left(x + \frac{b}{2a}\right) = 0$. Hence $x = -\frac{b}{2a}$, $y = \frac{-b + 4ac}{4c}$.

33. Since $a > 0$, the point in (7) is the lowest point. Since $(-b + 4ac) > 0$, every term in the expression for y is positive, and the lowest point is above the X-axis.

Problems 5.13

1. $x^2 - 4 = 0$.
3. $x^3 - 6x + 11x - 6 = 0$.
5. $x - 2 - 3i = 0$.
7. $x^4 - 5x^3 + 13x^2 - 19x + 10 = 0$.
9. $x^3 - 1 = 0$.
11. $x^3 + 3x^2 - 3x - 9 = 0$.
13. $x^3 - 7x^3 + 19x - 13 = 0$.

15. $x^3 - 2ax^2 - x^2 + a^2x + b^2x + 2ax - a^2 - b^2 = 0$.

17. $(x - 1)(ax^2 + bx + c) = 0, a \neq 0$.
19. $x^3 - 3x^2 + 3x - 1 = 0$.
21. 5.
23. 0.
25. 59.
27. $-1 + i$.
29. $2, -1 \pm \sqrt{3}\,i$.
31. $1, -1, 1, -1$.
33. 3; 3, they are $1, (-1 \pm \sqrt{3}\,i)/2$.
35. 4.
37. 3.
39. 2.

Problems 5.14

1. 3.
3. 91.
5. 3.31.
7. $2\frac{1}{8}$.
9. 0.
11. $(-4,4)$.
13. $(-2,3)$.
17. $k = 0$.
19. 34.

Problems 5.15

1. $\frac{1}{2}, -3.$ **3.** $-\frac{5}{2}, 0.$ **5.** $1, -1, -3.$
7. $\pm\frac{1}{2}.$ **9.** $0, \pm 1.$ **11.** $0, \pm 1, \pm i.$
13. $\pm 3, -2, 4.$

Problems 5.16

1. $-2, (-1 \pm \sqrt{3}\, i)/2.$ **3.** $-\frac{1}{3}, \pm i.$ **5.** $\frac{1}{2}, -1, (1 \pm \sqrt{3}\, i)/2.$
7. No rational roots. **9.** No rational roots, but $\sqrt{2}$ is a root.

Problems 5.17

1. $1.5 < x_0 < 1.6.$ **3.** $-1.5 < x_0 < -1.4.$ **5.** $0.6 < x_0 < 0.7.$
7. $0.7 < x_0 < 0.8.$ **9.** $1.4 < x_0 < 1.5.$ **11.** $1.1 < x_0 < 1.2.$

Problems 5.19

1. $\{1, -\frac{14}{9}\}.$ **3.** $\{3, -\frac{9}{5}\}.$ **5.** $\{2, \frac{9}{5}\}.$
7. $\{3, -3\}.$ **9.** $\{3\}.$ **11.** $\{2\}.$
13. $\emptyset.$ **15.** $\{-5\}.$ **17.** $\{0\}.$
19. $\{9\}.$

Problems 5.20

1. X-intercept: ± 2
Y-intercept: ± 3
Domain $= \{x \mid -2 \le x \le 2\}$
Range $= \{y \mid -3 \le y \le 3\}$
Symmetry in X-axis, Y-axis, and the origin
(Fig. 5.20.1)

3. Y-intercept: $\frac{2}{3}$
Domain $= \{x \mid x \neq -3\}$
Range $= \{y \mid y \neq 0\}$
Asymptotes: $x = -3$
$y = 0$
Symmetric about $(-3, 0)$
(Fig. 5.20.3)

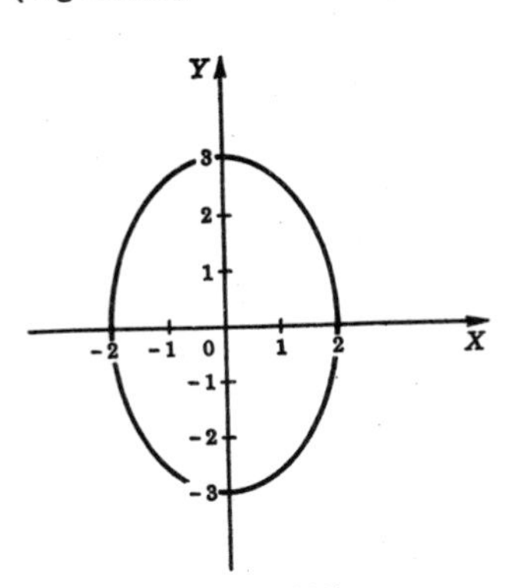

Figure **5.20.1**

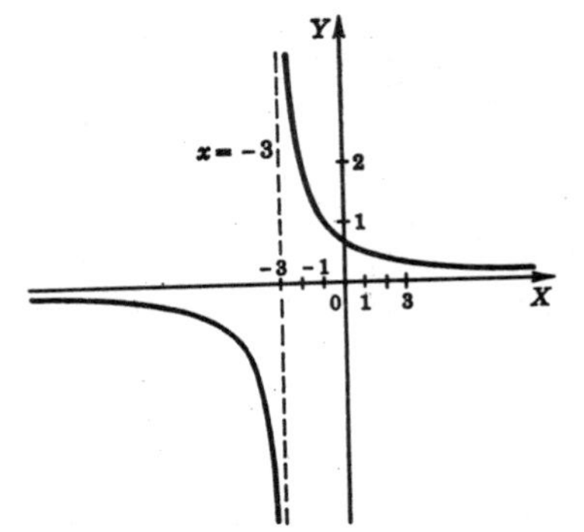

Figure **5.20.3**

5. X-intercept: 0, ± 1
Y-intercept: 0
Domain $= X$; range $= Y$
Symmetric about the origin
(Fig. 5.20.5)

7. Y-intercept: -2
X-intercept: $\pm 1, -2$
Domain $= X$; range $= Y$
(Fig. 5.20.7)

9. X-intercept: 0, ± 1
Y-intercept: 0
Domain $= X$
Range $= \{y \mid y \geq -\frac{1}{4}\}$
(cannot be calculated by methods of this section)
Symmetric about Y-axis
(Fig. 5.20.9)

11. X-intercept: 0, ± 1
Y-intercept: 0
Domain $= \{x \mid (0 \leq x \leq 1) \vee (x \leq -1)\}$
Range $= Y$
Symmetric about X-axis
(Fig. 5.20.11)

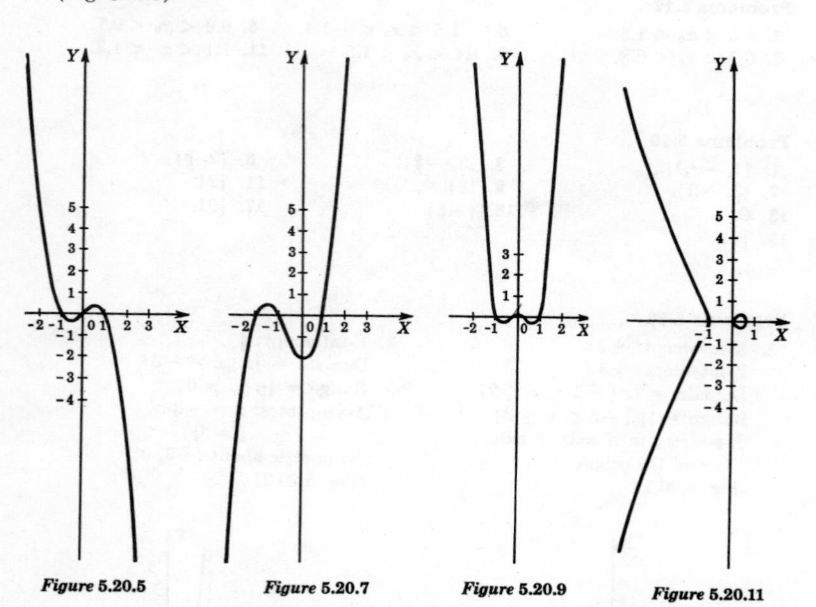

Figure 5.20.5 *Figure* 5.20.7 *Figure* 5.20.9 *Figure* 5.20.11

13. X-intercept: 0, ± 1
Y-intercept: 0
Domain $= \{x \mid (|x| \geq 1) \vee (x = 0)\}$
Range $= Y$
Symmetric about both axes
and the origin
(Fig. 5.20.13)

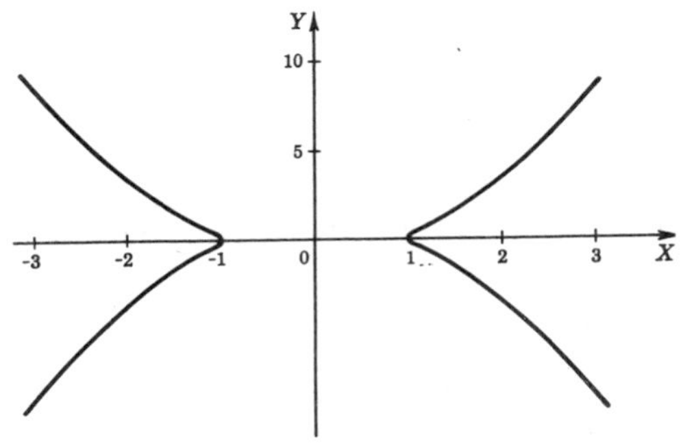

Figure **5.20.13**

15. X-intercept: 0
Y-intercept: 0
Domain $= \{x \mid x < 1\}$
Range $= Y$
Asymptote: $x = 1$
Symmetric about X-axis
(Fig. 5.20.15)

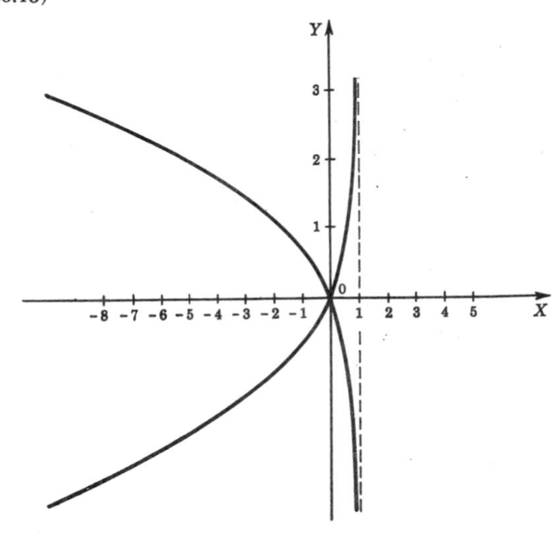

Figure **5.20.15**

17. Y-intercept: 2
Domain $= X$
Range $= \{y \mid 0 < y \leq 2\}$
Asymptote: $y = 0$
Symmetric about Y-axis
(Fig. 5.20.17)

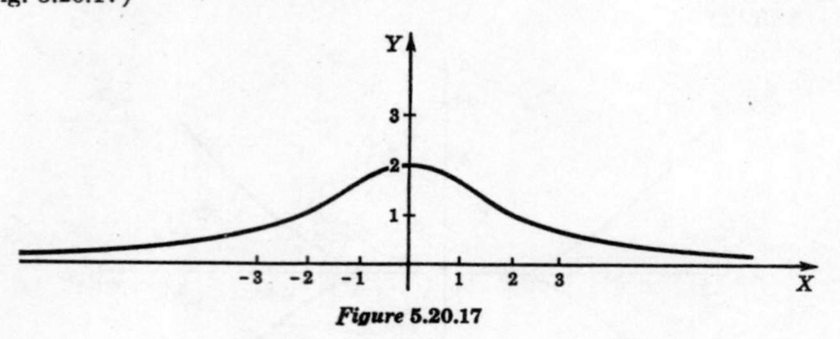

***Figure* 5.20.17**

19. X-intercept: 1
Y-intercept: 1
Domain $= \{x \mid 0 \leq x \leq 1\}$
Range $= \{y \mid 0 \leq y \leq 1\}$
(Fig. 5.20.19)

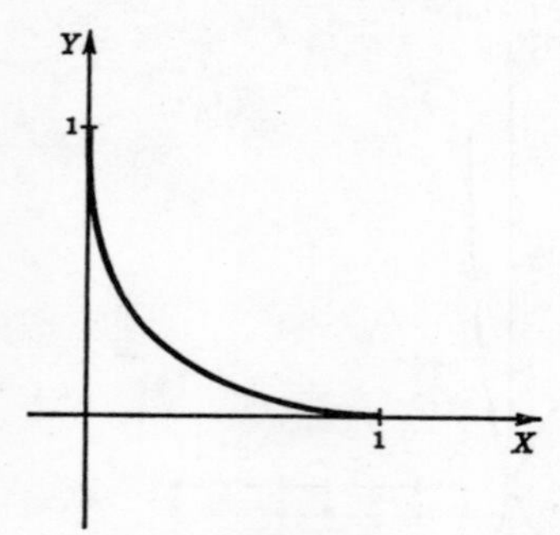

***Figure* 5.20.19**

21. X-intercept: ± 1
Y-intercept: ± 1
Domain $= \{x \mid -1 \leq x \leq 1\}$
Range $= \{y \mid -1 \leq y \leq 1\}$
Symmetric about both axes and the origin
(Fig. 5.20.21)

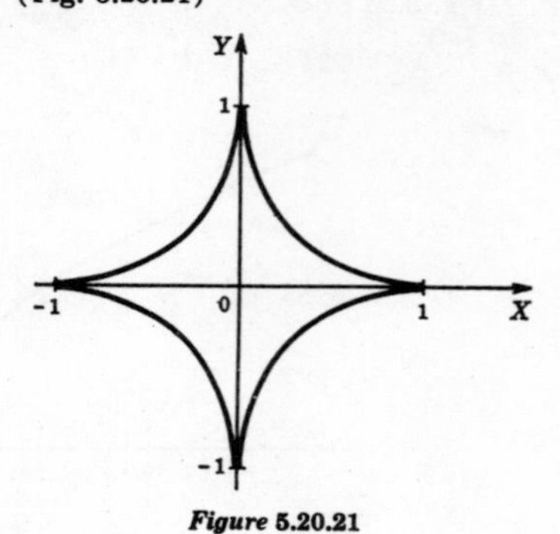

***Figure* 5.20.21**

23. X-intercept: -1
Y-intercept: $-\frac{1}{2}$
Domain $= \{x \mid x \neq 2\}$
Range $= \{y \mid y \neq 1\}$
Asymptote: $x = 2,\ y = 1$
(Fig. 5.20.23)

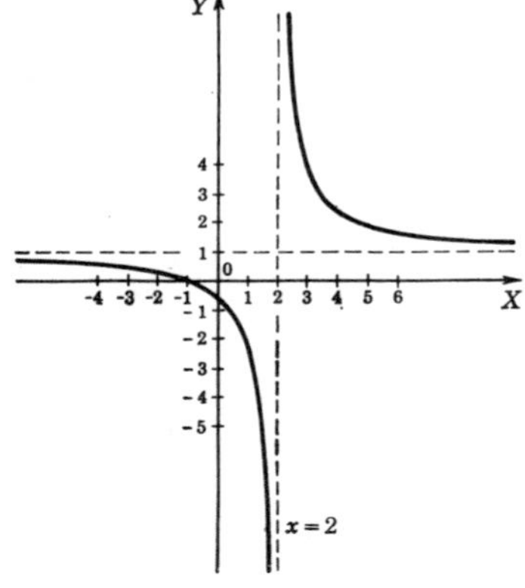

***Figure* 5.20.23**

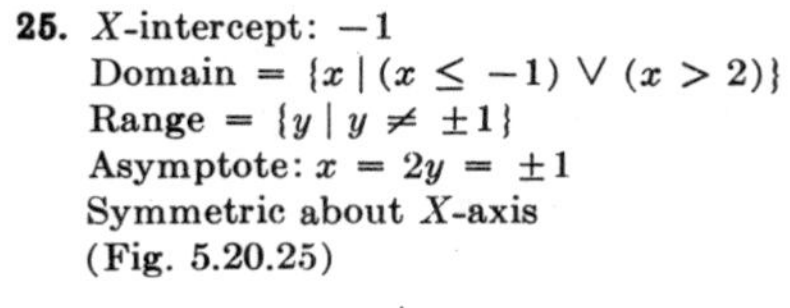

25. X-intercept: -1
Domain $= \{x \mid (x \leq -1) \vee (x > 2)\}$
Range $= \{y \mid y \neq \pm 1\}$
Asymptote: $x = 2y = \pm 1$
Symmetric about X-axis
(Fig. 5.20.25)

***Figure* 5.20.25**

27. X-intercept: 0, 1
Y-intercept: 0
Domain $= \{x \mid x \neq -2\}$
Range $= \{y \mid (y \leq -\sqrt{24} - 5) \vee (y \geq \sqrt{24} - 5)\}$
Asymptote: $x = -2$; $y = x - 3$ (cannot be calculated by the methods of this section)
(Fig. 5.20.27)

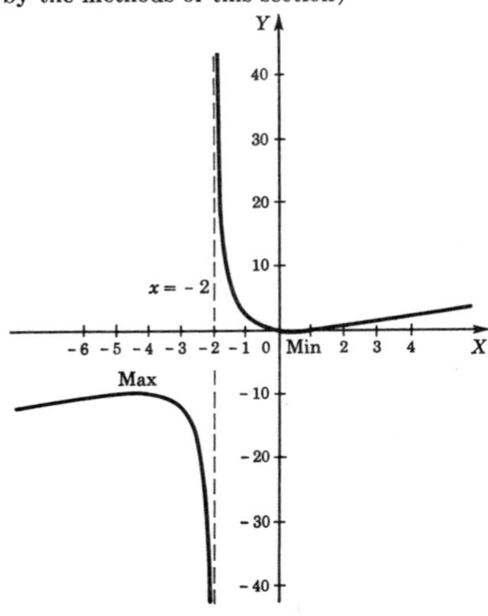

***Figure* 5.20.27**

29. (Fig. 5.20.29).
31. (Fig. 5.20.31).
33. (Fig. 5.20.33).
35. (Fig. 5.20.35).

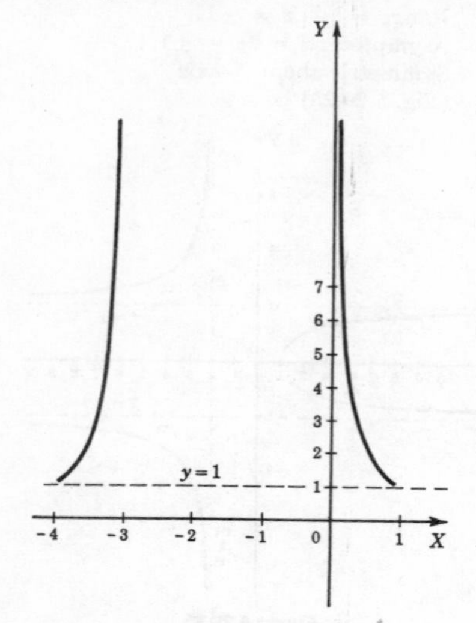

Figure 5.20.29

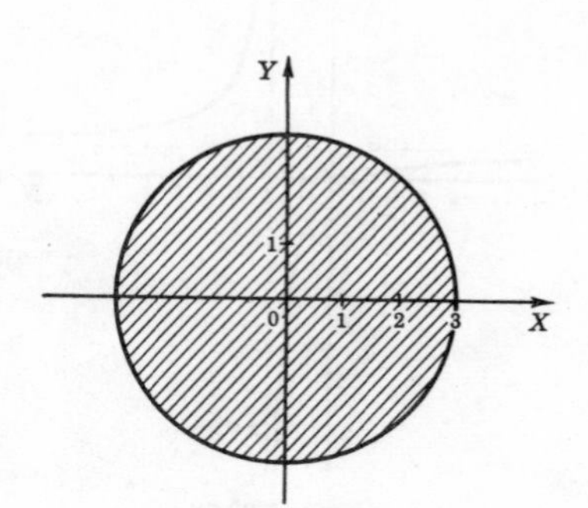

Figure 5.20.31

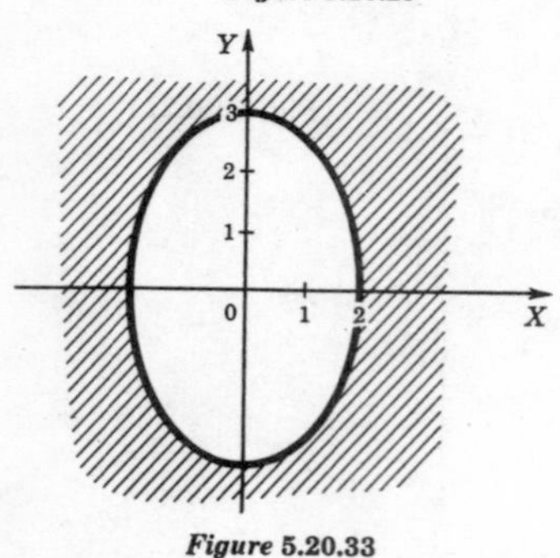

Figure 5.20.33

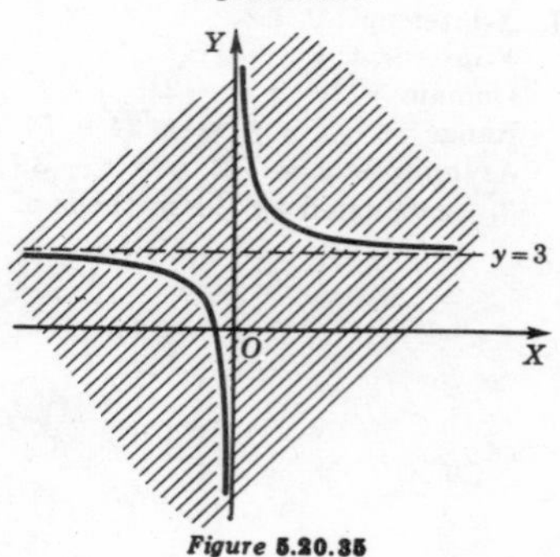

Figure 5.20.35

37. The region outside the curve for Prob. 21.
39. The origin only.

Problems 6.2

1. -4, 28, 1.
3. (1,0), (3,2), (4,3).
5. (b).
7. Yes; t.

9. $0, -\frac{1}{3}, \frac{\sqrt{2}-4}{3}, \frac{\sqrt{3}-6}{3}, \frac{2}{3}+\frac{1}{3}i, \frac{4}{3}+\frac{\sqrt{2}}{3}i.$

11. (a) $-4ah$; (b) 5.
13. $f(3)$ does not exist (6/0 is meaningless).

15. $f:(t, 200t)$, t hr. **17.** $\{y \mid y > -6\}$.

19. (**a**) $4 + h$; (**b**) $18 + 4h$; (**c**) 1; (**d**) 0.

Problems 6.4

1. (**b**) and (**d**) are functions; (**a**) and (**c**) are not.
3. (**a**) Domain, reals; range, reals.
(**c**) Domain, reals; range, reals.
(**d**) Domain, reals; range, $y \geq 1$.
(**e**) Domain, reals; range, reals.
Rule in each case is the equation.
5. Domain, reals; range, nonnegative reals.
7. $f(0) = 0$ $\quad f(-\sqrt{3}) = -2\sqrt{3} + 2$
$f(1) = -1$ $\quad f(-2) = -2$
$f(2.5) = -2$ $\quad f(-4.2) = -3.4$
9. In each case there are ordered pairs of the form (a,b), (a,c) with $b \neq c$.

Problems 6.5

1. Domain, reals; range, reals. **3.** No. **5.** No.

7. Range, $\{y \mid (y \leq -4) \vee (y > 0)\}$; domain, reals except $x = 0, 1$.

9. Reals. **11.** Reals, except 0. **13.** $0 \leq y \leq 4$.

15. For each item there is one and only one price.
Rule: price list
Domain: all items
Range: all prices
Elements: set of pairs (item, price)
17. Domain, reals; range, integers.
19. Domain, reals; range, nonnegative integers.

Problems 6.6

1. $\frac{1}{x} + \frac{1}{x+1} + x$; $\frac{1}{x} + \frac{1}{x+1} - x$; $\left(\frac{1}{x} + \frac{1}{x+1}\right)x$; $\left(\frac{1}{x} + \frac{1}{x+1}\right)\Big/x$. Domain, reals except $x = 0, -1$.

3. $\frac{1}{x+1} + x - 1$; $\frac{1}{x+1} - x + 1$; $\left(\frac{1}{x+1}\right)(x - 1)$; $\frac{1}{x+1}\Big/(x - 1)$. Domain, reals except $x = -1$; in quotient also exclude $x = 1$.

5. $x + 1 + \frac{1}{(x+1)^2}$; $(x + 1) - \frac{1}{(x+1)^2}$; $(x + 1)\frac{1}{(x+1)^2}$; $(x + 1)\Big/\frac{1}{(x+1)^2}$.
Domain, reals except $x = -1$.

7.
$$(f + g)(x) = \begin{cases} \frac{1}{x-1} + \frac{2}{(x-1)(x+4)}, & x \neq 1, -4 \\ 4 + \sqrt{7}, & x = 1 \\ \text{undefined}, & x = -4 \end{cases}$$

$$(f - g)(x) = \begin{cases} \frac{1}{x-1} - \frac{2}{(x-1)(x+4)}, & x \neq 1, -4 \\ 4 - \sqrt{7}, & x = 1 \\ \text{undefined}, & x = -4 \end{cases}$$

$$(f \times g)(x) = \begin{cases} \dfrac{1}{x-1} \times \dfrac{1}{(x-1)(x+4)}, \ x \neq 1, -4 \\ 4\sqrt{7}, \ x = 1 \\ \text{undefined}, \ x = -4 \end{cases}$$

$$(f/g)(x) = \begin{cases} \dfrac{1}{x-1} \Big/ \dfrac{1}{(x-1)(x+4)}, \ x \neq 1, -4 \\ 4/\sqrt{7}, \ x = 1 \\ \text{undefined}, \ x = -4 \end{cases}$$

9. $(g \circ f)(x) = 1/[1 + 1/(2 - x)]$. Reals, $x \neq 2, 3$.

11. $(g \circ f)(x) = (5 + 1/x + |x|)^2 + 4$. Reals, $x \neq 0$.
$(f \circ g)(x) = 5 + 1/(x^2 + 4) + |x^2 + 4|$. Reals.

13. $(g \circ f)(x) = [\![|x|]\!]$. Reals.
$(f \circ g)(x) = |[\![x]\!]|$. Reals.

15. $f(f(x)) = \||x|\| = |x| = f(x)$.

17. 9, 1, 5.

Problems 6.7

1. (Fig. 6.7.1). **3.** (Fig. 6.7.3). **5.** (Fig. 6.7.5). **7.** (Fig. 6.7.7).

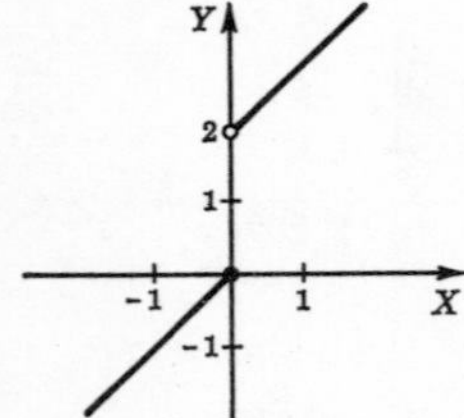

Figure **6.7.1**

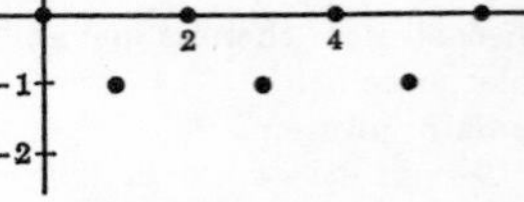

Figure **6.7.3**

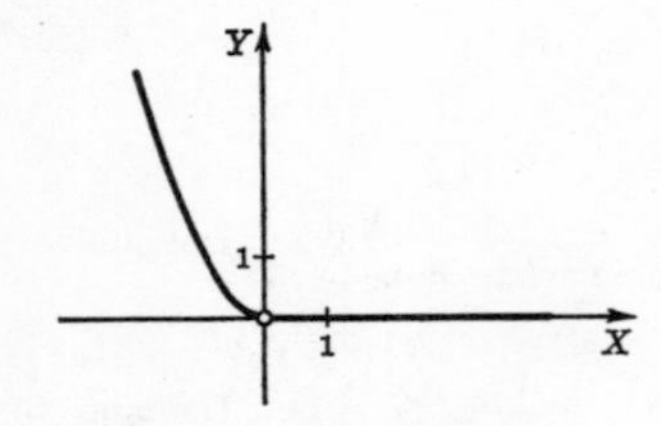

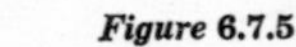

Figure **6.7.5**

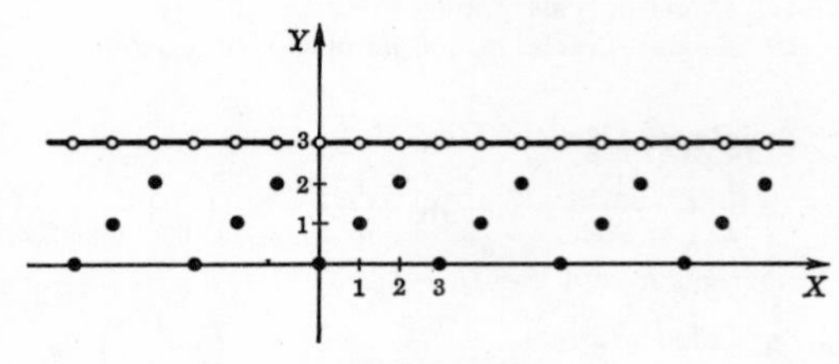

Figure **6.7.7**

9. (Fig. 6.7.9). **11.** (Fig. 6.7.11). **13.** (Fig. 6.7.13).

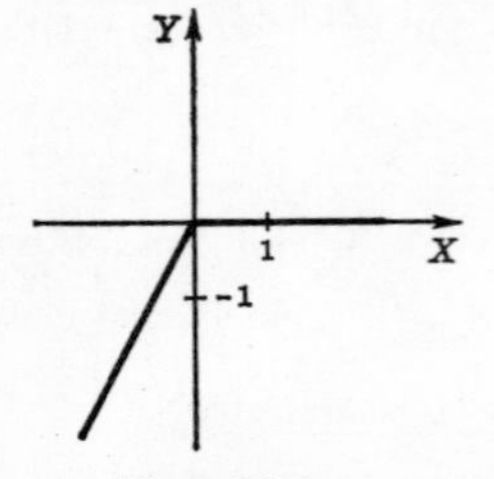

Figure **6.7.9**

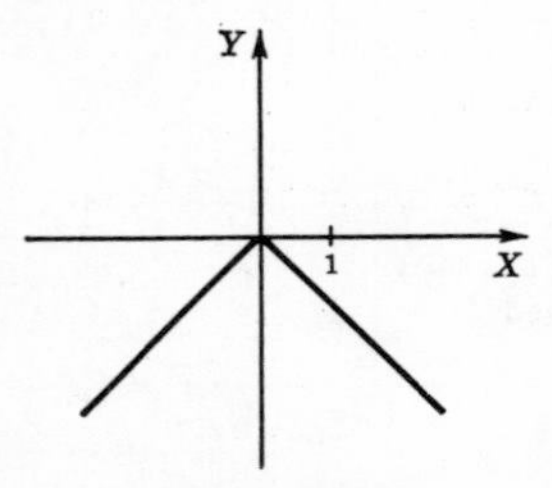

Figure **6.7.11**

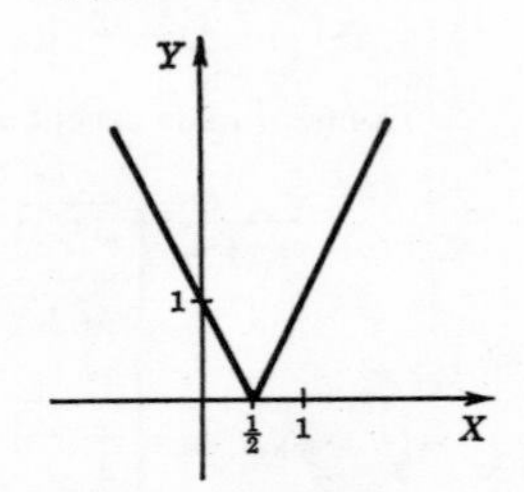

Figure **6.7.13**

15. (Fig. 6.7.15).

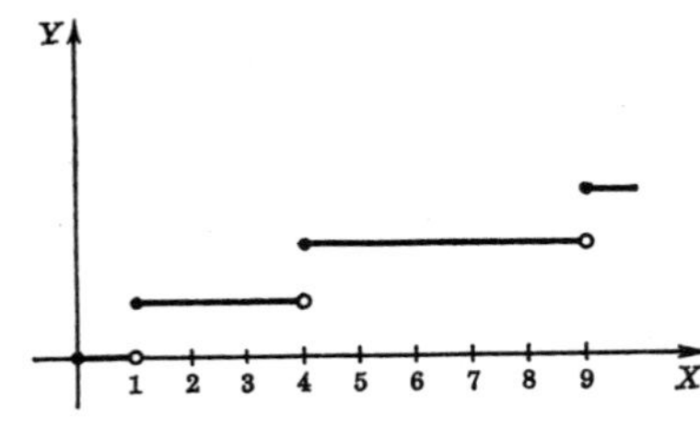

Figure **6.7.15**

17. (Fig. 6.7.17).

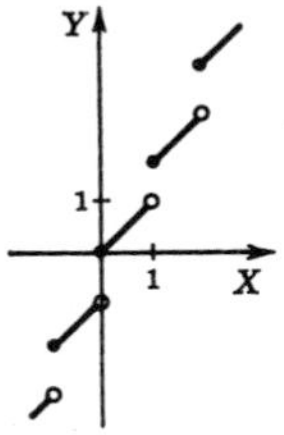

Figure **6.7.17**

19. (Fig. 6.7.19).

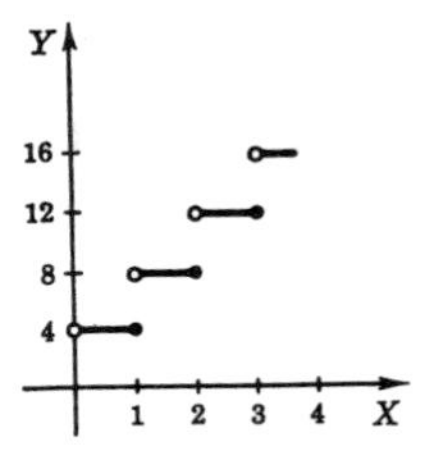

Figure **6.7.19**

21. (Fig. 6.7.21).

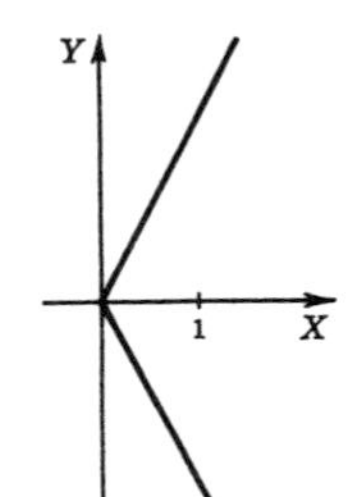

Figure **6.7.21**

23. (Fig. 6.7.23).

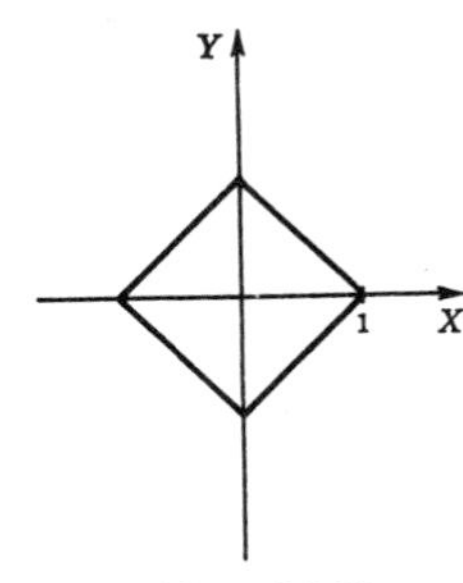

Figure **6.7.23**

25. (Fig. 6.7.25).

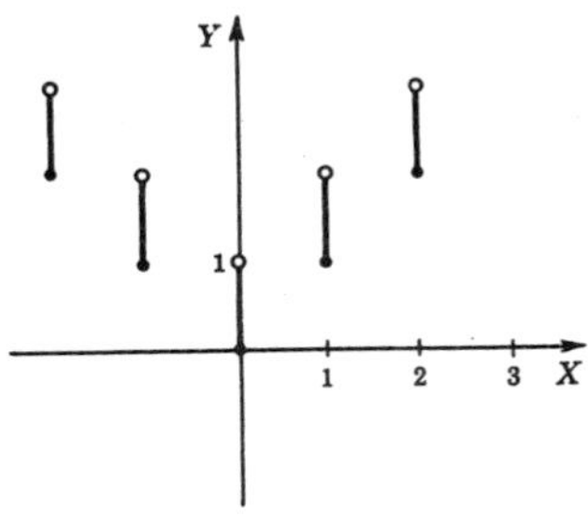

Figure **6.7.25**

Problems 6.8

1. Domain, reals; range, reals.

3. (**a**) Domain, reals; range, reals. (**b**) Domain, reals; range, reals.

5. $r_f = \{y \mid y \geq 0\}$. (Fig. 6.8.5)

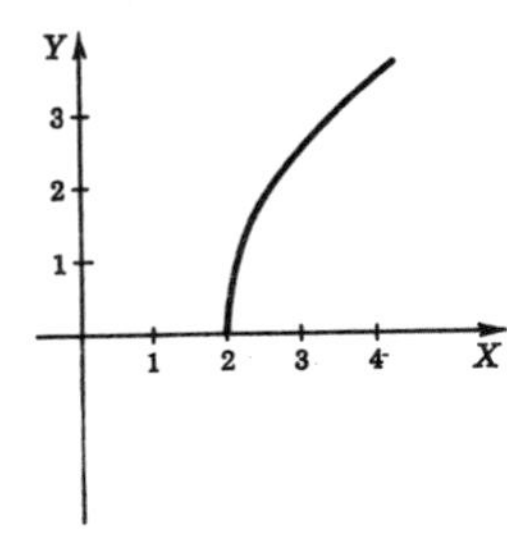

Figure **6.8.5**

7. $r_f = \{y \mid 0 \leq y \leq 2\}$. (Fig. 6.8.7)

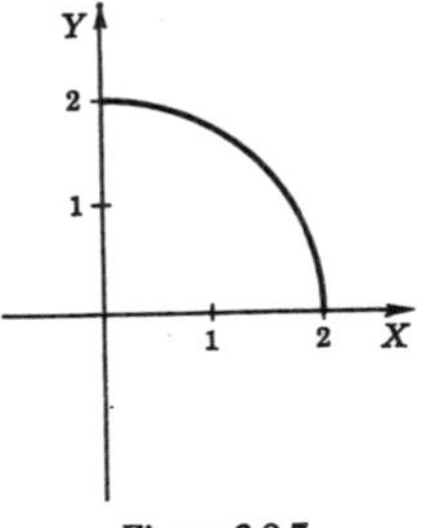

Figure **6.8.7**

9. $r_f = \{y \mid -2 \leq y \leq 0\}$. (Fig. 6.8.9)

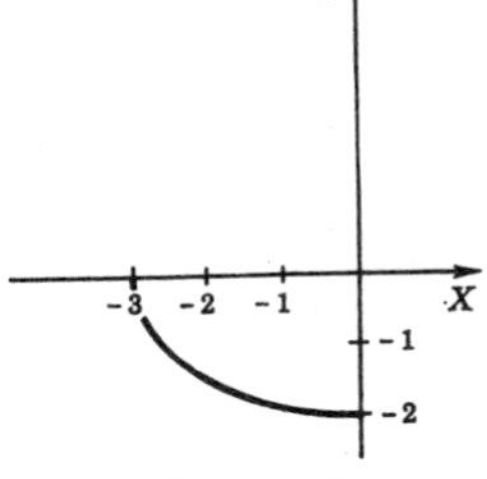

Figure **6.8.9**

11. For f defined by $f(x) = x$.

13. No.

Problems 6.9

1. $x = -\frac{3 + 3y}{4}$. Domain, reals; range, reals.

$y = -\frac{3 + 4x}{3}$. Domain, reals; range, reals.

3. $x = \sqrt{\frac{1 - y^2}{3}}$. Domain, $|y| \leq 1$; range, $0 \leq x \leq \frac{\sqrt{3}}{3}$.

$x = -\sqrt{\frac{1 - y^2}{3}}$. Domain, $|y| \leq 1$; range, $-\frac{\sqrt{3}}{3} \leq x \leq 0$.

$y = \sqrt{1 - 3x^2}$. Domain, $|x| \leq \frac{\sqrt{3}}{3}$; range, $0 \leq y \leq 1$.

$y = -\sqrt{1 - 3x^2}$. Domain, $|x| \leq \frac{\sqrt{3}}{3}$; range, $-1 \leq y \leq 0$.

5. $x = \sqrt{\frac{1 + y}{y}}$. Domain, $(y > 0) \vee (y \leq 1)$; range, positive reals except $x = 1$.

$x = -\sqrt{\frac{1 + y}{y}}$. Domain, $(y > 0) \vee (y \leq -1)$; range, negative reals except $x = -1$.

$y = \frac{1}{x^2 - 1}$. Domain, reals except $x = \pm 1$; range, $(y > 0) \vee (y \leq -1)$.

7. $x = 1 - |y|$. Domain, $|y| \leq 1$; range, $0 \leq x \leq 1$.
$x = |y| - 1$. Domain, $|y| < 1$; range, $-1 \leq x < 0$.
$y = 1 - |x|$. Domain, $|x| \leq 1$; range, $0 \leq y \leq 1$.
$y = |x| - 1$. Domain, $|x| < 1$; range, $-1 \leq y < 0$.

9. $s = -16t^2 + v_0t$. Domain, nonnegative reals; range, $0 \leq s \leq \frac{v_0^2}{64}$.

$t = \frac{v_0 + \sqrt{v_0^2 - 64s}}{32}$. Domain, $0 \leq s \leq \frac{v_0^2}{64}$; range, nonnegative reals.

$t = \frac{v_0 - \sqrt{v_0^2 - 64s}}{32}$. Domain, $0 \leq s \leq \frac{v_0^2}{64}$; range, nonnegative reals.

11. (Fig. 6.9.11). 13. (Fig. 6.9.13). 15. All points in the plane.

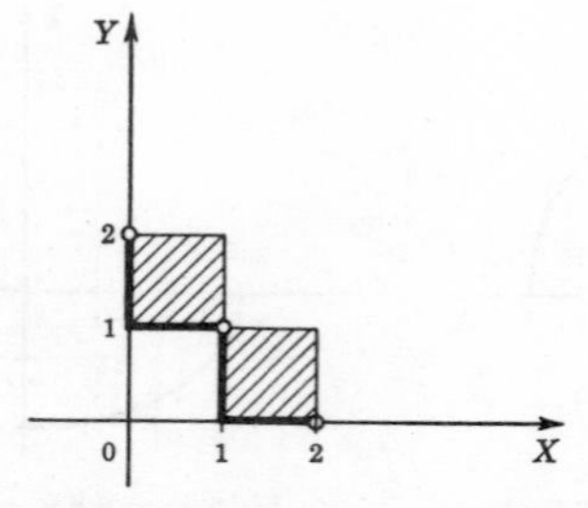

Figure 6.9.11

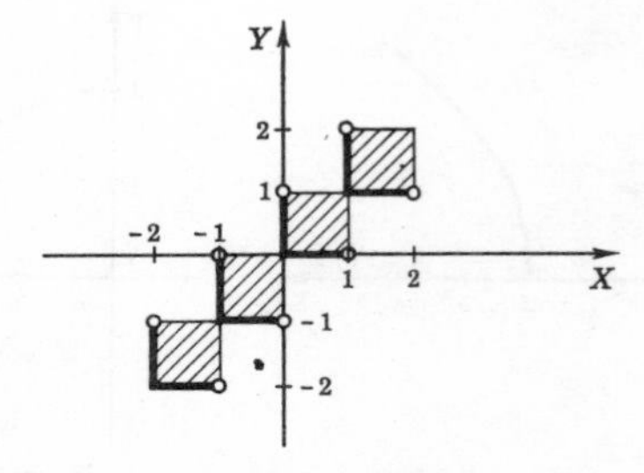

Figure 6.9.13

Problems 6.10

1. No. **3.** No. **5.** No.
7. Yes. (Over the complex field.) **9.** Yes.
11. Yes. **13.** No. **15.** Yes.
17. No. **19.** Yes. **21.** No.
23. No. **25.** Yes. **27.** No.
29. No. **31.** Polynomial function. Domain, reals.

33. Explicit algebraic function. Domain, $(0 \leq x \leq 1) \vee (x \leq -1)$.
35. Explicit algebraic function. Domain, nonnegative reals.
37. Rational function. Domain, reals except $x = 0, 1, -2$.
39. Explicit algebraic function. Domain, $x \geq 2$.

Problems 7.1

1. $\frac{9}{128}$. **3.** $\frac{4}{25}$. **5.** 4.
7. 4^{10}. **9.** 18. **11.** a^{-x}.
13. $c^{2x} + \frac{8}{3}$. **15.** 2^{3t-2}. **17.** 2^{-1}.

Problems 7.2

1. $e^{1.35} = 3.8574$.
$e^{-0.07} = 0.93239$.
$e^{\sqrt{2}} = 4.1132$.

3. (Fig. 7.2.3). **5.** (Fig. 7.2.5). **7.** (Fig. 7.2.7).

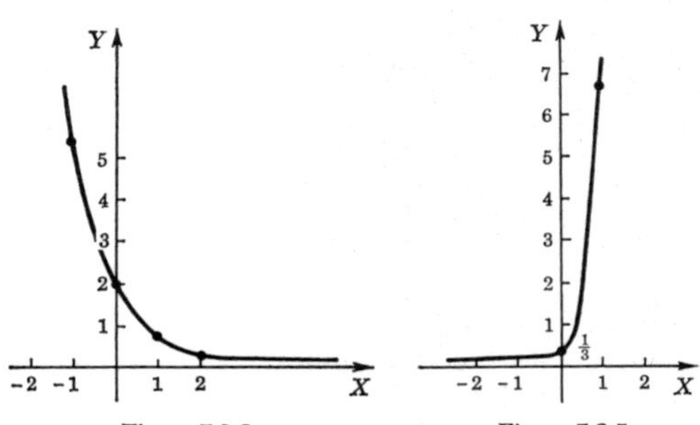

Figure 7.2.3 *Figure* 7.2.5

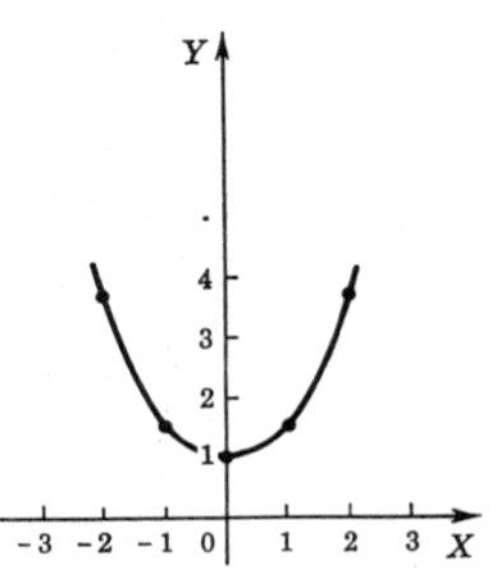

Figure 7.2.7

9. HINT: Add.
11. Use the infinite-sum formula for e^x, and assume that the usual processes of algebra are valid.

Problems 7.3

1. 3^{-x}, $\log_3(1/x)$. Domain of 3^{-x} = range of $\log_3(1/x)$ = the set of reals. Range of 3^{-x} = domain of $\log_3(1/x)$ = the set of positive reals. 4^{6x}, $\frac{1}{6}\log_4 x$. Domain of 4^{6x} = range of $\frac{1}{6}\log_4 x$ = the set of reals. Range of 4^{6x} = domain of $\frac{1}{6}\log_4 x$ = the set of positive reals.

3. (a) $3^{2.5} = 15.59$. (b) $10^{1.2} = 15.85$. (c) $e^{0.5} = 1.649$.
5. 17.3. **9.** $\sqrt{5}$. **13.** $z = x(\log_b a + 2)$.
15. x. **17.** x.

Problems 7.5

11. N_0. $\frac{1}{3}\log_e 2$. **13.** $\dfrac{1}{\log_{10} 2}$. **15.** $\log_e \dfrac{\sqrt{5}-1}{2}$.

17. $x = \dfrac{\log_5 2}{\log_3 4(1 - \log_5 3)}$

$y = \dfrac{\log_5 2}{1 - \log_5 3}$ **19.** 1.

Problems 8.2

1. 5. **3.** 2. **5.** $\sqrt{242}$.

Problems 8.3

1. (**a**) $\sin\theta = 0$, $\cos\theta = 1$, $\tan\theta = 0$. (**b**) $\sin\theta = 1$, $\cos\theta = 0$.
(**c**) $\sin\theta = 0$, $\cos\theta = -1$, $\tan\theta = 0$. (**d**) $\sin\theta = -1$, $\cos\theta = 0$.
(**e**) $\sin\theta = 0$, $\cos\theta = 1$, $\tan\theta = 0$. (**f**) $\sin\theta = 0$, $\cos\theta = -1$, $\tan\theta = 0$.
(**g**) $\sin\theta = 0$, $\cos\theta = 1$, $\tan\theta = 0$. (**h**) $\sin\theta = 0$, $\cos\theta = -1$, $\tan\theta = 0$.

Problems 8.4

1. $\sin\left(-\frac{\pi}{4}\right) = -\frac{\sqrt{2}}{2}$ $\sin\left(-\frac{3}{4}\pi\right) = -\frac{\sqrt{2}}{2}$

$\cos\left(-\frac{\pi}{4}\right) = \frac{\sqrt{2}}{2}$ $\cos\left(-\frac{3}{4}\pi\right) = -\frac{\sqrt{2}}{2}$

$\tan\left(-\frac{\pi}{4}\right) = -1$ $\tan\left(-\frac{3}{4}\pi\right) = 1$

$\sin\left(-\frac{5}{4}\pi\right) = \frac{\sqrt{2}}{2}$ $\sin\left(-\frac{7}{4}\pi\right) = \frac{\sqrt{2}}{2}$

$\cos\left(-\frac{5}{4}\pi\right) = -\frac{\sqrt{2}}{2}$ $\cos\left(-\frac{7}{4}\pi\right) = \frac{\sqrt{2}}{2}$

$\tan\left(-\frac{5}{4}\pi\right) = -1$ $\tan\left(-\frac{7}{4}\pi\right) = 1$

3. $\sin\left(\frac{\pi}{4} + 2n\pi\right) = \frac{\sqrt{2}}{2}$ $\sin\left(\frac{3}{4}\pi + 2n\pi\right) = \frac{\sqrt{2}}{2}$

$\cos\left(\frac{\pi}{4} + 2n\pi\right) = \frac{\sqrt{2}}{2}$ $\cos\left(\frac{3}{4}\pi + 2n\pi\right) = -\frac{\sqrt{2}}{2}$

$\tan\left(\frac{\pi}{4} + 2n\pi\right) = 1$ $\tan\left(\frac{3}{4}\pi + 2n\pi\right) = -1$

5. $\sin\left(\frac{\pi}{6} + 2n\pi\right) = \frac{1}{2}$ $\sin\left(\frac{5}{6}\pi + 2n\pi\right) = \frac{1}{2}$

$\cos\left(\frac{\pi}{6} + 2n\pi\right) = \frac{\sqrt{3}}{2}$ $\cos\left(\frac{5}{6}\pi + 2n\pi\right) = -\frac{\sqrt{3}}{2}$

$\tan\left(\frac{\pi}{6} + 2n\pi\right) = \frac{1}{\sqrt{3}}$ $\tan\left(\frac{5}{6}\pi + 2n\pi\right) = -\frac{1}{\sqrt{3}}$

$$\sin\left(\frac{7}{6}\pi + 2n\pi\right) = -\frac{1}{2} \qquad \sin\left(\frac{11}{16}\pi + 2n\pi\right) = -\frac{1}{2}$$

$$\cos\left(\frac{7}{6}\pi + 2n\pi\right) = -\frac{\sqrt{3}}{2} \qquad \cos\left(\frac{11}{6}\pi + 2n\pi\right) = \frac{\sqrt{3}}{2}$$

$$\tan\left(\frac{7}{6}\pi + 2n\pi\right) = \frac{1}{\sqrt{3}} \qquad \tan\left(\frac{11}{6}\pi + 2n\pi\right) = -\frac{1}{\sqrt{3}}$$

7. $\dfrac{1}{\cos 0} = 1$ $\qquad \dfrac{1}{\cos(\pi/2)}$ does not exist

$\dfrac{1}{\sin 0}$ does not exist $\qquad \dfrac{1}{\sin(\pi/2)} = 1$

$\dfrac{1}{\tan 0}$ does not exist $\qquad \dfrac{1}{\tan(\pi/2)}$ does not exist

$\dfrac{1}{\cos \pi} = -1$ $\qquad \dfrac{1}{\cos \frac{3}{2}\pi}$ does not exist

$\dfrac{1}{\sin \pi}$ does not exist $\qquad \dfrac{1}{\sin \frac{3}{2}\pi} = -1$

$\dfrac{1}{\tan \pi}$ does not exist $\qquad \dfrac{1}{\tan \frac{3}{2}\pi}$ does not exist

$\dfrac{1}{\cos 2\pi} = 1$

$\dfrac{1}{\sin 2\pi}$ does not exist

$\dfrac{1}{\tan 2\pi}$ does not exist

9. No.

11. 1.

Problems 8.5

1. $\sin 0.25 = 0.24740$.

3. $\sin 1.00 = 0.84147$.

5. $\sin(\pi + 0.20) = -0.19867$.

7. $\cos 0.30 = 0.95534$.

9. $\cos 1.57 = 0.00080$.

11. $\cos(\pi + 0.20) = -0.98007$.

13. $\tan 1.57 = 1255.8$.

15. $\tan(\pi + 0.90) = 1.2602$.

17. $\tan(\pi - 1.20) = -2.5722$.

19. $1 + \dfrac{\theta^6}{360} - \dfrac{\theta^8}{960} + \dfrac{\theta^{10}}{14{,}400}$.

Problems 8.6

1.

θ	$\cot\theta$	$\sec\theta$	$\csc\theta$
0	∞	1	∞
$\pi/6$	1.732	1.155	2
$\pi/4$	1	1.414	1.414
$\pi/3$	0.577	2	1.155
$\pi/2$	0	∞	1

3. (Fig. 8.6.3).

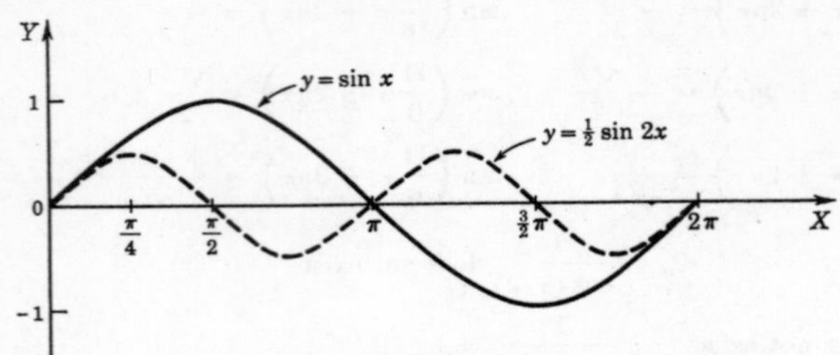

Figure **8.6.3**

5. (Fig. 8.6.5).

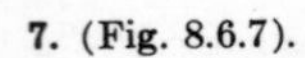
7. (Fig. 8.6.7).

9. (Fig. 8.6.9).

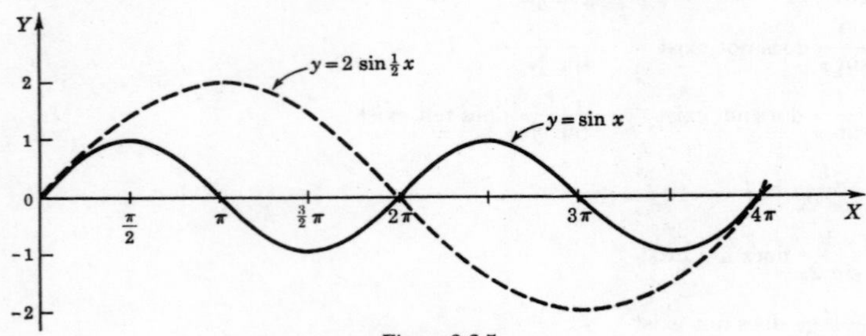

Figure **8.6.5**

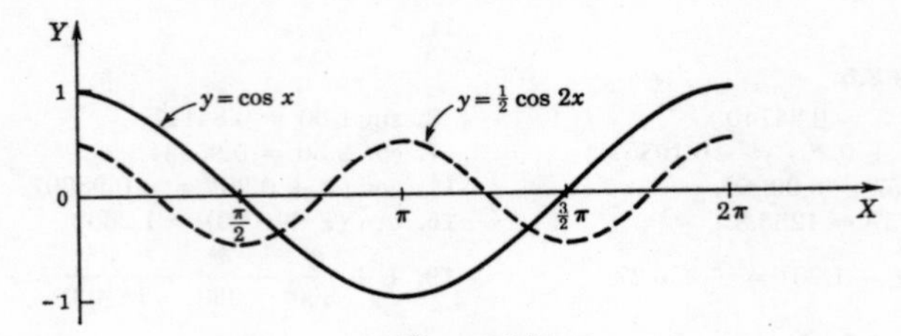

Figure **8.6.7**

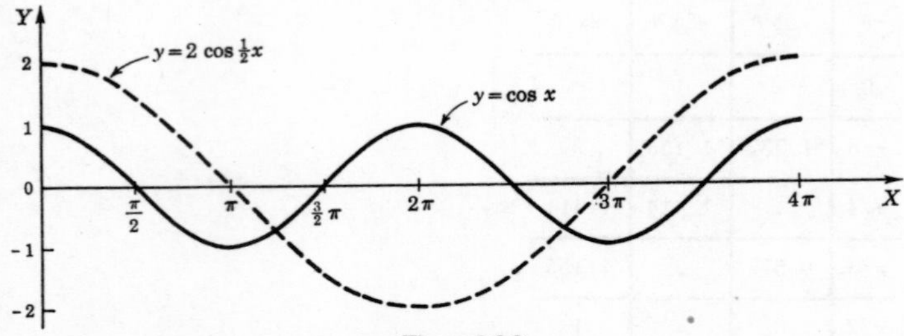

Figure **8.6.9**

11. (Fig. 8.6.11). **13.** (Fig. 8.6.13). **15.** (Fig. 8.6.15).

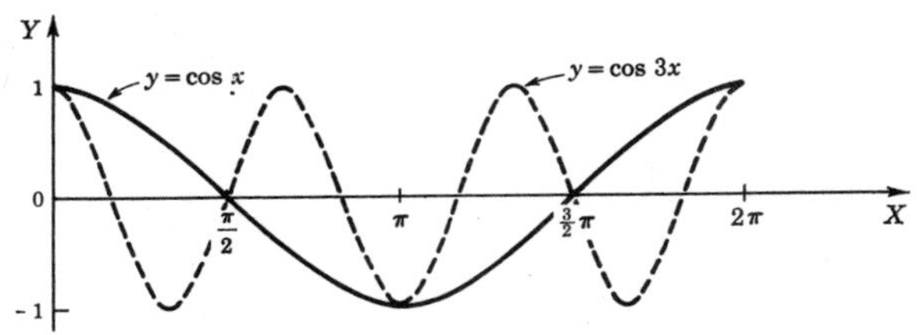

Figure **8.6.11**

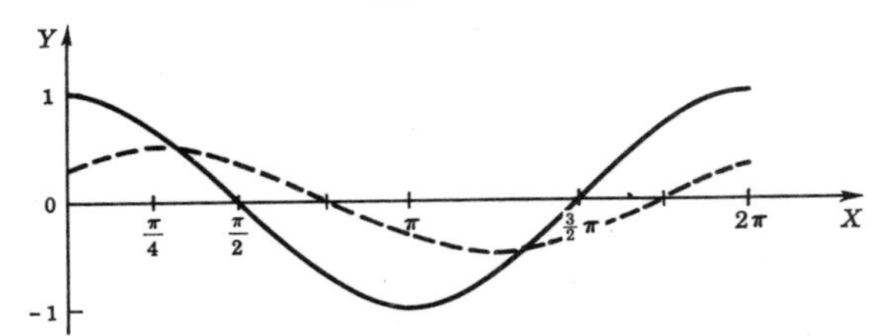

Figure **8.6.13**

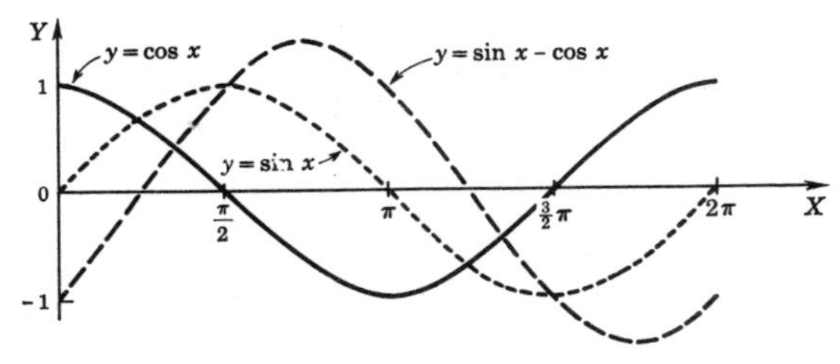

Figure **8.6.15**

17. (Fig. 8.6.17).

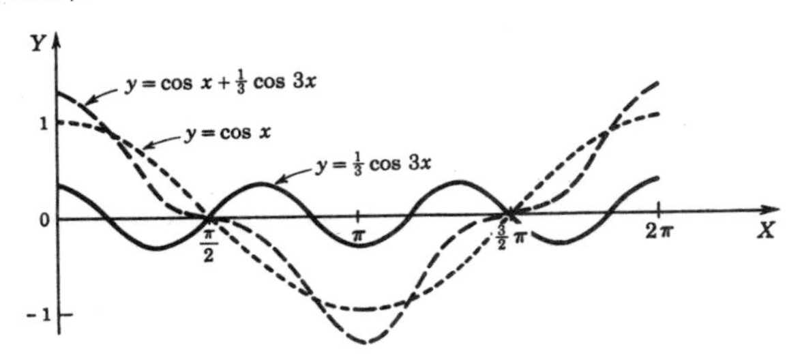

Figure **8.6.17**

19. (Fig. 8.6.19). **21.** (Fig. 8.6.21).

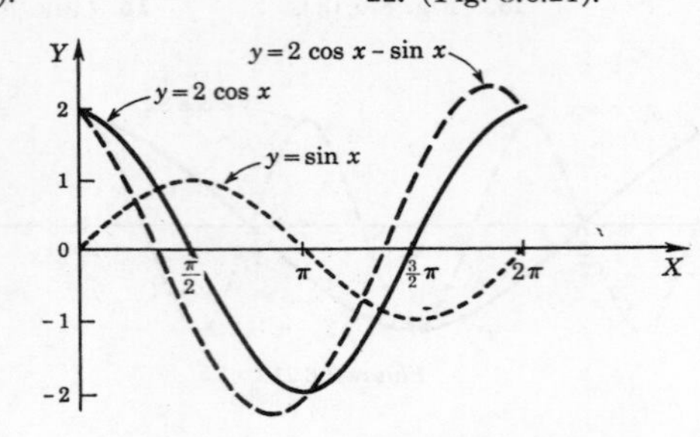

Figure **8.6.19**

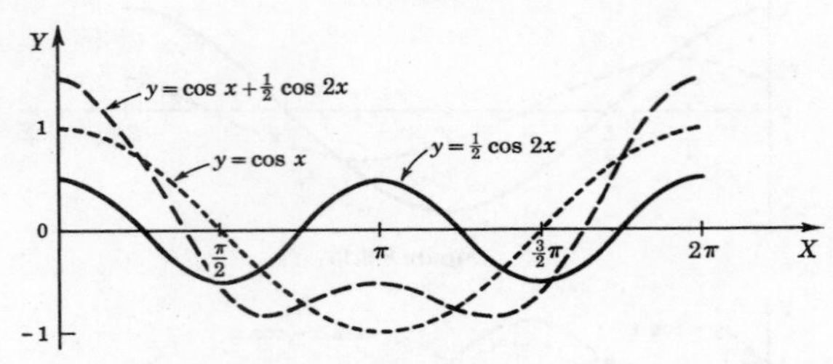

Figure **8.6.21**

23 and **25.**

Quadrant	As θ varies from:	sec θ varies from:	cot θ varies from:
I	0 to $\pi/2$	1 to ∞	∞ to 0
II	$\pi/2$ to π	$-\infty$ to -1	0 to $-\infty$
III	π to $\frac{3}{2}\pi$	-1 to $-\infty$	$-\infty$ to 0
IV	$\frac{3}{2}\pi$ to 2π	∞ to 1	0 to ∞

Problems 8.8

23. $\cos^3\theta - 3\sin^2\theta\cos\theta$

33. $\sin\frac{1}{12}\pi = \sqrt{\frac{1}{2} - \frac{1}{4}\sqrt{3}}$

$\cos\frac{1}{12}\pi = \sqrt{\frac{1}{2} + \frac{1}{4}\sqrt{3}}$

35. $\sin\frac{7}{12}\pi = \sqrt{\frac{1}{2} + \frac{1}{4}\sqrt{3}}$

$\cos\frac{7}{12}\pi = -\sqrt{\frac{1}{2} - \frac{1}{4}\sqrt{3}}$

37. $\sin\frac{13}{12}\pi = -\sqrt{\frac{1}{2} - \frac{1}{4}\sqrt{3}}$

$\cos\frac{13}{12}\pi = -\sqrt{\frac{1}{2} + \frac{1}{4}\sqrt{3}}$

39. $\sin\frac{19}{12}\pi = -\sqrt{\frac{1}{2} + \frac{1}{4}\sqrt{3}}$

$\cos\frac{19}{12}\pi = \sqrt{\frac{1}{2} - \frac{1}{4}\sqrt{3}}$

41. $\sin\frac{1}{8}\pi = \sqrt{\frac{1}{2} - \frac{1}{4}\sqrt{2}}$

$\cos\frac{1}{8}\pi = \sqrt{\frac{1}{2} + \frac{1}{4}\sqrt{2}}$

43. $\sin\frac{5}{8}\pi = \sqrt{\frac{1}{2} + \frac{1}{4}\sqrt{2}}$

$\cos\frac{5}{8}\pi = -\sqrt{\frac{1}{2} - \frac{1}{4}\sqrt{2}}$

45. $\sin\frac{9}{8}\pi = -\sqrt{\frac{1}{2} - \frac{1}{4}\sqrt{2}}$

$\cos\frac{9}{8}\pi = -\sqrt{\frac{1}{2} + \frac{1}{4}\sqrt{2}}$

47. $\sin\frac{13}{8}\pi = -\sqrt{\frac{1}{2} + \frac{1}{4}\sqrt{2}}$

$\cos\frac{13}{8}\pi = \sqrt{\frac{1}{2} - \frac{1}{4}\sqrt{2}}$

49. $\sin\frac{1}{24}\pi = \sqrt{\frac{1}{2} - \frac{1}{8}\sqrt{8 + 4\sqrt{3}}}$

$\cos\frac{1}{24}\pi = \sqrt{\frac{1}{2} + \frac{1}{8}\sqrt{8 + 4\sqrt{3}}}$

51. $\sin \frac{5}{24}\pi = \frac{1}{2}\sqrt{1 + \frac{1}{4}\sqrt{8 + 4\sqrt{3}}} - \frac{1}{2}\sqrt{1 - \frac{1}{4}\sqrt{8 + \sqrt{3}}}$

$\cos \frac{5}{24}\pi = \frac{1}{2}\sqrt{1 + \frac{1}{4}\sqrt{8 + 4\sqrt{3}}} + \frac{1}{2}\sqrt{1 - \frac{1}{4}\sqrt{8 + 4\sqrt{3}}}$

Problems 8.10

1. $n\pi, \frac{1}{6}\pi + 2n\pi, \frac{5}{6}\pi + 2n\pi.$
3. $\frac{1}{2}\pi + 2n\pi, \frac{1}{6}\pi + 2n\pi, \frac{5}{6}\pi + 2n\pi.$
5. $\frac{3}{4}\pi + n\pi.$
7. $\frac{1}{6}\pi + 2n\pi, \frac{5}{6}\pi + 2n\pi.$
9. $\frac{1}{4}\pi + n\pi.$
11. $0.26180 + n\pi, 1.30896 + n\pi.$
13. $\frac{1}{18}\pi + \frac{2}{3}\pi + n\pi.$
15. $\frac{1}{3}\pi + n\pi, \frac{2}{3}\pi + n\pi.$
17. $n\pi.$
19. $-0.20862 + 2n\pi, 3.35021 + 2n\pi.$

Problems 8.11

1. $\frac{1}{4}\pi.$
3. $\frac{1}{4}\pi.$
5. $\frac{1}{2}\pi.$
7. $-\frac{1}{3}\pi.$
9. $\pi.$
11. $\frac{2}{5}\sqrt{5}$
13. $\frac{4}{5}.$
15. $-\frac{1}{3}\sqrt{5}.$
17. $-0.80005.$

29. $y = \text{Csc}\, x$; domain, $-\frac{1}{2}\pi \le x \le \frac{1}{2}\pi$; range, $(-\infty < y \le -1) \vee (1 \le y < \infty)$.
$y = \text{Csc}^{-1} x$; domain, $(-\infty < x \le -1) \vee (1 \le x < \infty)$; range, $-\frac{1}{2}\pi \le y \le \frac{1}{2}\pi$

Problems 8.12

1. 0.1062.
3. 0.9218.
5. $-0.9426.$
7. $-0.7349.$
9. $-0.6115.$
11. 0.7030.
13. $-0.6517.$
15. 0.9075.
17. $-0.9150.$
19. 0.1069.
21. 2.3797.
23. 2.8258.
25. $-1.0839.$
27. $-0.7730.$

Problems 8.13

1. $2 \text{ cis } \frac{1}{4}\pi.$
3. $2 \text{ cis } \frac{1}{3}\pi.$
5. $\text{cis } (-\frac{1}{2}\pi).$
7. 1.
9. $-2.$
11. $-\frac{1}{2}\sqrt{3} - \frac{1}{2}i.$
13. $-i.$
15. $6 \text{ cis } 105°.$
17. $4 \text{ cis } 60°.$
19. cis 0.
21. $\frac{3}{2} \text{ cis } (-15°).$
23. cis 180°.
25. $\frac{3}{2} \text{ cis } \frac{5}{6}\pi.$
27. $\sqrt{3} \text{ cis } 75°, \sqrt{3} \text{ cis } 255°.$
29. $2 \text{ cis } \frac{1}{3}\pi, 2 \text{ cis } \pi, 2 \text{ cis } \frac{5}{3}\pi.$
31. $\pm 1, \pm i.$

33. $\text{cis}\left(\frac{\pi}{8} + n\frac{\pi}{4}\right), n = 0, 1, 2, 3.$ Or:

$$\begin{array}{r} 0.9239 + 0.3827i \\ -0.9239 - 0.3827i \\ -0.3827 + 0.9239i \\ 0.3827 - 0.9239i \end{array}$$

35. $\text{cis } n\frac{2}{5}\pi, n = 0, 1, 2, 3, 4.$ Or: $1, 0.3090 \pm 0.9511i, -0.8090 \pm 0.5878i$

37. $1, -1, \frac{1}{2} + \frac{i}{2}\sqrt{3}, -\frac{1}{2} + \frac{i}{2}\sqrt{3}, -\frac{1}{2} - \frac{i}{2}\sqrt{3}, \frac{1}{2} - \frac{i}{2}\sqrt{3}.$

Problems 9.4

1. (2,3).
3. (0,2).
5. $(a/2, a/2).$
7. $(\frac{1}{2}, a/2).$
9. **(a)** 2; **(c)** -2.
11. **(a)** 5; **(b)** 17; **(c)** $\frac{5}{17}$.
13. **(a)** 0; **(b)** 20; **(c)** 0.
15. **(a)** —; **(b)** —; **(c)** —.
17. **(a)** $\frac{b}{a}$; **(b)** $\theta = \text{arc Tan } (b/a).$
19. **(a)** $-\frac{5}{6}$; **(b)** $\theta = \text{arc Tan } (-\frac{5}{6}).$

21. (**a**) —; (**b**) $\theta = 90°$. **23.** (**a**) $\dfrac{d-b}{c-a}$; (**b**) $\theta = \text{arc Tan } \dfrac{d-b}{c-a}$.
25. (**b**) 2; (**c**) $5\sqrt{5}$. **27.** (**b**) 3; (**c**) 5. **29.** (**b**) 6; (**c**) $\sqrt{221}$.
31. $(10, -3)$. **33.** $(-4, 24)$. **35.** $\overline{AB}^2 = \overline{AC}^2 + \overline{BC}^2$.
37. $AC = BC \neq AB$. **39.** $m_{AB} = m_{CD} = -\frac{1}{12}$, $m_{BC} = m_{DA} = -\frac{7}{2}$.
41. $m_{AB} = m_{AC} = \frac{3}{2}$. **43.** $m_{AE} = \frac{11}{4}$, $m_{BF} = -1$, $m_{CD} = -16$.

45. $(5, -3)$. Use $m_{CD} = m_{AB}$ and $m_{CD} = -1/m_{BC}$.

Problems 9.6

1. $\lambda = \dfrac{1}{5\sqrt{2}}$, $\mu = \dfrac{-7}{5\sqrt{2}}$. **3.** $\lambda = \frac{1}{2}\sqrt{2}$, $\mu = \frac{1}{2}\sqrt{2}$.
5. $\lambda = \dfrac{2}{\sqrt{5}}$, $\mu = \dfrac{1}{\sqrt{5}}$. **7.** $\lambda = \dfrac{7}{\sqrt{53}}$, $\mu = \dfrac{-2}{\sqrt{53}}$.

9. (**a**) $\lambda_{AB} = \frac{1}{5}\sqrt{5}$, $\mu_{AB} = \frac{2}{5}\sqrt{5}$, $\lambda_{AC} = -\frac{2}{5}\sqrt{5}$, $\mu_{AC} = \frac{1}{5}\sqrt{5}$.
(**b**) $m_{AB} = 2$, $m_{AC} = -\frac{1}{2}$.
11. (**a**) $\lambda_{AB} = \dfrac{4}{\sqrt{52}}$, $\mu_{AB} = \dfrac{-6}{\sqrt{52}}$, $\lambda_{BC} = \dfrac{6}{\sqrt{52}}$, $\mu_{BC} = \dfrac{4}{\sqrt{52}}$.
(**b**) $m_{AB} = -\frac{3}{2}$, $m_{BC} = \frac{2}{3}$.
13. AB and CD are $\parallel$ to Y-axis, $m_{BC} = m_{AD} = \frac{1}{4}$.
15. $m_{AB} = m_{CD} = -\frac{1}{4}$, $m_{AD} = m_{BC} = \frac{6}{5}$.

17. $4/\sqrt{17}$. **19.** $\dfrac{32}{\sqrt{29}\sqrt{37}}$. **21.** $7/\sqrt{85}$.
23. $1/\sqrt{10}$. **25.** $1/\sqrt{17}$.
27. (**a**) 0.
(**b**) Does not exist.
29. (**a**) 1.
(**b**) -1.
31. Mid-points concide. **33.** $\left|\frac{3}{10} - \frac{4}{10}\sqrt{3}\right| = \frac{1}{10}\sqrt{3}$.
35. $\frac{1}{10}\sqrt{10}$. **37.** 0.
39. $\dfrac{22}{\sqrt{13}\sqrt{41}}$.

41. $\lambda_1 = -\lambda_2$, $\mu_1 = -\mu_2$, $\cos\theta = -\lambda_1^2 - \mu_1^2 = -1$, $\theta = 180°$.
43. $a = -\frac{2}{7}$.

Problems 9.8

13. $4x - 3y - 17 = 0$. **15.** $y = 3x$.
17. $5x - 2y - 26 = 0$. **19.** $2x + y + 1 = 0$.
21. $3x + 4y - 8 = 0$. **23.** Straight line.
25. Two straight lines. **27.** $y + 3 = \pm\sqrt{3}(x - 4)$.

31. Yes. Use only addition, subtraction, multiplication, and division in solving.
33. No. Rationals may be involved.
35. Point of intersection of first and second lines, $(-3, 2)$, lies on third.
37. $\dfrac{x}{\sqrt{2}} - \dfrac{y}{\sqrt{2}} - \dfrac{3}{\sqrt{2}} = 0$. **39.** $-\frac{4}{5}x + \frac{3}{5}y - 1 = 0$.
41. $-\frac{6}{10}x - \frac{8}{10}y - 2 = 0$. **45.** $\left|-\frac{3}{5}(2) + \frac{4}{5}(-5) - 2\right| = d = \frac{36}{5}$.
47. $\left|-\frac{1}{2}\sqrt{2}(-4) + \frac{1}{2}\sqrt{2}(2) - 2\right| = d = 3\sqrt{2} - 2$.

Problems 9.10

1. $(x - 3)^2 + (y - 5)^2 = 36$. (Fig. 9.10.1)

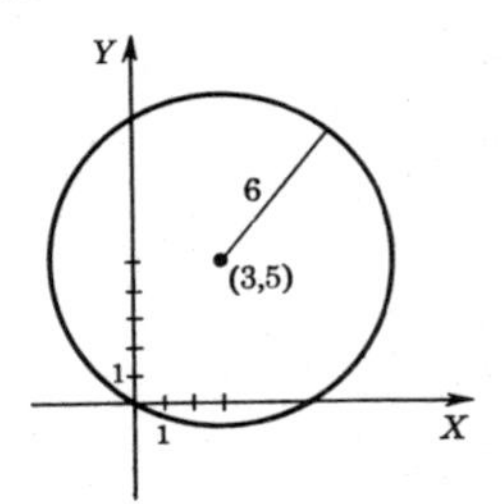

Figure **9.10.1**

3. $(x - 2)^2 + (y + 2)^2 = 25$. (Fig. 9.10.3)

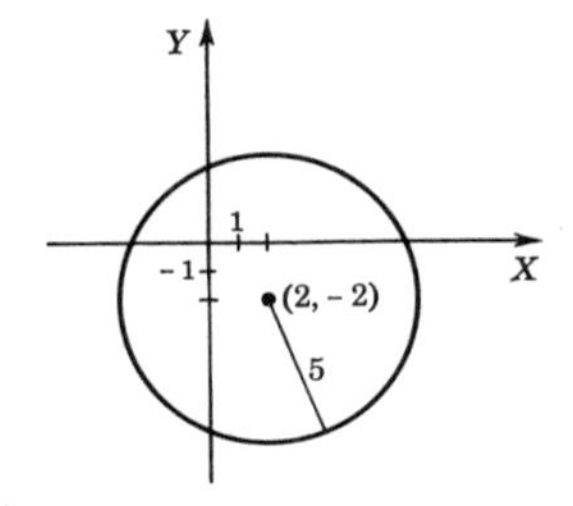

Figure **9.10.3**

5. $(x - \frac{1}{2})^2 + (y - \frac{3}{2})^2 = \frac{25}{2}$. (Fig. 9.10.5)

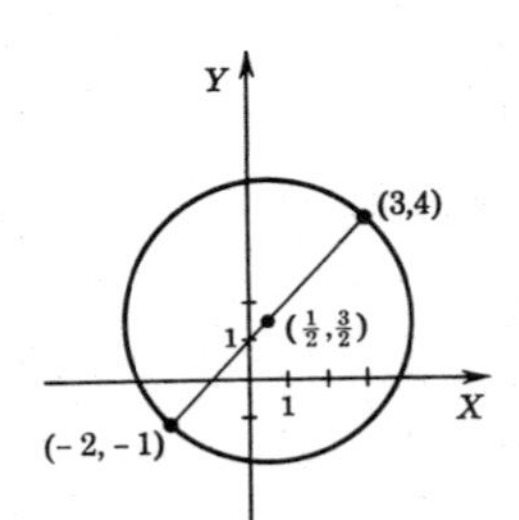

Figure **9.10.5**

7. $(x - 2)^2 + (y - 1)^2 = 1$. (Fig. 9.10.7)

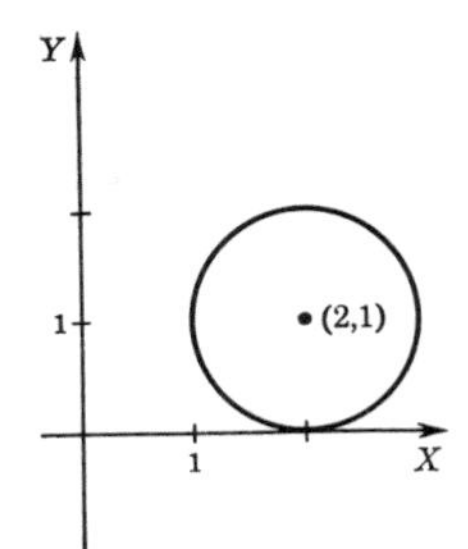

Figure **9.10.7**

9. $(x + \frac{1}{2})^2 + (y - \frac{1}{4})^2 = \frac{5}{16}$. (Fig. 9.10.9)

11. $(x - 4)^2 + (y - 4)^2 = 16$, $(x + 4)^2 + (y - 4)^2 = 16$, $(x + 4)^2 + (y + 4)^2 = 16$, $(x - 4)^2 + (y + 4)^2 = 16$. (Fig. 9.10.11)

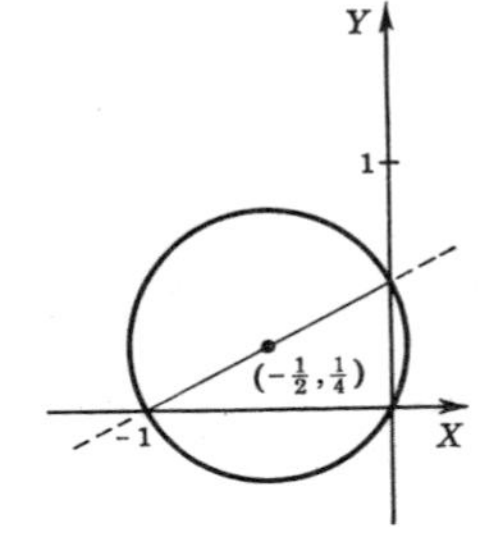

Figure **9.10.9**

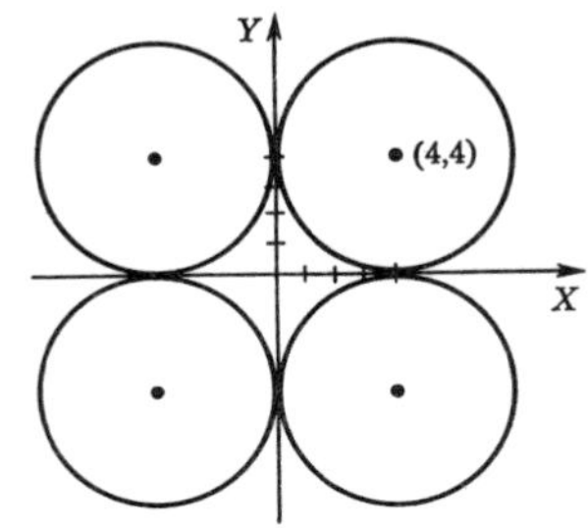

Figure **9.10.11**

13. $(2, -3)$, $r = \sqrt{13}$. (Fig. 9.10.13) **15.** $(3,0)$, $r = 5$. (Fig. 9.10.15)

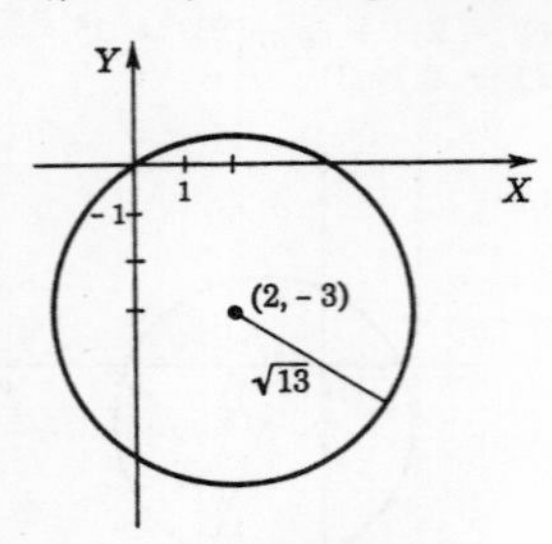

Figure 9.10.13

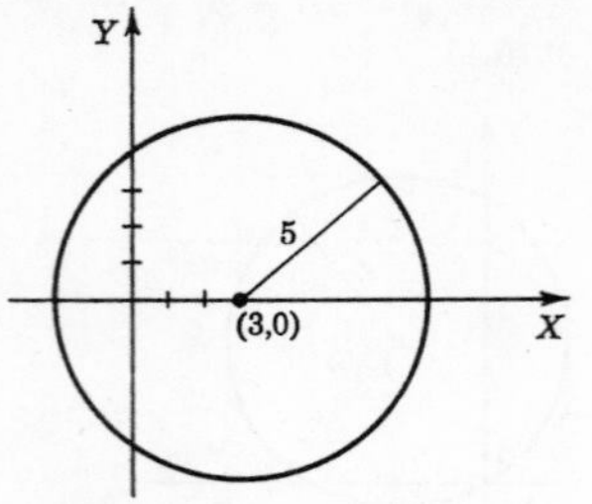

Figure 9.10.15

17. $(1, -2)$, r (not real). Equation is satisfied by no point.
19. $(-\frac{5}{4}, \frac{7}{4})$, $r = \frac{1}{4}\sqrt{154}$. (Fig. 9.10.19)

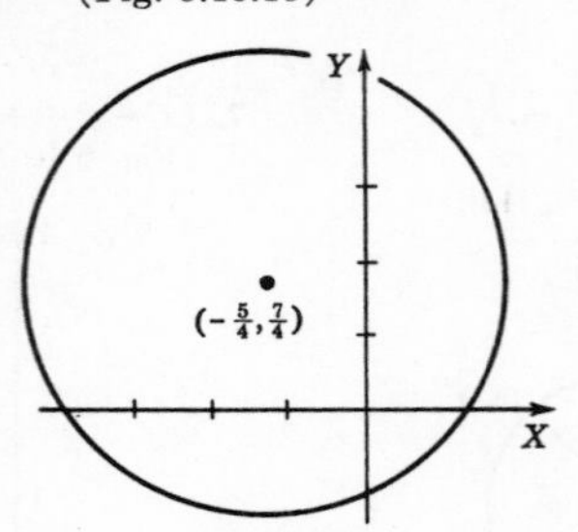

Figure 9.10.19

21. $(x - \frac{3}{2})^2 + (y + \frac{1}{2})^2 = 16$. **23.** $(x - h)^2 + (y - k)^2 = 1$.

25. $(x - h)^2 + (y - k)^2 = 1$, where $h^2 + k^2 = 4$.

Problems 9.11

1. $F(1,0)$, $x = -1$. (Fig. 9.11.1)
3. $F(-\frac{1}{4}, 0)$, $x = \frac{1}{4}$. (Fig. 9.11.3)

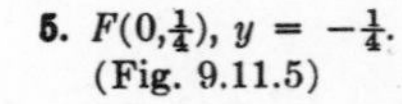

5. $F(0,\frac{1}{4})$, $y = -\frac{1}{4}$. (Fig. 9.11.5)

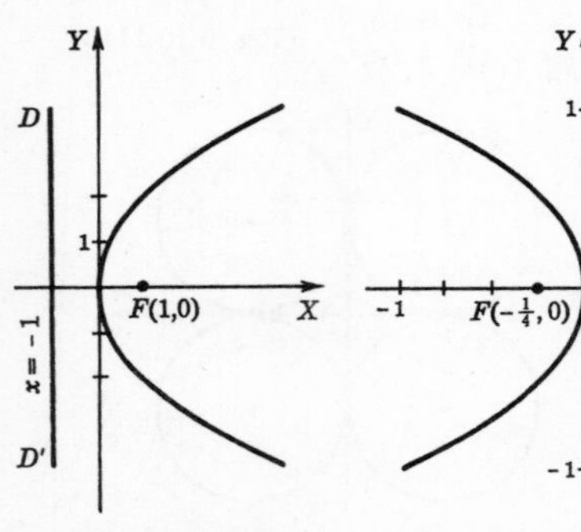

Figure 9.11.1 *Figure* 9.11.3

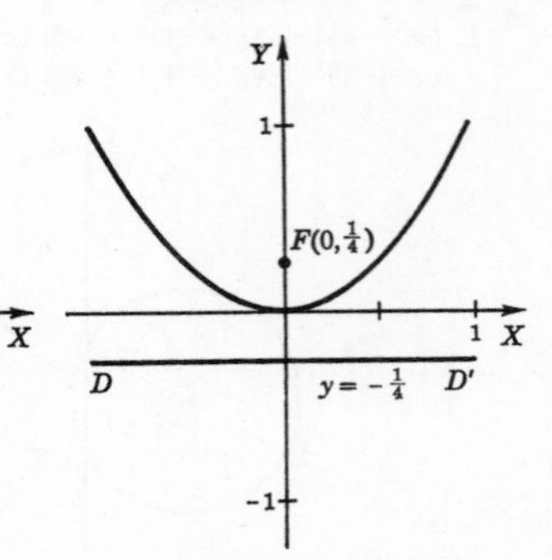

Figure 9.11.5

7. $F(0, -3)$, $y = 3$. (Fig. 9.11.7)

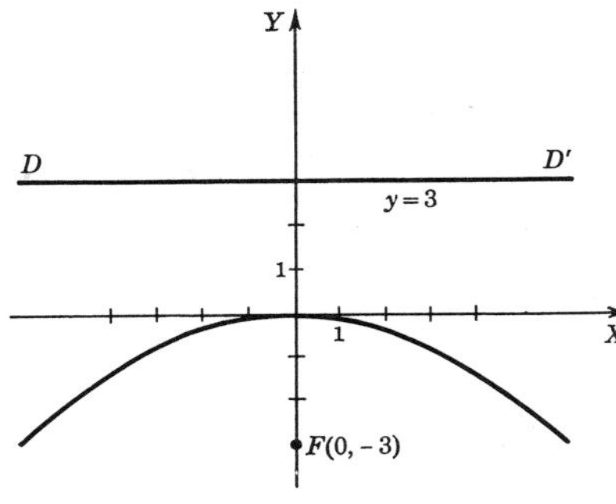

Figure **9.11.7**

9. $F(1,2)$, $x = -3$. (Fig. 9.11.9)

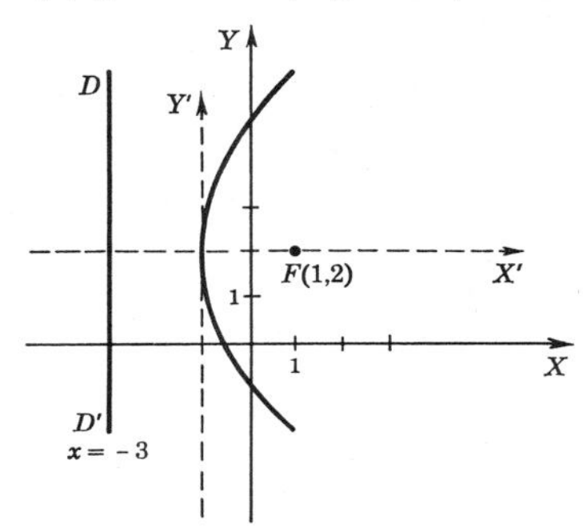

Figure **9.11.9**

11. $y^2 = 20x$. (Fig. 9.11.11)

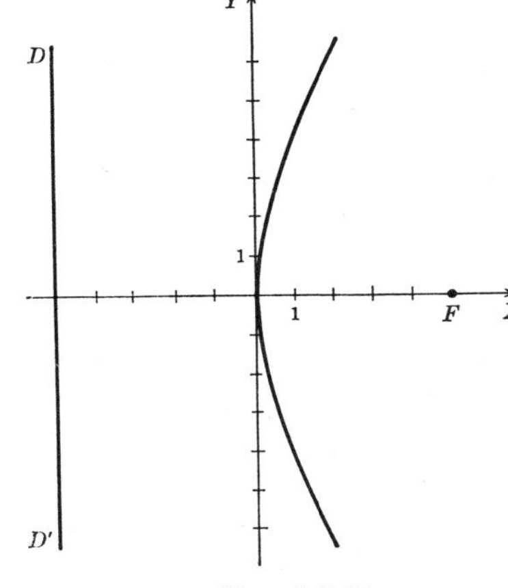

Figure **9.11.11**

13. $y^2 = -12x$. (Fig. 9.11.13)

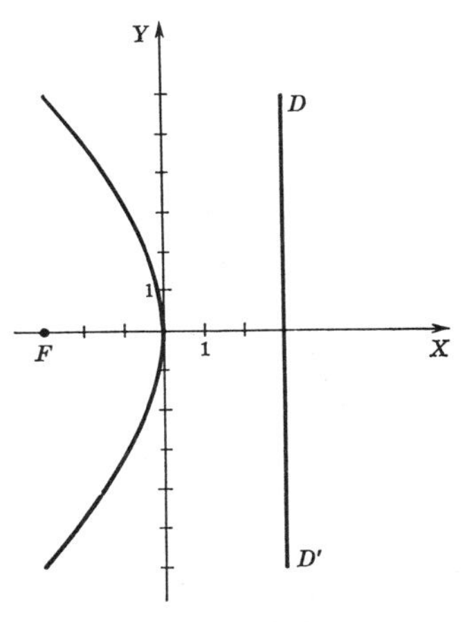

Figure **9.11,13**

15. $x^2 = 8y$. (Fig. 9.11.15)

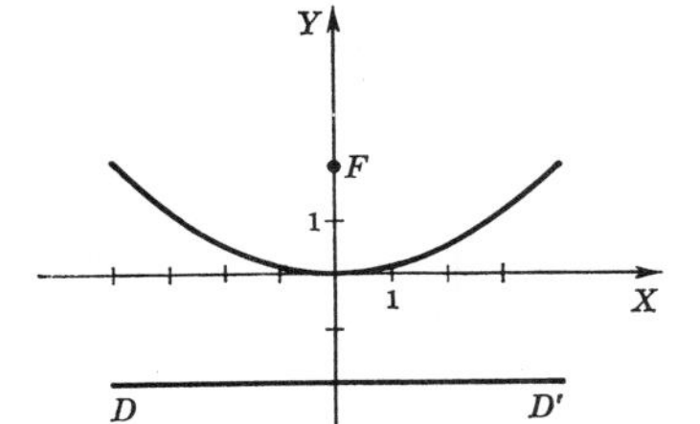

Figure **9.11.15**

17. $x^2 = -12y$. (Fig. 9.11.17)

Figure **9.11.17**

19. $x^2 = 4(y - 1)$. (Fig. 9.11.19)

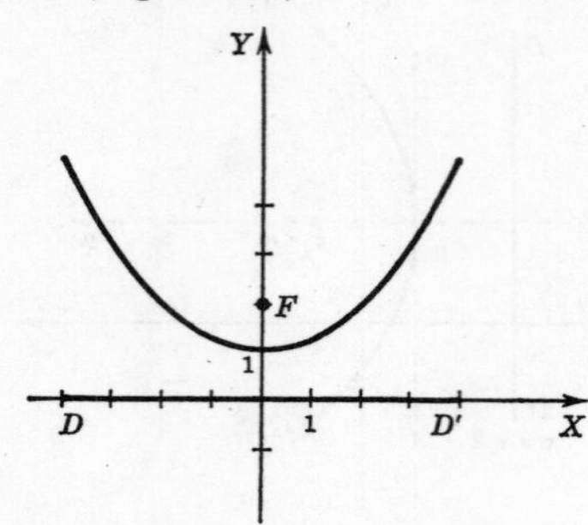

Figure 9.11.19

21. $(x - 3)^2 = 4(y - 1)$. (Fig. 9.11.21)

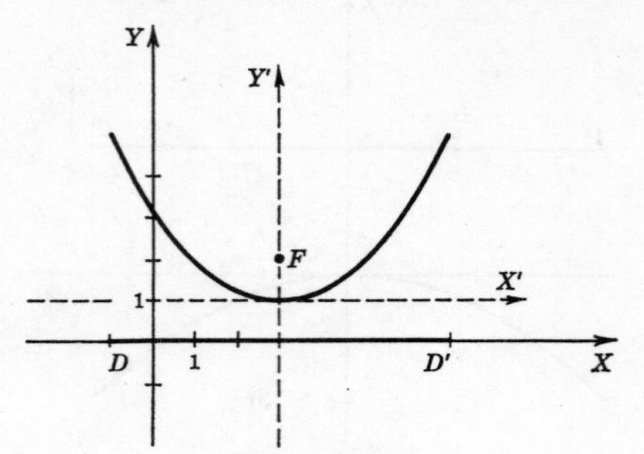

Figure 9.11.21

23. (0,0), (2,2).

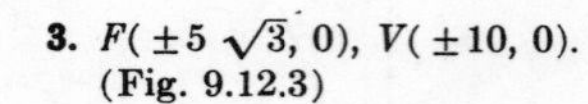

25. $\dfrac{x^2}{25} + \dfrac{(y - 1)^2}{9} = 1.$ **27.** $y^2 = 8(x - 1)$.

Problems 9.12

1. $F(\pm\sqrt{5}, 0)$, $V(\pm 3, 0)$. (Fig. 9.12.1)

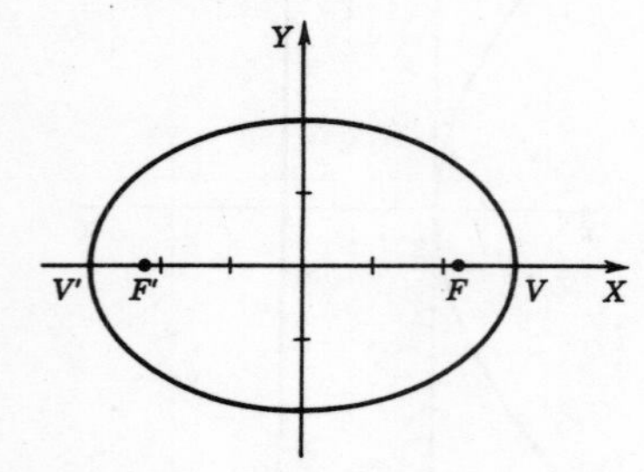

Figure 9.12.1

3. $F(\pm 5\sqrt{3}, 0)$, $V(\pm 10, 0)$. (Fig. 9.12.3)

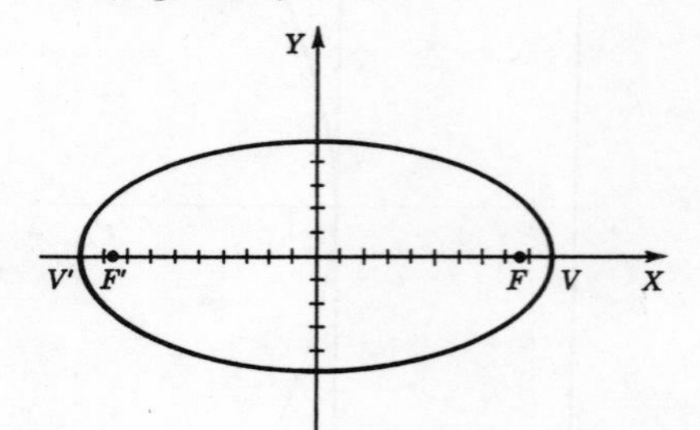

Figure 9.12.3

5. $F(\pm\sqrt{3}, 0)$, $V(\pm\sqrt{5}, 0)$. (Fig. 9.12.5)

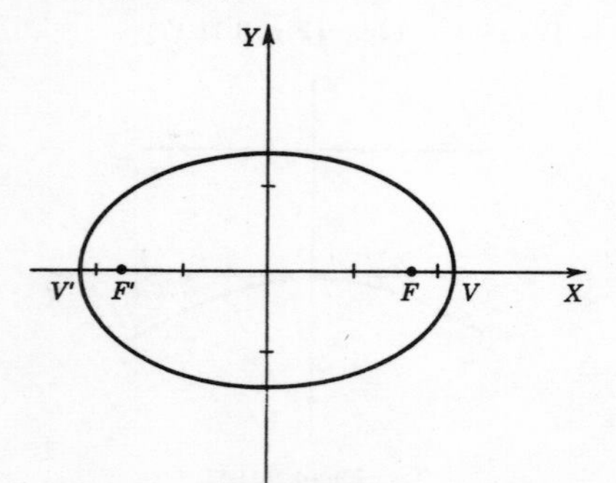

Figure 9.12.5

7. $F(0, \pm 2\sqrt{5})$, $V(0, \pm 6)$. (Fig. 9.12.7)

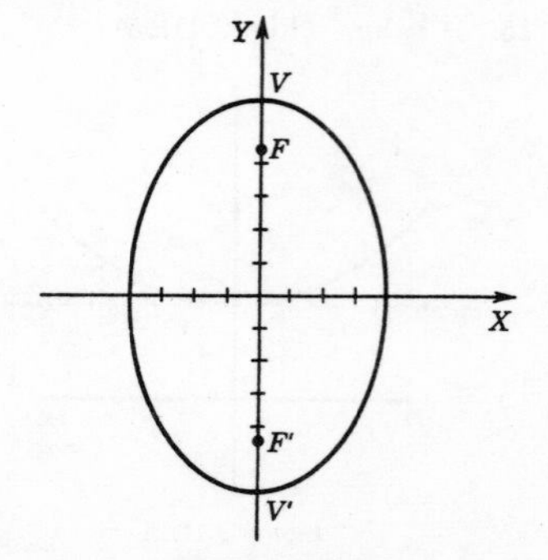

Figure 9.12.7

9. $F(2 \pm \sqrt{2}, -1)$, $V(2 \pm 2, -1)$. (Fig. 9.12.9)

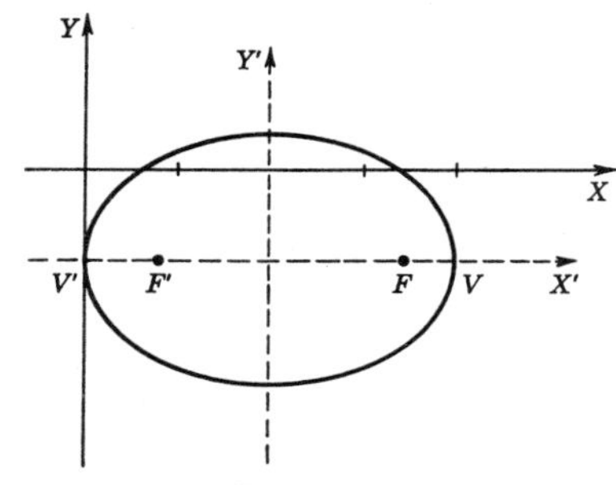

Figure **9.12.9**

11. $\dfrac{x^2}{9} + \dfrac{y^2}{8} = 1$. (Fig. 9.12.11)

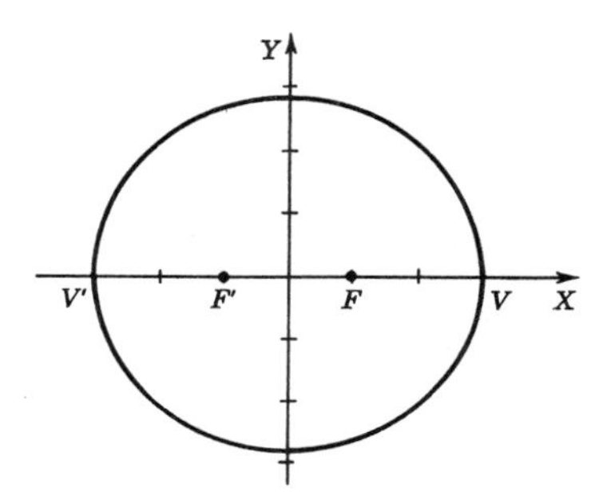

Figure **9.12.11**

13. $\dfrac{x^2}{10} + \dfrac{y^2}{9} = 1$. (Fig. 9.12.13)

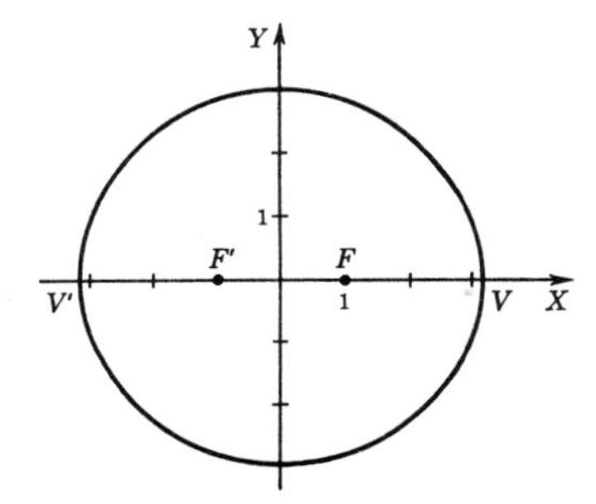

Figure **9.12.13**

15. $\dfrac{x^2}{25} + \dfrac{y^2}{9} = 1$. (Fig. 9.12.15)

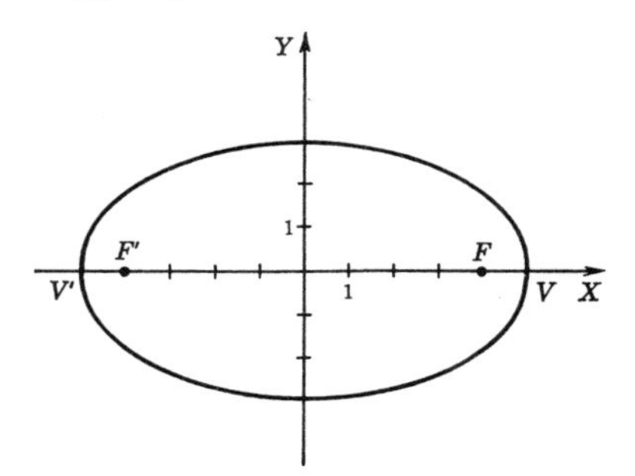

Figure **9.12.15**

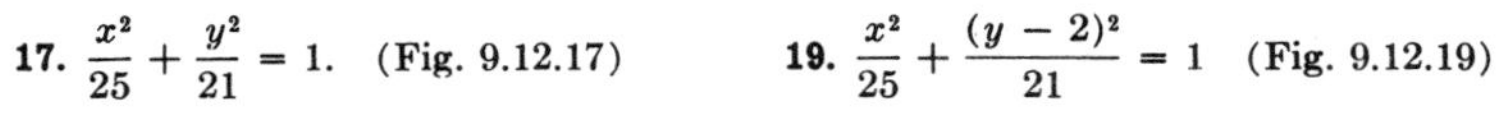

17. $\dfrac{x^2}{25} + \dfrac{y^2}{21} = 1$. (Fig. 9.12.17)

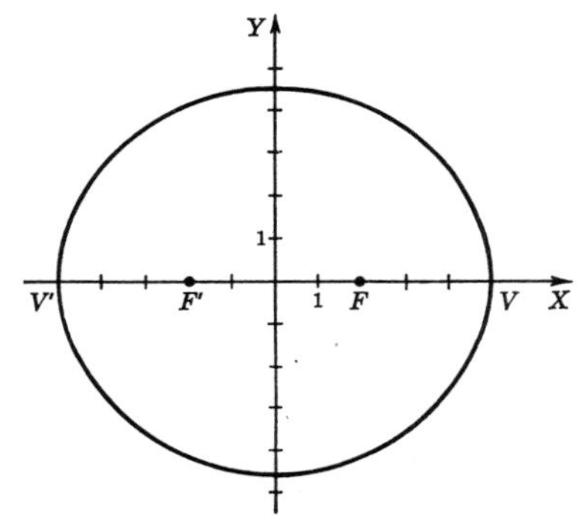

Figure **9.12.17**

19. $\dfrac{x^2}{25} + \dfrac{(y-2)^2}{21} = 1$ (Fig. 9.12.19)

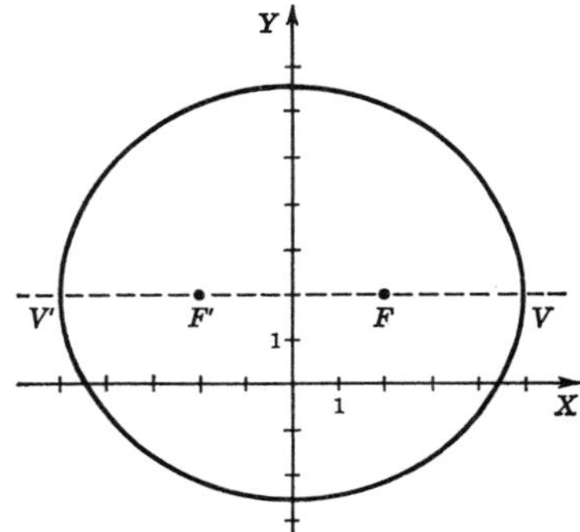

Figure **9.12.19**

21. $\dfrac{x^2}{84} + \dfrac{y^2}{100} = 1.$ (Fig. 9.12.21)

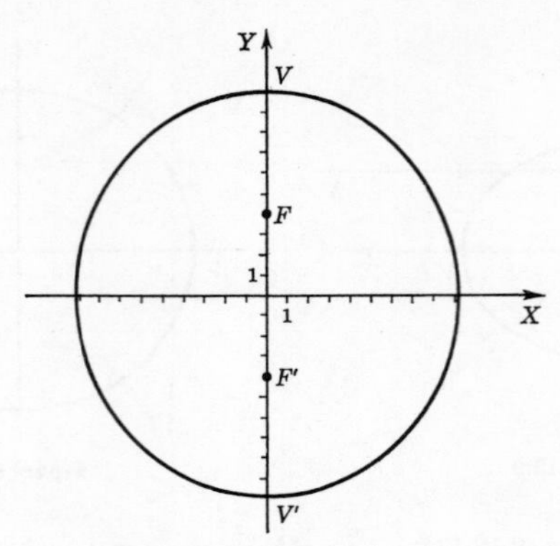

Figure 9.12.21

23. $\dfrac{x^2}{4} - \dfrac{(y-2)^2}{12} = 1.$

25. $(x - \frac{1}{2})^2 + y^2 = \frac{1}{4}.$

Problems 9.13

1. $(\frac{1}{2}, -1)$, $r = \frac{1}{2}\sqrt{85}$.

3. $y = \frac{3}{2}$, $V(-\frac{1}{2}, \frac{5}{4})$, $F(-\frac{1}{2}, 1)$.

5. $y = \pm(x + \frac{5}{2})$, $V(-\frac{5}{2} \pm \frac{1}{2}\sqrt{65}, 0)$, $F(-\frac{5}{2} \pm \sqrt{\frac{65}{2}}, 0)$. (Fig. 9.13.5)

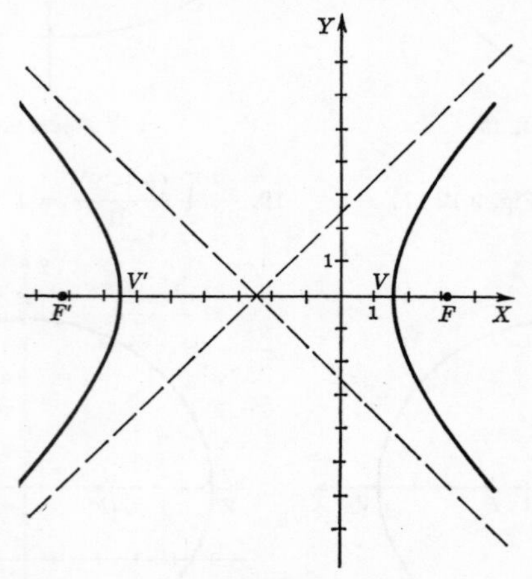

Figure 9.13.5

7. Two lines.

9. $y = \frac{1}{2}$, $V(\frac{1}{2}, -\frac{3}{4})$, $F(\frac{1}{2}, -1)$.

11. (Fig. 9.13.11).

13. $(x - h)^2 + (y - k) = 25$.

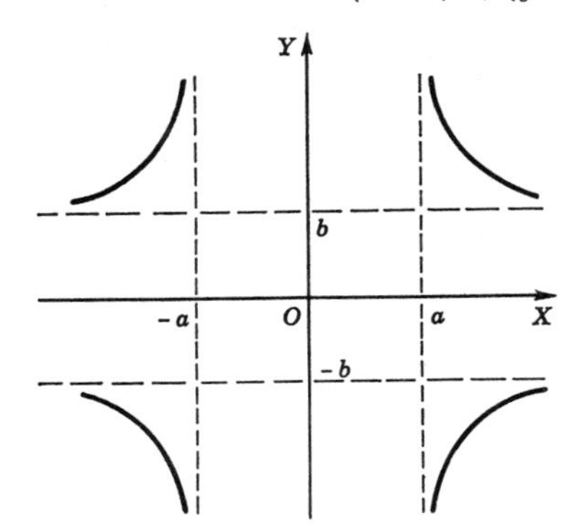

Figure **9.13.11**

15. $y^2 = -4(x - 2)$.

17. $(x^2/36) - (y^2/28) = 1$.

19. No. Two parallel lines. Cylindrical section.

21. Parabola. $V(h,k)$, $F(h + p, k)$, directrix $x = h - p$.

23. Hyperbola. Center (h,k), asymptotes $y - k = \pm \dfrac{b}{a}(x - h)$, $V(h \pm a, k)$, $F(h \pm \sqrt{a^2 + b^2}, k)$.

25. $x = m \pm \sqrt{m^2 - 1}$; $m^2 - 1 = 0$; each line is tangent to the parabola.

27. Two lines.

Problems 9.15

1. $r = 2 \cos \theta$.

3. $\theta = \text{Tan}^{-1} \frac{1}{2}$.

5. $r \sin^2 \theta = 8 \cos \theta$.

7. $r^2(1 + 3 \sin^2 \theta) = 16$.

9. $r^2\left(\dfrac{\cos^2 \theta}{25} - \dfrac{\sin^2 \theta}{9}\right) = 1$.

11. $r^2\left(-\dfrac{\cos^2 \theta}{4} + \sin^2 \theta\right) = 1$.

13. $(r - \cos \theta)^2 = 1$.

15. $x^2 + y^2 - 2y = 0$.

17. $x^2 + y^2 - 3x = 0$.

19. $(x^2 + y^2 - x)^2 = x^2 + y^2$.

21. $(x^2 + y^2 - 2y)^2 = x^2 + y^2$.

23. $y^2 - 2x - 1 = 0$.

25. $x^2 + y^2 = 16$.

27. $y = \sqrt{3}\, x$.

29. $(x^2 + y^2)^3 = (x^2 - y^2)^2$.

Problems 9.16

1. $3x^2 + 4y^2 - 14x - 49 = 0$. (Fig. 9.16.1)

3. $25x^2 + 24y^2 - 10y - 25 = 0$. (Fig. 9.16.3)

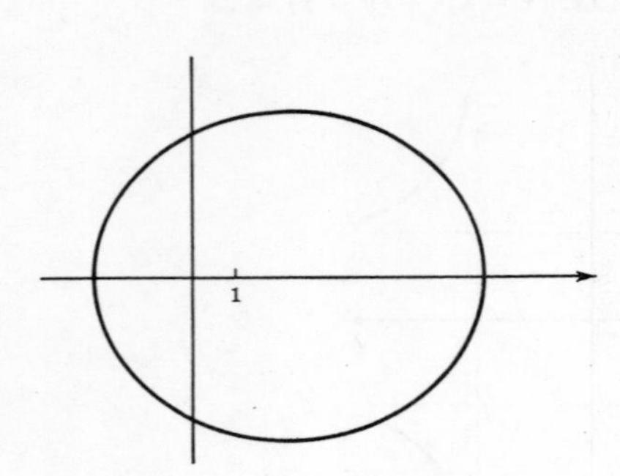

Figure 9.16.1

Figure 9.16.3

5. $x^2 + y^2 - x - \sqrt{3}\, y = 0.$
(Fig. 9.16.5)

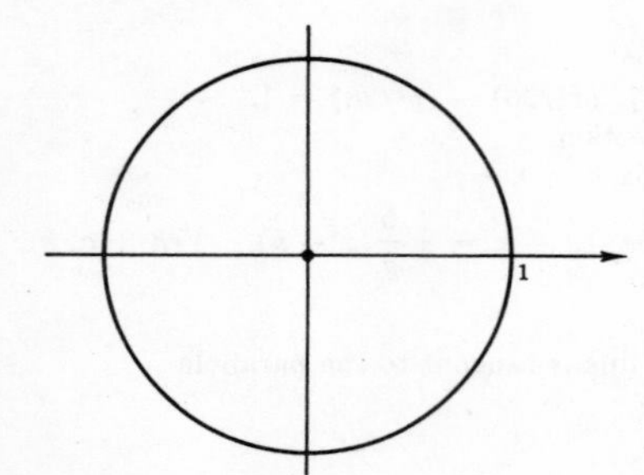

Figure 9.16.5

7. $x^2 + y^2 - y = 0.$
(Fig. 9.16.7)

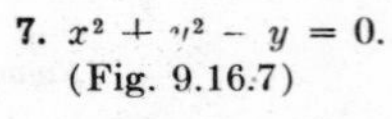

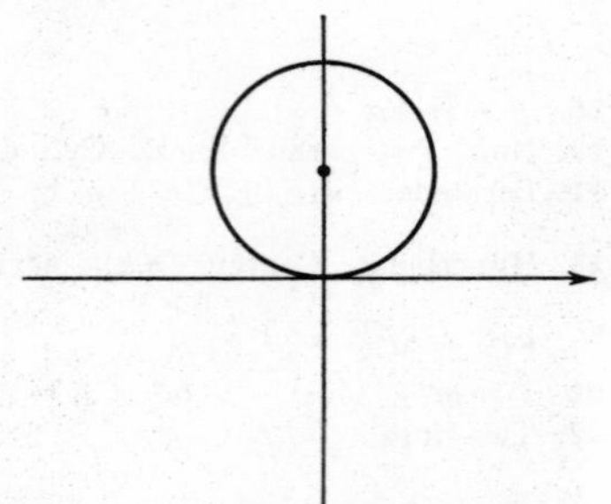

Figure 9.16.7

9. $(x^2 + y^2)^5 = 16x^2y^2(x^2 - y^2)^2.$
(Fig. 9.16.9)

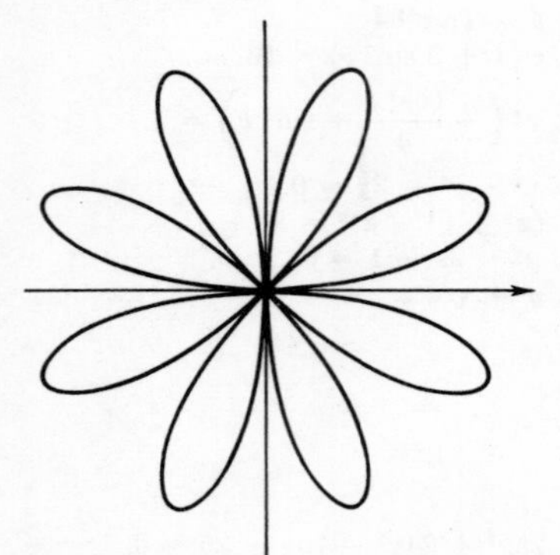

Figure 9.16.9

11. $(x^2 + y^2)^3 = (x^2 - y^2)^2.$
(Fig. 9.16.11)

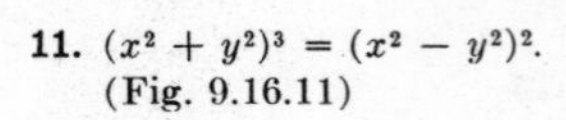

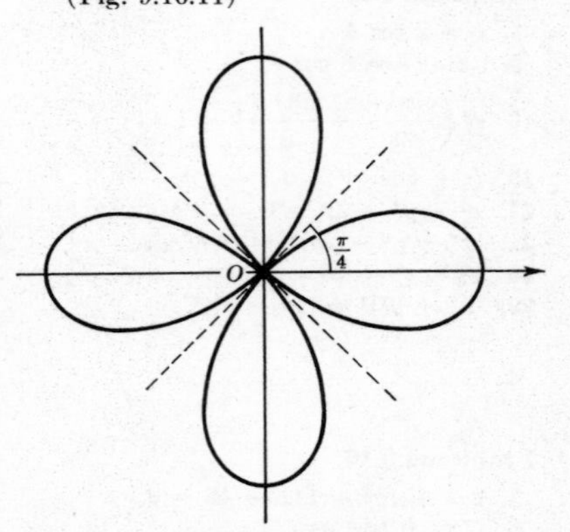

Figure 9.16.11

13. $(x^2 + y^2)^5 = (x^4 - 6x^2y^2 + y^4)^2$.
(Fig. 9.16.13)

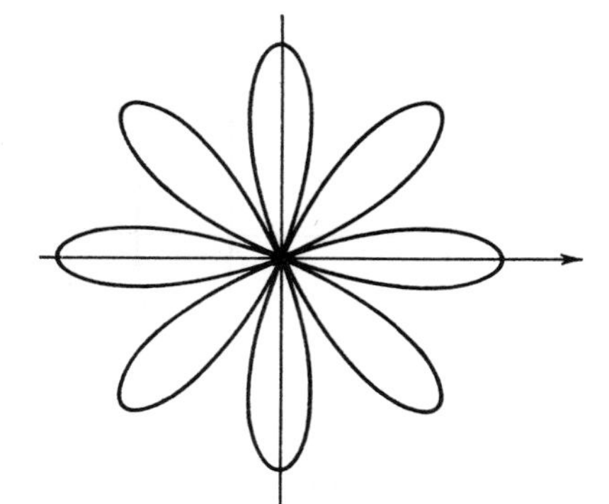

Figure **9.16.13**

15. $(x^2 + y^2 + x)^2 = x^2 + y^2$.
(Fig. 9.16.15)

Figure **9.16.15**

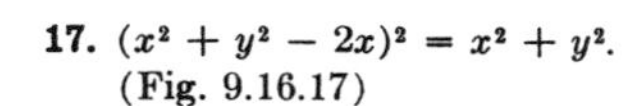

17. $(x^2 + y^2 - 2x)^2 = x^2 + y^2$.
(Fig. 9.16.17)

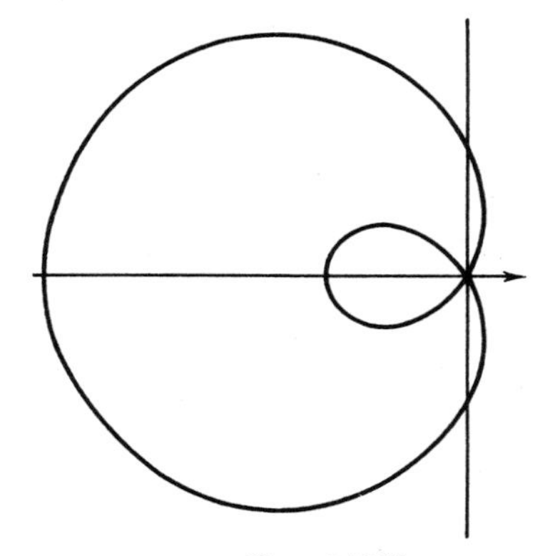

Figure **9.16.17**

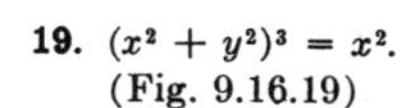

19. $(x^2 + y^2)^3 = x^2$.
(Fig. 9.16.19)

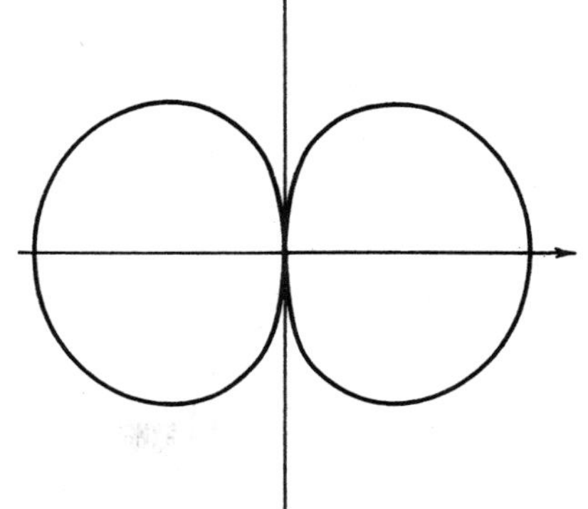

Figure **9.16.19**

21. $(x^2 + y^2)^2 = x^2 - y^2$.
(Fig. 9.16.21)

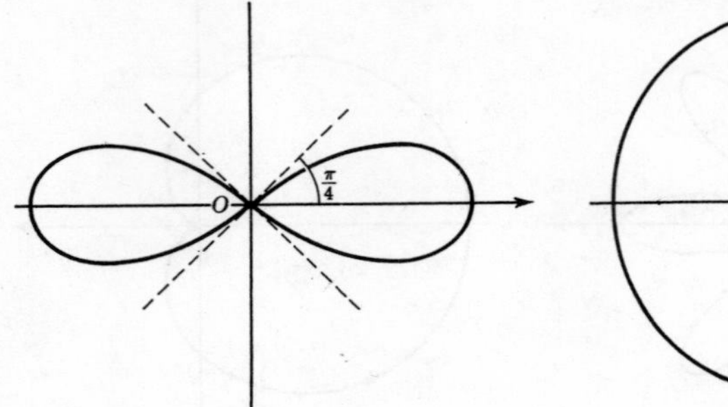

Figure **9.16.21**

23. $(x^2 + y^2 + x)^2 = 4(x^2 + y^2)$.
(Fig. 9.16.23)

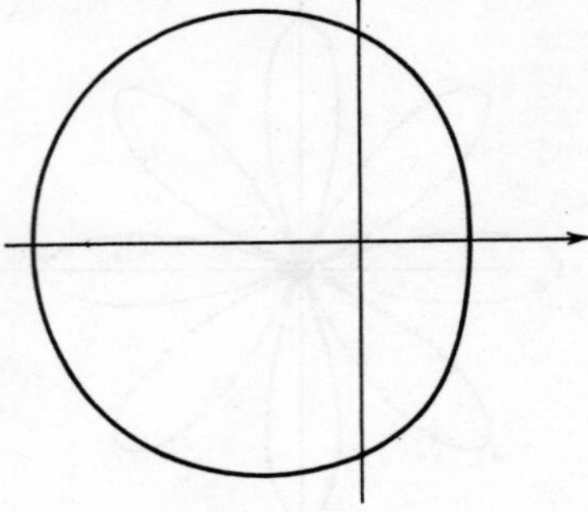

Figure **9.16.23**

25. $y = x \tan\left(\frac{-\pi}{\sqrt{x^2 + y^2}}\right)$.
(Fig. 9.16.25)

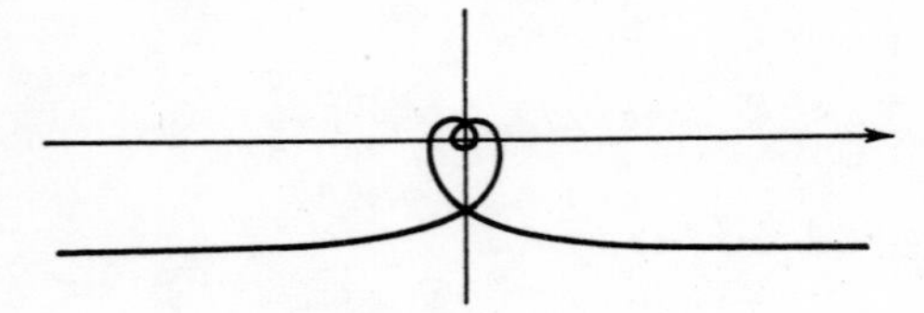

Figure **9.16.25**

27. $y = x \tan\left(\frac{-\pi}{x^2 + y^2}\right)$.
(Fig. 9.16.27)

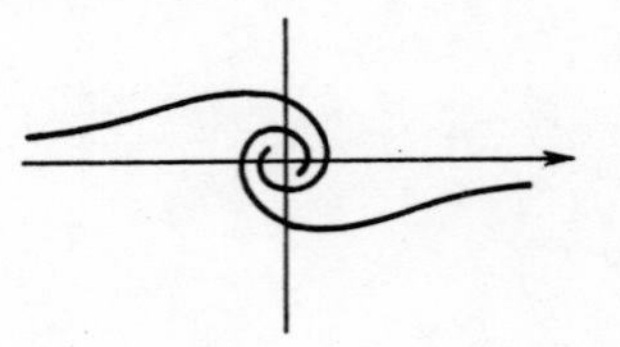

Figure **9.16.27**

29. $x = 1$.

31. $(\frac{1}{2}, \pi/6)$, $(\frac{1}{2}, 5\pi/6)$ and the origin. (Fig. 9.16.31)

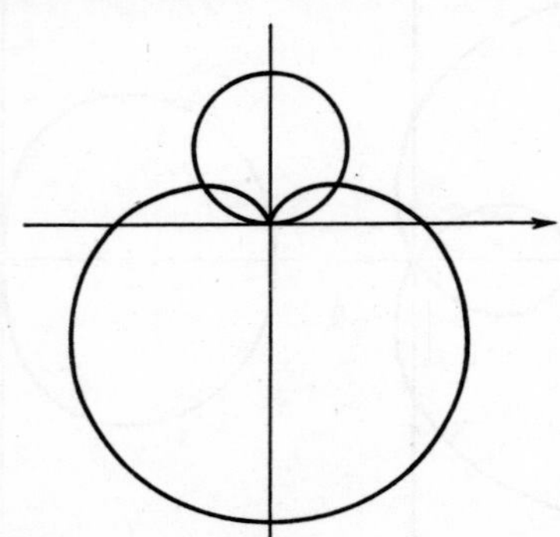

Figure **9.16.31**

33. $r = a \cos \theta$.

Problems 9.17

1. $x - 2y + 5 = 0$, straight line.
3. $y^2 = \frac{1}{2}x$, parabola.
5. $y = x^2 - 2x$, parabola.
7. $(x - y)^2 - 9(x + y) = 0$, parabola.
9. $(x - 1)^2 + (y - 1)^2 = 1$, circle.
11. $x^2 + y^2 = 2$, circle.
13. $xy = 1$, hyperbola.
15. $y = x$, straight line.
17. $(x - 2)^2 = (y + 1)$, parabola.
19. $(x^2/4) + (y^2/1) = 1$, ellipse.
21. $y^2 = (x - 4)$, parabola.
23. $-(x^2/1) + (y^2/1) = 1$, hyperbola.
25. $y^2 = -\frac{1}{2}(x - 1)$, parabola.

Problems 10.3

1. $\frac{1}{2}, \frac{2}{3}, \frac{3}{4}, \frac{4}{5}, \frac{5}{6}$.
3. $\frac{1}{2}, \frac{1}{5}, \frac{1}{10}, \frac{1}{17}, \frac{1}{26}$.
5. $\frac{1}{2}, \frac{4}{3}, \frac{9}{4}, \frac{16}{5}, \frac{25}{6}$.
7. 3, 9, 27, 81, 243.
9. $-1, \frac{1}{2}, -\frac{1}{3}, \frac{1}{4}, -\frac{1}{5}$.
11. $x, -\frac{x^3}{3!}, \frac{x^5}{5!}, -\frac{x^7}{7!}, \frac{x^9}{9!}$.
13. $\sin x, \sin 2x, \sin 3x, \sin 4x, \sin 5x$.

Problems 10.4

1. (a) $1, \frac{1}{4}, \frac{1}{9}, \frac{1}{16}, \frac{1}{25}$.
(b) $S = 0$.
(c) $n_0(\epsilon) = \left[\sqrt{\frac{1}{\epsilon}}\right] + 1$.

3. (a) $1, \frac{2}{3}, \frac{3}{5}, \frac{4}{7}, \frac{5}{9}$.
(b) $S = \frac{1}{2}$.
(c) $n_0(\epsilon) = \left[\frac{1}{4\epsilon} + \frac{1}{2}\right] + 1$.

5. (a) $\frac{3}{2}, 1, \frac{3}{4}, \frac{3}{5}, \frac{1}{2}$.
(b) $S = 0$.
(c) $n_0(\epsilon) = \left[\frac{3 - \epsilon}{\epsilon}\right] + 1$.

7. (a) $3, 2, \frac{5}{3}, \frac{3}{2}, \frac{7}{5}$.
(b) $S = 1$.
(c) $n_0(\epsilon) = \left[\frac{2 + 2\epsilon}{\epsilon}\right] + 1$.

9. (a) $1, 2, \frac{7}{3}, \frac{5}{2}, \frac{13}{5}$.
(b) $S = 3$.
(c) $n_0(\epsilon) = \left[\frac{2}{\epsilon}\right] + 1$.

11. (a) $\frac{3}{2}, \frac{3}{4}, \frac{3}{8}, \frac{3}{16}, \frac{3}{32}$.
(b) $S = 0$.
(c) $n_0(\epsilon) = \left[\frac{\log 3 - \log \epsilon}{\log 2}\right] + 1$.

13. (a) $\frac{1}{3}, \frac{1}{7}, \frac{1}{13}, \frac{1}{21}, \frac{1}{31}$.
(b) $S = 0$.
(c) $n_0(\epsilon) = \left[\frac{\sqrt{4 - 3\epsilon^2}}{2\epsilon} - \frac{1}{2}\right] + 1$.

15. (a) $\frac{2}{3}, \frac{3}{8}, \frac{4}{15}, \frac{5}{24}, \frac{6}{35}$.
(b) $S = 0$.
(c) $n_0(\epsilon) = \left[\frac{1 + \sqrt{1 + 4\epsilon^2}}{2\epsilon}\right] + 1$.

Problems 10.6

1. 1.
3. 0.
5. 0.
7. 2.
9. 0.
11. 0.

13. $n^{n-1} > n$.
15. $\cos 2k\pi = 1$ for all k; $\cos (2k + 1)\pi = -1$ for all k.
17. For $n = a^k - 1$, k a positive integer, $\log_a (n + 1) = \log_a a^k = k$.
19. $0, 1, \frac{1}{2}, 2, \frac{1}{3}, 3, \frac{1}{4}, \frac{2}{3}, \frac{3}{2}, 4, \frac{1}{5}, 5, \ldots$. No limit.
21. Let L_n be the length of one side of a regular polygon of n sides inscribed in a circle of diameter 1. Then $\pi = \lim_{n \to \infty} nL_n$.

Problems 10.7

1. 1.
3. $\frac{1}{3}$.
5. $-\frac{3}{4}$.
7. 5.
9. 3^7.
13. $\frac{3}{20}$.

15. $\frac{2,132}{999}$.

17. 0.995004. Value from table is 0.99500.

19. No need to consider $x > \frac{\pi}{4}$, and $\frac{\pi}{8!8} = 0.0000097$.

Problems 10.8

1. 8.
3. -1.
5. 3.
7. 6.
9. 27.
11. $\frac{1}{2}$.
13. 1.
15. 0.
17. 0.
19. 1.
21. 3.

Problems 10.9

1. 1.
3. -16.
5. 0.
7. 11.
9. $1 - \frac{1}{3!} + \frac{1}{5!} - \frac{1}{7!}$.
11. $\frac{\pi}{6} + \frac{1}{2}$.
13. $9/2\pi$.
15. 1.
17. 12.
19. $-\frac{1}{2}$.
21. (Fig. 10.9.21).
23. (Fig. 10.9.23).

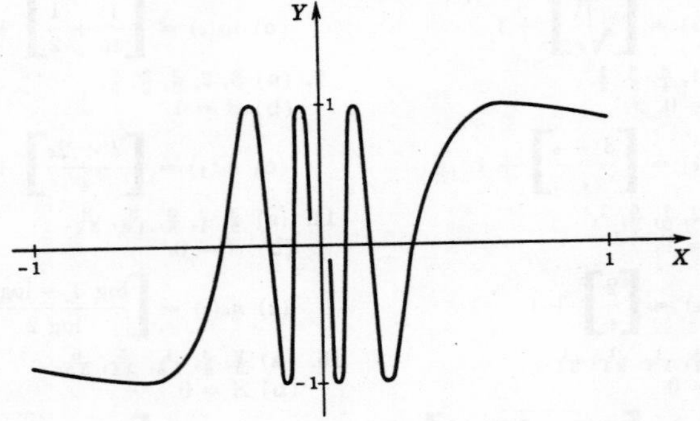

Figure **10.9.21**

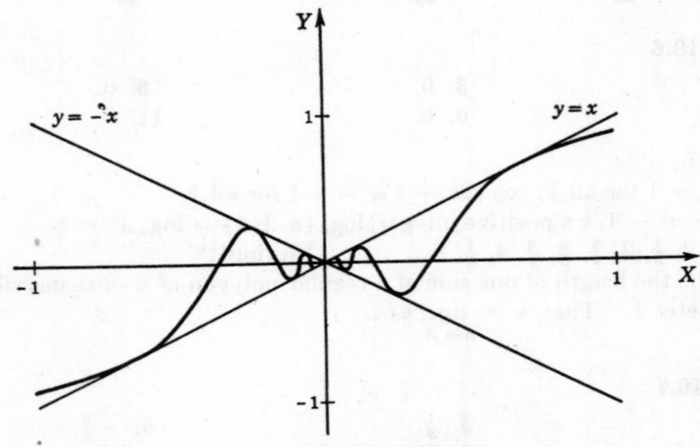

Figure **10.9.23**

25. (Fig. 10.9.25). **27.** No limit. **29.** 0.

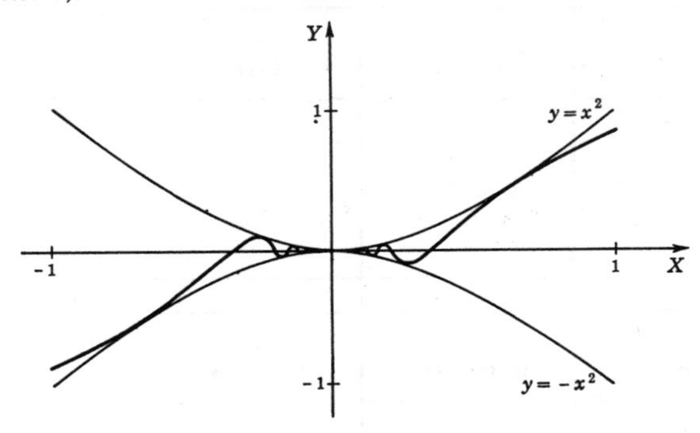

Figure **10.9.25**

31. 10¢, 5¢.

Problems 10.10

3. $\frac{3}{2}$. **5.** No limit. **7** 0.
9. $\frac{1}{2}$. **11.** $\pi/3$. **13.** 0.
15. 0. **17.** $\log(\pi/2)$. **19.** Continuous.
21. Discontinuous. **23.** Not defined at $x = 2$. **25.** $x = 1$.
27. $x = (2k + 1)(\pi/6)$. **29.** $x = 1, 2$.

Problems 10.11

1. $A = \lim_{n\to\infty} \sum_{i=1}^{n} \frac{a}{n} \cdot \left[\frac{i}{n} a\right]^2 = \frac{a^3}{3}$

3. $A = \lim_{n\to\infty} \sum_{i=1}^{n} \frac{a}{n} \cdot \left[\frac{i}{n} a\right]^3 = \frac{a^4}{4}$

5. $A = \lim_{n\to\infty} \sum_{i=1}^{n} \frac{1}{n} \cdot \left[\frac{i}{n}\right]^4 = \frac{1}{5}$

7. $A = \lim_{n\to\infty} \sum_{i=1}^{n} \frac{a}{n} \cdot \left[\frac{i}{n} a\right]^4 = \frac{a^5}{5}$

9. $A = \lim_{n\to\infty} \sum_{i=1}^{n} \frac{b}{n} \cdot \left[\frac{h}{b} \cdot \frac{i}{n} b\right] = \frac{1}{2}bh$

Problems 10.12

	(a) ft/sec	(b) ft/sec	(c) ft/sec
1.	16	48	96
3.	112	128	160
5.	9/2	3/2	0
7.	$-16t_1$	$-16(t_1 + t_2)$	$-32t_1$
9.	b	b	b

	(a) ft/sec	(b) ft/sec/sec	(c) ft/sec/sec
11.	5	3	3
13.	32	32	32
15.	4	-32	-32
17.	$-16{,}000$	-32	-32
19.	a	b	b

	(a) ft	(b) ft/sec	(c) ft/sec/sec
21.	3	$2t_1 + 1$	2
23.	12	$8t_1 - 2$	8
25.	$a + bt_1 + ct_1^2$	$b + 2ct_1$	$2c$
27.	at_1^3	$3at_1^2$	$6at_1$
29.	t_1^n	nt_1^{n-1}	$n(n-1)t_1^{n-2}$

Problems 10.13

1. $2x - y + 5 = 0.$ **3.** $7x - y - 2 = 0.$ **5.** $y = 3x.$
7. $3x - y + 5 = 0.$ **9.** $x + 9y - 6 = 0.$ **11.** $x - 4y + 1 = 0.$
13. $3ax_1^2x - y - 2ax_1^3 + b = 0.$ **15.** $ax - y + b = 0.$
17. $4x_1^3x - y - 3x_1^4 = 0.$ **19.** $nx_1^{n-1}x - y - (n-1)x^n = 0.$

Problems 11.1

1. $\frac{1}{6}$. (Fig. 11.1.1) **3.** 0. **5.** $-\frac{1}{6}$.

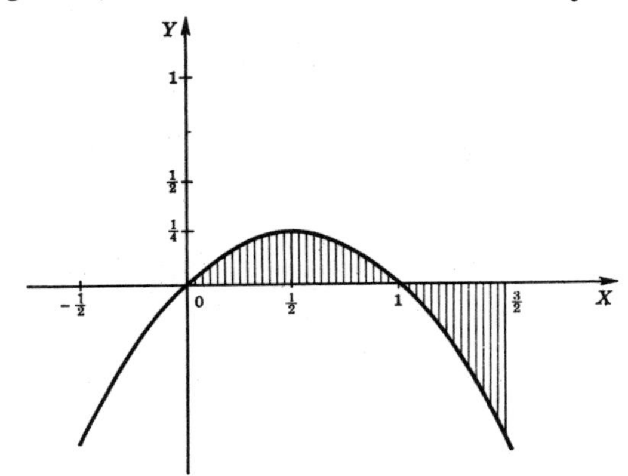

Figure 11.1.1

7. $\frac{1}{6}$. (Fig. 11.1.7)

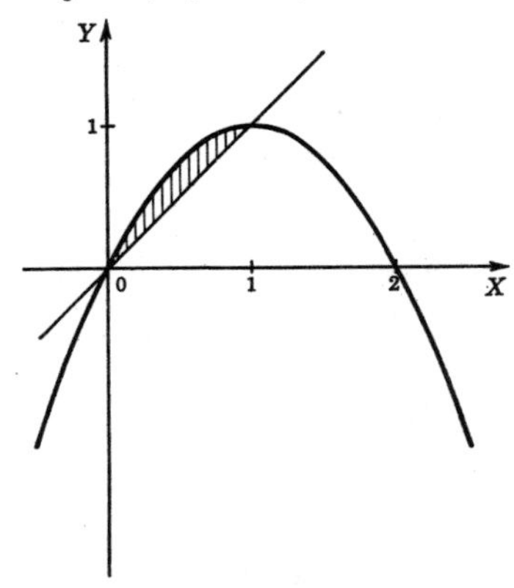

Figure 11.1.7

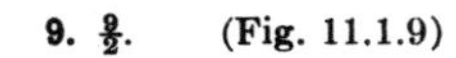

9. $\frac{9}{2}$. (Fig. 11.1.9)

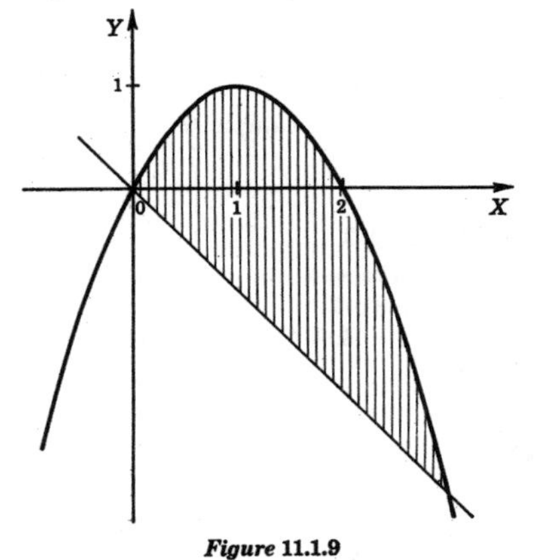

Figure 11.1.9

11. $\frac{3}{5}$. (Fig. 11.1.11)

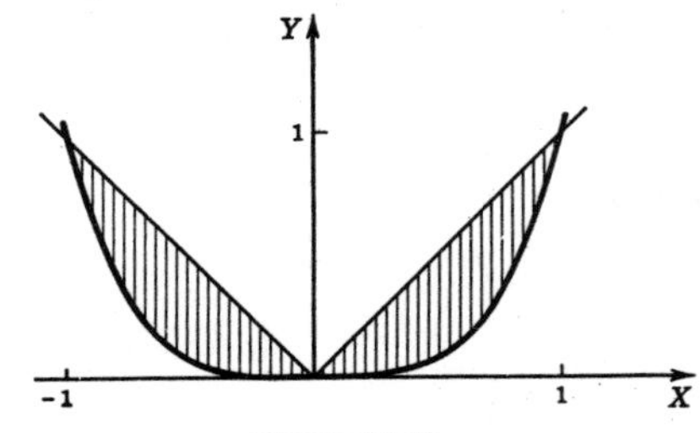

Figure 11.1.11

13. $\frac{1}{2}$. (Fig. 11.1.13)

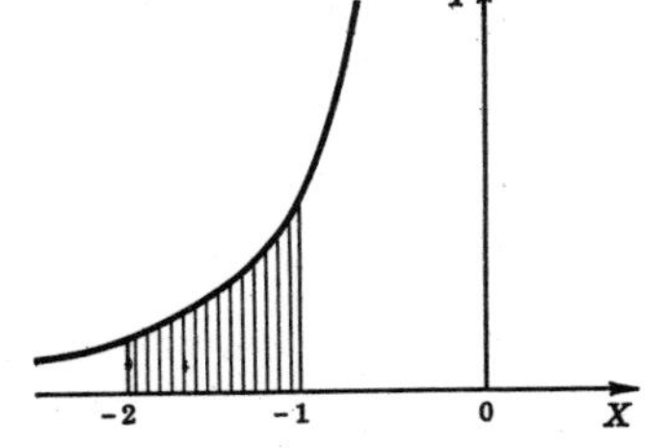

Figure 11.1.13

15. $-\frac{3}{8}$. (Fig. 11.1.15) **17.** $\frac{9}{3}$.

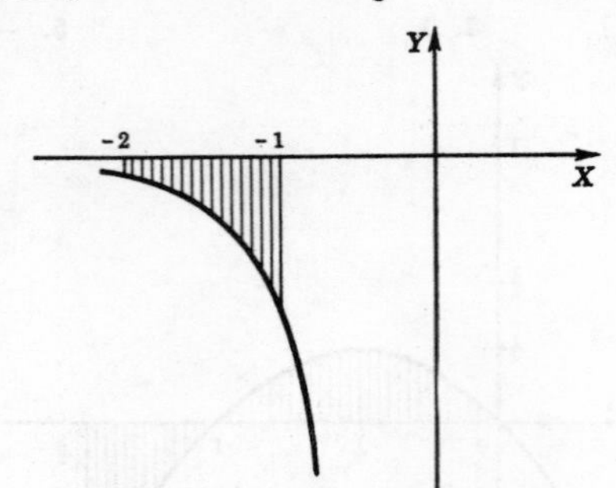

Figure **11.1.15**

19. $\frac{2}{3}$. **21.** -10. **23.** $7(e - 1)$.

25. (a) $1 - (1/n)$.
(b) 1.
(c) Area from 1 to n: area from 1 to ∞.

27. $a_0x + \frac{a_1x^2}{2} + \frac{a_2x^3}{3} + \cdots + \frac{a_nx^{n+1}}{n+1}$. $\int_0^x kt^n\,dt = \frac{kt^{n+1}}{n+1}$, $n \neq -1$. The integral of a sum is the sum of the integrals.

29. No. Counterexample: $\int_0^x \frac{u}{u^2}\,du$.

Problems 11.2

1. $3x^2 - 12x + 3$.
3. $1 + 2x$.
5. $\frac{1}{2}x^{-\frac{1}{2}} + \frac{3}{2}x^{-\frac{3}{2}}$.
7. $-2x^{-2} - 2x$.
9. $1 + 2x + 3x^2 + \cdots + nx^{n-1}$.
11. $1/x$.
13. $1/x$
15. $(2/x)\log_{10} e$.
17. 1.
19. $3e^{x-5}$.

21. $2x - 3y - 3 = 0$. (Fig. 11.2.21)

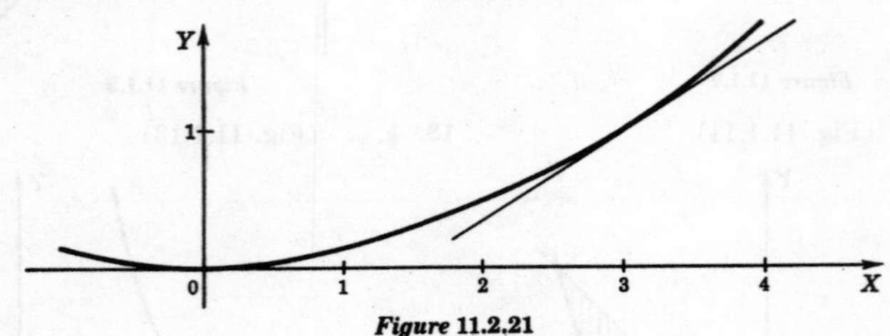

Figure **11.2.21**

23. $x - y - 2 = 0.$
(Fig. 11.2.23)

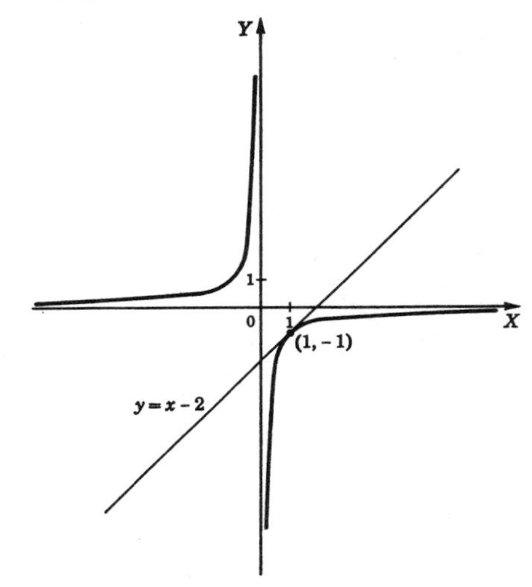

Figure 11.2.23

25. $4x + y = 0.$
(Fig. 11.2.25)

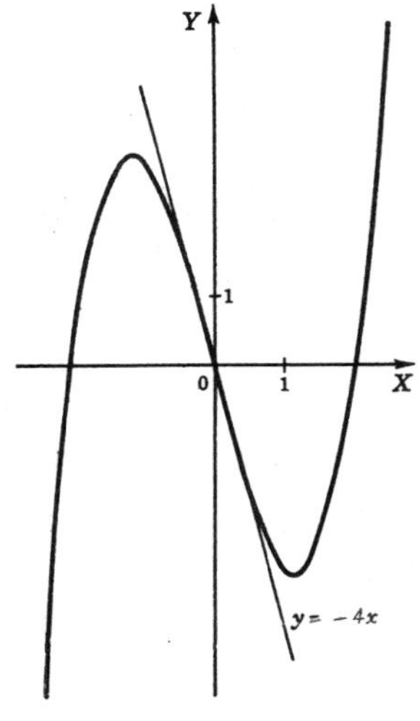

Figure 11.2.25

27. $x - y - 1 = 0.$

29. $\sqrt{3}\,x + 2y - 1 - \dfrac{\sqrt{3}}{3}\pi = 0.$

31. $x + y - 3 = 0.$

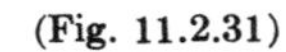
(Fig. 11.2.31)

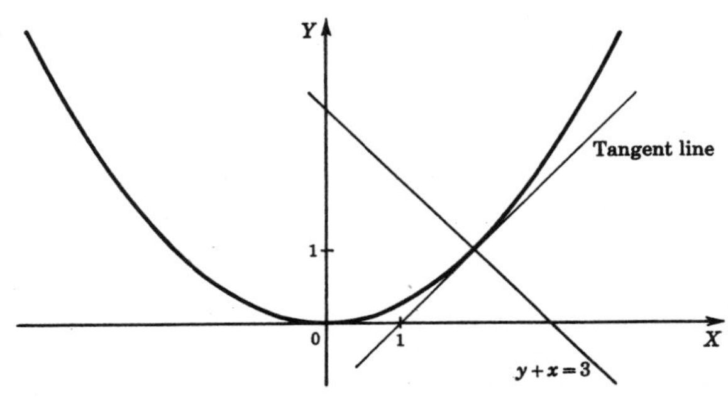

Figure 11.2.31

33. $8x - 2y + 15 = 0.$

35. $x + y - 4 = 0.$ (Fig. 11.2.35)

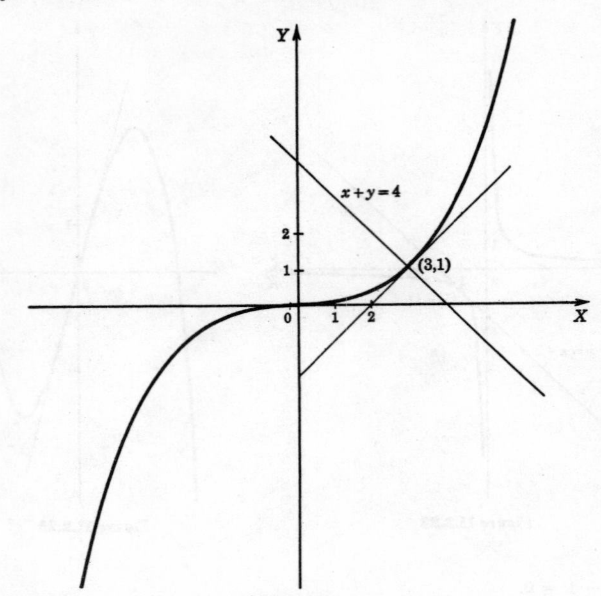

Figure 11.2.35

37. $x + 3y - 1 = 0.$

39. $x = 0.$

41. $$\Delta y = \frac{1}{(x + \Delta x)^2} - \frac{1}{x^2}$$

$$D_x y = \lim_{\Delta x \to 0} \frac{-2x - \Delta x}{x^2(x + \Delta x)^2} = -2x^{-3}$$

43. Define $f(x) = g(x) = x$; then $D(f \cdot g) = D(x^2) = 2x$
But $D(f) \cdot D(g) = D(x) \cdot D(x) = 1$

Problems 11.3

1. (**a**) $G(x) = \frac{1}{3}x^3 + \frac{3}{2}x^2 - 5x + C.$
(**b**) $F(x) = G(x) - G(a).$

3. (**a**) $G(x) = -x^{-1} + \log_e x + C.$
(**b**) $F(x) = G(x) - G(a), (x > 0, a > 0).$

5. (**a**) $G(x) = \frac{3}{5}x^{\frac{5}{3}} + 2^{\frac{2}{3}}x + C.$
(**b**) $F(x) = G(x) - G(a).$

7. (**a**) $G(x) = 16 \log_e x - 8x + \frac{1}{2}x^2 + C.$
(**b**) $F(x) = G(x) - G(a), (x > 0, a > 0).$

9. (**a**) $G(x) = 7e^x + C.$
(**b**) $F(x) = G(x) - G(a).$

11. $f(x) = 0$. **13.** $f(x) = 3x^2 - 1$. **15.** $f(x) = -2x^{-3} + x^{-2}$.
17. $f(x) = -(1/x^2)(2x + 3)(3 - x)^2$. **19.** $f(x) = e^x$.
21. (a) $v = 4t^3 + 2t - 2$.
(b) -2.
23. (a) $v = 3t^2 + 4t - 3$.
(b) $a = 6t + 4$.
27. $f(x) = 2x$.

Problems 11.4

1. $2x(1 - x^2)^4(1 - 6x^2)$.
3. $\dfrac{3(-t^4 + 4t^3 - 2t) + t^2}{(3 - t)^2}$.
5. $\frac{2}{3}(z^2 + 2z)^{-\frac{1}{3}}(z + 1)$.
7. $-2n(1 - 2\theta)^{n-1}$.
9. $e^x(\cos x + \sin x)$.
11. $\frac{1}{15}(x^3 + 1)^5 + C$.
13. $-\frac{1}{8}(1 - x - x^2)^8 + C$.
15. $\frac{3}{4}(1 + x)^{\frac{4}{3}} + C$.
17. $\frac{2}{7}e^{7x} + \sin x + C$.
19. $x^3(1 + x)^4 + C$.
21. $2a$.
23. $-\frac{1}{4}(1 - x)^{-\frac{4}{3}}$.
25. $-1/x^2$.
27. $\dfrac{ax^4}{12} + \dfrac{bx^3}{6} + \dfrac{cx^2}{2} + px + q$.
29. $\frac{4}{15}(4 - x)^{\frac{5}{2}} + px + q$.
33. $\dfrac{xx_1}{a^2} + \dfrac{yy_1}{b^2} = 1$.

Problems 11.6

1. Min. $(-3, -1)$. (Fig. 11.6.1)
3. Min. $\left(\frac{1}{2}, \frac{-9}{4}\right)$. (Fig. 11.6.3)

Figure 11.6.1

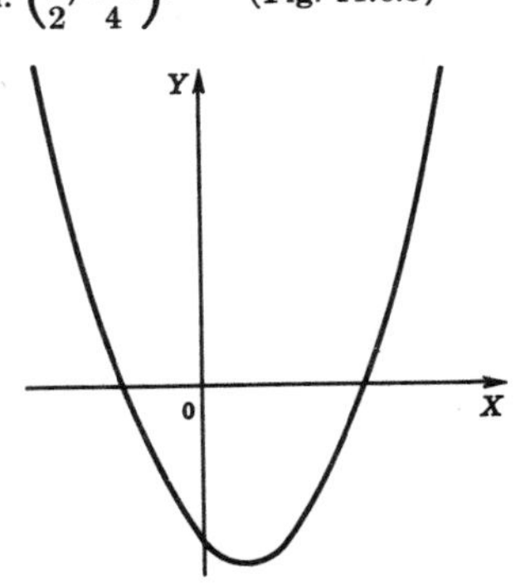

Figure 11.6.3

5. Min. $\left(\dfrac{1+\sqrt{7}}{3}, -\dfrac{20+14\sqrt{7}}{27}\right)$.

Max. $\left(\dfrac{1-\sqrt{7}}{2}, \dfrac{-20+14\sqrt{7}}{27}\right)$.

(Fig. 11.6.5)

7. Max. $(-2-\frac{2}{3}\sqrt{3}, \frac{16}{9}\sqrt{3})$.
Min. $(-2+\frac{2}{3}\sqrt{3}, -\frac{16}{9}\sqrt{3})$.
(Fig. 11.6.7)

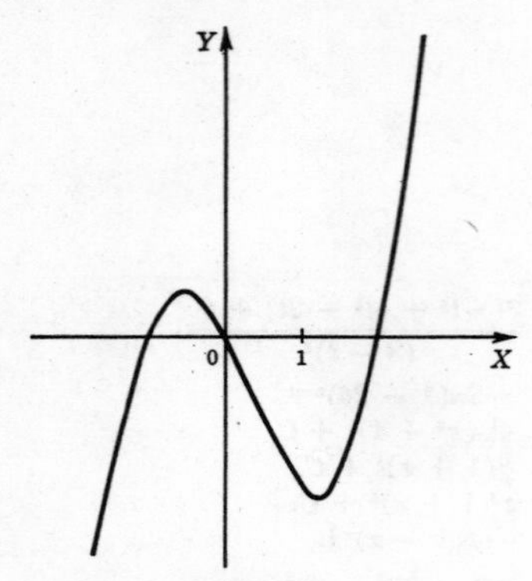

Figure **11.6.5**

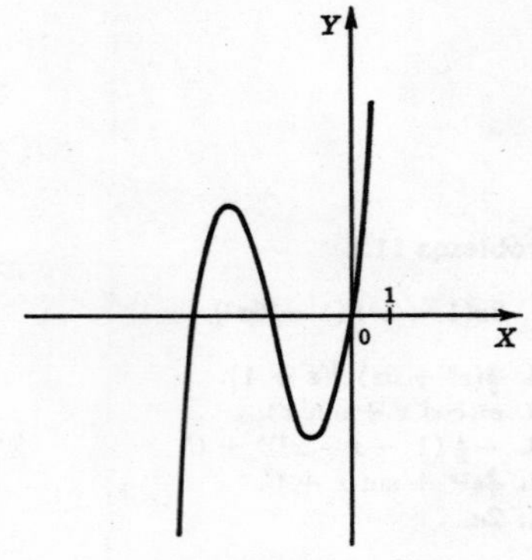

Figure **11.6.7**

9. Min. (0,0).
Max. $(1/\sqrt{2}, \frac{1}{4})$.
Max. $(-1/\sqrt{2}, \frac{1}{4})$.
(Fig. 11.6.9)

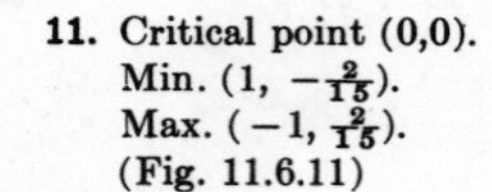

11. Critical point (0,0).
Min. $(1, -\frac{2}{15})$.
Max. $(-1, \frac{2}{15})$.
(Fig. 11.6.11)

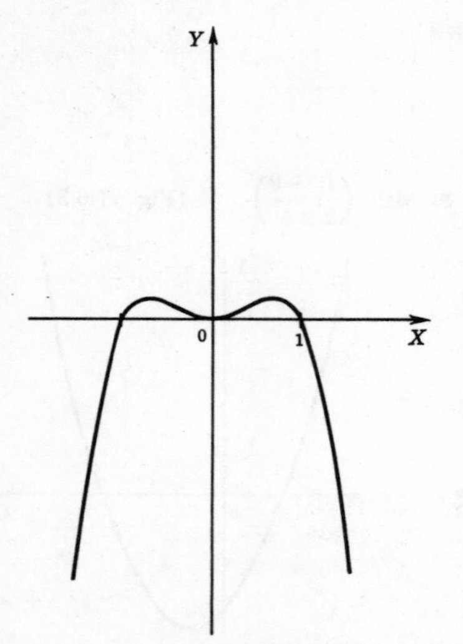

Figure **11.6.9**

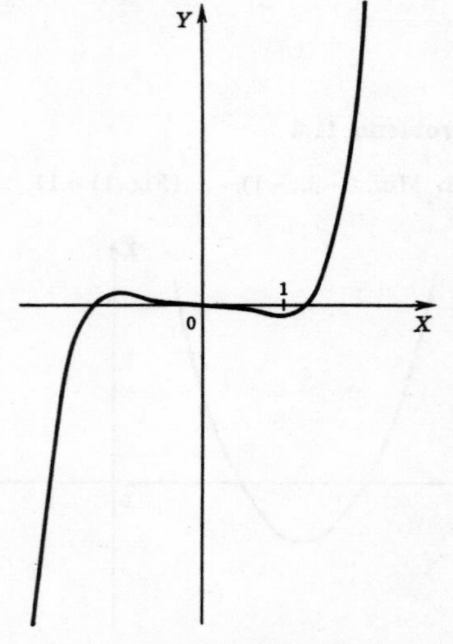

Figure **11.6.11**

13. Max. $\left(\frac{1 + \sqrt{33}}{4}, \frac{4(\sqrt{33} - 3)}{93 + 13\sqrt{33}}\right)$.

Max. $\left(\frac{1 - \sqrt{33}}{4}, \frac{4(-\sqrt{33} - 3)}{93 - 13\sqrt{33}}\right)$.

(Fig. 11.6.13)

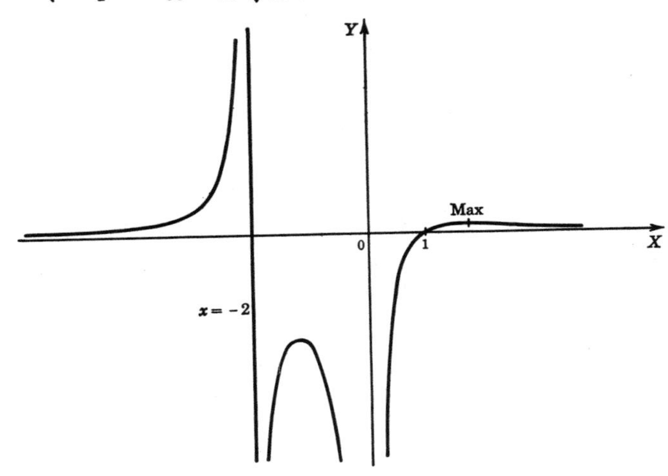

Figure 11.6.13

15. Min. $(-3, -5)$.
(Fig. 11.6.15)

17. Min. $(0,0)$.
Max. $(-2, 4e^{-2})$.
(Fig. 11.6.17)

19. Min. acceleration $= -2$.

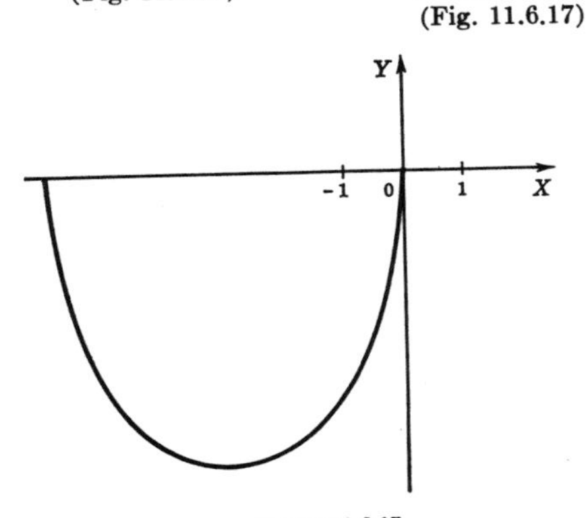

Figure 11.6.15

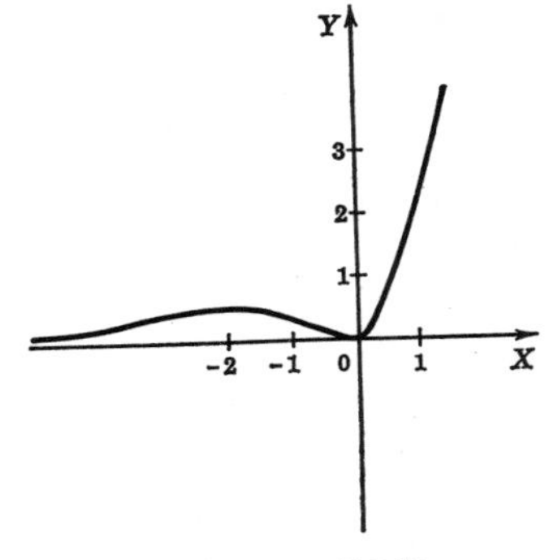

Figure 11.6.17

21. (**a**) $t_1 = 1$.
(**b**) $t_2 = 2$.
(**c**) 0, 0.
(**d**) -6, 6.
(**e**) $t = \frac{3}{2}$.

23. (**a**) 48,400 ft.
(**b**) 100 sec afterward.
(**c**) 1,760 ft/sec.

27. 5 weeks.

29. 1.

Problems 11.7

1. $30/\sqrt{2}$ ft/sec.

3. (a) 3 ft/sec.
   (b) 9 ft/sec.

5. (a) 0.01548384 cc/min. (1 in. = 2.54 cm.)
   (b) 0.012192 cm²/min.

7. $\dfrac{36{,}040}{3605} \approx 599.5$ mi/hr.

9. $\dfrac{27}{103.2256\pi} \approx \dfrac{0.2616}{\pi}$ cm/min.

Problems 12.2

1. $\{HH, HT, TH, TT\}$.
3. $\{H1, H2, H3, H4, H5, H6, T1, T2, T3, T4, T5, T6\}$.
5. $\{1, 2, 3, 4\}$.
7. $\{1, 2, 3, 4, 5, 6, 7, 8\}$.
9. $\{1, 2, 3, \ldots, 20\}$.
11. $\{1, 2, 3, 4, 5\}$. We assume no interest in sides marked 1 other than the mark "1".
13. $\{GFP, GFP', GF'P, G'FP, GF'P', G'FP', G'F'P, G'F'P'\}$.
15. $\{MFF, FMF, FFM\}$.
17. Permissible: (a), (c), (d). Not permissible: (b), not exhaustive; (e), not mutually exclusive.
19. (a), (b), (c), (d).

Problems 12.4

1. $\{H, T\}$, $P(H) = P(T) = \frac{1}{2}$.
3. $\{1, 2, 3, 4\}$, $P(i) = \frac{1}{4}$, $i = 1, 2, 3, 4$!
5. $\{1, 2, 3\}$, $P(i) = \frac{1}{6}$, $P(2) = \frac{2}{6}$, $P(3) = \frac{3}{6}$.
7. $\{1, 2, \ldots, 8\}$, $P(i) = \frac{1}{8}$, $i = 1, 2, \ldots, 8$.
9. $\{1, 2, \ldots, 12\}$, $P(i) = \frac{1}{12}$, $i = 1, 2, \ldots, 12$.
11. $\{1, 2, \ldots, 20\}$, $P(i) = \frac{1}{20}$, $i = 1, 2, \ldots, 20$.

Problems 12.5

1. $P(\text{odd}) = P(\text{even}) = \frac{1}{2}$.
3. $P(\text{odd}) = P(\text{even}) = \frac{1}{2}$.
5. $P(\text{odd}) = P(\text{even}) = \frac{1}{2}$.
7. $P(1) = \frac{1}{12}$, $P(\text{prime}) = \frac{5}{12}$, $P(\text{composite}) = \frac{6}{12}$.
9. $P(\text{multiple of } 3) = \frac{6}{20}$, $P(\text{otherwise}) = \frac{14}{20}$.
11. $P(\text{multiple of } 6) = \frac{3}{20}$, $P(\text{multiple of } 7) = \frac{2}{20}$, $P(\text{otherwise}) = \frac{15}{20}$.
13. $P(c) = P(d) = P(h) = P(s) = \frac{1}{4}$.
15. $P(i) = \frac{1}{13}$, $i = 2, 3, \ldots, 10$; $P(J) = P(Q) = P(K) = P(A) = \frac{1}{13}$.
17. $P(\text{red}) = \frac{5}{12}$, $P(\text{green}) = \frac{4}{12}$, $P(\text{black}) = \frac{3}{12}$.
19. $P(\text{odd}) = \frac{7}{12}$, $P(\text{even}) = \frac{5}{12}$.

Problems 12.6

1. $U = U_1 \times U_1$ where $U_1 = \{1, 2, 3, 4\}$
   $P(ij) = \frac{1}{16}$ $i, j = 1, 2, 3, 4$
3. $U = U_1 \times U_1$ where $U_1 = \{1, 2, \ldots, 12\}$
   $P(ij) = \frac{1}{144}$ $i, j = 1, 2, \ldots, 12$
5. $U = \{1, 2, 3, 4\} \times \{1, 2, 3, 4, 5, 6, 7, 8\}$
   $P(ij) = \frac{1}{32}$ $i = 1, 2, 3, 4, j = 1, 2, \ldots, 8$
7. $U = \{1, 2, 3, 4\} \times \{1, 2, \ldots, 20\}$
   $P(ij) = \frac{1}{80}$ $i = 1, 2, 3, 4; j = 1, 2, \ldots, 20$
9. $U = \{1, 2, \ldots, 8\} \times \{1, 2, \ldots, 20\}$
   $P(ij) = \frac{1}{160}$ $i = 1, 2, \ldots, 8; j = 1, 2, \ldots, 20$

11. $U = \{1, 2, 3\} \times \{1, 2, 3\}$

$P(11) = \frac{2}{16}$ $P(12) = \frac{4}{16}$ $P(13) = \frac{2}{16}$

$P(21) = \frac{1}{16}$ $P(22) = \frac{2}{16}$ $P(23) = \frac{1}{16}$

$P(31) = \frac{1}{16}$ $P(32) = \frac{2}{16}$ $P(33) = \frac{1}{16}$

13. $U = \{1, 2\} \times \{1, 2\}$

$P(11) = \frac{121}{144}$ $P(12) = \frac{11}{144}$

$P(21) = \frac{11}{144}$ $P(22) = \frac{1}{144}$

15. $U = \{1, 2, \ldots, n\} \times \{1, 2, \ldots, n\}$

$P(ij) = \dfrac{1}{n^2}$ $i, j = 1, 2, \ldots, n$

A "deck" of cards marked $1, 2, \ldots, n$

17. $U = \{HT\} \times \{HT\} \times \{HT\}$

$= \{HHH, HHT, HTH, THH, HTT, THT, TTH, TTT\}$

$P(HHH) = P(HHT) = \cdots = P(TTT) = \frac{1}{8}$.

19. $U = \{1, 2\} \times \{1, 2\} \times \{1, 2\}$

$= \{111, 112, 121, 211, 122, 212, 221, 222\}$

$P(111) = \frac{27}{64}$ $P(112) = P(121) = P(211) = \frac{9}{64}$

$P(122) = P(212) = P(221) = \frac{3}{64}$ $P(222) = \frac{1}{64}$

21. $U = \{1, 2, \ldots, 52\} \times \{1, 2, \ldots, 52\}$

$P(ij) = (\frac{1}{52})^2$ $i, j = 1, 2, \ldots, 52$

23. $U = \{\text{red, black}\} \times \{\text{red, black}\}$

$P(rr) = \frac{16}{81}$ $P(rb) = \frac{20}{81}$

$P(br) = \frac{20}{81}$ $P(bb) = \frac{25}{81}$

25. $U = \{\text{red, black}\}$.

$P(r) = \frac{4}{9}$ $P(b) = \frac{5}{9}$

27. $P(\text{all correct}) = (\frac{1}{5})^{50}$.

$U = \{\text{correct, not correct}\} \times \cdots \times \{\text{correct, not correct}\}$, 50 times. This nonuniform joint distribution has 2^{50} simple events. No.

29. $\frac{49}{100} \times \frac{999}{1{,}000}$.

31. $U = \{\text{I}s, \text{II}s, \text{II}g, \text{III}g\}$

$P(\text{I}s) = \frac{1}{3}$ $P(\text{II}s) = P(\text{II}g) = \frac{1}{6}$ $P(\text{III}g) = \frac{1}{3}$

Problems 12.7

1. $\frac{1}{2}$. **3.** $\frac{1}{6}$. **5.** $\frac{1}{4}$.
7. $\frac{1}{10}$. **9.** $\frac{9}{80}$. **11.** $\frac{1}{4}$.
13. $\frac{11}{72}$. **15.** $\frac{1}{8}$. **17.** $\frac{1}{8}$.
19. $\frac{1}{8}$. **21.** $\frac{1}{8}$. **23.** $\frac{4}{21}$.
25. $(\frac{1}{26})^4$. **27.** $(\frac{1}{365})^2$. **29.** $\frac{1}{64}$.
31. 1/1,728. **33.** $\frac{1}{364}$. **35.** $\frac{1}{2}$.
37. (a) $\dfrac{1{,}381}{3{,}861}$. (b) $\left(\dfrac{1{,}381}{3{,}861}\right)^2$. (c) $\left(\dfrac{1{,}381}{3{,}861}\right)^n$.

39. Note that the intersection of $(A \cap B')$ and $(A \cap B)$ is the null set and the union is A. Hence $P(A \cap B') + P(A \cap B) = P(A)$, etc.

Problems 12.8

1. $\frac{1}{2}$. **3.** $\frac{3}{4}$. **5.** $\frac{1}{2}$.
7. $\frac{5}{8}$. **11.** $\frac{7}{8}$. **13.** $\frac{1}{12}$.
15. $\frac{11}{12}$. **17.** $\frac{7}{36}$. **19.** $\frac{1}{6}$.
21. $P(A \cap B) = \frac{1}{4}$ $P(A \cup B) = \frac{3}{4}$. **23.** $\frac{35}{36}$.
25. $\frac{17}{18}$. **27.** $\frac{3}{7}$.
29. $\frac{215}{216}$. **31.** $\frac{21}{41}$.
33. (a) $\frac{17}{25}$. (b) $\frac{17}{25}$. **35.** (a) 1.00. (b) 0.80.

Problems 12.9

1. $\frac{1}{2}$.
3. $\frac{3}{7}$.
5. $\frac{3}{4}$.
7. $\frac{1}{4}$.
9. $\frac{2}{3}$.
11. $\frac{2}{3}$.
13. $\frac{2}{5}$.
15. $\frac{1}{3}$.
17. $\frac{1}{2}$.
19. $\frac{3}{5}$.

Problems 12.11

1. (a) 20.
   (b) 380.
3. 5,040.
5. (a) 720.
   (b) 46,656.
7. 125.
9. (a) 90.
   (b) 45.
11. 3,780.
13. $11 \cdot 10 \cdot 9 \cdot 7 \cdot 5 - 7 \cdot 6 \cdot 5 \cdot 4 = 33{,}810$.
15. $11 \cdot 10 \cdot 9 \cdot 8 \cdot 7 \cdot 6$.
17. $\dfrac{15!}{4!}$.
19. $24^2 \cdot 15 \cdot 10^4$.
21. $4 + 4^2 + 4^3 + 4^4 - 27 = 3 \cdot 4^3 + 2 \cdot 4^2 + 5 = 313$.
23. $\dfrac{(n-1)!}{2}$.
25. $45 \cdot 28 \cdot 15 \cdot 6$.
27. 30.
29. m^n.

Problems 12.12

3. 1.
5. (a) 120.
   (b) 120.
7. $\dfrac{20!}{3!}$.
9. 8.
11. 19.
13. (a) 12; (b) 24.
15. 15.
17. 39.
19. $\binom{30}{6}!$
21. (a) 120.
    (b) 252.
    (c) 126.
23. 22.
25. $\binom{40}{5} + 60\binom{40}{4} + 1{,}770\binom{40}{3}$.
27. $4n^4$.
29. $3\binom{33}{3} + 33^3$.

Problems 12.13

1. 1, 6, 15, 20, 15, 6, 1.
3. $15 + 20 = 35$.
5. $1 + 4 + 6 + 4 + 1 = 16 = 2^4$.
7. With $a = 1$, $b = 1$, $(1+1)^n = 2^n = 1 + \binom{n}{1} + \binom{n}{2} + \cdots + \binom{n}{n}$.
9. $a^5 - 10a^4b + 40a^3b^2 - 80a^2b^3 + 80ab^4 - 32b^5$.
11. $x^6 + 6x^5\,\overline{\Delta x} + 15x^4\,\overline{\Delta x}^2 + 20x^3\,\overline{\Delta x}^3 + 15x^2\,\overline{\Delta x}^4 + 6x\,\overline{\Delta x}^5 + \overline{\Delta x}^6$.
13. $\dfrac{32}{x^5} - \dfrac{80}{x^2} + 80x - 40x^4 + 10x^7 - x^{10}$.
15. 1.08243216.
17. -286.
19. 60.

Problems 12.14

1. (a) $\frac{1}{4}$.
   (b) $\frac{9}{64}$.
   (c) $\frac{5}{32}$.
3. $\dfrac{53}{5^5}$.
5. (a) $\frac{25}{216}$.
   (b) $\dfrac{171}{6^4}$.

7. $\frac{20}{6^5}$.

9. $[1 - (\frac{1}{2})^{500} - 500(\frac{1}{2})^{500}]$.

11. $\sum_{i=2}^{5} \binom{500}{i} (\frac{1}{2})^{500}$.

13. $\frac{11}{3^5}$.

15. $\frac{200}{3^5}$.

17. $\frac{3}{16}$.

19. $1 - \frac{131}{6^5}$.

21. $\frac{1}{4}$.

23. $\binom{25}{x} (\frac{3}{4})^{25-x}(\frac{1}{4})^x$.

25. $\frac{17}{30}$.

27. $$\frac{\frac{1}{\binom{64}{2}} \cdot 7 \cdot 7 \cdot 2}{\frac{1}{\binom{64}{2}} \cdot \binom{64}{2}} = \frac{7}{144}.$$

29. (a) $$\frac{\binom{52}{1}\binom{48}{1}\binom{44}{1}\binom{40}{1}\binom{36}{1}}{\binom{52}{5}}.$$

(b) $$\frac{\binom{4}{2}\binom{48}{3} + \binom{4}{3}\binom{48}{2} + \binom{4}{4}\binom{48}{1}}{\binom{52}{5}}.$$

31. (a) $\frac{5}{72}$.

(b) $$\frac{\binom{11}{1} + 2\binom{11}{2} + \binom{11}{3}}{6^{11}}.$$

33. $$\frac{15! \cdot 14 \cdot 13 \cdot 12 \cdot 2}{19!}.$$

35. $\frac{2}{9}$.

Problems 12.15

1. (a) Reject; (b) accept; (c) accept; (d) accept; (e) accept.

3. Accept.

5. Quit playing, but don't shoot him.

7. Reject.

9. Reject.

Problems 13.3

1. Show that $A' \cap A = A \cap A' = O$ and $A' \cup A = A \cup A' = I$. Then apply B.5.

3. $A \subseteq B$ is equivalent to $A \cap B' = O$. Also, $A \cap (B \cup B') = A \cap I = A$. Expanding: $(A \cap B) \cup (A \cap B') = A$; $(A \cap B) \cup O = A$; $A \cap B = A$. Conversely: $A \cap B' = (A \cap B) \cap B' = A \cap (B \cap B') = A \cap O = O$.

5. Use the definition and De Morgan's Law.

7. By hypothesis, $A \cap B' = O$. Thus $A \cap (B \cup C)' = A \cap (B' \cap C') = (A \cap B') \cap C' = O$. Hence $A \subseteq (B \cup C)$.

9. $A \cup (A' \cap B) = (A \cup A') \cap (A \cup B) = I \cap (A \cup B) = A \cup B$.

11. $(A \cup B) \cap (C \cup D) = [(A \cup B) \cap C] \cup [(A \cup B) \cap D] = \cdots$

13. $(A \cup B) \cap (A \cup B') = A \cup (B \cap B') = A \cup O = A$.

15. Follow instructions.

17. $(A \cap B) \cup (A' \cap C) = [A \cap (A' \cup B)] \cup [A' \cap (A \cup C)] = (A \cup A') \cap (A \cup A \cup C) \cap (A' \cup A' \cup B) \cap (A' \cup B \cup A \cup C) = I \cap (A \cup C) \cap (A' \cup B) \cap I = (A \cup C) \cap (A' \cup B)$.

19. This is the dual of Prob. 18.

Problems 13.6

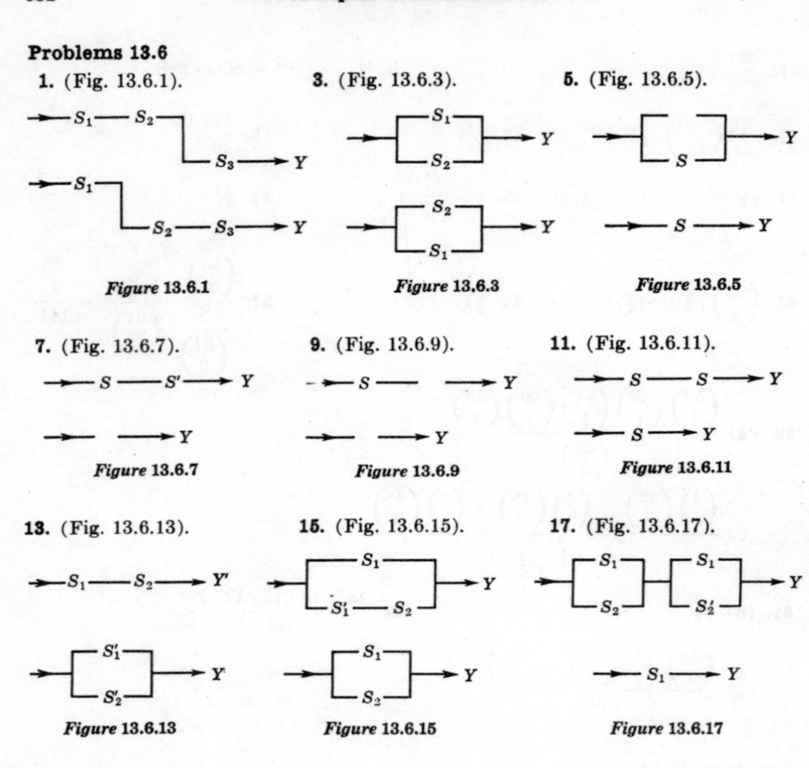

Figure **13.6.1** *Figure* **13.6.3** *Figure* **13.6.5**

Figure **13.6.7** *Figure* **13.6.9** *Figure* **13.6.11**

Figure **13.6.13** *Figure* **13.6.15** *Figure* **13.6.17**

19. (Fig. 13.6.19).

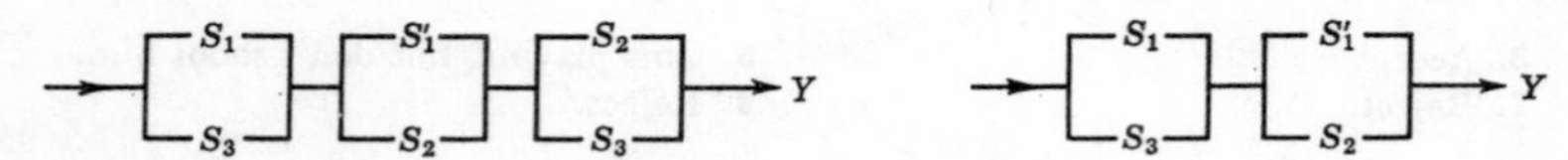

Figure **13.6.19**

21. $(S_1 \times S_2 \times S_3') + (S_1' \times S_2) + (S_2 \times S_3)$

$$\begin{aligned} &= S_2 \times [(S_1 \times S_3') + S_1' + S_3] \\ &= S_2 \times (S_3 + S_1 + S_1') && \text{[Prob. 15]} \\ &= S_2 \times (1 + S_1) && \text{[Prob. 6]} \\ &= S_2 \times 1 && \text{[Prob. 10]} \\ &= S_2 && \text{[Prob. 4]} \end{aligned}$$

23. $(S_1 + S_2) \times (S_1 + S_2') \times (S_1' + S_2') = S_1 \times (S_1' + S_2')$ [Prob. 17]

$$\begin{aligned} &= (S_1 \times S_1') + (S_1 \times S_2') \\ &= 0 + (S_1 \times S_2') \\ &= S_1 + S_2' \end{aligned}$$

25. $(S_1 \times S_2) + (S_1' \times S_3) + (S_2 \times S_3) = (S_1 \times S_2) + (S_1' \times S_3)$ [Prob. 20]

Problems 13.7

1. $(A \times B \times C) + (A' \times B' \times C) + (A \times B' \times C') + (A' \times B \times C') =$
$A \times [(B \times C) + (B' \times C')] + A' \times [(B' \times C) + (B \times C')]$.
3. $CO \times (A_1 + A_2)$.
5. $(M_1 \times M_2) + [(W_1 \times W_2) \times (M_1 + M_2)]$.
7. Pointing to one of R, R' and one of L, L', he asks, "Is it safe to drink these two vials?" If the answer is "yes", he drinks them; if the answer is "no", he drinks the R vial to which he is pointing and the L vial to which he is not pointing.
9. $p_1p_2 + p_3p_4 - p_1p_2p_3p_4$.

Index

The symbols used in this book–and the pages on which they are defined

Page	Symbol	Meaning
8	$a \in A$	a is an element of set A
9	$\{\ \}$	notation for a set
9	$\{3\}$	the set of which 3 is the only element
9	$\{2, 7, 15, 36\}$	the set whose elements are 2, 7, 15, 36
9	$X = \{x \mid x \text{ is a real number}\}$	the set of all real numbers
9	$X \times Y$	the set of all ordered pairs (x,y) of real numbers
10	$\emptyset$	the empty set
9	U	the universal set
9	$A \subseteq B$	set inclusion
9	$A \subset B$	proper set inclusion
16	$A \cap B$	intersection of A and B
18	$A \cup B$	union of A and B
20	A'	complement of A
7	p, q, r	propositions
13	p_x, q_x, r_x	open sentences
13	P, Q, R	truth sets of p_x, q_x, r_x
14	$\forall_x$	for all x
14	$\exists_x$	for some x
15	$p \wedge q$	conjunction of p and q
17	$p \vee q$	inclusive disjunction of p and q
18	$p \underline{\vee} q$	exclusive disjunction of p and q
20	$\sim p$	negation of p
22	$p \rightarrow q$	the implication: if p then q
23	$p \leftrightarrow q$	p is equivalent to q
52	$\lvert a \rvert$	absolute value
61	$a + bi$	complex number in usual notation
63	(a,b)	complex number in ordered pair notation
271	$r(\cos\theta + i \sin\theta) = r \text{ cis } \theta$	complex number in polar form
67	$a \equiv b, \bmod m$	a congruent to b modulo m
75	$>$	greater than
75	$\geq$	greater than or equal to
75	$<$	less than
75	$\leq$	less than or equal to
81	$1.\overline{142857}$	repeating decimal
91, 429	$n!$	n factorial